ISNM
International Series of Numerical Mathematics
Vol. 134

# Ginzburg-Landau Phase Transition Theory and Superconductivity

K.-H. Hoffmann
Q. Tang

Springer Basel AG

Authors:

Karl-Heinz Hoffmann
Caesar
Friedensplatz 16
53111 Bonn
Germany
and

Qi Tang
SMS
University of Sussex
Brighton BN2 8LR
UK

2000 Mathematics Subject Classification 35-02, 35Qxx

A CIP catalogue record for this book is available from the Library of Congress, Washington D.C., USA

Deutsche Bibliothek Cataloging-in-Publication Data
Hoffmann, Karl-Heinz:
Ginzburg-Landau phase transition theory and superconductivity /
K.-H. Hoffmann ; Q. Tang. - Basel ; Boston ; Berlin : Birkhäuser, 2001
(International series of numerical mathematics ; Vol. 134)
ISBN 978-3-0348-9499-9 ISBN 978-3-0348-8274-3 (eBook)
DOI 10.1007/978-3-0348-8274-3

ISBN 978-3-0348-9499-9


Originally published by Birkhäuser Verlag in 2001
Softcover reprint of the hardcover 1st edition 2001

Printed on acid-free paper produced of chlorine-free pulp. TCF ∞

ISBN 978-3-0348-9499-9

# Contents

# Preface

The theory of complex Ginzburg-Landau type phase transition and its applications to superconductivity and superfluidity has been a topic of great interest to theoretical physicists and has been continuously and persistently studied since the 1950s. Today, there is an abundance of mathematical results spread over numerous scientific journals. However, before 1992, most of the studies concentrated on formal asymptotics or linear analysis. Only isolated results by Berger, Jaffe and Taubes and some of their colleagues touched the nonlinear aspects in great detail.

In 1991, a physics seminar given by Ed Copeland at Sussex University inspired Q. Tang, the co-author of this monograph, to study the subject. Independently in Munich, K.-H. Hoffmann and his collaborators Z. Chen and J. Liang started to work on the topic at the same time. Soon it became clear that at that time, groups of mathematicians at Oxford and Virginia Tech had already studied the subject for a couple of years. They inspired experts in interface phase transition problems and their combined effort established a rigorous mathematical framework for the Ginzburg-Landau system.

At the beginning Q. Tang collaborated with C.M. Elliott and H. Matano. When they met at the University of Sussex in the Summer of 1991, they discussed what kind of problems they should study. They found that the mathematical justification of the vortex structure and the relation between the applied magnetic field and the number of vortices were the two fundamental mathematical problems that the physicists failed to address, despite the long period they had been working on the subject. The trio also intended to look at a problem which was mathematically interesting: the limit behavior of the energy minimizers of the following Landau problem

$$\inf_{u|_{\partial B_1}=e^{in\theta}} \int_{B_1} \left(|\nabla u|^2 + \frac{1}{\varepsilon^2}(|u|^2-1)^2\right)dx$$

as $\varepsilon \to 0$. Here $B_1$ is the unit ball in $\mathbb{R}^2$ and $n$ is any integer. But during Matano's stay at Sussex, they didn't find a solution to the problem.

Back in Tokyo, Matano talked to H. Brezis, who visited Japan at the time, about the limit problem (in a Japanese café, according to H. Brezis), and that led to the famous work of [BBH 94]. Elliott, Matano and Tang justified the 2-d vortex structure rigorously.

By this time, many papers in this field treating modeling, asymptotic analysis and numerical analysis had already appeared. The research field is becoming even more active today with many new results being presented at every conference. However, to the best of our knowledge, the fundamental question of revealing the relation between the applied magnetic field and the number of vortices remains open. Hoffmann and his research team (as well as many other physicists and mathematicians around the world) carried out various numerical computations trying to gain insight into this relationship. The most spectacular numerical simulation we have seen so far was a 3-d film by H. Kappa who had the necessary superb computing facilities in his Argonne laboratory. Recently, H.-J. Bauer in his PhD thesis (from TU Munich) obtained progress in handling the problem with less computer power by using new concepts in scientific computing.

In this monograph, we try to collect the recent research results in the complex G-L theory with or without immediate applications to the theory of superconductivity. The purpose is to present as many mathematically sound results on various aspects of the PDE system as possible and provide a good reference for researchers who are interested in studying mathematical and physical problems in this field.

To fulfill this purpose, we include rigorous mathematical analysis, formal asymptotics as well as numerical analysis for the PDE system.

We start with some material on the physical background and point out some of the weaknesses in the modelling and theoretical studies of physicists. We then treat the mathematical scaling in a systematic way and analyze the implications on various limit problems. After that we address the mathematical foundation and formal asymptotic analysis of vortex motion. Then we concentrate on rigorous mathematics: we present results on existence, regularity and long time behavior of solutions and discuss the rigorous results on vortex location and law of motion. Furthermore, we look at various ways of deriving lower dimensional models from higher dimensional models and study rigorous results for the pinning of vortices.

We would like to thank the Alexander von Humboldt Foundation, the Center of Advanced European Studies and Research (caesar), Deutsche Forschungsgemeinschaft (DFG), Sussex University and the Technical University Munich for providing financial and moral support for the writing of this monograph.

We also would like to thank numerous colleagues for their interest in our work and in particular, Professor C.M. Elliott for his careful reading of the manuscript.

Q. Tang is also grateful to his wife, Xiaoyin, for her sincere and continuous support during his long absence from their beloved home in Brighton, England while staying in Bonn and Munich.

# Chapter 1
# Introduction

Before starting formally our introduction, we define some mathematical symbols that will be used throughout this book.

$\Omega$ is a bounded, piecewise $C^2$ Lipschitz domain in $I\!R^n$ with $n = 2$ or 3. If there are additional requirements such as simple connectedness, $C^2$ boundary etc., we will specify in the specific places concerned.

$\partial\Omega$ is the boundary of $\Omega$.

$\boldsymbol{n}$ is the outward unit normal vector of $\Omega$ along $\partial\Omega$ which is defined almost everywhere on $\partial\Omega$. When the space dimension is 2, $\boldsymbol{\tau}$ is the tangential vector of $\partial\Omega$ obtained by rotating $\boldsymbol{n}$ anti-clockwise for 90 degrees.

$\nabla$ is the standard gradient operator.

div is the standard divergence operator.

**curl** is the standard curl operator. It has several definitions depending on the dimension of the domain of the function and the dimension of the image space: (i) If $\boldsymbol{f} : I\!R^3 \to I\!R^3$, $\mathbf{curl}\boldsymbol{f} = \nabla \times \boldsymbol{f}$ where $\times$ is the formal cross (vector) product. (ii) If $f : I\!R^2 \to I\!R$, $\mathbf{curl} f = (\partial_2 f, \; -\partial_1 f)^T$. In this particular case, the notation $\mathbf{curl} = \nabla^\perp$ is also frequently used. If we put this into the 3-d perspective, we have taken the scalar function $f$ as the third component of the vector $(0, 0, f)$. (iii) If $\boldsymbol{f} : I\!R^2 \to I\!R^2$, $\mathbf{curl}\boldsymbol{f} = \partial_1 f_2 - \partial_2 f_1$. Sometimes, it is also possible to regard **curl** as the 3-d vector $(0, 0, \partial_1 f_2 - \partial_2 f_1) = \vec{\nabla} \times (f_1, f_2, 0)$.

Partial derivatives (for example the partial derivative of $f$ with respect to $t$) can be denoted by any of the following symbols: $\frac{\partial f}{\partial t}$, $\partial_t f$ or $f_t$. In addition, if $\boldsymbol{n}$ is unit vector, we use $\frac{\partial f}{\partial \boldsymbol{n}} = \partial_{\boldsymbol{n}} f = \nabla f \cdot \boldsymbol{n}$ to denote the derivative of $f$ along the $\boldsymbol{n}$ direction.

Throughout this book, bold letters will be used for vectors, normal letters for real valued scalars and curly letters for complex valued scalars.

## 1.1 Brief history

The discovery of the phenomenon of superconductivity was made by Kamerlingh-Onnes in 1911. Since then, various efforts were made to find a mathematical description of superconductivity. At the beginning, superconductivity was simply regarded as a vanishing of all electrical resistance below the transition temperature.

In the early days, people had only lower temperature superconductors available. These superconductors exhibit superconducting behaviour at temperatures very close to absolute zero, usually between 0 and 5 K. These superconductors are also called hard superconductors or type I superconductors. In 1933, W. Meissner and R. Ochsenfeld discovered the so-called Meissner effect: if a not very strong magnetic field is applied to a sample of lower temperature superconductor, magnetic flux is excluded from the interior of the superconductor body. Near the surface of the sample, however, a thin penetration region appears. That is, the magnetic field penetrated into the sample and destroyed superconductivity only in a thin region near the surface.

Not long afterwards in 1935, H. London and F. London proposed a phenomenological theory of the electromagnetic properties for superconductors. Many other scientists also made contributions to the establishment of a consistent mathematical theory in the subsequent years. However, the common problem with a phenomenological theory is that it needs a microscopic theory to make it valid and consistent.

In 1957, J. Bardeen, L.N. Cooper and J.R. Schrieffer developed the BCS theory that is the most satisfactory microscopic theory for describing lower temperature superconductivity phenomena. For an introduction to the BCS theory and the many earlier related works, we refer to [BCS 57] and the references quoted there. The main achievement of the BCS theory is to present a convincing argument that the mean field approach (averaging the behaviour of large amount of superconducting electrons) can be used to describe superconductivity. It therefore gave theoretical support to phenomenological theories such as those of London and Ginzburg-Landau.

The London theory is based on the Maxwell system by interpreting the Meissner effect across the boundary of the superconductor samples appropriately (see Section 1.2 in the following).

By taking into account the microscopic electron motion, Landau and Ginzburg-Landau developed the so-called Ginzburg-Landau phenomenological theory (referred to as G-L theory in the following) of superconductivity over a period of 13 years (1937–1950). This theory can be used to describe both type I and type II superconducting behaviour by adjusting the values of certain parameters involved in the system. It also has been shown that some relation between the London theory and the G-L theory exists. In the following, we will discuss the settings of the London theory and G-L theory in detail.

In the remainder of this section, we give a brief introduction to the various physical phenomena accompanying superconductivity behaviour and give a qualitative description on how various parameters are used to distinguish type I and type II superconductors.

### 1.1.1 Meissner effect – diamagnetism

The Meissner effect was discovered by Meissner and Ochsenfeld in 1933. This effect illustrates that a superconductor is not simply a perfect conductor. In the presence of an applied magnetic field, a superconductor behaves rather differently from a perfect conductor (or more realistically, a very good conductor). To explain this better, we consider a concrete example.

Let a long solid cylindrical shaped superconducting sample be placed in a gradually increasing applied magnetic field of strength $H_e$ with direction parallel to the axis of the cylinder. As the field is increased from zero to a certain magnitude $H_c$, the field creates electric currents flowing on the surface of the superconductor sample. The surface currents are induced currents, so they will create a magnetic field to oppose the penetration of the applied field. Consequently, the magnetic induction $B$ inside the cylinder remains 0 except in a thin layer near the boundary which the magnetic field decreases rapidly from the outside to the inside of the sample (see Figure 1). Up to now, the superconductor behaves in exactly the same way as a perfect conductor.

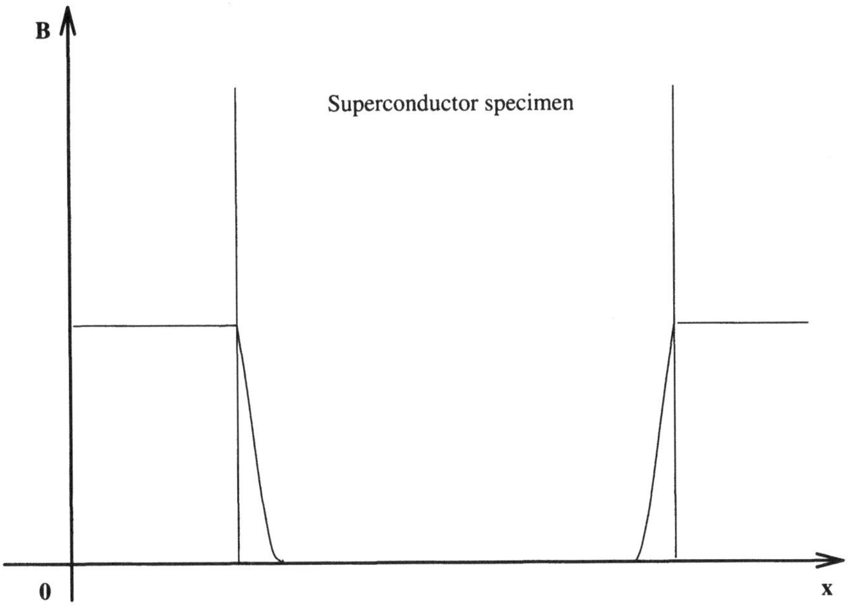


Figure 1: Variation of flux density of the magnetic field across a specimen

At $H_e = H_c$, the superconductor becomes normal, the magnetic field penetrates and the induction $B$ becomes equal to $H_c$. Let (i.e., non-superconducting) the applied field now be lowered below $H_c$, if the superconductor were a perfect conductor, the surface current (which is not disappearing) would maintain the induction inside so that $B = H_c$. However, experiments show that for a superconductor, as soon as $H_e$ is below $H_c$ again, $B = 0$.

The conclusion is that (hard) superconductors behave like a perfect diamagnet and their magnetic property is independent of their history.

As this phenomenon was discovered in 1933, it doesn't apply to the so called type II superconductors discovered later whose behaviour is described by the G-L phenomenological theory with parameters satisfying certain conditions (For detailed discussion, see Section 1.1.4.). In the G-L theory, when the applied magnetic field is decreased from $H_c$, the induced magnetic field remains nonzero in the superconductor. This is rather different from the phenomenon described by the Meissner effect. However, the Meissner effect is still an important phenomenon to our understanding of superconductivity.

In order to classify type I and type II superconductors, we need some concrete mathematical quantities to serve as indicators distinguishing their behaviour. This is our main concern in the rest of this section.

### 1.1.2 The London equation and the penetration depth

As we have pointed out, the London theory attempted to use the Maxwell system to describe the Meissner effect across the boundary of a superconductor.

We denote the supercurrent by $\boldsymbol{j}_s$ and the magnetic field by $\boldsymbol{h}$. The term "supercurrent" is used because the current is produced without the presence of a (electrical) potential difference. The Maxwell equation inside the superconductor $\Omega \subset I\!R^3$ has the form

$$\mathbf{curl}\boldsymbol{h} = \frac{4\pi}{c}\boldsymbol{j}_s \tag{1.1}$$

where $\boldsymbol{h}$ denotes the magnetic field and $c$ is the velocity of light.

The free energy of the system has the form

$$F = F_s + E_{kin} + E_{mag} \tag{1.2}$$

where $F_s$ is the energy of the electrons in the condensed state at rest. $E_{kin}$ is the kinetic energy associated with the permanent currents:

$$n_s e\boldsymbol{v}(\boldsymbol{r}) = \boldsymbol{j}_s(\boldsymbol{r}), \quad \text{for all } \boldsymbol{r} \in \Omega \tag{1.3}$$

where $e$ is the electron charge, $n_s$ is the number of superconducting electrons per unit volume and $\boldsymbol{v}$ is velocity of the electrons. If we assume that $\boldsymbol{v}(\boldsymbol{r})$ varies very slowly (almost a constant), we have

$$E_{kin} = \frac{1}{2}\int_\Omega mn_s|\boldsymbol{v}|^2 d\boldsymbol{r} = \frac{1}{8\pi}\int_\Omega \lambda_L^2|\mathbf{curl}\boldsymbol{h}|^2 d\boldsymbol{r} \tag{1.4}$$

where $m$ is the effective mass of the electrons. $\lambda_L$ is a length scale called penetration depth and is defined by

$$\lambda_L = \sqrt{\frac{mc^2}{4\pi n_s e^2}}. \tag{1.5}$$

$E_{mag}$ is the magnetic energy:

$$E_{mag} = \int_\Omega \frac{h^2}{8\pi} d\boldsymbol{r} \tag{1.6}$$

The free energy is then

$$F = F_s + \frac{1}{8\pi}\int_\Omega (h^2 + \lambda_L^2 |\mathbf{curl}\boldsymbol{h}|^2) d\boldsymbol{r}. \tag{1.7}$$

Assuming that $F_s$ is a constant, the energy minimizers satisfy the following equation

$$\boldsymbol{h} + \lambda_L^2 \mathbf{curlcurl}\boldsymbol{h} = 0. \tag{1.8}$$

This is the London equation derived by F. and H. London in 1935.

Let us now investigate a semi-infinite domain $\Omega$ which has its surface lying on the $x-y$ plane and whose bulk occupies the region $z > 0$. We also assume that the half space $z < 0$ is vacuum. Since $\boldsymbol{h}$ is the magnetic field, it must satisfy the equation

$$\text{div}\boldsymbol{h} = 0. \tag{1.9}$$

Because of the symmetry of the problem, we assume that $\boldsymbol{h}$ is a function of $z$ only. Equation (1.9) gives

$$\frac{dh_3}{dz} = 0. \tag{1.10}$$

Using the third component of equation (1.8), we derive that $h_3 = 0$.

This implies that $\boldsymbol{h}$ has no normal component to the surface of the specimen. If we simplify the problem further by assuming that $\boldsymbol{h} = (h, 0, 0)^T$, then (1.8) can be written as

$$\frac{d^2h}{dz^2} = \frac{h}{\lambda_L^2}.$$

This equation has the solution $h(z) = h(0)\exp(-z/\lambda_L)$. The solution tells us that $h$ decays exponentially inside the superconductor on the length scale $\lambda_L$. This is the reason that $\lambda_L$ is called the **penetration depth**. In practice, $\lambda_L$ is a small number. Its value varies in the range of a few hundreds of angstroms.

### 1.1.3 The coherence length

In establishing the London equation, we have assumed that $\boldsymbol{v}(\boldsymbol{r})$ is a slowly varying function of $\boldsymbol{r}$. It is under this assumption that the solution $h$ varies on the length scale $\lambda_L$.

In the condensed state, the correlation between the superconducting electrons is strong. If we use $\xi_0$ to denote the range of this correlation, then the condition that $\boldsymbol{v}(\boldsymbol{r})$ is a slowly varying function of $\boldsymbol{r}$ can be quantified as $\lambda_L \gg \xi_0$. The magnitude of this coherence length $\xi_0$ is of the order

$$\frac{\hbar v_F}{k_B T_0} \tag{1.11}$$

where $\hbar$ is the Planck constant, $v_F$ is the Fermi velocity, and $k_B T_0$ is the distance between the condensed electrons and the Fermi surface.

### 1.1.4 Classification of superconductors

Using the coherence length $\xi_0$ and the London penetration depth $\lambda_L$ we can classify the superconductors into two types:

(a) When $\lambda_L \ll \xi_0$, we say that the superconductor is a Type I superconductor. The London equation is not valid in this case because the penetration region is a very thin one. The relation implies that correlation of electrons is the dominant factor and the Meissner effect governs the behaviour of the material in response to an exterior magnetic field.

Most type I materials are pure metals and low concentration alloys. They usually require very low temperature (a few K) to exhibit superconducting behaviour. For this reason, they are also called low temperature superconductors or hard superconductors.

To prepare the discussion of type II superconductivity, we give a description of the concept of vortex. Mathematically speaking, a vortex is a co-dimension two object. For example, a point in the plane, or a curve in the space. This concept has its origin in fluid mechanics. A typical example of vortex is the circulating line void formed in bath water as it sinks into the drain and around which, the water circulates, clockwise or anti-clockwise. This illustrates the concept of a vortex and its orientation. In the context of superconductivity, the superconducting electron flow corresponds to the water flow. When a void is formed, the magnetic flux will fill into it just as the air does to the fluid vortex. In the vortex core, there is an absence of superconducting electrons (i.e., the material is in a normal state) just as in the fluid vortex, there is no water. This is an ideal picture. The real situation near a vortex core is more complicated because the quantities vary continuously. We are now ready to discuss type II superconductors:

(b) When $\lambda_L \gg \xi_0$, we say that the superconductor is a Type II superconductor. The London equation is valid only in low fields when there is a relatively mild magnetic penetration across the boundary of the sample.

Most type II materials are chemical compounds of higher concentration superconducting alloys. Type II material becomes superconducting at a relatively high temperature (about 30 K or above).

Type II superconductors behave differently from type I material. Instead of generalising, we take the example of a macroscopic cylinder sample of type II superconductor. When it is placed in an applied magnetic field $H$ parallel to its axis, there are three ranges of values of $H$ that are of interesting.

i) There is a value $H_{c1}$ such that if $H < H_{c1}$, then only a relatively mild Meissner effect is observed on the boundary. The value of $H_{c1}$ depends on the temperature, geometric shape and other properties of the material involved. In this regime, the property of the superconductor can be described by the London equation or some particular form of the G-L equation to be introduced later.

ii) There is another critical value $H_{c2} > H_{c1}$ such that for $H \in (H_{c1}, H_{c2})$, the magnetic flux gradually penetrates into the sample. However, the superconducting property is not fully destroyed and consequently, the magnetic flux can still not fully penetrate into the sample. Within this region, the penetrated flux strength is smaller than that of the applied field. In this state, the partially penetrated magnetic flux gather around a number of quantized flux-enclosing supercurrent vortices inside the body. This state is often called the mixed state or the vortex state.

iii) For $H > H_{c2}$, the specimen becomes completely normal and magnetic flux penetrates thoroughly into the body. The value of $H_{c2}$ can be very high and its value depends on a number of factors including temperature and geometry.

iv) When the value of $H$ is decreased from above $H_{c2}$ to just under $H_{c2}$, we see surface superconductivity appearing and nucleation of vortices inside the body.

v) It can also be observed that the creation of vortices requires a stronger magnetic field than that required to maintain vortices. To be more precise, if $H_1$ is the minimum value required to create a vortex in the sample, then after the vortex has been created, if we start lowering the applied magnetic field, to just under $H_1$, we can still observe the persistent presence of this vortex in the domain. It will not disappear until the magnetic field is lowered a bit further. This phenomenon is known as hysteresis. This has also been observed in numerical analysis of the problem (see [CHo 95]).

Finally, we emphasise that the values of $H_{c1}$ and $H_{c2}$ are situation dependent. That is, even if the material is the same, the value distribution and direction of the applied magnetic field, the temperature, the geometric shape of the sample and the history of the sample may all affect their values (see Figure 2).

### 1.1.5 Vortices

We already know that for a type II superconductor, when $H_{c1} < |\boldsymbol{H}| < H_{c2}$, we have the mixed state or vortex state and the magnetic field gradually penetrates into the material. The most interesting physical observation is that the penetrated magnetic field does not distribute evenly inside the superconductor. It instead con-

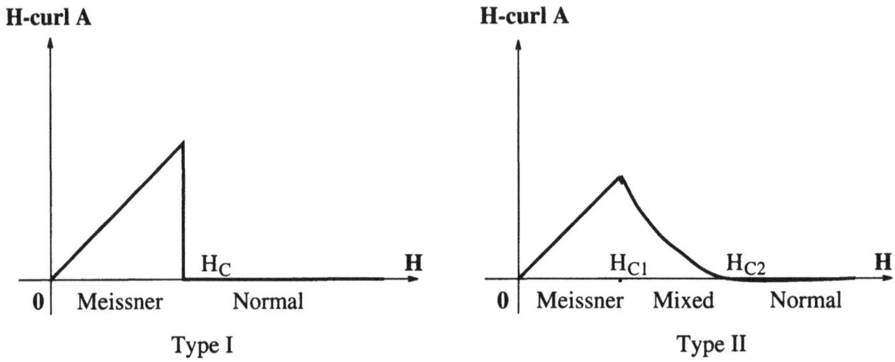


Figure 2. Relationship between induced and applied magnetic fields for types I and II superconductors

centrates along some co-dimension two objects inside the superconductor sample. Co-dimension two objects are points if the space dimension is 2 and curves when the space dimension is three. These co-dimension two objects are the **vortices**. In experiments, vortices are observed to meet the boundary of the superconductor sample perpendicularly.

Around the core of a vortex, there is a supercurrent circulating. This supercurrent produces a strong magnetic field (usually with strength much larger than $H_{c2}$) in the tangential direction of the vortex along the vortex core (or perpendicular to the superconductor in which the vortex is situated if the space dimension is two). So near the vortex core, the material becomes normal instead of superconducting. This induced magnetic field will also radiate into the surrounding region and affect the magnetic field perpendicular to the surface of the superconductor sample.

The mathematical interests in studying superconductivity related mathematical models lies in the study of the structure of co-dimension 2 phase transition domains. The word phase transition in this context means the change from the normal phase to the superconducting phase or vice versa.

Apart from the co-dimension 2 feature of the vortices, another quantity is orientation. This is decided by the direction of the circulating supercurrent around the vortex core. It is easy to understand that there are only two possibilities for orientation. Viewed from a fixed direction at a cross section perpendicular to the vortex core curve, the supercurrent can circulate either clockwise or anti-clockwise.

Therefore, it is mathematically possible to fix a choice, say a vortex with clockwise circulating supercurrent to be a positive vortex and a vortex with anti-clockwise circulating supercurrent to be a negative vortex. This distinction is only useful in the 2-d case. When the space is 3-d, it is not possible to define positive and negative vortices (see Figure 3, where $|\psi|^2$ represents the density of superconducting electron pairs).

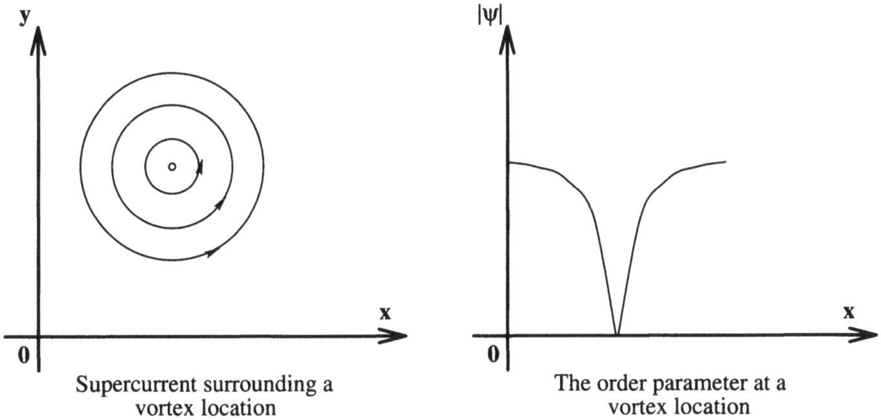


Figure 3: Supercurrent and order parameter near a vortex

### 1.1.6 Summary

In this monograph, we report on some of the most recent research progress on the G-L superconductivity model. We will consider the mathematics justifications of the phenomena mentioned above and many other related issues. We are not going to discuss the thermodynamic aspects of the theory as we do not know any satisfactory mathematical thermodynamic model. So if temperature is involved in our arguments, it will be regarded as a constant given in the predetermined circumstances.

The primary mathematical interest in studying the G-L theory of superconductivity has been the understanding of co-dimension two phase transition problems (vortex structured phase transition problems). Of particular interest will be those properties with possible physical applications. Some of the interesting topics are, for example: are necessarily the phase transition domains described by the G-L system of co-dimension 2 structure? Do they actually move in space-time? What is the law of motion? In the equilibrium state, can vortices of different orientation co-exist (this question may only be relevant in 2-d case)? How do the vortices interact? What is the relation between the vortices and the applied magnetic field? How are they created by increase of magnetic field? These are all interesting physical as well as mathematical problems.

In this book, we will give the mathematical answer to some of these questions. Many of the problems we asked here are still open problems.

In the rest of this chapter, we will introduce the G-L system, explain the scaling involved, study some of the basic concepts of co-dimension two-phase transition problems and some relevant mathematical tools associated to pave the way for further studies.

## 1.2 The G-L phenomenological theory

In this section, we introduce the main object of our study, the G-L system. The key issue at this stage is to show how to rescale the system to certain familiar forms and discuss some of the outcomes of the scaling involved. We discuss the mathematical implications of some of the conditions which the founders of the G-L theory regarded as important for the model to remain valid. We also give a brief introductory discussion on the influence of rescaling when we send certain parameters in the rescaled G-L system to zero or infinity in order to derive some limit equations. The derivation of these limits is becoming an important part of recent mathematical research. However, it is very difficult to find a systematic discussion in the existing literature on these topics.

### 1.2.1 The free energy and the G-L equations

In 1937, Landau proposed a general theory of second order phase transitions based on the following fundamental assumptions:

a) There exists a (complex) scalar order parameter $\psi$ which goes to zero at the transition.

b) There is an homogeneous free energy density which may be expanded in powers of $\psi$.

c) The coefficients of the expansion are regular functions of the temperature $T$.

Following these assumptions, Landau wrote down the free energy per unit volume as follows ($\cdots$ represents higher powers of $|\psi|$)

$$F = F_n + \alpha(T)|\psi|^2 + \frac{\beta(T)}{2}|\psi|^4 + \cdots . \tag{2.1}$$

He pointed out that the temperature $T$ is an important factor intervening in the phase transition process. He claimed that the model is valid when $T$ is close to a particular critical value $T_0$. This critical temperature $T_0$ is defined in the following sense: when $T < T_0$, superconductivity property appears, when $T > T_0$, the superconductivity property disappears.

Mathematically, this property is expressed through the following expressions:

$$\alpha(T) \sim (T - T_0)\alpha_1, \ \beta(T) \sim \beta_0.$$

In the above expressions, $\alpha_1$ and $\beta_0$ are positive constants. When $T < T_0$, the free energy (2.1) has an absolute minimum for $|\psi|^2 = -\frac{\alpha}{\beta}$. This implies that the energy minimizer $\psi$ should not be identically zero. Obviously, when $T \geq T_0$, the energy minimizer $\psi$ satisfies $|\psi| = 0$.

In the context of superconductivity, $|\psi|^2$ will be used as the density of superconducting electron pairs. Following the above discussions, when $T < T_0$, the material prefers to take the superconducting state. By contrast, when $T > T_0$, the material prefers to take the normal state.

We have to point out that the expression (2.1) applies only when $\psi$ is a constant in the domain of consideration. If $\psi$ is not a constant, the term $|\nabla\psi|^2$ has to be added and an integration over the domain of consideration has to be performed. To consider the full picture, the energy of the magnetic field should also be taken care of.

In 1950, Ginzburg and Landau proposed an extension of the Landau energy to describe the superconductor in the presence of a magnetic field. They proposed that $\psi$ is taken as the wave function for a particle of charge $e^*$ and mass $m^*$. In order to ensure gauge invariance, they wrote the free energy as ($c$ is the light speed, $\hbar$ is the Planck constant)

$$E(\psi, \boldsymbol{A}) = \int_\Omega F_n + \alpha|\psi|^2 + \frac{\beta}{2}|\psi|^4 + \frac{1}{2m^*}\left|\left(-i\hbar\nabla - \frac{e^*\boldsymbol{A}}{c}\right)\psi\right|^2 + \frac{|\mathbf{curl}\boldsymbol{A}|^2}{8\pi} \tag{2.2}$$

where $\boldsymbol{A}$ is the vector potential for the magnetic field $\boldsymbol{h}$, i.e.,

$$\boldsymbol{h} = \mathbf{curl}\boldsymbol{A}. \tag{2.3}$$

This is the original G-L energy for superconductivity taking into consideration the effect of the magnetic field.

It is also possible to study the Gibbs energy, which takes into account the applied magnetic field more explicitly. The Gibbs energy has the following form:

$$E(\psi, \boldsymbol{A}) = \int_\Omega F_n + \alpha|\psi|^2 + \frac{\beta}{2}|\psi|^4 + \frac{1}{2m^*}\left|\left(-i\hbar\nabla - \frac{e^*\boldsymbol{A}}{c}\right)\psi\right|^2 + \frac{|\mathbf{curl}\boldsymbol{A} - \boldsymbol{H}|^2}{8\pi} \tag{2.4}$$

where $\boldsymbol{H}$ is the applied field from outside. The difference in these two models will be seen when we try to derive the PDE (partial differential equation) system with associated boundary conditions. Physically, we believe this concerns how to deal with the applied, total and induced field although no one states the reason explicitly.

As a matter of fact, we now understand that the energy expression (2.4) is not the whole story. If we consider the superconductivity problem on a domain $\Omega$ which is not the entire space, we will need the Maxwell system in $\mathbb{R}^n \setminus \Omega$. It can be proved, however, that under some reasonable assumptions, (2.4) is valid when the space dimension is 2. We will discuss this in greater detail later in Section 8. Here we limit our discussions temporarily to the 2-d case. Hence, $\boldsymbol{A} = (A_1, A_2)$, $\mathbf{curl}\boldsymbol{A} = \partial_1 A_2 - \partial_2 A_1$ and $\boldsymbol{H}$ becomes a scalar function $H$.

Consequently, if we calculate the first order variation and let it be zero (the idea behind it is to minimize the energy), we obtain, in the traditional G-L case (2.2),

$$
\begin{aligned}
0 = \delta E &= \operatorname{Re} \int_\Omega \delta\psi^* \Big[ \alpha\psi + \beta|\psi|^2\psi + \frac{1}{2m^*}\Big( -i\hbar\nabla - \frac{e^*\boldsymbol{A}}{c}\Big)^2 \psi \Big] \\
&+ \operatorname{Re} \int_\Omega \delta\boldsymbol{A} \cdot \Big( \frac{\mathbf{curl}^2\boldsymbol{A}}{8\pi} - \frac{e^*\psi^*}{2m^*c}(-i\hbar\nabla - \frac{e^*}{c}\boldsymbol{A})\psi \Big) \\
&+ \operatorname{Re} \int_{\partial\Omega} \delta\psi^*(-i\hbar\nabla - \frac{e^*}{c}\boldsymbol{A})\psi \cdot \boldsymbol{n} + \frac{1}{8\pi}(\delta\boldsymbol{A}\cdot\boldsymbol{\tau})\mathbf{curl}\boldsymbol{A}
\end{aligned}
$$

and the corresponding G-L equations are

$$
\begin{cases}
\frac{1}{2m^*}(-i\hbar\nabla - \frac{e^*\boldsymbol{A}}{c})^2\psi + \alpha\psi + \beta|\psi|^2\psi = 0, \\
\frac{\mathbf{curl}^2\boldsymbol{A}}{4\pi} = \frac{\boldsymbol{j}}{c} = \frac{e^*\hbar}{2im^*c}(\psi^*\nabla\psi - \psi\nabla\psi^*) - \frac{e^{*2}}{m^*c^2}|\psi|^2\boldsymbol{A}
\end{cases}
\tag{2.5}
$$

with corresponding boundary conditions

$$
\begin{cases}
(-i\hbar\nabla - \frac{e^*}{c}\boldsymbol{A})\psi \cdot \boldsymbol{n} = 0, & x \in \partial\Omega, \\
\mathbf{curl}\boldsymbol{A} = 0, & x \in \partial\Omega.
\end{cases}
\tag{2.6}
$$

It is clear that the boundary condition for the magnetic field on the boundary is not satisfactory because it states that the applied field is not playing a natural role in the mathematics formulation and the model is becoming isolated from outside. Of course, we can impose the following physically more reasonable boundary conditions

$$
\begin{cases}
(-i\hbar\nabla - \frac{e^*}{c}\boldsymbol{A})\psi \cdot \boldsymbol{n} = 0, & x \in \partial\Omega, \\
\mathbf{curl}\boldsymbol{A} = H, & x \in \partial\Omega.
\end{cases}
\tag{2.7}
$$

But then, the solutions of the PDE problem are neither energy minimizers nor stationary points of the energy.

However, with the Gibbs energy (2.4), there is no such inconvenience. The first variation leads to the following PDE system:

$$
\begin{cases}
\frac{1}{2m^*}(-i\hbar\nabla - \frac{e^*\boldsymbol{A}}{c})^2\psi + \alpha\psi + \beta|\psi|^2\psi = 0, \\
\frac{\mathbf{curl}^2\boldsymbol{A} - \mathbf{curl}\boldsymbol{H}}{4\pi} = \frac{\boldsymbol{j}}{c} = \frac{e^*\hbar}{2im^*c}(\psi^*\nabla\psi - \psi\nabla\psi^*) - \frac{e^{*2}}{m^*c^2}|\psi|^2\boldsymbol{A}
\end{cases}
\tag{2.8}
$$

with natural boundary conditions

$$
\begin{cases}
(-i\hbar\nabla - \frac{e^*}{c}\boldsymbol{A})\psi \cdot \boldsymbol{n} = 0, & x \in \partial\Omega, \\
\mathbf{curl}\boldsymbol{A} = H, & x \in \partial\Omega.
\end{cases}
\tag{2.9}
$$

We want the condition $\mathbf{curl}\boldsymbol{A} = H$ when $x \in \partial\Omega$ because we want the applied magnetic field and the induced magnetic field to be the same along the tangential direction of the boundary of the superconductor sample.

These conditions are for a sample placed in the vacuum. If the outside of the superconductor sample is filled with some kind of conductor, De Gennes [Gen 66] proposed that we should impose the following boundary conditions:

$$\begin{cases} (i\hbar\nabla + \frac{e^*}{c}\boldsymbol{A})\psi \cdot \boldsymbol{n} = -i\eta\psi, & x \in \partial\Omega, \\ \mathbf{curl}\boldsymbol{A} = H, & x \in \partial\Omega \end{cases} \tag{2.10}$$

for some (preferably nonnegative for mathematics convenience) real valued function $\eta$. This is a more complicated issue. The energy expression needs to be changed correspondingly as well.

When the space dimension is 3, the Maxwell equation in the outside domain should also be taken into account (see [CDG 96]). For example, when the outside domain is vacuum, we should use the energy

$$\begin{aligned} E(\psi, \boldsymbol{A}) &= \int_\Omega \Big( F_n + \alpha|\psi|^2 + \frac{\beta}{2}|\psi|^4 \\ &+ \frac{1}{2m^*}\Big|\Big( - i\hbar\nabla - \frac{e^*\boldsymbol{A}}{c}\Big)\psi\Big|^2\Big) + \int_{\mathbb{R}^n} \frac{|\mathbf{curl}\boldsymbol{A} - \boldsymbol{H}|^2}{8\pi}. \end{aligned}$$

**Remark 1.2.1** *It is formally known that the vortices tend to meet the boundary of the superconductor sample perpendicularly. As vortices are formed by the cores of circulating supercurrents, the supercurrent generates a magnetic field along the tangential direction of the core, and subsequently, perpendicular to the sample boundary. It is this observation that leads to the thinking of using Maxwell system in the outside domain.*

In the following, when we discuss the G-L energy, we are referring to the Gibbs energy most of the time. The Gibbs energy leads to much better understanding of the problem and enables us to have sound mathematical formulations.

In principle, we can now decide the order parameter (wave function) $\psi$ and the magnetic potential $\boldsymbol{A}$ with the help of the PDE system and the boundary conditions. However, the PDE system (2.5) and associated conditions (2.6), (2.7) or (2.10) satisfy also the so-called gauge invariance property. As there are too many physical constants involved in our systems, we postpone the discussion of gauge invariance after rescaling has been carried out.

**Remark 1.2.2** *From the mathematical meaning of the order parameter, the vortex (discussed in Section 1.5) locations are given by*

$$\{x \in \Omega, \ \psi(x) = 0.\}$$

### 1.2.2 Rescaling and the values of the constants

For the rescaling of the problem, we take the 2-d case as an example. The whole process extends to 3-d case automatically. The burden there is that we have to take care of the Maxwell system at the same time.

Firstly, let us specify the values of $e^*$ and $m^*$. They are the values of the electric charge and mass of the Cooper pairs, respectively. It was shown by Gorkov in 1959 that $e^* = 2e$ where $e$ is the standard electronic charge. He also showed that the value of $m^*$ can be arbitrary. The physics convention is then to let it be $m$ (not $2m$ !), the standard electron mass. So the G-L system can be rewritten as

$$\begin{cases} \frac{1}{2m}(-i\hbar\nabla - \frac{2e\boldsymbol{A}}{c})^2\psi + \alpha\psi + \beta|\psi|^2\psi = 0, & x \in \Omega, \\ \frac{\mathbf{curl}h}{4\pi} = \frac{\dot{\jmath}}{c} = \frac{2e\hbar}{2imc}(\psi^*\nabla\psi - \psi\nabla\psi^*) - \frac{4e^2}{mc^2}|\psi|^2\boldsymbol{A}, & x \in \Omega \end{cases} \tag{2.11}$$

where depending on the choice of energy, $h = \mathbf{curl}\boldsymbol{A}$ or $h = \mathbf{curl}\boldsymbol{A} - H$.

As to the quantities $\alpha$ and $\beta$, there are various values assigned to them. Gorkov showed that for pure metal, we have

$$\alpha = 1.83\frac{\hbar^2}{2m}\frac{1}{\xi_0^2}\frac{T - T_0}{T_0}, \quad \beta = 0.35\frac{1}{N(0)}\left(\frac{\hbar^2}{2m}\frac{1}{\xi_0^2}\right)^2\frac{1}{(k_B T_0)^2} \tag{2.12}$$

where $\xi_0$ is the ground state coherence length and $N(0)$ is the density of states at the Fermi level. In the BCS theory, they have respective values

$$\alpha = N(0)\frac{T - T_0}{T_0}, \quad \beta = 0.098\frac{N(0)}{(k_B T_0)^2}. \tag{2.13}$$

Now we are in a position to discuss the scaling of the G-L system. Let us introduce the following physical parameters:

$$\begin{cases} |\psi_0|^2 = |\alpha|/\beta, \\ H_c = (4\pi|\alpha|^2/\beta)^{1/2}, \\ \lambda = \lambda(T) = (mc^2\beta/16\pi e^2|\alpha|)^{1/2}, \\ \xi = \xi(T) = \hbar/(2m|\alpha|)^{1/2}, \\ \kappa = \lambda/\xi. \end{cases} \tag{2.14}$$

Physically, $|\psi_0|^2$ stands for the equilibrium density, $H_c$ for the thermodynamic critical field, $\lambda$ for the penetration depth, $\xi$ for the coherence length. The ratio of the two characteristic lengths $\kappa = \lambda/\xi$ is the G-L parameter of the substance and is a dimensionless constant.

We now introduce the following new non-dimensional variables (those with prime):

$$\begin{cases} x = \lambda x', \\ \psi = \psi_0\psi', \\ \boldsymbol{A} = \sqrt{2}H_c\lambda\boldsymbol{A}', \\ H = \sqrt{2}H_c H' \end{cases} \tag{2.15}$$

Then we have the following non-dimensional G-L equations (we henceforth drop the primes):

$$\begin{cases} (\frac{i}{\kappa}\nabla + \boldsymbol{A})^2\psi + (|\psi|^2 - 1)\psi = 0, \\ \mathbf{curl}h + \frac{i}{2\kappa}(\psi^*\nabla\psi - \psi\nabla\psi^*) + |\psi|^2\boldsymbol{A} = 0. \end{cases} \tag{2.16}$$

Here again $h$ could be $\mathbf{curl}\boldsymbol{A}$ or $\mathbf{curl}\boldsymbol{A} - H$ depending on which energy we use. The corresponding boundary conditions are

$$(\frac{i}{\kappa}\nabla\psi + \boldsymbol{A}\psi)\cdot\boldsymbol{n} = 0, \quad \mathbf{curl}\boldsymbol{A} = H. \tag{2.17}$$

This is one particular form of the rescaled system. The corresponding rescaled energy is

$$E(\psi, \boldsymbol{A}) = \int_\Omega (|\psi|^2 - 1)^2 + \left|(\frac{i}{\kappa}\nabla + \boldsymbol{A})\psi\right|^2 + |h|^2 dx \tag{2.18}$$

As a matter of fact, if we scale again to introduce new dimensionless quantities (those with prime)

$$\boldsymbol{A} = \frac{1}{\kappa}\boldsymbol{A}', \; H = \frac{1}{\kappa}H', \; \psi = \psi', \tag{2.19}$$

we obtain (drop the primes again)

$$\begin{cases} (i\nabla + \boldsymbol{A})^2\psi + \kappa^2(|\psi|^2 - 1)\psi = 0, \\ \mathbf{curl}h + \frac{i}{2}(\psi^*\nabla\psi - \psi\nabla\psi^*) + |\psi|^2\boldsymbol{A} = 0. \end{cases} \tag{2.20}$$

The corresponding boundary conditions are

$$(i\nabla\psi + \boldsymbol{A}\psi)\cdot\boldsymbol{n} = 0, \quad \mathbf{curl}\boldsymbol{A} = H. \tag{2.21}$$

The corresponding rescaled energy is

$$E(\psi, \boldsymbol{A}) = \int_\Omega \kappa^2(|\psi|^2 - 1)^2 + \left|(i\nabla + \boldsymbol{A})\psi\right|^2 + |h|^2 dx. \tag{2.22}$$

(2.16) and (2.20) are two of the most popularly used rescaled forms of the G-L system.

**Remark 1.2.3** *From now on, we mainly use the rescaled forms (2.16) and (2.20). We do not give a clear preference because depending on the circumstances, one form may be preferred to the other.*

### 1.2.3 Gauge invariance

In the energy expression (2.18), it is easy to see that for any given smooth function $\theta$,

$$E(\psi, \boldsymbol{A}) = E(\psi\exp(i\kappa\theta), \boldsymbol{A} + \nabla\theta) \tag{2.23}$$

and in the energy form (2.22), it is easy to see that

$$E(\psi, \boldsymbol{A}) = E(\psi\exp(i\theta), \boldsymbol{A} + \nabla\theta). \tag{2.24}$$

Let us take the form in (2.22) as an example. The transform

$$(\psi, \boldsymbol{A}) \to (\psi \exp(i\theta), \boldsymbol{A} + \nabla\theta) \tag{2.25}$$

is called a **gauge transformation**. The fact that the energy is invariant under gauge transformation is called **gauge invariance**. This invariance property will help us to establish the existence of solutions later on.

## 1.3 Some considerations arising from scaling

In the previous section, we have seen the introduction of the G-L system and associated boundary conditions. However, some boundary conditions may couse mathematical difficulties. For example,

$$(i\nabla + \boldsymbol{A})\psi \cdot \boldsymbol{n} = -i\eta\psi \tag{3.1}$$

is one such condition. If the superconductor sample is very thin, the superconductivity property might be destroyed by the presence of $\eta$ (An informal communication by J. Rubinstein). De Gennes showed that (see [Gen 66] (3.1) holds for a superconductor-normal metal interface with nonzero $\eta$ and holds for super-insulator (vacuum) interface with zero $\eta$. At the first sight, we seem to have difficulty in explaining this phenomenon from a physical point of view. This might have, as a matter of fact, more to do with the validity of the G-L system as a mathematical model as we will briefly discuss at the end of this section.

### 1.3.1 The two characteristic lengths $\xi(T)$ and $\lambda(T)$

The G-L system has two particular solutions:

1) $\psi \equiv 0$ and $\textbf{curl}\boldsymbol{A} = H$ when $H$ is large (we will give a qualitative confirmation of this result later). Here $H$ is the applied field. This solution describes the normal state.

2) $\psi = \psi_0 = \sqrt{|\alpha|/\beta}$ and $\boldsymbol{A} = 0$ when $H = 0$. This is the pure superconducting state without outside applied field.

When the applied field is very weak, $\psi$ will vary slowly, close to $\psi_0$. We can let $\boldsymbol{A} = 0$ and $f = \psi/\psi_0$, then $f$ satisfies

$$-\frac{\hbar^2}{2m}\Delta f + \alpha(f - f^3) = 0.$$

This leads to the definition of

$$-\frac{\hbar^2}{2m\alpha} = \xi^2(T) \tag{3.2}$$

as the temperature dependent coherence length (firstly introduced in (2.14)). It represents the length scale on which $f$ varies between 0 and 1.

Assuming again that the field is low and $\psi$ is almost $\psi_0$, we write $\psi = \psi_0 e^{i\chi}$ and we have (by (2.11))

$$\boldsymbol{j} = \frac{2e\hbar}{m}\psi_0^2 \nabla\chi - \frac{4e^2}{mc}\psi_0^2 \boldsymbol{A}. \tag{3.3}$$

Taking the curl of $\boldsymbol{j}$, one obtains

$$\mathbf{curl}\boldsymbol{j} = -\frac{4e^2}{mc}\psi_0^2 \boldsymbol{h}. \tag{3.4}$$

This is equivalent to the London equation (1.8) with the penetration depth

$$\lambda(T) = \sqrt{mc^2/16\pi e^2 \psi_0^2}. \tag{3.5}$$

**Remark 1.3.1** *(3.2) and (3.5) give some physical insight on how $\xi(T)$ and $\lambda(T)$ are defined. In contrast, in (2.14), they were introduced as if for the convenience of scaling only.*

*It is clear that $\xi(T)$ and $\lambda(T)$ are different from the quantities $\xi_0$ and $\lambda_L$ we encountered in Section 1. $\xi_0$ and $\lambda_L$ are called the ground state coherence length and penetration depth, respectively.*

### 1.3.2 The validity of the G-L theory

There are quite a few issues concerning the validity of the G-L theory; for example, from the physical side, the validity of the order parameter, the justification of the coupling relation between the order parameter and the magnetic field. In this section, we only discuss those that are regarded as mathematically relevant issues (however, they all come from physical considerations). From the existing physics literature, we identified some of the conditions that have to be imposed to mathematically validate the G-L theory as in the following:

$$\xi(T) \gg \xi_0, \tag{3.6}$$

$$\lambda(T) \gg \xi_0. \tag{3.7}$$

Looking at (3.2) and (3.5) and also at (2.12)–(2.13), both conditions imply that $|T - T_0|$ is small. The physical consideration behind these assumptions is that the wave function should vary SLOWLY over a range of distance $\xi_0$.

In the rescaled problem, as the equations depend on $\kappa$ only and as $\kappa$ is the ratio between $\lambda(T)$ and $\xi(T)$ which is independent of $T$, validity is not an issue when the value of $\kappa$ is kept constant. However, when we let $\kappa$ tend to 0 or infinity to investigate the limits, we will have to justify that the validity arguments are maintained. Now we investigate a few situations that will be of interests later on in the book.

**1) Large $\kappa$ case – general observation**

The typical large $\kappa$ limits are so far investigated under the rescaled energy (2.22). The mathematical research has been concentrated in the cases where $\boldsymbol{A} = 0$ (see [BBH 94]) or $\boldsymbol{H}$ is proportional in magnitude to $\kappa$ (see [CDGP 95]). In the scaling scheme (2.22), it is clear that large $\kappa$ tends to push the vortices out of the domain of consideration. So in the case $\boldsymbol{A} = 0$, vortices are maintained by imposing a Dirichlet boundary condition. In the large $\kappa$, large $\boldsymbol{H}$ case, the effect of large $\boldsymbol{H}$ should maintain the presence of the vortices.

The question to ask is what does large $\kappa$ mean from the scaling point of view? There are many possibilities but two of them are of particular interests:

**The first scenario is** based on the facts that

1) $\xi(T) \gg \xi_0$,

2) $\xi_0$ is a constant for a given material irrespective how we adjust temperature and other factors,

we know that $\xi(T)$ can not tend to zero and $\lambda$ has to tend to $\infty$ if $\kappa$ were to go to infinity. Looking at the scaling formulae (2.15) and (2.19), we have $x = \lambda x'$ where $x'$ is the new variable. In order to work on a finite domain in the new variable space, the original physical space has to be extended (eventually to an unbounded domain).

If we maintain a finite number of vortices in the new domain and discuss the limit, then going back to the original space, we are looking at some far away tiny, gently varying vortices. On the other hand, the relations (2.15) and (2.19) imply that the scaling does not affect the order of the magnetic field. Consequently, it is not a good physical setting except for the pure Landau energy case (where $\boldsymbol{A}$ and $\boldsymbol{H}$ are all set to be 0). Works in this area include [BBH 94] and [Lin 96].

**The second scenario is** to fix $\lambda$ and decrease $\xi$ to 0. We are consequently implying that $\xi_0 \to 0$. That is, the coherence length tends to zero, or equivalently, the electron interaction range tends to zero. The electrons are no longer in a condensed state, they really behave like a group of non-interacting particles (but without physical size). This can be interpreted as extreme type II behaviour although we do not particularly anticipate that $\xi_0$ is extremely small (in comparison to $\lambda$).

This way, we achieve the limit $\kappa \to \infty$ without shrinking the domain but it also implies that the original magnetic field is very small (see (2.15) and (2.19)). In this kind of limit, we expect the effect of the magnetic field is very small on the equation of motion for vortices. Works in this field include [BRi 95]. The research in this context demonstrates again that vortices need to be maintained by Dirichlet type boundary conditions.

It seems that when $\kappa \to \infty$, the problem is associated with weak magnetic field. One way to overcome this is to introduce an imposed large magnetic field (of the order of $\kappa$), then the limit problem is of a complete different nature. Works in this field include, for example, [CDPG 95].

Many other combination of $\xi$ and $\lambda$ lead to the limit $\kappa \to \infty$. We obviously can not discuss all of them here. However, once we understand that $\lambda$ is the distance needed for magnetic field to come through into the sample and $\xi$ is the "bonding" distance among the electrons (to counter the penetration of magnetic field), it is not difficult to make a tailor made discussion for each individual case when necessary.

**2) Large $\kappa$ case – mean field model**

Worth mentioning is the recently studied **mean field model** (see [Cha2 95]). This fits the image of non-interacting particles in our second scenario above. The mean field theory is based on the following arguments: according to (3.3), we have

$$\frac{c}{4\pi}\mathbf{curl}\boldsymbol{h} = \frac{2e\hbar}{m}\psi_0^2\nabla\chi - \frac{4e^2}{mc}\psi_0^2\boldsymbol{A}. \tag{3.8}$$

Assume that $\kappa$ is large under the context that $\xi$ is small, $\lambda$ remains constant. As the vortices are non-interacting particles, they do not interact and $\mathbf{curl}\nabla\chi$ is a combination of approximate Dirac functions supported by the vortex locations. Taking **curl** on both sides of (3.8), we have

$$\frac{c}{4\pi}\mathbf{curl}^2\boldsymbol{h} = 2\pi\frac{2e\hbar}{m}\psi_0^2\sum_k\delta(x - x_k) - \frac{4e^2}{mc}\psi_0^2\boldsymbol{h} \tag{3.9}$$

where the sum is taken over $k$ and $\{x_k\}$ are the locations of vortices. If there are many vortices, we can define a vorticity term $\omega$ to replace $2\pi\frac{2e\hbar}{m}\psi_0^2\sum_k\delta(x - x_k)$ to obtain

$$\frac{c}{4\pi}\mathbf{curl}^2\boldsymbol{h} + \frac{4e^2}{mc}\psi_0^2\boldsymbol{h} = \omega. \tag{3.10}$$

Imposing an additional conservation law on the vorticity, we have the mean field model (see Section 3, Chapter 3 for more details).

This seems to be an interesting physically relevant result coming out of large $\kappa$ limit of the G-L model. There are several remarks we would like to make concerning possible future rigorous mathematical justification (open problem so far as we know) of the model:

**The first remark** is that the mean field model is two limits away from the G-L model: the first limit is a large $\kappa$ limit with (bounded-large – depending on how scaling is interpreted) magnetic field. In this case, both [BBH 94] and [BRi 95] seemed to have to maintain the vortex presence by using some "non-physical" boundary conditions (this is the only technical difficulty to overcome, as far as we know). The second limit is a large $\mathbf{curl}\boldsymbol{A}$, large vortex number limit (where $\mathbf{curl}\boldsymbol{A}$ is proportional to the number of vortices). The second limit, by contrast, is much easier to justify rigorously.

**The second remark** concerning the model is the imposition of a conservation law on the vorticity. This assumption needs many physical requirements such as vortices could not merge and could not move relative to each other. Weather this can be derived from a phenomenological theory using mathematical methods remains to be seen.

**The third remark** is somehow related to the first one: the derivation of the Dirac functions on the right-hand side is based on the assumption that

$$\mathbf{curl}(|\psi|^2\nabla\chi) \sim 2\pi|\psi|^2\delta,\ \mathbf{curl}(\psi_0^2\boldsymbol{A}) \sim \psi_0^2\boldsymbol{h} \tag{3.11}$$

near each vortex. This can be justified when non-physical Dirichlet type boundary condition is imposed. If we have Neumann type boundary conditions, it presents a serious mathematical challenge.

Despite lacking rigorous justification, the mean field model gives a fresh and innovative perspective on the modelling of practical problems where many non-interacting vortices seem to exist in a superconductor sample (many thousands at least) and move in a well organized way.

**3) Small physical domains**

From a pure mathematical point of view, the G-L expansion is only valid when $\psi$ varies slowly on the distance range $\xi_0$, which is a constant for a given material.

If we investigate problems such as a 3-d plate shrinking in thickness to a two dimensional film, the question can be asked if the G-L model is still valid. For this kind of problems, we need to make sure that the gradient of the order parameter is uniformly bounded in proportion to the size of the domain. For example, if the plate occupies $\Omega_\varepsilon = \Omega_0 \times (-\varepsilon, \varepsilon)$ where $\Omega_0$ is a fixed domain in $\mathbb{R}^2$ and $\varepsilon \to 0$, we have to make sure that the solution $\psi_\varepsilon$ on $\Omega_\varepsilon$ satisfies

$$||\partial_3\psi_\varepsilon|| \le cf(\varepsilon) \tag{3.12}$$

for some appropriate norm $||\cdot||$ and appropriate function $f(\cdot)$ (as we can usually work out higher regularity estimates from $L^2$ estimates, $L^2$ norm and $f(\varepsilon) = \sqrt{\varepsilon}$ in (3.12) are normally enough). As long as this criterion is fulfilled, at least from a mathematical perspective, there is no problem in applying the G-L theory.

**Answer to our introductory example**

After having seen some analysis on how scaling argument could affect the validity of the G-L model and how to analyse various situations, we now go back to the question raised at the beginning of this section.

Now we can offer a reasonable explanation as to why (3.1) might cause difficulty for analysing a thin plate model. Let the middle surface of the thin plate be placed in the plane $x_3 = 0$ with thickness $2\varepsilon$. Then (3.1) implies that there is a change of value of $\psi_{x_3}$ (of a finite magnitude) from $\eta\psi$ to $-\eta\psi$ over a distance $2\varepsilon$ in the $x_3$ direction. If we estimate $||\partial_3\psi_\varepsilon||$, the best we can now say is

$$||\partial_3\psi_\varepsilon|| \le c/\sqrt{\varepsilon}. \tag{3.13}$$

This is definitely disproportionate to the size of the domain (unless $\psi$ is identically zero on the upper and lower surface – $\psi$ would then be identically zero as the thickness of the plate shrinks). Hence, if we were to expect nonzero solutions, the assumption that the order parameter varies slowly would be broken. This is a very interesting situation where the G-L energy ceases to describe a physical situation due to validity issues.

Finally, what is wrong with this physics setting? We believe that if we place conductors around a superconductor sample, there must be a surface current along the junction, caused by electromagnetic interactions. If the thickness of the superconducting plate is getting smaller, the effect of the surface current will become more pronounced. If we include the surface current in the boundary conditions, the validity problem will be solved.

## 1.4 The evolutionary G-L system – 2-d case

In this section, because the problem is 2-d, we take the magnetic field in the outside domain as the applied field $H$. By doing so, we implicitly assumed that the outside field is quasi-static.

### 1.4.1 The system

The evolutionary G-L system was first derived by Gor'kov and Eliashberg in 1968. We simply write down the equations in the following standard form:

$$\begin{aligned} &\tfrac{\hbar^2}{2mD}\Big(\tfrac{\partial}{\partial t}+\tfrac{2ie}{\hbar}\phi\Big)\psi-|\alpha|\psi+\beta|\psi|^2\psi+\tfrac{1}{2m}\Big(-\tfrac{\hbar}{i}\nabla+\tfrac{2e}{c}\boldsymbol{A}\Big)^2\psi=0,\\ &\mathbf{curl}^2\boldsymbol{A}-\mathbf{curl}\boldsymbol{H}=-\tfrac{4\pi\sigma}{c}\Big(\tfrac{1}{c}\boldsymbol{A}_t+\nabla\phi\Big)+\\ &\qquad\tfrac{4\pi}{c}\Big[\tfrac{e\hbar}{im}(\psi^*\nabla\psi-\psi\nabla\psi^*)-\tfrac{4e^2}{mc}|\psi|^2\boldsymbol{A}\Big]. \end{aligned} \tag{4.1}$$

Compared with the steady state case, the newly introduced constants are $\sigma$, the conductivity of the normal phase; $D$, the diffusion constant and the function $\phi$, the electric potential.

It is easy to verify that for any given smooth function $\theta$, the following transformation

$$(\psi,\ \phi,\ \boldsymbol{A})\ \to\ (\psi\exp(i\theta),\ \phi-\frac{\hbar}{2e}\theta_t,\ \boldsymbol{A}-\frac{\hbar c}{2e}\nabla\theta) \tag{4.2}$$

does not change the equation. This transformation is the **gauge transformation** for the evolutionary problem.

### 1.4.2 Mathematical scaling

Similar to the steady state case, we introduce some physical parameters

$$\begin{aligned} |\psi|^2&=|\alpha|/\beta, & \kappa&=\lambda/\xi,\\ H_c&=\sqrt{4\pi|\alpha|^2/\beta}, & \eta&=4\pi\sigma D/c^2,\\ \lambda&=\sqrt{mc^2\beta/16\pi e^2|\alpha|}, & \tau&=\lambda^2/D\\ \xi&=\hbar/\sqrt{2m|\alpha|}, \end{aligned}$$

and non-dimensional variables

$$\begin{aligned} x &= \lambda x', \\ t &= \tau t', \\ \psi &= \psi_0 \psi', \\ \boldsymbol{A} &= \sqrt{2} H_c \lambda \boldsymbol{A}', \\ \phi &= \frac{\sqrt{2} D H_c}{c} \phi', \\ H &= \sqrt{2} H_c H' \end{aligned}$$

to arrive at (dropping the primes in the new equations)

$$\begin{cases} \frac{1}{\kappa^2}(\psi_t + i\kappa\phi\psi) + (\frac{i}{\kappa}\nabla + \boldsymbol{A})^2\psi + (|\psi|^2 - 1)\psi = 0, \\ \eta(\boldsymbol{A}_t + \nabla\phi) + \frac{i}{2\kappa}(\psi^*\nabla\psi - \psi\nabla\psi^*) \\ \qquad + |\psi|^2\boldsymbol{A} + \mathbf{curl}^2\boldsymbol{A} - \mathbf{curl}H = 0. \end{cases} \tag{4.3}$$

The corresponding boundary and initial conditions are

$$\begin{cases} (\frac{i}{\kappa}\nabla\psi + \boldsymbol{A}\psi) \cdot \boldsymbol{n} = -i\eta_0\psi, & \text{on } \partial\Omega \times (0,T), \\ \mathbf{curl}\boldsymbol{A} = H, & \text{on } \partial\Omega \times (0,T), \\ \psi(x,0) = \psi_0(x), & \text{in } \Omega, \\ \boldsymbol{A}(x,0) = \boldsymbol{A}_0(x) & \text{in } \Omega. \end{cases} \tag{4.4}$$

Here $\eta_0$, $\psi_0$ and $\boldsymbol{A}_0$ are given functions.

If we let

$$\boldsymbol{A} = \frac{1}{\kappa}\boldsymbol{A}', \ \phi = \frac{1}{\kappa}\phi', \ H = \frac{1}{\kappa}H', \ \psi = \psi',$$

we obtain

$$\begin{cases} \psi_t + i\phi\psi + (i\nabla + \boldsymbol{A})^2\psi + \kappa^2(|\psi|^2 - 1)\psi = 0, \\ \eta(\boldsymbol{A}_t + \nabla\phi) + \frac{i}{2}(\psi^*\nabla\psi - \psi\nabla\psi^*) \\ \qquad + |\psi|^2\boldsymbol{A} + \mathbf{curl}^2\boldsymbol{A} - \mathbf{curl}H = 0. \end{cases} \tag{4.5}$$

The corresponding boundary and initial conditions are

$$\begin{cases} (i\nabla\psi + \boldsymbol{A}\psi) \cdot \boldsymbol{n} = -i\eta_0\psi, & \text{on } \partial\Omega \times (0,T), \\ \mathbf{curl}\boldsymbol{A} = H, & \text{on } \partial\Omega \times (0,T), \\ \psi(x,0) = \psi_0(x), & \text{in } \Omega, \\ \boldsymbol{A}(x,0) = \boldsymbol{A}_0(x), & \text{in}\Omega. \end{cases} \tag{4.6}$$

Here $\eta_0$, $\psi_0$ and $\boldsymbol{A}_0$ are given functions.

### 1.4.3 The G-L functional as a Lyapunov functional

In the following, to have a fixed idea for our discussions, we will use the simpler formulation (4.5)–(4.6) unless otherwise stated.

It is also noticed that in much of the physics literature, there is no **curl**$H$ in the evolutionary system (see e.g. [GEl 68]). The reason is probably that they consider either the problem on the whole space (there is no place to send an applied field then) or the case where $H$ is a constant, so **curl**$H = 0$. However, we intend to keep **curl**$H$ in the equation because as we will see from the next theorem, it is a mathematically reasonable term to be present in the equations to relate the evolutionary system and the steady state system.

**Theorem 1.4.1** *Let $u = (\psi,\ \phi,\ \boldsymbol{A})$ be a triple of solutions of (4.5)–(4.6), $H$ be independent of time $t$ and $\eta_0 = 0$. Defining*

$$E(u) = \int_\Omega \frac{1}{2}|(i\nabla + \boldsymbol{A})\psi|^2 + \frac{1}{4}\kappa^2(|\psi|^2 - 1)^2 + \frac{1}{2}|\mathbf{curl}\boldsymbol{A} - H|^2 dx, \tag{4.7}$$

*we have*

$$\frac{d}{dt}E(u) = -|\psi_t + i\phi\psi|^2 - \eta|\boldsymbol{A}_t + \nabla\phi|^2 \leq 0. \tag{4.8}$$

*Proof.* Introduce the following gauge transformation (which affects neither the G-L system nor the G-L (or equivalently Gibbs) energy):

$$\begin{cases} \xi = \psi \exp(i\chi), \\ \boldsymbol{B} = \boldsymbol{A} + \nabla\chi, \\ \chi = \int_0^t \phi(x,s)ds. \end{cases} \tag{4.9}$$

Noticing that

$$\begin{cases} \xi_t = (\psi_t + i\phi\psi)\exp(i\chi), \\ \boldsymbol{B}_t = \boldsymbol{A}_t + \nabla\phi, \\ (i\nabla + \boldsymbol{B})\xi = ((i\nabla + \boldsymbol{A})\psi)\exp(i\chi), \end{cases} \tag{4.10}$$

we multiply the complex conjugate of (4.5-1) by $\psi_t + i\phi\psi$ and take the real part, then multiply (4.5-2) by $\boldsymbol{A}_t + \nabla\phi$. Adding up the results and using gauge invariance, we obtain,

$$\begin{aligned} &-|\psi_t + i\phi\psi|^2 - \eta|\boldsymbol{A}_t + \nabla\phi|^2 \\ &= \int_\Omega \Big[\mathrm{Re}[(\psi_t + i\phi\psi)((-i\nabla + \boldsymbol{A})^2\psi^* + \kappa^2(|\psi|^2 - 1)\psi^*)] \\ &\quad +(\boldsymbol{A}_t + \nabla\phi)\cdot((\mathbf{curl}^2\boldsymbol{A} - \mathbf{curl}H) \\ &\quad +\tfrac{i}{2}(\psi^*\nabla\psi - \psi\nabla\psi^*) + |\psi|^2\boldsymbol{A})\Big]dx \\ &= \int_\Omega \Big[\mathrm{Re}[((i\nabla + \boldsymbol{B})\xi)_t(-i\nabla + \boldsymbol{B})\xi^* + \kappa^2(|\xi|^2 - 1)\xi^*\xi_t)] \\ &\quad +(\mathbf{curl}\boldsymbol{B} - H)(\mathbf{curl}\boldsymbol{B} - H)_t + (\mathbf{curl}\boldsymbol{B} - H)H_t\Big]dx \\ &= \tfrac{dE}{dt}(\xi, \boldsymbol{B}) + \int_\Omega(\mathbf{curl}\boldsymbol{B} - H)H_t dx \\ &= \tfrac{dE}{dt}(u) + \int_\Omega(\mathbf{curl}\boldsymbol{A} - H)H_t dx. \end{aligned} \tag{4.11}$$

The boundary terms have all vanished due to boundary conditions. The theorem is now proved.

This theorem indicates that the phase transition phenomenon described by the G-L system under the conditions specified is energy dissipative, therefore the process is irreversible.

### 1.4.4 Gauge invariance

We have noticed that the G-L energy (system) is gauge invariant. This implies that the associated problems usually have infinite many solutions. In order to have a fixed idea of the mathematical formulation, it is therefore important to fix a choice of gauge when discussing a given problem. As we can imagine, there are many ways of fixing a gauge. However, there are two main choices that are particularly favoured by the mathematicians. We only discuss the choice for the evolutionary problem. For the steady state problem, by just ignoring the electric potential $\phi$ and the time derivatives involved, we can make the same claims and give similar proofs. Moreover, we only give the gauge choice in one particular scaling scheme. The results are similar in other scaling schemes.

The first choice is

**Lemma 1.4.1** **(Lorentz Gauge)** *Let $(\psi',\ \phi',\ \boldsymbol{A}')$ satisfy (4.5)–(4.6), we can choose a function $\Theta$ such that*

$$(\psi,\ \phi,\ \boldsymbol{A}) = (\psi' \exp(i\Theta),\ \phi' - \Theta_t,\ \boldsymbol{A}' + \nabla\Theta) \tag{4.12}$$

*satisfy*

$$\begin{cases} \operatorname{div}\boldsymbol{A} + \phi = 0, \\ \int_\Omega \phi dx = 0, \\ \boldsymbol{A}\cdot\boldsymbol{n} = 0. \end{cases} \tag{4.13}$$

*Proof.* First, we let

$$\begin{aligned} a &= \frac{1}{|\Omega|}\int_\Omega \phi' dx, \\ \phi'' &= \phi' - a, \\ \psi'' &= \psi' \exp(i\int_0^t a dt), \\ \boldsymbol{A}'' &= \boldsymbol{A}'. \end{aligned}$$

Then we solve the standard heat equation problem

$$\begin{aligned} -\Theta'_t + \Delta\Theta' + \operatorname{div}\boldsymbol{A}'' + \phi'' = 0, \\ \partial_{\boldsymbol{n}}\Theta' + \boldsymbol{A}''\cdot\boldsymbol{n} = 0 \end{aligned}$$

to decide $\Theta'$. Letting $\Theta = \Theta' + \int_0^t a dt$, we easily verify that the formula (4.12) gives the new $(\psi,\ \phi,\ \boldsymbol{A})$ satisfying (4.13).

The second choice is

**Lemma 1.4.2 (Coulomb Gauge)** *Let $(\psi', \phi', \boldsymbol{A}')$ satisfy (4.5)–(4.6), we can choose a function $\Theta$ such that*

$$(\psi, \phi, \boldsymbol{A}) = (\psi' \exp(i\Theta), \phi' - \Theta_t, \boldsymbol{A}' + \nabla\Theta) \tag{4.14}$$

*satisfy*

$$\begin{cases} \mathrm{div}\boldsymbol{A} = 0, \\ \boldsymbol{A} \cdot \boldsymbol{n} = 0. \end{cases} \tag{4.15}$$

The proof is not difficult to see with the help of that of Lemma 4.1.

In this book, we may use either of the above gauge choices depending on the circumstances. As a matter of fact, we will simply add either (4.13) or (4.15) into our system without commenting again. Results proved in one gauge choice can be obtained in another gauge choice with some minor technical modification of the proofs.

### 1.4.5 A uniform bound on $|\psi|$

As we have noticed that $|\psi|^2$ is a density function (density of superconducting electron pairs), it is desirable to show that the $\psi$ solution of the system (4.5)–(4.6) (or in the steady state case, (2.20)–(2.21)) satisfies the estimate

$$|\psi|(x,t) \leq 1 \text{ for almost all } x \in \Omega,\ t > 0. \tag{4.16}$$

We again discuss this in the evolutionary case only. The discussion for the steady state case is similar.

**Theorem 1.4.2** *Let $(\psi, \phi, \boldsymbol{A})$ be a triple of solutions of (4.5)–(4.6). Assuming that $|\psi_0|(x) \leq 1$, $\eta_0 \geq 0$, we have (4.16).*

*Proof.* Writing $\psi = f \exp(i\chi)$ with $f = |\psi|$, substituting into the equation satisfied by $\psi$ and taking the real part, we obtain the equation satisfied by $f$:

$$\begin{aligned} &f_t - \Delta f + |\boldsymbol{A} - \nabla\chi|^2 f + \kappa^2(f^2 - 1)f = 0, \\ &\partial_{\boldsymbol{n}} f = -\eta_0 f, \\ &f(x,0) = |\psi_0(x)|. \end{aligned}$$

Multiplying both sides of the equation by $(f-1)_+$, the positive parts of $f - 1$, integrating by parts over $\Omega$, we obtain

$$\int_\Omega \Big(\frac{1}{2}\partial_t((f-1)_+)^2 + |\nabla(f-1)_+|^2 + |\boldsymbol{A} - \nabla\chi|^2 f(f-1)_+$$
$$+\kappa^2 f(f+1)((f-1)_+)^2\Big)dx + \int_{\partial\Omega} \eta_0 f(f-1)_+ d\Gamma = 0.$$

Noticing that $f \geq 0$, we have

$$\int_\Omega \frac{1}{2}\partial_t((f-1)_+)^2 dx \leq 0.$$

Using the initial condition, we obtain

$$\int_\Omega ((f-1)_+)^2(x,t)dx = 0$$

which implies (4.16).

## 1.5 Exterior evolutionary Maxwell system

In practice, the supercurrent in a superconductor sample produces a magnetic field (especially at the places where vortices meet the boundary of the superconductor) and this induced magnetic field, together with the applied magnetic field, will act back to influence the superconductor. The effect of this action depends on the geometric shape of the sample and the direction and value distribution of the applied field. In this context, we regard the superconductor as an integral part of the whole space. That is, in the domain that is occupied by the superconductor, we use the G-L system to describe the electromagnetic properties. In the outside domain, instead of assuming that the applied magnetic field is given, we assume that the outside medium responds to the situation in the superconductor sample and its corresponding electromagnetic behaviour changes accordingly. In this case, the electromagnetic properties of the outside domain are governed by the Maxwell system.

We recognize that most classical physics literature on superconductivity do not give a clear statement about what kind of boundary conditions should be used. This idea of using Maxwell system in the exterior domain is first advocated clearly in [CDG 96] in a rigorous mathematics context. The reason for the physicists to be obscure on this point is rather unclear. For example, in [TTi 86], pp304, apart from establishing the Neumann boundary condition for the order parameter, the authors claimed: "In general, we shall also have to use the boundary condition of equation (8.21) (the equation for the magnetic potential – remark by Hoffmann and Tang) for any free surface which are present." In some others (for example in [Tin 75]), the issue was simply ignored. In [SST 69], it was put as $\mathbf{curl}\boldsymbol{A} \times \boldsymbol{n} = \boldsymbol{H} \times \boldsymbol{n}$, the natural boundary condition coming out of the variational principle.

As we can see, the boundary conditions should be imposed to satisfy physical requirements. In the following, we take the approach of using G-L system in the superconductor domain and using the Maxwell system in the outside domain. Across the boundary of the superconductor sample, we assume that the magnetic field is continuous.

### 1.5.1 Review of the Maxwell system

Let $\boldsymbol{E}$ be the electric field, $\boldsymbol{H}$ be the magnetic field, $\varepsilon_0$ and $\mu_0$ be the free space dielectric constant and permeability, $\chi_e$ be the relative dielectric constant, $\chi_m$ be the relative permeability $\boldsymbol{B}$ be the magnetic induction, $\boldsymbol{D}$ be the dielectric displacement, $\boldsymbol{j}$ be the current density, $\rho$ be the charge density. To simplify notation, we introduce the parameters $\tilde{\varepsilon} = \varepsilon_0\chi_e$ and $\mu = \mu_0\chi_m$. So our parameters are the same as in [CHO 92] and we will be able to use their scaling results without alteration. Then in SI units with $c$ denoting the light speed, the evolutionary Maxwell system can be summarized as

$$\begin{aligned} &\boldsymbol{B} = \mu\boldsymbol{H}, \\ &\boldsymbol{D} = \tilde{\varepsilon}\boldsymbol{E}, \\ &\mathrm{div}\boldsymbol{D} = \rho, \\ &\mathrm{div}\boldsymbol{B} = 0, \\ &\mathbf{curl}\boldsymbol{H} = \boldsymbol{j} + \boldsymbol{D}_t, \\ &\mathbf{curl}\boldsymbol{E} = -\boldsymbol{B}_t, \\ &\boldsymbol{j} = \sigma\boldsymbol{E}. \end{aligned} \tag{5.1}$$

Here $\sigma$ is the conductivity, it is large for conductors and small for insulators.

Since only the equation concerning the magnetic field will be of interest to us, we try to find out the corresponding equation from the system. First, the Maxwell system can be reduced to the following system involving electrical field and magnetic field only:

$$\begin{aligned} &\mathrm{div}\boldsymbol{E} = (1/\tilde{\varepsilon})\rho, \\ &\mathrm{div}\boldsymbol{H} = 0, \\ &\mathbf{curl}\boldsymbol{H} = \sigma\boldsymbol{E} + \tilde{\varepsilon}\boldsymbol{E}_t, \\ &\mathbf{curl}\boldsymbol{E} = -\mu\boldsymbol{H}_t. \end{aligned} \tag{5.2}$$

Since $\mathrm{div}\boldsymbol{H} = 0$, there exists a vector field $\boldsymbol{A}$, unique up to an added gradient term, such that $\boldsymbol{H} = \mathbf{curl}\boldsymbol{A}$. Using $\boldsymbol{A}$ instead of $\boldsymbol{H}$ in $\mathbf{curl}\boldsymbol{E} = -\mu\boldsymbol{H}_t$, we obtain

$$\mathbf{curl}(\boldsymbol{E} + \mu\boldsymbol{A}_t) = 0,$$

which leads to

$$\mu\boldsymbol{A}_t + \nabla\phi + \boldsymbol{E} = 0 \tag{5.3}$$

for some scalar function $\phi$ called the electric potential. Using the equation $\mathbf{curl}\boldsymbol{H} = \sigma\boldsymbol{E} + \tilde{\varepsilon}\boldsymbol{E}_t$ and (2.3), we have

$$\mathbf{curl}\boldsymbol{H} = \mathbf{curl}^2\boldsymbol{A} = -\sigma(\mu\boldsymbol{A}_t + \nabla\phi) - \tilde{\varepsilon}(\mu\boldsymbol{A}_{tt} + \nabla\phi_t). \tag{5.4}$$

In physics literature, we see the following approximation arguments:
1) *When the conductor is very poor*, we have approximately

$$\mathbf{curl}^2\boldsymbol{A} = -\tilde{\varepsilon}(\mu\boldsymbol{A}_{tt} + \nabla\phi_t).$$

2) *When the conductor is very good*, we have approximately

$$\mathbf{curl}^2 \boldsymbol{A} = -\sigma(\mu \boldsymbol{A}_t + \nabla\phi).$$

To avoid extreme cases, we use the intermediate equation (5.4) in the medium external to the superconductor. Using the non-dimensionalization transform introduced in [CHO 92], we have

$$\nu(\boldsymbol{A}_{tt} + \nabla\phi_t) + (\boldsymbol{A}_t + \nabla\phi) + \mathbf{curl}^2 \boldsymbol{A} = 0. \tag{5.5}$$

Here $\nu$ is a dimensionless constant which depends only on the property of the material.

### 1.5.2 The G-L superconductivity model

The evolutionary G-L superconductivity model we use is already presented in Section 4.

Now, nondimensionalizing the system as in [CHO 92], we obtain

$$\begin{aligned} &\alpha\xi^2\psi_t + \tfrac{\alpha\xi}{\lambda} i\psi\phi + |\psi|^2\psi - \psi - (\xi\nabla - i\tfrac{\gamma}{\lambda}\boldsymbol{A})^2\psi = 0, \\ &\lambda^2(\boldsymbol{A}_t + \nabla\phi + \mathbf{curl}^2\boldsymbol{A}) + \tfrac{\xi\lambda}{2\gamma} i(\psi^*\nabla\psi - \psi\nabla\psi^*) + |\psi|^2\boldsymbol{A} = 0. \end{aligned} \tag{5.6}$$

This system looks slightly more complicated than the scaled systems given in Section 4. However, it can be shown that they are equivalent. The reason we use this particular scaling is that (5.6) has the advantage that it matches the Maxwell equations in scaling and therefore, there is no hidden "time gap" between the G-L model and the Maxwell system. As a consequence, we can put the two systems together in the same context.

### 1.5.3 The setting of the problem

In this section, $\Omega$ is the domain occupied by the superconductor sample. $B_M$ is a large ball in $I\!R^3$ containing $\Omega$. We explain the reason of making this choice in Remark 5.3 in the following. Our final remark is about the choice of the boundary conditions: similar to the 2-d case, we impose $\mathbf{curl}\boldsymbol{A} \times \boldsymbol{n} = \boldsymbol{H} \times \boldsymbol{n}$ and the gauge choice condition $\boldsymbol{A} \cdot \boldsymbol{n} = 0$ on $\partial B_M$. We will also give more discussions about this choice in Remark 5.3.

From this setting, we want the G-L system in $\Omega$ to describe the superconducting electromagnetic properties. By contrast, we want the Maxwell system outside $\Omega$ with matching boundary conditions across $\partial\Omega$. The equations are

$$\begin{cases} \alpha\xi^2\psi_t + \frac{\alpha\xi}{\lambda} i\psi\phi + |\psi|^2\psi - \psi - (\xi\nabla - i\frac{\gamma}{\lambda}\boldsymbol{A})^2\psi = 0, \\ \qquad \boldsymbol{x} \in \Omega,\ t > 0, \\ \gamma^2(\boldsymbol{A}_t + \nabla\phi + \mathbf{curl}^2\boldsymbol{A}) + \frac{\xi\gamma}{2\lambda} i(\psi^*\nabla\psi - \psi\nabla\psi^*) + \frac{\gamma^2}{\lambda^2}|\psi|^2\boldsymbol{A} = 0, \\ \qquad \boldsymbol{x} \in \Omega,\ t > 0, \\ \gamma^2\nu(\boldsymbol{A}_{tt} + \nabla\phi_t) + \gamma^2(\boldsymbol{A}_t + \nabla\phi + \mathbf{curl}^2\boldsymbol{A}) = 0, \\ \qquad \boldsymbol{x} \in B_M \setminus \overline{\Omega},\ t > 0. \end{cases} \tag{5.7}$$

With the gauge choice made in Section 4.4, we can fix

$$\operatorname{div}\boldsymbol{A} + \phi = 0 \quad \text{for } \boldsymbol{x} \in B_M,\ t > 0. \tag{5.8}$$

On the boundary $\partial\Omega$, we impose

$$\frac{\partial \psi}{\partial \boldsymbol{n}_0} - i\boldsymbol{A} \cdot \boldsymbol{n}_0 \psi = -\eta_0 \psi, \quad \text{on } \partial\Omega,\ t > 0. \tag{5.9}$$

Here $\boldsymbol{n}_0$ is the outward unit normal vector of $\Omega$ along $\partial\Omega$, $\eta_0$ is a given smooth (nonnegative) function.

On $\partial B_M$, we impose, using gauge invariance and imposing natural boundary conditions,

$$\begin{array}{lll} \boldsymbol{A} \cdot \boldsymbol{n} = 0, & \text{on } \partial B_M, & t > 0, \\ \mathbf{curl}\boldsymbol{A} \times \boldsymbol{n} = \boldsymbol{H} \times \boldsymbol{n} & \text{on } \partial B_M, & t > 0. \end{array} \tag{5.10}$$

The initial conditions are given as

$$\begin{cases} \psi(\boldsymbol{x}, 0) = \psi_0(\boldsymbol{x}), & \boldsymbol{x} \in \Omega, \\ \boldsymbol{A}(\boldsymbol{x}, 0) = \boldsymbol{A}_0(\boldsymbol{x}), & \boldsymbol{x} \in B_M, \\ \boldsymbol{A}_t(\boldsymbol{x}, 0) = \boldsymbol{A}_1(\boldsymbol{x}), & \boldsymbol{x} \in B_M \setminus \overline{\Omega}, \\ \phi(\boldsymbol{x}, 0) = \phi_0(\boldsymbol{x}), & \boldsymbol{x} \in B_M \setminus \overline{\Omega}. \end{cases} \tag{5.11}$$

**Remark 1.5.1** *We need to point out that the 3rd and 4th conditions in (5.11) are not entirely independent of what is going on in $\Omega$ where the same information is not given. As a matter of fact, as we will see from the following full setting of the problem, on $\partial\Omega \times \{0\}$, we need the following compatibility conditions:*

$$\begin{cases} \phi_0(\boldsymbol{x}) = -\operatorname{div}\boldsymbol{A}_0(\boldsymbol{x}), \\ \boldsymbol{A}_1(\boldsymbol{x}) = -\Big( - \nabla\operatorname{div}\boldsymbol{A}_0 + \mathbf{curl}^2\boldsymbol{A}_0 \\ \qquad + \frac{\xi i}{2\lambda\gamma}(\psi_0^*\nabla\psi_0 - \psi_0\nabla\psi_0^*) + \frac{1}{\lambda^2}|\psi_0|^2\boldsymbol{A}_0\Big), \\ \operatorname{div}\boldsymbol{A}_1(\boldsymbol{x}) = -\operatorname{div}\Big( - \nabla\operatorname{div}\boldsymbol{A}_0 + \mathbf{curl}^2\boldsymbol{A}_0 \\ \qquad + \frac{\xi i}{2\lambda\gamma}(\psi_0^*\nabla\psi_0 - \psi_0\nabla\psi_0^*) + \frac{1}{\lambda^2}|\psi_0|^2\boldsymbol{A}_0\Big). \end{cases} \tag{5.12}$$

*As there are compatibility conditions on the given initial data, we have to make strong assumptions on the smoothness of the initial data in $\Omega$ ($H^4$, for example) to make them meaningful.*

The problem consisting of (5.7)–(5.12) plus some continuity condition on $\boldsymbol{A}$ across $\partial\Omega$ will form our full problem. To simplify notation, we set all the physical constants in (5.7)–(5.12) to be 1 which will not affect the mathematics argument. For application purposes however, it is important to put the constants back again and specify their values. The simplified problem can be summarized as follows (need to add compatibility conditions (5.12) for completeness):

$$\begin{cases} \psi_t + i\psi\phi + |\psi|^2\psi- & & \\ \qquad \psi - (\nabla - i\boldsymbol{A})^2\psi = 0, & \boldsymbol{x} \in \Omega, & t > 0, \\ \boldsymbol{A}_t + \nabla\phi + \mathbf{curl}^2\boldsymbol{A} + \frac{i}{2}(\psi^*\nabla\psi & & \\ \qquad -\psi\nabla\psi^*) + |\psi|^2\boldsymbol{A} = 0, & \boldsymbol{x} \in \Omega, & t > 0, \\ \boldsymbol{A}_{tt} + \nabla\phi_t + \boldsymbol{A}_t + \nabla\phi & & \\ +\mathbf{curl}^2\boldsymbol{A} = 0, & \boldsymbol{x} \in B_M \setminus \overline{\Omega}, & t > 0, \\ \mathrm{div}\boldsymbol{A} + \phi = 0, & \boldsymbol{x} \in B_M, & t > 0, \\ \frac{\partial\psi}{\partial\boldsymbol{n}_0} - i\boldsymbol{A}\cdot\boldsymbol{n}_0\psi = -\eta_0\psi, & \boldsymbol{x} \in \partial\Omega, & t > 0, \\ [\mathbf{curl}\boldsymbol{A}] = 0, & \text{on } \partial\Omega, & t > 0, \\ [\boldsymbol{A}] = 0, & \text{on } \partial\Omega, & t > 0, \\ \boldsymbol{A}\cdot\boldsymbol{n} = 0, & \boldsymbol{x} \in \partial B_M, & t > 0, \\ \mathbf{curl}\boldsymbol{A} \times \boldsymbol{n} = \boldsymbol{H} \times \boldsymbol{n} & \boldsymbol{x} \in \partial B_M, & t > 0, \\ \psi(\boldsymbol{x},0) = \psi_0(\boldsymbol{x}), & \boldsymbol{x} \in \Omega, & \\ \boldsymbol{A}(\boldsymbol{x},0) = \boldsymbol{A}_0(\boldsymbol{x}), & \boldsymbol{x} \in B_M, & \\ \boldsymbol{A}_t(x,0) = \boldsymbol{A}_1(x), & \boldsymbol{x} \in B_M \setminus \overline{\Omega}, & \\ \phi(\boldsymbol{x},0) = \phi_0(\boldsymbol{x}), & \boldsymbol{x} \in B_M \setminus \overline{\Omega}. & \end{cases} \tag{5.13}$$

Here $[\cdot]$ is the jump of the quantities concerned across $\partial\Omega$. We stress again here the presence of the set $B_M$ is for pure technical reasons (see Remark 5.3 in the following).

**Remark 1.5.2** *We notice that while all the boundary restrictions on $\psi$ is imposed on $\partial\Omega$, all the boundary restrictions on $\boldsymbol{A}$ is given on $B_M$. Only some continuity conditions on $\boldsymbol{A}$ are imposed across $\partial\Omega$.*

**Remark 1.5.3** *According to [CDG 96], where the explicit idea of using Maxwell's equation in the exterior of a superconductor obeying the G-L equations was first clearly proposed, the mathematics setting is as follows:*

*In the domain $\Omega$, we have the G-L system to model the superconductor.*

*In the exterior domain $I\!R^3 \setminus \Omega$, we have the Maxwell system for the magnetic field.*

*Across the boundary of the superconductor $\partial\Omega$, the magnetic potential and magnetic field satisfy continuity conditions.*

*In our setting, however, we inserted an intermediate layer. That is, in $\Omega$, we have the G-L system to describe the superconductivity behaviour.*

*Then we proposed a larger but finite domain (a large ball) $B_M$, in which we have the Maxwell system describing the magnetic field.*

*Finally, we imposed that the magnetic field is equal to the applied magnetic field in $I\!R^3 \setminus B_M$.*

*If we stick to the proposal of [CDG 96], the domain $B_M$ is $I\!R^3$ and the boundary condition*

$$\mathbf{curl}\boldsymbol{A} \times \boldsymbol{n} = \boldsymbol{H} \times \boldsymbol{n}, \quad x \in \partial B_M, \ t > 0 \tag{5.14}$$

*should be replaced by*

$$\mathbf{curl}\boldsymbol{A} \to \boldsymbol{H} \ \textit{as} \ |x| \to \infty, \ t > 0. \tag{5.15}$$

*As (5.15) has one more condition than (5.14), the condition* $\boldsymbol{A} \cdot \boldsymbol{n} = 0$ *on* $\partial\Omega$ *is dropped as a compensation.*

*The reason we choose an intermediate layer* $B_M$ *to use Maxwell system is for technical conveniences. When proving existence and uniqueness of solutions, we usually use some kind of discretization method. It is technically simpler to work on a bounded domain rather than on an unbounded one.*

*As a matter of fact, if we take* $\Omega = B_M(0)$ *and let* $M \to \infty$*, we can easily obtain the convergence of the models on* $\Omega$ *to the model on* $\mathbb{R}^3$*.*

*Finally, the use of the jump condition*$[\mathbf{curl}\boldsymbol{A}] = 0$ *across* $\partial\Omega_0$ *is also unusual because the standard variational principle would imply only that* $[\mathbf{curl}\boldsymbol{A} \times \boldsymbol{n}] = 0$*. However, it is worth noting that* $[\mathbf{curl}\boldsymbol{A} \times \boldsymbol{n}] = 0$ *and* $[\boldsymbol{A}] = 0$ *on* $\partial\Omega_0$ *imply that* $[\mathbf{curl}\boldsymbol{A} \cdot \boldsymbol{n}] = 0$ *and consequently,* $[\mathbf{curl}\boldsymbol{A}] = 0$*.*

*Now we look at yet another formulation of the problem with fewer jump conditions involved: the PDEs of the magnetic potential can be rewritten in the following form to hold in the sense of distributions:*

$$\begin{aligned} & \chi(x)(\boldsymbol{A}_{tt} + \nabla\phi_t) + \boldsymbol{A}_t + \nabla\phi + \mathbf{curl}^2\boldsymbol{A} \\ + \quad & (1 - \chi(x))(\frac{i}{2}(\psi^*\nabla\psi - \psi\nabla\psi^*) + |\psi|^2\boldsymbol{A}) = 0 \end{aligned}$$

*where* $\chi(x) = 1$ *if* $x \in \mathbb{R}^3 \setminus \Omega$*, and* $\chi(x) = 0$ *otherwise. From this unified formulation, taking into account* $\mathrm{div}\boldsymbol{A} = 0$*, using standard elliptic PDE theory, an* $H^2$ *in space solution* $\boldsymbol{A}$ *would be expected to satisfy* $[\mathbf{curl}\boldsymbol{A}] = 0$ *across* $\partial\Omega$*. However, this is only a formal argument and is not a rigorous mathematical proof of the fact.*

## 1.6 Exterior steady-state Maxwell system

With the same notation as in the evolutionary case, a non-dimensionalized consistent steady state G-L superconductivity system consists of the following equations (we take $B_M = \mathbb{R}^3$):

$$\begin{cases} (i\nabla + \boldsymbol{A})^2\psi + \kappa^2(|\psi|^2 - 1)\psi = 0, & \text{in } \Omega, \\ \mathbf{curl}^2\boldsymbol{A} - \mathbf{curl}\boldsymbol{H} + \frac{i}{2}(\psi^*\nabla\psi & \\ \quad -\psi\nabla\psi^*) + |\psi|^2\boldsymbol{A} = 0, & \text{in } \Omega, \\ \mathrm{div}\boldsymbol{A} = 0, & \text{in } \mathbb{R}^3, \\ \mathbf{curl}^2\boldsymbol{A} = \mathbf{curl}\boldsymbol{H}, & \text{in } \mathbb{R}^3 \setminus \Omega, \\ [\mathbf{curl}\boldsymbol{A}] = 0, & \text{on } \partial\Omega, \\ [\boldsymbol{A}] = 0, & \text{on } \partial\Omega, \\ (i\nabla + \boldsymbol{A})\psi \cdot \boldsymbol{n}_0 = 0, & \text{on } \partial\Omega, \\ \mathbf{curl}\boldsymbol{A} \to \boldsymbol{H} \text{ uniformly} & \text{as } |x| \to \infty. \end{cases} \tag{6.1}$$

The uniform convergence in (6.1-8) is a purely technical assumption.

## 1.7 Surface energy, superconductor classification

It is now interesting to see how type I and type II superconductivity phenomena can be classified by using the G-L system. We discuss the problem in the steady state case only.

We point out that the classic physics approach outlined in this section (see [SST 69]) has its own defects and we should treat it with caution from the mathematical point of view.

London observed that the total exclusion of the field (Meissner effect) would not lead to a state of lowest energy unless a boundary (surface) energy exists. The key point is to introduce a surface energy $\sigma_{ns}$ per unit surface between the superconducting layer and the normal layer in the case when the Meissner effect dominates. The magnitude of the surface energy is such that its contribution exceeds the gain in magnetic energy.

The superconductor will exhibit two different behaviours with respect to the surface energy depending on the parameter values in the G-L energy which gives us the criterion to distinguish type I and type II superconductors.

In the G-L theory, the concept of surface energy is simply introduced by solving the following problem: consider an infinitely long sample which has the following boundary conditions:

$$\begin{aligned} \psi = 0,\ h = H_c, \quad &\text{when } z = -\infty, \\ \psi = \psi_0,\ h = 0, \quad &\text{when } z = +\infty. \end{aligned}$$

The order parameter $\psi$ and the field $h$ solving the G-L system will vary gradually along the z-direction changing from value 0 to $\psi_0$ and from $H_c$ to 0, respectively. A phase changing region is thus produced.

The surface energy is the difference between the energy $F_p$ of the above situation and the condensation (perfect superconducting) energy $\int (H_c^2/8\pi) dx$. We use the original non-scaled form of the energy

$$F_p = \int_{-\infty}^{\infty} \left( \alpha|\psi|^2 + \frac{\beta}{2}|\psi|^4 + \frac{1}{2m}|(i\hbar\nabla + \frac{2e}{c}\boldsymbol{A})\psi|^2 + \frac{h^2}{8\pi} + \frac{H_c(H_c - h)}{4\pi} \right) dz \qquad (7.1)$$

where the last term is a new term representing the 1-d magnetic energy contribution due to penetration of the field. The surface energy per unit area is then defined as

$$\begin{aligned} \sigma_{ns} &= \int_{-\infty}^{\infty} \left( \alpha|\psi|^2 + \frac{\beta}{2}|\psi|^4 + \frac{1}{2m}|(i\hbar\nabla + \frac{2e}{c}\boldsymbol{A})\psi|^2 + \frac{h^2}{8\pi} + \frac{H_c(H_c - h)}{4\pi} - \frac{H_c^2}{8\pi} \right) dz \\ &= \int_{-\infty}^{\infty} \left( \alpha|\psi|^2 + \frac{\beta}{2}|\psi|^4 + \frac{\hbar^2}{2m}|\frac{d\psi}{dz}|^2 + \frac{2e^2}{mc^2}|\boldsymbol{A}\psi|^2 + \frac{h^2}{8\pi} + \frac{H_c^2}{8\pi} - \frac{H_c h}{4\pi} \right) dz \end{aligned} \qquad (7.2)$$

Rescaling to (2.16) and simplifying the equations, we obtain

$$-\frac{1}{\kappa^2} f_{zz} + \frac{1}{f^3}(\frac{dh}{dz})^2 = f - f^3, \tag{7.3}$$

$$f^2 h = -\frac{2}{f} f_z h_z + h_{zz}, \tag{7.4}$$

$$A f^2 = h_z, \tag{7.5}$$

where $h$ and $A$ are the moduli of the reduced field and of the reduced vector potential and $f$ is the modulus of the reduced order parameter. As a matter of fact, from

$$(\frac{i}{\kappa}\nabla + \boldsymbol{A})^2 \psi = \psi - |\psi|^2 \psi,$$
$$\mathbf{curl}^2 \boldsymbol{A} = \mathbf{curl}\boldsymbol{h} = -\frac{i}{2\kappa}(\psi^* \nabla \psi - \psi \nabla \psi^*) - |\psi|^2 \boldsymbol{A},$$
$$\boldsymbol{h} = \mathbf{curl}\boldsymbol{A},$$

writing $\psi = f \exp(i\chi)$ and $\boldsymbol{A}_0 = \boldsymbol{A} - \frac{1}{\kappa}\nabla\chi$, we have

$$-\frac{1}{\kappa^2}\Delta f + |\boldsymbol{A}_0|^2 f = f - f^3,$$
$$\mathbf{curl}^2 \boldsymbol{A}_0 = \mathbf{curl}\boldsymbol{h} = f^2 \boldsymbol{A}_0,$$
$$\boldsymbol{h} = \mathbf{curl}\boldsymbol{A}_0$$

Eliminating $\boldsymbol{A}_0$ from the system, we obtain (7.3)–(7.5).

The surface energy takes the following form (using the rescaled variables):

$$\sigma_{ns} = 2\lambda \frac{H_c^2}{8\pi} \int_{-\infty}^{\infty} (\frac{1}{2}(1 - f^2)^2 + \frac{1}{f^2} h_z^2 + \frac{1}{\kappa^2} f_z^2 + h^2 - \sqrt{2}h) dz. \tag{7.6}$$

Using (7.3)–(7.5), we have

$$\begin{aligned} \sigma_{ns} &= 2\lambda \frac{H_c^2}{8\pi} \int_{-\infty}^{\infty} (\frac{1}{2}(1 - f^4) + \frac{1}{\kappa^2}\frac{d}{dz}(f f_z) + h^2 - \sqrt{2}h) dz \\ &= 2\lambda \frac{H_c^2}{8\pi} \int_{-\infty}^{\infty} (\frac{1}{2}(1 - f^4) + h^2 - \sqrt{2}h) dz \end{aligned} \tag{7.7}$$

where we have used the conditions

$$f = 0 \ \text{ for } \ z = -\infty,$$
$$f_z = 0, \ f = 1 \ \text{ for } \ z = \infty.$$

We now introduce the length

$$\delta = 2I\lambda \tag{7.8}$$

where

$$I = \int_{-\infty}^{\infty} (\frac{1}{2}(1 - f^4) + h^2 - \sqrt{2}h)dz$$

so that the surface energy is given by

$$\sigma_{ns} = \delta \frac{H_c^2}{8\pi}. \tag{7.9}$$

**Remark 1.7.1** *If we observe carefully, the integral in (7.1) is unbounded although $\sigma_{ns}$ is bounded. Hence, all the related calculations we carried out are formal subjecting to this unsatisfactory point. This issue will not rise if the domain we are considering is a bounded interval. But if we adopt this assumption, later discussions can not be carried out.*

*The sign of the value of $\sigma_{ns}$ or equivalently, the sign of the value of $I$, will give us the classification of the superconductors.*

### 1.7.1 The sign of $\sigma_{ns}$ when $\kappa \ll 1$

When $\kappa \ll 1$, the penetration depth is very small compared with the coherence length $\xi$. It can be assumed that $h$ is zero as soon as $f$ is different from 0. Equation (7.3) reduces to

$$-\frac{1}{\kappa^2} f_{zz} = f - f^3. \tag{7.10}$$

This gives, when $z \geq 0$,

$$-\frac{1}{2\kappa^2} f_z^2 = \frac{1}{2} f^2 - \frac{1}{4} f^4 + c. \tag{7.11}$$

If we assume that $f = 1$ implies $f_z = 0$, we obtain $c = -1/4$. Hence

$$\frac{1}{\kappa^2} f_z^2 = \frac{1}{2}(1 - f^2)^2. \tag{7.12}$$

The convenient solution is then (approximate to a $\kappa$ order jump in $f_z$ at 0)

$$\begin{array}{lll} f = 0, & h = 1/\sqrt{2}, & \text{for } z < 0, \\ f = \tanh(z\kappa/\sqrt{2}), & h = 0, & \text{for } z > 0 \end{array} \tag{7.13}$$

and the energy integral yields

$$I = \frac{1}{2} \int_0^{\infty} (1 - f^4)dz = \frac{2\sqrt{2}}{3\kappa}. \tag{7.14}$$

Hence, when $\kappa \ll 1$, the surface energy is positive.

### 1.7.2 The sign of $\sigma_{ns}$ when $\kappa \gg 1$

In this case we neglect the term $(1/\kappa^2)f_{zz}$ in (7.3) so that

$$h_z^2 = f^4(1 - f^2). \tag{7.15}$$

As $h$ must decrease with increasing $z$,

$$h_z = -f^2\sqrt{1 - f^2}. \tag{7.16}$$

From (7.4), we obtain

$$h = \frac{d}{dz}(\frac{1}{f^2}h_z) = -\frac{d}{dz}\sqrt{1 - f^2}. \tag{7.17}$$

Hence

$$\frac{d^2}{dz^2}\sqrt{1 - f^2} = f^2\sqrt{1 - f^2} = \sqrt{1 - f^2} - (\sqrt{1 - f^2})^3$$

or

$$u_{zz} = u - u^3 \tag{7.18}$$

where $u = \sqrt{1 - f^2}$. Integrating, we have

$$u_z^2 = u^2\sqrt{1 - \frac{u^2}{2}} + c. \tag{7.19}$$

The constant of integration is obviously 0 (using the condition at $-\infty$) and since $du/dz$ should be negative, we have

$$u_z = -u\sqrt{1 - u^2/2}. \tag{7.20}$$

We can then evaluate the integral:

$$\begin{aligned} I &= \int_{-\infty}^{\infty} \left(2u^2(1 - u^2/2) - \sqrt{2}u\sqrt{1 - u^2/2}\right)dz \\ &= \int_0^1 (2u\sqrt{1 - u^2/2} - \sqrt{2})\frac{du}{dz}dz = -\frac{4}{3}(\sqrt{2} - 1). \end{aligned} \tag{7.21}$$

Thus when $\kappa \gg 1$, the surface energy is negative.

### 1.7.3 The case $\kappa = 1/\sqrt{2}$

It is easy to see that for the surface integral

$$I = \int_{-\infty}^{\infty} [\frac{1}{2}(1 - f^4) + h^2 - \sqrt{2}h]dz \tag{7.22}$$

to be zero, we can set

$$h = \frac{1}{\sqrt{2}}(1 - f^2). \tag{7.23}$$

In this case, (7.3) and (7.4) take the form

$$-\frac{1}{\kappa^2} f_{zz} + \frac{2}{f} f_z^2 = f - f^3, \tag{7.24}$$

$$f f_{zz} + f_z^2 = \frac{1}{2} f^2 (1 - f^2). \tag{7.25}$$

These two equations are identical for $\kappa = 1/\sqrt{2}$. So $I = 0$ when $\kappa = 1/\sqrt{2}$

The implication of $I = 0$ has profound effect on the behaviour of the solutions. In [JTa 80], a large part of the book is devoted to this case and the conclusion is that we can have any number of vortices, at anywhere we want them to be. This is certainly not the case when $\kappa \neq 1/\sqrt{2}$.

### 1.7.4 Conclusion

Now it is clear how to classify the type I and type II superconductors from the calculation carried out so far. Following Abrikosov (1952),

Type I superconductivity: $\kappa < 1/\sqrt{2}$, positive surface energy.

Type II superconductivity: $\kappa > 1/\sqrt{2}$, negative surface energy.

**Remark 1.7.2** *From a pure mathematical point of view, the approximations made in the computations are rather crude. There are also unsatisfactory points in the formulation of surface energy and the solution of the equations. As a consequence, we do not believe this is a definitely good classification. We present the theory here because it is interesting. More about this issue will be discussed later.*

## 1.8 Difference between 2-d and 3-d models

At first glance, it is rather obvious that should we use Maxwell system in the outside domain when the space dimension is 3, no exception should be made when the space dimension is 2. But so far, we have simply taken the outside magnetic field as the applied magnetic field.

Through the following result, we prove that in SOME circumstances, the Maxwell system in $\mathbb{R}^2 \backslash \Omega$ gives $\mathbf{curl} \boldsymbol{A} = H$. Hence, it is correct to take the magnetic field as the applied magnetic field in the outside domain in these circumstances. Now we specify the conditions:

a) $\Omega \subset \mathbb{R}^2$ is a simply connected domain with smooth boundary.

b) 1) $\Omega$ is a bounded domain or

2) $\Omega$ is not bounded and the area of $\mathbb{R}^2 \setminus \Omega$ is infinity.

**Proposition 1.8.1** : *Under a) and b) above, let $(\psi, \boldsymbol{A})$ be the solution of the steady state 2-d G-L system (note that here we do not require energy minimizers) coupled with the Maxwell system in the outside domain*

$$\begin{cases} (i\nabla + \boldsymbol{A})^2\psi + \kappa^2(|\psi|^2 - 1)\psi = 0, & \text{in } \Omega, \\ \mathbf{curl}^2\boldsymbol{A} - \mathbf{curl}H + \frac{i}{2}(\psi^*\nabla\psi - \psi\nabla\psi^*) & \\ \qquad +|\psi|^2\boldsymbol{A} = 0, & \text{in } \Omega, \\ \mathrm{div}\boldsymbol{A} = 0, & \text{in } I\!R^3, \\ \mathbf{curl}^2\boldsymbol{A} = \mathbf{curl}H, & \text{in } I\!R^3 \setminus \Omega, \\ \left[\mathbf{curl}\boldsymbol{A}\right] = 0, & \text{on } \partial\Omega, \\ \left[\boldsymbol{A}\right] = 0, & \text{on } \partial\Omega, \\ (i\nabla + \boldsymbol{A})\psi \cdot \boldsymbol{n}_0 = 0, & \text{on } \partial\Omega, \\ \mathbf{curl}\boldsymbol{A} \to H \ \ \text{uniformly} & \text{as } |x| \to \infty \end{cases}$$

*such that*

$$\int_\Omega |(i\nabla + \boldsymbol{A})\psi|^2 + \frac{\kappa^2}{2}(|\psi|^2 - 1)^2 + \int_{I\!R^2} |\mathbf{curl}\boldsymbol{A} - H|^2 < \infty.$$

*Then, in $I\!R^2 \setminus \Omega$, we have*

$$\mathbf{curl}\boldsymbol{A} = H.$$

*Proof.* Rewriting the equation

$$\mathbf{curl}^2\boldsymbol{A} = \mathbf{curl}H, \quad \text{in } I\!R^3 \setminus \Omega \tag{8.1}$$

component by component, we have

$$\partial_2(\mathbf{curl}\boldsymbol{A} - H) = 0, \quad -\partial_1(\mathbf{curl}\boldsymbol{A} - H) = 0.$$

This clearly imply that

$$\mathbf{curl}\boldsymbol{A} - H = \text{constant}$$

Since the energy is finite and since the area of $I\!R^2 \setminus \Omega$ is infinity, the constant must be 0. Hence the proposition is proved.

**Remark 1.8.1** *1) This proposition is only true if we make the assumption that the problem is 2-d everywhere. That is $\boldsymbol{A}$ has only two components $(A_1, A_2)$ and all the variables depend on two independent real variables $(x_1, x_2)$ (or one complex variable).*

*2) This proposition is generally NOT true for evolutionary problems because (8.1) does not hold. However, if we take the evolutionary G-L system in $\Omega$, and use the steady state Maxwell system in $I\!R^2 \setminus \Omega$, then the result of this proposition also applies.*

*3) The assumption that $\Omega$ is simply connected is very important.*

*It is easy to see that if the domain $\Omega$ is not simply connected, or if $\Omega \subset I\!R^2$ is unbounded but the area of $I\!R^2 \setminus \Omega$ is finite, we cannot conclude that the constant obtained in the proof of Proposition 8.1 is zero.*

In the rest of this book, when discussing the 2-d problems, unless otherwise stated, we will assume that the domain $\Omega$ satisfies the properties required by Proposition 8.1.

Similarly, when discussing 2-d evolutionary systems, we always take the outside magnetic field static, so that $\mathbf{curl}\boldsymbol{A} = H$ in the outside domain.

## 1.9 Bibliographical remarks

The general theory of type II superconductivitity and the rationale of using the G-L system as the mathematics model have been widely discussed in the physics literature for many years. Some of the interesting references include [Gen 66], [SST 69], [Tin 75] and [TTi 86].

The theory of G-L has been developed in [GLa 50] and [GEl 68] for steady state and evolutionary problems respectively. The microscopic theory of superconductivity has been studied in [BCS 57], giving a theoretical foundation to the mean field approach for developing the G-L theory.

In the early days, the G-L theory was developed mainly for the superconductor sample itself with little attention been paid to the effect of the surrounding magnetic field. Most of the studies assumed simply that the outside space is dominated by the given applied field. In recent years, it was realized that in the surrounding region of the superconductor sample, it is appropriate to set up the Maxwell system and regard the whole physical space as one entity. This coupling between the G-L system and the Maxwell system raised some interesting mathematical issues which haven't been observed before. Some of the issues concerned have been discussed in [CDG 96] and [HTa3] and of course, more systematically in this monograph.

The mathematical scaling of the G-L system has also received considerable attention. It has been discussed in [DGP 92], [CHO 92] and more recently, in [TWa 95]. In [TWa 95], we started looking at problems related to domain size and the inherent physical implications that might be caused by various scaling. This is important especially when various parameters in the rescaled G-L system have been sent to zero or infinity in order to study various asymptotic limits. This has been a very active research area over the last few years and many outstanding mathematics works have been carried out. Some of the most exciting works in this direction have been summarized in [BBH 94]. In this monograph, we have had an opportunity to look at the matter of scaling systematically and give more extensive comments on the physical implications of the various asymptotic limits.

# Chapter 2
# Mathematical Foundation

In this chapter, we discuss a simplified Landau phase transition model by setting the magnetic potential in the superconductivity problem to be zero. The aim is to concentrate on the mathematical issues involved in describing the phase transition phenomena associated with the model. The G-L energy we look at takes the form

$$E(u) = \frac{1}{2}\int_{\Omega} |\nabla u|^2 + \frac{1}{2}(|u|^2 - 1)^2 dx. \tag{0.1}$$

The associated steady state PDE is

$$-\Delta u + (|u|^2 - 1)u = 0 \tag{0.2}$$

and the associated evolutionary PDE is

$$u_t - \Delta u + (|u|^2 - 1)u = 0 \tag{0.3}$$

Because the solutions to this equation change their value slowly, to see rapid phase changes around the points where the solutions take up the value 0, the energy is supposed to take the form

$$E_\varepsilon(u) = \frac{1}{2}\int_{\Omega} |\nabla u|^2 + \frac{1}{2\varepsilon^2}(|u|^2 - 1)^2 dx$$

with $\varepsilon$ a small parameter. Here $\Omega$ is a bounded, smooth domain in $I\!R^n$ with $n = 2$ or 3. The phase transition phenomenon appears over a length scale $O(\varepsilon^{1/2})$ because when $u$ takes up the value zero, the small $\varepsilon$ will force it rapidly back to the unit circle. This $\varepsilon$ dependent equation can be obtained from the original equation via a scaling of the form (disregard the change of the underlying domain)

$$t' = \varepsilon^2 t, \qquad x' = \varepsilon x.$$

The fact that the rescaled model is effective in deriving the asymptotic equations of phase changes suggests that the original Landau model describes slow moving and slow varying phase transition phenomenon.

Along with the discussions, we cover the mathematical description of the co-dimension two phase transition phenomenon and its formal asymptotic analysis. For the two space dimensional case, topological degree is used to explain some of the terms appearing in the G-L superconductivity system and how they are related to vortices. In three space dimensional case, however, there is no such topological quantity for the description of vortices because such a topological measure of the presence of a line is not known.

We then carry out formal asymptotic analysis on the co-dimension two vortex structures. More details are given in the two space dimension case than in the three space dimension case because in the latter case, the computation becomes very tedious.

The main message we wish to communicate through this formal asymptotic analysis is that near a particular given vortex (a point in 2-d, a curve in 3-d) with multiplicity $n$, the solution of our PDE takes the form $U(r)e^{in(\theta+\theta_0)}$. Here $U$ is a nonnegative real valued function, $(r, \theta)$ is the polar co-ordinate system in the plane (containing the vortex in 2-d, intersecting the vortex curve perpendicularly in 3-d) with the origin at the vortex (in 2-d) location or at the intersection point between the plane and the vortex curve (in 3-d).

This compares interestingly to the co-dimension one phase transition problems where the signed distance function from the phases' interface plays an important role. The co-dimension one phase transition problem is briefly presented in the next section.

## 2.1 Co-dimension one phase transition problems

In this section, we briefly describe an effective mathematical method used to describe co-dimension one phase transition problems. We start with some intuitive ideas on how such a phase transition takes place and then verify them with formal asymptotic analysis computations.

We discuss both steady state and evolutionary cases. Then we link them through a brief discussion of long time behaviour of the evolutionary system and show that as $t \to \infty$, the evolutionary solutions converge to the steady state solutions.

In the following, unless otherwise specified, $\Omega$ will be a bounded, open, simply connected subset of $\mathbb{R}^n$ with smooth boundary. Here $n = 1, 2$ or $3$ and $u$ is a real scalar valued function

### 2.1.1 Steady state problems

The standard G-L type phase transition theory involves the introduction of a real valued order parameter function $u$ and the minimization of the following energy over the domain $\Omega$:

$$E_\varepsilon(u) = \frac{1}{2}\int_\Omega |\nabla u|^2 + \frac{1}{2\varepsilon^2}(u^2-1)^2 dx. \tag{1.1}$$

Here $\varepsilon$ is a small constant. The energy term $(u^2-1)^2$ will favour the order parameter $u$ to take values $\pm 1$. The term $|\nabla u|^2$ is an approximation of the surface energy which makes the transition of $u$ between the favoured values $\pm 1$ smooth.

If we just minimize the energy $E_\varepsilon(u)$ over $H^1(\Omega)$ without conditions attached, we end up with $u_\varepsilon \equiv 1$ or $u_\varepsilon \equiv -1$. So it is desirable to impose some conditions so that the energy minimizer takes values close to both 1 and $-1$. In the one-dimensional case for example, assuming that $\Omega = (0,1)$, imposing the boundary condition $u(0) = -1$ and $u(1) = 1$ and letting $u_\varepsilon$ be a minimizer of (1.1), then there exists a unique $\xi_0 \in (0,1)$ such that, as $\varepsilon \to 0$, $u_\varepsilon(x) \to -1$ when $x \in (0,\xi_0)$ and $u_\varepsilon(x) \to 1$ when $x \in (\xi_0, 1)$. The fact that the energy minimizers $u_\varepsilon$ do not oscillate between $-1$ and 1 implies that the limiting behaviour of the minimizers as $\varepsilon \to 0$ is predictable (even boring). This can be justified by the following simple arguments. The energy minimizers must satisfy the following problem, for fixed $\varepsilon$,

$$\begin{cases} -u_{xx} + \frac{1}{\varepsilon^2}(u^2-1)u = 0, \\ u(0) = -1, \ u(1) = 1. \end{cases} \tag{1.2}$$

It is standard that the solution to (1.2) satisfies $-1 \le u \le 1$. It is also easy to verify that the solutions are real analytic in the interval $(0,1)$.

Multiplying the differential equation by $u_x$ and integrating from 0 to $x \in (0,1)$, we obtain

$$u_x^2(0) + \frac{1}{2\varepsilon^2}(u^2-1)^2 = u_x^2(x).$$

If $u_x^2(0) \ne 0$, then $u_x^2(x)$ never vanishes in $(0,1)$, so the minimizer must be monotonically increasing because of the boundary conditions.

If $u_x^2(0) = 0$, then there are two possibilities: i) $|u|$ never equals to 1 in $(0,1)$, then $u_x^2(x)$ never vanishes in $(0,1)$, so the minimizer must be monotonically increasing because of the boundary conditions. ii) $|u|(\xi) = 1$ for some $\xi \in (0,1)$. If $u(\xi) = -1$, we redefine

$$\tilde{u}(x) = \begin{cases} -1 & x \in (0,\xi) \\ u(x) & x \in (\xi, 1). \end{cases}$$

Then $\tilde{u}$ has smaller energy than $u$ unless $\tilde{u} \equiv u$, the first scenario contradicts the fact that $u$ is an energy minimizer, the second scenario implies that $\tilde{u} \equiv -1$ on $(0,1)$ because it is real analytic. Consequently, it can not satisfy the boundary condition $u(1) = 1$. So neither is possible. The only possibility now is $u(x) > -1$ for all $x \in (0,1)$.

If $u(\xi) = 1$ for some $\xi \in (0,1)$, a symmetric argument to the case $u(\xi) = -1$ leads also to contradiction. We have therefore justified the monotonicity of the energy minimizers. The existence of $\xi_0$ follows straightway.

In the two-dimensional case, a Dirichlet type boundary condition is no longer appropriate because we can not accommodate both $\pm 1$ on $\partial\Omega$ in a manner that

$u|_{\partial\Omega}$ is still continuous, independent of $\varepsilon$ and friendly to the limit $\varepsilon \to 0$. We then have to impose other more flexible conditions such as

$$\int_\Omega u(x)dx = m \int_\Omega dx$$

for some constant $m \in (-1, 1)$. The problem then becomes

$$\begin{cases} -\Delta u + \frac{1}{\varepsilon^2}(u^2 - 1)u = c, \\ \int_\Omega u(x) = m \int_\Omega dx, \\ \frac{\partial u}{\partial \boldsymbol{n}} = 0 \end{cases} \tag{1.3}$$

for some unknown constant $c$.

In this case, the analogous result to the one space dimensional case is (cf. [Mod 87]): as $\varepsilon \to 0$, the minimizers$u_\varepsilon$ of (1.1) tend to a function $u_0$ almost everywhere such that $\Omega$ is divided into two domains $\Omega_1$ and $\Omega_2$ and

$$u_0(x) = \begin{cases} 1 & x \in \Omega_1, \\ -1 & x \in \Omega_2. \end{cases}$$

Moreover, (for any curve $C$ in $I\!R^2$, we use $|C|$ to denote the length of the curve)

$$|\partial\Omega_1 \cap \Omega| = \min |\partial A \cap \Omega|$$

where the minimum is taken over all the subsets $A$ of $\Omega$ with measurable boundary such that

$$\int_A dx - \int_{\Omega\setminus A} dx = m \int_\Omega dx.$$

The proof of this result goes beyond the scope of this book and we do not include the details.

### 2.1.2 Evolutionary problems

The evolutionary scalar G-L phase transition theory leads to the motion by mean curvature law. We take the N-dimensional Allen-Cahn equation as an example for our discussions. Let $u : I\!R^N \times (0, T) \to I\!R$ be a solution of

$$\varepsilon u_t = \varepsilon \Delta u - \frac{1}{\varepsilon}(u^2 - 1)u. \tag{1.4}$$

It is clear that the equation favours the solution taking the values 1 or $-1$.

First, let's make some basic mathematical assumptions when $N = 2$. Assume that there is a closed evolving curve $\Gamma_t$ enclosing an interior domain $\Omega_t$. Let $d(x, t)$ be the signed distance function to $\Gamma_t$ so that $\Omega_t = \{x : \ d(x, t) < 0\}$ and $I\!R^N \setminus \overline{\Omega_t} = \{x : \ d(x, t) > 0\}$. It follows that $\boldsymbol{n} = \nabla d$ is the unit outward normal vector of $\Omega_t$,

$-d_t$ is the normal velocity of $\Gamma_t$ in the direction of $\boldsymbol{n}$ and $\kappa = \Delta d|_{\Gamma_t}$ is the mean curvature.

We then work in the spirit that the solution $u$ is almost $-1$ in $\Omega_t$, almost 1 in $I\!R^N \setminus \overline{\Omega_t}$. The rapid phase transition zone of $u$ going from $-1$ to 1 is concentrated, therefore, near $\Gamma_t$.

To realize the idea, we define a one-dimensional function $\psi_1(z)$ which changes from $-1$ at $z = -\infty$ to 1 at $z = \infty$. The function $\psi_1(d/\varepsilon)$ will then have the property that $\psi_1$ is almost $-1$ in $\Omega_t$, almost 1 in $I\!R^N \setminus \overline{\Omega_t}$ with a rapid phase transition zone near $\Gamma_t$.

Setting

$$P(u) = \varepsilon u_t - \varepsilon \Delta u + \frac{1}{\varepsilon}(u^2 - 1)u, \tag{1.5}$$

we find that for $\psi_1 : I\!R \to I\!R$,

$$P(\psi_1(\frac{d}{\varepsilon})) = \psi_1'(d_t - \Delta d) - \frac{1}{\varepsilon}(\psi_1'' - (\psi_1^2 - 1)\psi_1). \tag{1.6}$$

Let $\psi_1 = \psi_1(z) = \tanh \frac{z}{\sqrt{2}}$ be the unique solution to

$$-\psi_1'' + (\psi_1^2 - 1)\psi_1 = 0, \ z \in I\!R,$$
$$\psi_1(0) = 0, \ \psi_1'(z) > 0, \ \psi_1(z) \to \pm 1 \text{ as } z \to \pm\infty,$$

then in order that $P(\psi_1(\frac{d}{\varepsilon})) = 0$, we should choose $(d_t - \Delta d)|_{\Gamma_t} = 0$. Since $\psi_1'(\frac{d}{\varepsilon})$ is only large in an $\varepsilon$-neighbourhood of $\Gamma_t$ and tends to 0 away from $\Gamma_t$, it follows from (1.6) that $\psi_1(\frac{d}{\varepsilon})$ is now close to being a solution of the Allen-Cahn equation. It can be proved that this is indeed the case. The real solution is $\psi_1(\frac{d}{\varepsilon})$ plus small terms. This yields, as $\varepsilon \to 0$, that the equation of motion of the interface is the equation of motion by mean curvature

$$d_t - \Delta d = 0 \text{ or equivalently } v - \kappa = 0.$$

Pure motion by mean curvature would of course, shrink all closed curves to a point. This implies that the phase transition region (where the order parameter changes rapidly from the value $-1$ to 1) disappears altogether within a finite period of time. This is largely due to the fact that in the model discussed, no restrictions were imposed to maintain the presence of the phase transition region. The analogue in the steady state case is that we get only trivial solutions.

Similar to the steady state case, restrictions which maintain the presence of a phase transition region can be imposed on evolutionary cases and modified motion by mean curvature equations can be derived. For example, equations of the type

$$v = f(\kappa, \boldsymbol{n}, x, t)$$

for some given functions $f$. This is again beyond the scope of this book.

### 2.1.3 Long time behaviour

It is clear that from the physical background of the problems that as $t \to \infty$, it is important that we can derive from the mathematical formulation that the solutions of the evolutionary problems converge to the solutions of the corresponding steady state problems. This is luckily true in most of the cases.

We take the following problem as an example: (note that we have dropped the parameter $\varepsilon$ because it plays no role in the mathematical analysis)

$$\begin{cases} u_t - \Delta u + (u^2-1)u = c(t), & \text{in } \Omega \times (0,\infty) \\ \int_\Omega u(x,t)dx = a = m\int_\Omega dx, & \text{on } (0,\infty) \\ \frac{\partial u}{\partial \boldsymbol{n}} = 0, & \text{on } \partial\Omega \times (0,\infty) \\ u(x,0) = u_0(x), & \text{in } \Omega \end{cases} \tag{1.7}$$

where $c(t)$ is a function of $t$ only, $a$ is a constant independent of time, $u_0(x) \in H^1(\Omega)$ and $|u_0(x)| \leq 1$ for almost every $x \in \Omega$. We prove that as $t \to \infty$, the solutions of (1.7) converge to the set of solutions of

$$\begin{cases} -\Delta u + (u^2-1)u = c, \\ \int_\Omega u(x,t)dx = a = m\int_\Omega dx, \\ \frac{\partial u}{\partial \boldsymbol{n}} = 0 \end{cases} \tag{1.8}$$

for some suitable constant $c$.

In order to show this, we first observe that

$$E(u)(t) = \frac{1}{2}\int_\Omega |\nabla u|^2 + \frac{1}{2}(u^2-1)^2 dx \tag{1.9}$$

is the Lyapunov functional for the problem (1.7) because

$$\begin{aligned} \frac{d}{dt}E(u)(t) &= \int_\Omega u_t(-\Delta u + (u^2-1)u)dx \\ &= \int_\Omega (-u_t + c(t))u_t dx \\ &= \int_\Omega -u_t^2 dx + c(t)a_t \\ &= \int_\Omega -u_t^2 dx \leq 0. \end{aligned}$$

By integrating the equation in (1.7), using associated restrictions and boundary conditions, it is easy to verify that

$$c(t) = \frac{\int_\Omega (u^2-1)u dx}{\int_\Omega dx}.$$

We have therefore the following conclusions: for any given smooth initial data $u_0$, there exists a constant $C > 0$ independent of time $t$,

$$||u||_{H^1(\Omega)}(t) \leq C,$$
$$|c(t)| \leq C.$$

Because $|u_0(x)| \leq 1$ for almost every $x \in \Omega$, it is easy to show that $|u(x,t)| \leq 1$ for almost every $x \in \Omega$ and $t \in (0, \infty)$. We then have

$$u_t - \Delta u = f \in L^\infty(0, \infty; L^2(\Omega)).$$

Using standard parabolic PDE theory (see for example, the arguments in Section 3, Chapter 5), we have

$$||u||_{H^2(\Omega)}(t) \leq C.$$

Using standard $\omega$-convergence theory (see for example [Hen 81], we obtain that $u(x,t)$ converges to the set of solutions of (1.8) as $t \to \infty$ (with the corresponding convergence of $c(t)$ to $c$ as a consequence).

**Remark 2.1.1** *The scalar G-L phase transition theory describes the situation where the material has two possible phases (characterized by the values $\pm 1$ of the order parameter) which are separated by a thin boundary between them.*

*In recent years, more complicated phase transition problems have been studied. For example, the phase transition model for crystal structures, which concerns the minimization of energy of the form*

$$\int_\Omega (|\nabla u|^2 - 1)^2 dx$$

*or its penalized form*

$$\int_\Omega \left(|Hu|^2 + \frac{1}{\varepsilon^2}(|\nabla u|^2 - 1)^2\right) dx$$

*where $Hu$ denotes the Hessian matrix of $u$.*

*Another example is the already discussed complex valued G-L phase transition theory with typical form of energy*

$$\int_\Omega \left(|\nabla u|^2 + \frac{1}{\varepsilon^2}(|u|^2 - 1)^2\right) dx.$$

So far, we have outlined the general philosophy and some particular approaches of dealing with *interface phase transition problems* or *co-dimension one phase transition problems.*

Our *vortex structured phase transition problems* or *co-dimension two phase transition problems,* will be studied by following the same kind of general philosophy.

## 2.2 Co-dimension two phase transition problems

In this section, we describe the philosophy of constructing co-dimension two phase transition domains. The steady state problem is discussed on both bounded and unbounded domains. The evolutionary problems and their long time behaviour are also looked at. Co-dimension two-phase transition domains are also called **vortices**. In plain language, these are points in $I\!R^2$ and curves in $I\!R^3$.

### 2.2.1 Steady state problems on bounded domains

By a co-dimension two phase transition, we mean that the phase change region is concentrated in regions which are of co-dimension two. For example, in the plane, these are just points. In three space dimensions, these are curves. In order to describe such transitions, we need two functions in the energy. The simplest natural extension using co-dimension one phase transition theory is to have two real valued functions $u_1$ and $u_2$ minimizing the energy

$$\frac{1}{2}\int_\Omega |\nabla u_1|^2 + |\nabla u_2|^2 + \frac{1}{2\varepsilon^2}((u_1^2 - 1)^2 + (u_2^2 - 1)^2)dx.$$

From the analysis made in Section 1, each of $u_1$ and $u_2$ has a co-dimension one transition region. It is easy to imagine that the intersection of these two co-dimension one regions may well form a co-dimension two phase transition region which fits our aim.

However, the problem is not so simple. From the fact that both $u_1$ and $u_2$ have to take up values $\pm 1$ near the transition zone, in the neighbourhood of the intersecting co-dimension two transition region, we would normally encounter four phases: $(1,1)$, $(1,-1)$, $(-1,1)$ and $(-1,-1)$. To bring 4 different phases to a single location in addition to dealing with all the secondary phase changing interfaces between each pair of different phases to form a co-dimension two phase transition region in practice is a mathematically daunting, if not impossible task.

As a consequence, a more subtle approach should be adopted. We see then the introduction of a complex valued order parameter $u$ and the corresponding energy of the form

$$\frac{1}{2}\int_\Omega |\nabla u|^2 + \frac{1}{2\varepsilon^2}(|u|^2 - 1)^2 dx \tag{2.1}$$

which has

$$-\Delta u + \frac{1}{\varepsilon^2}(|u|^2 - 1)u = 0 \tag{2.2}$$

as its Euler-Lagrange equation.

The homogeneous energy term $(|u|^2 - 1)^2$ is obviously pushing the order parameter to take up values on the unit circle (this is a continuum, not the isolated values of $\pm 1$ as in the previous section).

The other extreme for the value of the order parameter is now obviously 0. However, 0 is a local **maximum** value of the homogeneous energy instead of a minimum. So the order parameter does not take up this value easily because it is highly unstable. It should instead, be forced to take up this value by the boundary conditions or other additional quantities we are going to add to the energy functional. Once the order parameter is forced to take up the value zero, it will leave as quickly as possible, leaving a rapidly phase changing region around it. As a consequence, the order parameter will minimize the region on which it has to take the value zero because of instability, so the rationale of co-dimension two phase transition is realized. Another advantage of using a complex order parameter is that near the sites where rapid phase changes take place, there are no other secondary phase changing interfaces characterized by the 4-phase problem at the beginning of the section. The complex valued parameter smoothed out this effect.

As to boundary conditions, one possibility is to minimize the energy in the whole space $H^1$ without any restrictions, we then obtain the Neumann condition associated with the PDE problem (2.2):

$$\frac{\partial u}{\partial n} = 0.$$

However, this leads us, as in the interface problems, to the trivial solution $u = \exp(i\alpha)$ for some constant $\alpha$. So other methods have to be used to force the order parameter to take up value 0. For example, we can impose conditions of the type

$$u = g \text{ on } \partial\Omega, \tag{2.3}$$

or in the two space dimensional case,

$$|u| = 1, \quad \mathrm{Im} \int_{\partial\Omega} u \partial_\tau u^* d\Gamma = 2\pi k \tag{2.4}$$

where $u^*$ is the complex conjugate of $u$, $k$ is some integer fixed in advance. The reason for these choices will be specified later. Another method is of course to add additional terms into the energy which will force the order parameter to take up value 0. For example, coupling the G-L system with magnetic field, as we briefly discussed in Chapter 1.

### 2.2.2 Steady state problems on $\mathbb{R}^2$

The steady state problem on the whole of $\mathbb{R}^2$ is different to that posed on a bounded domain. In order that the energy

$$E_\varepsilon(u) = \frac{1}{2} \int_{\mathbb{R}^2} |\nabla u|^2 + \frac{1}{2\varepsilon^2}(u^2 - 1)^2 dx \tag{2.5}$$

is finite, we should have $|u|$ sufficiently close to 1 and $|\nabla u|$ sufficiently close to 0 as $|x| \to \infty$. Such solutions do not admit vortices. If we want to discuss solutions with vortices, they must be stationary points rather than energy minimizers. It is common practice to consider solutions having vortices in a bounded domain and we do not have the need to impose special boundary conditions to force the presence of vortices. A typical vortex steady state solution of the equation (2.2) on $\mathbb{R}^2$ takes the form

$$u(x) = U(r)e^{in\theta}, \quad U(0) = 0, \quad U(\infty) = 1 \tag{2.6}$$

where $(r, \theta)$ are the standard polar co-ordinates in $\mathbb{R}^2$, $n$ is an integer and $U$ is a nonnegative real valued function. Substituting into the energy and equation, we obtain respectively

$$E(u) = \pi \int_0^\infty \left((U')^2 + \frac{n^2}{r^2}U^2 + \frac{1}{2\varepsilon^2}(U^2-1)^2\right) r dr \tag{2.7}$$

and

$$-\frac{1}{r}(rU')' + \frac{n^2}{r^2}U + \frac{1}{\varepsilon^2}(U^2-1)U = 0 \tag{2.8}$$

which gives a solution that vanishes at the origin when $n \neq 0$.

The structure of $U$ is revealed in greater detail by using the shooting method – an important method in studying ordinary differential equations on bounded or unbounded domains (see [CET 94] for more details).

Now we make an intuitive remark about the multiplicity (or the winding) number $n$: The number $n$ appears in the energy as $\pi n^2 \int_0^\infty \frac{U^2}{r^2} r dr$. If $|n| > 1$, a vortex with multiplicity number $n$ seems to produce much higher energy than $|n|$ vortices each with multiplicity number 1. Of course when arguing this, we ignored energy contribution from possible vortex interaction and the vortex spreading out effect. (Each vortex will take up certain space. When space is scarce, there may be competition among them.) But later we will see this is indeed true when $\varepsilon$ is small and when the vortices have very small interaction force. Under these assumptions, the energy minimizerstake vortices of winding number $\pm 1$ only.

### 2.2.3 Evolutionary problems

The setting of the evolutionary problem is similar to the interface problem:

$$u_t = \Delta u - \frac{1}{\varepsilon^2}(|u|^2 - 1)u. \tag{2.9}$$

However, we now face the problem of deriving the equation of motion of a point in the 2-d plane and the equation of motion of a line in the 3-d space.

The asymptotic equations for the motion of vortices are very different from that in the interface case. We will look at the issue of formal asymptotic approximation in the next section.

### 2.2.4 Long time behaviour

Similar kind of arguments as in Section 1 link the evolutionary equation to the steady state equation: let $u(x,t)$ be a solution of

$$u_t = \Delta u - \frac{1}{\varepsilon^2}(|u|^2 - 1)u,$$

then as $t \to \infty$, the distance (measured by $H^1$ norms) between $u$ and the set of solutions of (2.2) approaches 0 when appropriate boundary/restriction conditions are imposed.

## 2.3 Mathematical description of vortices in $\mathbb{R}^2$

As we pointed out in Section 2, a major feature of the problem is the vortex structure of the phase transition region. In planar domains, these are points and in three dimensional domains, these are lines. In order to describe them mathematically, we need proper tools.

We introduce degree theory for the purpose of relating boundary conditions to the number of vortices in the two space dimensional case.

For a given *continuous* mapping $\phi : \Omega \subset \mathbb{R}^n \to \mathbb{R}^n$ with $n = 2$ or 3, the degree $d(\phi, \Omega, y)$ is a tool that describes the number of solutions of the equation $\phi(x) = y$ for $x \in \Omega$. When describing co-dimension two regions, this is only effective in two space dimensions. So we will concentrate on the 2-d case which is of interest to us and will not give a general description of the degree theory

When the space dimension is two, we can regard the mapping $\phi$ as from $\Omega \subset \mathbb{C} \to \mathbb{C}$ and then the quantity $d(\phi, \Omega, y)$ is also called the *winding number* of $\phi$ around $y$ in the context of complex analysis.

In the following, we use $z$ for the complex independent variable and use $x = (x_1, x_2)$, the real two-dimensional vector formed with the real and complex parts of $z$.

According to the definition, when regarded as a complex mapping with $\partial\Omega$ being parameterized by $s \in (0, S)$, the winding number of the curve $\phi(\partial\Omega)$ with respect to $y$ is given by,

$$d(\phi, \Omega, y) = \frac{1}{2\pi i} \int_{\partial\Omega} \frac{d\phi(z(s))}{\phi(z(s)) - y} \tag{3.1}$$

provided $y \notin \phi(\partial\Omega)$. Here $z = z(s)$ is the parameterization of $\partial\Omega$.

From the theory of complex analysis and definition (3.1), we can draw the following conclusions when $y \notin \phi(\partial\Omega)$:

1) $d(\phi, \Omega, y)$ is an integer.

2) $d(\phi, \Omega, y)$ depends on $\phi|_{\partial\Omega}$ only. Consequently, we can alter the definition of $\phi$ inside $\Omega$ arbitrarily without affecting the degree.

3) Using (3.1), we deduce that

$$d(\phi, \Omega, y) = \frac{1}{2\pi i} \int_{\partial\Omega} \frac{d\phi(z(s))}{\phi(z(s)) - y} = \frac{1}{2\pi i} \int_{\partial\Omega} \frac{d(\phi(z(s)) - y)}{\phi(z(s)) - y}.$$

So it doesn't lose generality if we assume that $y = 0$ and just investigate $d(\phi, \Omega, 0)$ in the following. Since $\phi(z(s)) \neq 0$ for all $s \in (0, S)$, we can use the expression $\phi = |\phi| e^{i\theta(z(s))}$ in the degree formula

$$d(\phi, \Omega, 0) = \frac{1}{2\pi i} \int_0^S \frac{d\phi(z(s))}{\phi(z(s))}$$

to obtain (notice that notation of complex valued functions and real vector valued functions are mixed up in the following computation for the convenience of expression)

$$\begin{aligned}
d(\phi, \Omega, 0) &= \frac{1}{2\pi i} \int_0^S \frac{e^{i\theta}|\phi|' + i|\phi|\theta' e^{i\theta}}{|\phi| e^{i\theta}} ds \\
&= \frac{1}{2\pi i} \int_0^S \frac{|\phi|'}{|\phi|} + i\theta' ds \\
&= \frac{1}{2\pi} \int_0^S \theta' ds = \int_{\partial\Omega} \partial_\tau \theta d\Gamma \\
&= \frac{1}{2\pi} \int_{\partial\Omega} (\nabla\theta \cdot \boldsymbol{\tau}) d\Gamma = \int_{\partial\Omega} \frac{1}{2i} \frac{\phi^* \nabla\phi - \phi \nabla\phi^*}{|\phi|^2} \cdot \tau d\Gamma \qquad (3.2) \\
&= \frac{1}{2\pi} \int_\Omega \frac{1}{2i} \mathbf{curl} \Big( \frac{\phi^* \nabla\phi - \phi \nabla\phi^*}{|\phi|^2} \Big) d\Omega \\
&= \frac{1}{2\pi} \int_\Omega \frac{1}{2i} \frac{\mathbf{curl}\phi^* \cdot \nabla\phi - \mathbf{curl}\phi \cdot \nabla\phi^*}{|\phi|^2} d\Omega \\
&= \frac{1}{2\pi} \int_\Omega \frac{\partial_1\phi_1 \partial_2\phi_2 - \partial_1\phi_2\partial_2\phi_1}{|\phi|^2} dx_1 dx_2 \qquad (3.3) \\
&= \frac{1}{2\pi} \int_\Omega \frac{\det \nabla(\phi_1, \phi_2)^T}{|\phi|^2} dx_1 dx_2 \qquad (3.4)
\end{aligned}$$

where $\phi_1$ and $\phi_2$ are the real and complex parts of $\phi$ respectively. According to the standard definitions, $\det\nabla\phi$ is a null Lagrangian of $\phi$ in the sense that $\int_\Omega \det\nabla\phi$ depends on $\phi|_{\partial\Omega}$ only (see [FGa 95]). The above computation reveals an interesting relationship among the topological degree, the null Lagrangian $\det \nabla\phi$ of $\phi$ and the expression $\phi^*\nabla\phi - \phi\nabla\phi^*$ which appeared in our superconductivity equations in (2.5) or (4.1) in Chapter 1. However, the quantity $|\phi|^2$ in the denominator of (3.4) is a barrier to an intuitive geometrical explanation of the integration of $\det\nabla\phi$. To simplify the matter, let us assume that

$$\phi : \Omega \subset I\!R^2 \to B(0, r) \subset I\!R^2$$

is a smooth mapping satisfying

$$|\phi(z)| = r \quad \forall z \in \partial\Omega,$$

then we have

$$d(\phi,\Omega,0)=\frac{1}{\pi r^2}\int_\Omega \det\nabla\phi dx$$

which leads to

$$\int_\Omega \det\nabla\phi dx = d(\phi,\Omega,0)\pi r^2 = \text{ degree } \times \text{ area of image.} \tag{3.5}$$

This is a particular case of the following general change of variable formula:

$$\int_\Omega f(\phi(x))\det\nabla\phi dx = \int_{\mathbb{R}^n} f(y)d(\phi,\Omega,y)dy.$$

This equality implies that if both $\partial\Omega$ and $\phi(\partial\Omega)$ are simple closed curves isomorphic to a circle, then for any $y\in\phi(\Omega)\setminus\phi(\partial\Omega)$, we have

$$\begin{aligned}\int_\Omega \det\nabla(\phi_1,\phi_2)^T dx &= \int_\Omega \mathbf{curl}(\frac{i}{2}(-\phi^*\boldsymbol{A}\phi+\phi\nabla\phi^*))\\ &= d(\phi,\Omega,y)\cdot \text{ area of } \phi(\Omega).\end{aligned} \tag{3.6}$$

**Remark 2.3.1** *The notion that the degree $d(\phi,\Omega,y)$ represents the "number" of solutions of*

$$\phi(z)=y,\ z\in\Omega \tag{3.7}$$

*is not entirely correct. For example, let $B(0,1)$ be the unit disc in the complex plane with centre 0, the equation*

$$\phi(z)=z(z-0.5)(\bar z-0.5)=0$$

*has three solutions: $z=0$, $z=0.5$ and $\bar z=0.5$. However, if we express $\phi=|z||z-0.5||\bar z-0.5|e^{i\theta}$, then $\theta$ is the standard polar angle variable and we have*

$$\begin{aligned}d(\phi,B(0,1),0) &= \frac{1}{2\pi i}\int_{|z|=1}\frac{d\phi}{\phi}=\frac{1}{2\pi i}\int_{|z|=1}\frac{d(|z||z-0.5|^2e^{i\theta})}{|z||z-0.5|^2e^{i\theta}}\\ &= \frac{1}{2\pi}\int_0^{2\pi}\theta' ds=\frac{1}{2\pi}\int_0^{2\pi}1ds=1.\end{aligned}$$

*As a matter of fact, if we write*

$$z=|z|e^{i\theta_1},\quad z-0.5=|z-0.5|e^{i\theta_2},\quad \bar z-0.5=|\bar z-0.5|e^{i\theta_3},$$

*then it is easy to see that $\theta=\theta_1+\theta_2+\theta_3$ and $\theta_1=\theta$, $\theta_2=-\theta_3$. Consequently, it is not difficult to conclude that*

$$\frac{1}{2\pi}\int_0^{2\pi}\theta_1' ds=1,\ \frac{1}{2\pi}\int_0^{2\pi}\theta_2' ds=1,\ \frac{1}{2\pi}\int_0^{2\pi}\theta_3' ds=-1.$$

*So the winding number produced by $\theta$ is the sum of the winding numbers produced by $\theta_1$, $\theta_2$ and $\theta_3$. If some of them are of opposite signs, then the final results will be an integer with absolute value smaller than the number of solutions of the equation.*

An even more extreme example is as follows: for any real valued function $f(z)$ satisfying $f|_{\partial B(0,1)} = 1$, define $\phi(z) = zf(z)$. Then $d(\phi, B(0,1), 0) = 1$, but $f(z)$ (consequently $\phi(z)$) could vanish on many points, on some curves and even maybe on some open subsets of $B(0,1)$.

**Remark 2.3.2** *In the definition of degree $d(\phi, \Omega, y)$, we assumed that $y \notin \phi(\partial\Omega)$. What happens if $y \in \phi(\partial\Omega)$? The answer is, there is no answer, it depends on the situation! The integral (3.1) is now singular and the following example shows that we can expect anything:*

*Let $\phi : B(0,1) \to B(0,1)$ such that $\phi|_{\partial B(0,1)} = (z-1)^n e^{i\alpha\theta}$ where $\theta$ is the standard polar angle variable, $\alpha$ is an arbitrary real number and $n$ is large positive integer, then a formal computation using (3.1) leads to*

$$d(\phi, \Omega, 0) = \alpha.$$

**Remark 2.3.3** *Here we give some other frequently used properties of the degree:*

*1) Let $y_1$ and $y_2$ be two points in a connected domain of $\phi(\Omega) \setminus \phi(\partial\Omega)$, then*

$$d(\phi, \Omega, y_1) = d(\phi, \Omega, y_2).$$

*2) Let $\phi_1$ and $\phi_2$ be two functions from $\Omega$ to $\mathbb{R}^2$ (or $\mathbb{C}$) such that $\phi_1|_{\partial\Omega} = \phi_2|_{\partial\Omega}$, then $\forall y \in \mathbb{R}^2 \setminus \phi_1(\partial\Omega)$,*

$$d(\phi_1, \Omega, y) = d(\phi_2, \Omega, y).$$

**Remark 2.3.4** *In carrying out the computations in (3.2)–(3.4), we have used the following identities: let $\psi = |\psi| e^{i\chi}$,*

$$\frac{i}{2}(\psi^* \nabla \psi - \psi \nabla \psi^*) = -|\psi|^2 \nabla \chi, \tag{3.8}$$

$$\mathbf{curl} \frac{i}{2}(\psi^* \nabla \psi - \psi \nabla \psi^*) = -\det \nabla (\psi_1, \psi_2)^T. \tag{3.9}$$

*Here $\psi_1$ and $\psi_2$ are the real and complex parts of $\psi$.*

Finally, we explain how the boundary conditions (2.3) and (2.4) impose vortices on the order parameter solution of the complex G-L equation.

First, if $u|_{\partial\Omega} = g$ and if $g$ has a nonzero winding number $m$ with respect to 0, then any solution $u$ of the G-L equation with this boundary data must have the same winding number $m$ and consequently, $u(z) = 0$ has at least $|m|$ solutions (including multiplicity).

Second, the condition (2.4) means simply that the winding number of $u$ with respect to 0 is $k$ (see (3.2-4)), so there are again at least $|k|$ vortices in $\Omega$.

## 2.4 Asymptotic methods for describing vortices in $I\!R^2$

In this section and through Chapter 3, the exceptional notation we use is $\boldsymbol{x}$ ($= (x_1, x_2)$ or $(x_1, x_2, x_3)$) depending on the situation. We have used $x = (x_1, x_2)$ so far. The reason is that in the context of formal asymptotic analysis, we want to distinguish vector and scalar quantities clearly because of the complicated notation used for formal asymptotic analysis. We will however use $x = (x_1, x_2)$ again when not dealing with formal asymptotic analysis methods. We follow mainly the results of [Neu 90] in the following.

### 2.4.1 Steady state case in $I\!R^2$:

Let $u$ be a complex valued function defined in $I\!R^2$ and solve the equation

$$-\Delta u + (|u|^2 - 1)u = 0. \tag{4.1}$$

The trivial uniform solution is then $u = e^{i\theta_0}$ with $\theta_0$ =constant. However, the equation also admits nontrivial vortex solutions of the form

$$u(\boldsymbol{x}) = U(r)e^{i(n\theta+\theta_0)} \tag{4.2}$$

where $(r, \theta)$ are polar co-ordinates of $I\!R^2$, $\theta_0$ is an arbitrary constant and $n$ is an integer. Substituting (4.2) into (4.1), we obtain

$$\begin{cases} U_{rr} + \frac{1}{r}U_r - \frac{n^2}{r^2}U + (1 - U^2)U = 0, \ r > 0 \\ U(0) = 0, \ U(\infty) = 1. \end{cases} \tag{4.3}$$

The continuity of $u$ at $r = 0$ forces $U(r) = 0$. The condition $U(\infty) = 1$ is imposed because we expect finite phase change energy $(\int_0^\infty (|U|^2 - 1)^2 r dr)$.

We can verify, using appropriate power expansion, that

$$U(r) \sim \begin{cases} ar^{|n|} + \mathcal{O}(r^{|n|+2}) & \text{as } r \to 0 \\ 1 - \frac{n^2}{2r^2} + \mathcal{O}(1/r^4) & \text{as } r \to \infty. \end{cases}$$

For example, when $r$ is small, we assume that $(U(0) = 0)$

$$U(r) = a_1 r + a_2 r^2 + \cdots .$$

Substituting into the equation in (4.3-1), we obtain

$$\begin{aligned} & 1 \times 2a_2 + 2 \times 3a_3 r + 3 \times 4a_4 r^2 + \cdots \\ + \ & \frac{1}{r}(a_1 + 2a_2 r + 3a_3 r^2 + 4a_4 r^3 + \cdots) \\ - \ & \frac{n^2}{r^2}(a_1 r + a_2 r^2 + a_3 r^3 + \cdots) \\ + \ & (1 - (a_1 r + a_2 r^2 + \cdots)^2)(a_1 r + a_2 r^2 + \cdots) = 0 \end{aligned}$$

which leads to

$$\begin{aligned}
&-n^2 a_1 + a_1 = 0,\\
&2a_2 + 2a_2 - n^2 a_2 = 0,\\
&6a_3 + 3a_3 - n^2 a_3 + a_1 = 0,\\
&\dots\\
&m(m-1)a_m + ma_m - n^2 a_m + f_m(a_1, \dots, a_{m-2}) = 0,\\
&\dots
\end{aligned}$$

where $f_m(0, \dots, 0) = 0$. The first nonzero coefficient appears when

$$m = |n|.$$

Moreover, $a_{|n|}$ is not determined by the equation itself and therefore, is an arbitrary constant which can be chosen to satisfy the behaviour of $U$ as $r \to \infty$. The method of making such choice is called the shooting method (c.f. [CET 94]).

The next nonzero coefficient is obviously $a_{|n|+2}$.

Similar computation works for the case when $r \to \infty$ by simply expanding

$$U(r) = 1 + \frac{a_1}{r} + \frac{a_2}{r^2} + \cdots .$$

Finally, the equation

$$-\Delta u_\varepsilon + \frac{1}{\varepsilon^2}(|u_\varepsilon|^2 - 1)u_\varepsilon = 0$$

has a corresponding solution

$$u_\varepsilon(\boldsymbol{x}) = u(\frac{\boldsymbol{x}}{\varepsilon}) = U(\frac{r}{\varepsilon})e^{i(n\theta+\theta_0)}.$$

### 2.4.2 Evolutionary case in $\mathbb{R}^2$ – Introduction:

Since the G-L equation

$$u_t - \Delta u + (|u|^2 - 1)u = 0 \tag{4.4}$$

is of 1st order in $t$ and 2nd order in $\boldsymbol{x}$, the asymptotic natural time scale corresponding to the length scale $1/\varepsilon$ is $1/\varepsilon^2$. The space length scale $1/\varepsilon$ is the typical expected distance between the vortices. The scaled G-L equation is (2.9). The natural representation of a vortex trajectory is then

$$\boldsymbol{X} = \boldsymbol{Q}(T, \varepsilon) \tag{4.5}$$

where $\boldsymbol{X}$ and $T$ are scaled space and time variables

$$\boldsymbol{X} = \varepsilon \boldsymbol{x}, \qquad T = \varepsilon^2 t. \tag{4.6}$$

This scaling ensures that the rapid phase change takes place in a $\varepsilon$ neighbourhood of the actual vortex point and we can therefore derive the vortex point behaviour from the equation (2.9).

Differentiating both sides of (4.5) and going back to the original variables $(x, t)$, we see that

$$\dot{\boldsymbol{x}} = \varepsilon \dot{\boldsymbol{Q}}(\varepsilon^2 t, \varepsilon). \tag{4.7}$$

Therefore, under our scaling, the velocity of the vortex is $\varepsilon \dot{\boldsymbol{Q}}$.

We now construct asymptotic expansions of $u(\boldsymbol{x}, t, \varepsilon)$. As usual with asymptotic analysis, there is a core expansion which is valid in neighbourhoods of radii $\mathcal{O}(1)$ (with respect to the real space variable $\boldsymbol{x}$) and a far field expansion which is valid at distances $\mathcal{O}(1/\varepsilon)$ from a vortex.

The core expansion about $\boldsymbol{x} = \boldsymbol{Q}(T, \varepsilon)/\varepsilon$ takes the form

$$u = u(\boldsymbol{r}, T, \varepsilon) \text{ with } \boldsymbol{r} = \boldsymbol{x} - \boldsymbol{Q}(T, \varepsilon)/\varepsilon. \tag{4.8}$$

The far field expansion takes the form

$$u = U(\boldsymbol{X}, T, \varepsilon). \tag{4.9}$$

**Remark 2.4.1** *We have noticed that in $\boldsymbol{X} = \boldsymbol{Q}(T, \varepsilon)$, there is an additional dependence on $\varepsilon$ in $\boldsymbol{Q}$ which we didn't specify. The reason is that our approach assumes that $\boldsymbol{Q}(T, \varepsilon)$ is given and we can not decide a priori what form this $\varepsilon$ dependence is.*

*As a matter of fact, 2-dimensional singularity asymptotic expansion involves not only terms such as $\varepsilon$, $\varepsilon^2$ and so on, they also involve $|\ln \varepsilon|$ and its powers. We will look at the reason of involving $\ln \varepsilon$ later through the fundamental solutions of the 2-d Laplacian in the rigorous 2-d asymptotic analysis. Here we treat temporarily unity and $\ln \varepsilon$ as formally equal orders of magnitude. As a consequence, the dependence on $\ln \varepsilon$ is treated as an implicit dependence through the dependance on $\varepsilon$ in $\boldsymbol{Q}(T, \varepsilon)$.*

Under this scaling scheme, the velocity of vortices is of the order $\mathcal{O}(\varepsilon)$ (because $\dot{\boldsymbol{x}} = \varepsilon \dot{\boldsymbol{Q}}$), the vortices are separated by distances of the order $\mathcal{O}(1/\varepsilon)$ (because $\boldsymbol{x} = \boldsymbol{X}/\varepsilon$). So the leading order term in the core expansion should be

$$u^0 = U(r) \exp\Big(i(n\theta + \theta_0(T, \varepsilon))\Big) \tag{4.10}$$

where $r$ and $\theta$ are the standard polar co-ordinates in the $\boldsymbol{x}$ plane. The winding number $n$ is assumed to be $\pm 1$ for stability reasons (as we have explained in Section 2.2). $\theta_0(T, \varepsilon)$ is a slowly varying phase shift.

The ultimate aim of asymptotic analysis is to produce a *compatibility condition* linking the evolution of the slowly varying phase $\Theta$ and the vortex position $\boldsymbol{Q}(T, \varepsilon)$ in the limit $\varepsilon \to 0$.

The compatibility condition is also known as *the matching of the core and far field expansions.*

### 2.4.3 Evolutionary case in $\mathbb{R}^2$ – far field expansion:

Substituting $U(\boldsymbol{X},T,\varepsilon) = F(\boldsymbol{X},T,\varepsilon)\exp(i\Theta(\boldsymbol{X},T,\varepsilon))$ into (2.9), we obtain

$$\begin{cases} (F^2-1)F = -\varepsilon^2 F_T + \varepsilon^2(\Delta F - |\nabla\Theta|^2 F), \\ F\Theta_T = F\Delta\Theta - 2\nabla F\cdot\nabla\Theta \end{cases} \tag{4.11}$$

From the first equation, it follows straightforwardly that if $F$ is bounded away from 0, then $F = 1 + \mathcal{O}(\varepsilon^2)$ (another possibility is $F = \mathcal{O}(\varepsilon^2)$ but this is not desirable for the far field). Looking at the higher order terms, we easily see that

$$F = 1 - \frac{1}{2}\varepsilon^2|\nabla\Theta|^2 + \mathcal{O}(\varepsilon^4), \quad \Theta_T - \Delta\Theta = \mathcal{O}(\varepsilon^2). \tag{4.12}$$

So the leading order approximation $\Theta^0$ of the far field phase $\Theta$ satisfies the heat equation

$$\Theta^0_T = \Delta\Theta^0 \tag{4.13}$$

and the leading order far field approximation $U^0$ of the order parameter $U$ is

$$U^0 = e^{i\Theta^0}. \tag{4.14}$$

Near the vortex position $\boldsymbol{X} = \boldsymbol{Q}$, bearing mind that the winding number of the order parameter is $n$, we can write

$$\Theta^0 \sim n\theta(\boldsymbol{X}-\boldsymbol{Q}) + \theta_0. \tag{4.15}$$

Here $\theta(\boldsymbol{X}-\boldsymbol{Q})$ is the standard polar angle variable with respect to $\boldsymbol{X}-\boldsymbol{Q}$.

Given suitable initial and topological boundary conditions around each vortex, $\Theta^0$ can now be determined entirely with the help of the heat equation it satisfies.

### 2.4.4 Evolutionary case in $\mathbb{R}^2$ – local structure of the far field solution near a vortex

For the far field expansion, we use the spatial variable $\boldsymbol{R} = \boldsymbol{X}-\boldsymbol{Q}$ and we deal with the region where $|\boldsymbol{R}| \geq c > 0$. In this co-ordinate system, $\Theta^0 = \Theta^0(\boldsymbol{R},T) = \Theta^0(\boldsymbol{X}-\boldsymbol{Q},T)$ and the heat equation (4.13) becomes

$$\Delta\Theta^0 = -\dot{\boldsymbol{Q}}\cdot\nabla\Theta^0 + \Theta^0_T. \tag{4.16}$$

Note that spatial derivatives here are with respect to $\boldsymbol{R}$ instead of $\boldsymbol{x}$.

In a neighbourhood of $\boldsymbol{R} = 0$, $\Theta^0$ has a local expansion

$$\Theta^0 = n\theta(\boldsymbol{R}) + \theta_0(T,\varepsilon) + H(\boldsymbol{R},T,\varepsilon). \tag{4.17}$$

The first two terms in the right-hand side of (4.17) are the leading order expansion for $\Theta^0$ (see (4.15)), $H$ is the remainder and must satisfy $H(0, T, \varepsilon) = 0$. Substituting (4.17) into (4.16), we obtain (note that $\dot{\boldsymbol{R}} = -\dot{\boldsymbol{Q}}$) an equation for $H$:

$$H_T - \Delta H - \dot{\boldsymbol{Q}} \cdot \nabla H = n \frac{\dot{\boldsymbol{Q}} \cdot \boldsymbol{R}^\perp}{|\boldsymbol{R}|^2} + \Theta_{0,T}. \tag{4.18}$$

Looking at the leading order terms, $H$ should take the form ($R = |\boldsymbol{R}|$)

$$H = (\ln R)\boldsymbol{A} \cdot \boldsymbol{R} + \boldsymbol{K} \cdot \boldsymbol{R} + \mathcal{O}(R^2 \ln R) \tag{4.19}$$

where $\boldsymbol{A}$ and $\boldsymbol{K}$ are vector functions of $T$ but not of $\boldsymbol{R}$. Substituting (4.19) into (4.18), we are able to determine

$$\boldsymbol{A} = \frac{1}{2} n \dot{\boldsymbol{Q}}^\perp \tag{4.20}$$

but we are unable to determine $\boldsymbol{K}$ uniquely. As a matter of fact, by neglecting contributions from higher order terms, we obtain, for $\boldsymbol{K}$,

$$\begin{aligned} &-2\frac{\boldsymbol{R}}{R^2} \cdot \boldsymbol{A} + [\ln R(\dot{\boldsymbol{A}} \cdot \boldsymbol{R} - \dot{\boldsymbol{Q}} \cdot \boldsymbol{A}) + \frac{\dot{\boldsymbol{R}} \cdot \boldsymbol{R}}{R^2} \boldsymbol{A} \cdot \boldsymbol{R} + \dot{\boldsymbol{K}} \cdot \boldsymbol{R}] \\ &+\boldsymbol{K} \cdot \dot{\boldsymbol{R}} = -n \frac{\dot{\boldsymbol{R}} \cdot \boldsymbol{R}}{R^2} - \dot{\Theta}_0. \end{aligned}$$

The first terms of both the left-hand side and the right-hand side match under the choice (4.19). The terms in the square bracket in the left-hand side come from the contribution of high order terms and the last term of the left-hand side matches with $-\dot{\Theta}_0$ but does not decide $\boldsymbol{K}$ uniquely. This is because the higher order terms of the form $B(T)R^2$ could also contribute to the matching between $\dot{\Theta}_0$ and $\boldsymbol{K} \cdot \dot{\boldsymbol{R}}$. It we trace up the expansion, we see that infinite many terms will be needed before deciding $\boldsymbol{K}$. This is technically impossible.

Hence

$$\Theta^0 = n\theta(\boldsymbol{R}) + \theta_0 + \frac{1}{2} n \ln R \dot{\boldsymbol{Q}}^\perp \cdot \boldsymbol{R} + \boldsymbol{K} \cdot \boldsymbol{R} + \mathcal{O}(R^2 \ln R).$$

Redefining new $\boldsymbol{K}$ to be the old $\boldsymbol{K}$ plus $\frac{1}{2} n \ln \varepsilon \dot{\boldsymbol{Q}}^\perp$ and choosing $a = 1/\varepsilon$, we have

$$\Theta^0 = n\theta(\boldsymbol{R}) + \theta_0 + \frac{1}{2} n \ln(R/\varepsilon) \dot{\boldsymbol{Q}}^\perp \cdot \boldsymbol{R} + \boldsymbol{K} \cdot \boldsymbol{R} + \mathcal{O}(R^2 \ln R). \tag{4.21}$$

The corresponding expansion of $U^0 = e^{i\Theta^0}$ as $R \to 0$ is then

$$U^0 = \exp\{i(n\theta(\boldsymbol{R}) + \theta_0)\}\Big(1 + \frac{in}{2} \ln(\frac{R}{\varepsilon}) \dot{\boldsymbol{Q}}^\perp \cdot \boldsymbol{R} + i\boldsymbol{K} \cdot \boldsymbol{R} + \mathcal{O}(R^2 \ln R)\Big). \tag{4.22}$$

The quantity $\boldsymbol{K}$ is called the *locally uniform component of the phase gradient* at $\boldsymbol{X} = \boldsymbol{Q}$ and it is related to the moving speed of the vortex (see Section 4.7 in the following). As a consequence, the moving speed of the vortices can not be determined via formal asymptotic analysis this way.

### 2.4.5 Evolutionary case in $\mathbb{R}^2$ – Core expansion

In order to deal with the core domain, we make a change of variables from $(\boldsymbol{x}, t)$ to $(\boldsymbol{r}, T) = (\boldsymbol{x} - \boldsymbol{Q}/\varepsilon, \varepsilon^2 t)$ in (4.4). This rescaling makes the core domain more pronounced by making the vortices artificially quasi-static (time $T$ is rescaled so that the time dependent terms become very small whereas the space scaling means finite $\boldsymbol{x}$ is very close to the vortex core in the real physical space). As a consequence, we obtain

$$\Delta u + (1 - |u|^2)u = \varepsilon^2 u_T - \varepsilon \dot{\boldsymbol{Q}} \cdot \nabla u. \tag{4.23}$$

The first two terms for the core expansion take the form $u \sim u^0 + \varepsilon u^1$. Because the terms involving time derivatives have been rescaled to be very small, the leading order term $u^0 = U(r)\exp(i(n\theta + \theta_0))$ is the vortex state solution of the following steady-state problem:

$$-\Delta u^0 + (|u^0|^2 - 1)u^0 = 0 \tag{4.24}$$

and $u^1$ is the solution of following quasi-static problem:

$$Lu^1 = -\Delta u^1 + (2|u^0|^2 - 1)u^1 + (u^0)^2 u^{1*} = -\dot{\boldsymbol{Q}} \cdot \nabla u^0. \tag{4.25}$$

### 2.4.6 Evolutionary case in $\mathbb{R}^2$ – Matching of the core and far field expansions

Having studied the far and core fields solutions which are scaled by $\varepsilon$ apart and $\varepsilon$ is tending towards 0, it is important that we find a mid-field in which they are going to match each other. These matching conditions are the so-called compatibility conditions. Our job is to match the leading order far field expansion $U^0$ and the two term core expansion $u^0 + \varepsilon u^1$.

First, we have to define the mid-field or the "matching zone". Understandably, the far field and the core field should match when $\boldsymbol{R} \to 0$ and $\boldsymbol{r} \to \infty$. Notice that $\boldsymbol{R} = \varepsilon \boldsymbol{r}$, it is mathematically convenient to consider the matching zone to be $r = |\boldsymbol{r}| \in (\varepsilon^{-1}, \varepsilon^{-p})$ for some $0 < p < 1$ and with $\varepsilon r \to 0$ as $\varepsilon \to 0$. We make the choice that $r \sim \varepsilon^p$ with $-1 < p < -1/3$.

In the matching zone, noticing that for the core expansion,

$$u^0 = \exp(i(n\theta(r) + \theta_0)) + \mathcal{O}(1/r^2) \tag{4.26}$$

and for the far field expansion

$$U^0 = \exp(i(n\theta(\boldsymbol{R}) + \theta_0))\Big(1 + \frac{in}{2}\ln(\frac{R}{\varepsilon})\dot{\boldsymbol{Q}}^{\perp} \cdot \boldsymbol{R} + i\boldsymbol{K} \cdot \boldsymbol{R} + \mathcal{O}(R^2 \ln R)\Big), \tag{4.27}$$

we have

$$\begin{aligned} & u^0(\boldsymbol{r}, T, \varepsilon) + \varepsilon u^1(\boldsymbol{r}, T, \varepsilon) - U^0(\varepsilon \boldsymbol{r}, T, \varepsilon) \\ & = \varepsilon u^1(\boldsymbol{r}, T, \varepsilon) - i\varepsilon\Big(\frac{n}{2}\ln r\dot{\boldsymbol{Q}}^{\perp} + \boldsymbol{K}\Big) \cdot \boldsymbol{r}\exp(i(n\theta + \theta_0)) \\ & + \mathcal{O}(\varepsilon^2 r^2 \ln(\varepsilon r) + \frac{1}{r^2}). \end{aligned} \tag{4.28}$$

It is obvious then that the matching condition is

$$u^1(\boldsymbol{r}, T, \varepsilon) = i\Big(\frac{n}{2}\ln r\dot{\boldsymbol{Q}}^\perp + \boldsymbol{K}\Big)\cdot \boldsymbol{r}\exp(i(n\theta+\theta_0)) + \circ(r). \tag{4.29}$$

It is easy to verify that this expression of $u^1$ is consistent with the leading order solution of (4.25).

### 2.4.7 Vortex motion equation

After making a consistent solution (the far and core field solutions match), we are going to show that the leading order term of the vortex motion velocity is proportional to $\boldsymbol{K}^\perp$ in the limit $\varepsilon \to 0$. As we have explained that $\boldsymbol{K}$ can not be determined from the formal asymptotic expansion, we are unable to decide how the vortices move without making further assumptions. Interestingly enough, this formal asymptotic indecisiveness for the vortex is matched by the fact that in the rigorous setting, we are also unable to prove the equation of motion of vortices. We do have a rigorous result as to where the vortices will end up with after the transitional time period but the actual transition process seems to be less straightforward. We refer to Chapter 7 for the rigorous analysis.

By straightforward integration by parts, we have the identity

$$\int_D \mathrm{Re}(u(Lv)^* - v(Lu)^*)d\boldsymbol{x} = \int_{\partial D} \mathrm{Re}(u(\partial_{\boldsymbol{n}}v)^* - v(\partial_{\boldsymbol{n}}u)^*)ds \tag{4.30}$$

where $L$ is the operator defined in (4.25), $u$ and $v$ are arbitrary smooth complex valued functions and $D$ is any smooth, bounded domain in $\mathbb{R}^2$.

Multiplying both sides of (4.25) by $\boldsymbol{e}\cdot\nabla u^0$ for some constant unit vector $\boldsymbol{e}$, integrating by parts and using (4.30)), we derive

$$\begin{aligned}&\int_{r<r_0} \mathrm{Re}\Big(-(\boldsymbol{e}\cdot\nabla u^0)(\dot{\boldsymbol{Q}}\cdot\nabla u^{0*})\Big)d\boldsymbol{r}\\ &= \int_{r=r_0} \mathrm{Re}\Big((\boldsymbol{e}\cdot\nabla u^0)\partial_r u^{1*} - u^1\partial_r(\boldsymbol{e}\cdot\nabla u^{0*})\Big)dl.\end{aligned} \tag{4.31}$$

To obtain the equation of motion, we evaluate (4.31) for the far-field and core solutions in the overlapping domain $r_0 = \mathcal{O}(\varepsilon^p)$ for some $p \in (-1, -1/3)$ and then take the limit $\varepsilon \to 0$.

Substituting the far field solution $u^0 = U(r)\exp(i(n\theta+\theta_0))$ into the left-hand side of (4.31), we obtain (note that $n = \pm 1$)

$$\text{left-hand side} = -\int_{r<r_0} U_r^2(\boldsymbol{e}\cdot\frac{\boldsymbol{r}}{r})(\dot{\boldsymbol{Q}}\cdot\frac{\boldsymbol{r}}{r}) + \frac{U^2}{r^2}(\boldsymbol{e}\cdot\frac{\boldsymbol{r}^\perp}{r})(\dot{\boldsymbol{Q}}\cdot\frac{\boldsymbol{r}^\perp}{r})d\boldsymbol{x}$$

$$\begin{aligned}
&= -|\boldsymbol{e}||\dot{\boldsymbol{Q}}|\int_0^{2\pi}\int_0^{r_0}(U_r^2\cos(\theta_{\boldsymbol{e}}-\theta)\cos(\theta_{\dot{\boldsymbol{Q}}}-\theta)\\
&\quad +\frac{U^2}{r^2}\sin(\theta_{\boldsymbol{e}}-\theta)\sin(\theta_{\dot{\boldsymbol{Q}}}-\theta))rdrd\theta\\
&= -|\boldsymbol{e}||\dot{\boldsymbol{Q}}|\int_0^{r_0}\int_0^{2\pi}(U_r^2\frac{\cos(\theta_{\boldsymbol{e}}-\theta_{\dot{\boldsymbol{Q}}})+\cos(\theta_{\dot{\boldsymbol{Q}}}+\theta_{\boldsymbol{e}}-2\theta)}{2}\\
&\quad +\frac{U^2}{r^2}\frac{\cos(\theta_{\boldsymbol{e}}-\theta_{\dot{\boldsymbol{Q}}})-\cos(\theta_{\dot{\boldsymbol{Q}}}+\theta_{\boldsymbol{e}}-2\theta)}{2})d\theta\, rdr\\
&= -|\boldsymbol{e}||\dot{\boldsymbol{Q}}|\int_0^{r_0}\int_0^{2\pi}(U_r^2\frac{\cos(\theta_{\boldsymbol{e}}-\theta_{\dot{\boldsymbol{Q}}})}{2}+\frac{U^2}{r^2}\frac{\cos(\theta_{\boldsymbol{e}}-\theta_{\dot{\boldsymbol{Q}}})}{2})d\theta\, rdr\\
&= -\pi(\boldsymbol{e}\cdot\dot{\boldsymbol{Q}})\int_0^{r_0}(U_r^2+\frac{U^2}{r^2})rdr.
\end{aligned}$$

Since $U(r) = 1 + \mathcal{O}(1/r^2)$ as $r \to \infty$, we have, as $r_0 \to \infty$,

$$\text{left-hand side} = -\pi(\boldsymbol{e}\cdot\dot{\boldsymbol{Q}})(\ln r_0 + \alpha) + \mathcal{O}(1/r_0^2)$$

where

$$\alpha = \lim_{r_0\to\infty}\Big(\int_0^{r_0}(U_r^2+\frac{U^2}{r^2})rdr - \ln r_0\Big).$$

Substituting the expressions of $u^1$ and $u^0$ into the right-hand side of (4.31), we have

$$\text{right-hand side} = -\pi(\ln r_0 + 1)(\boldsymbol{e}\cdot\dot{\boldsymbol{Q}}) + 2\pi n\boldsymbol{e}\cdot\boldsymbol{K}^\perp + \mathrm{o}(1).$$

Since the left-hand side and the right-hand side are equal, we obtain

$$((1-\alpha)\dot{\boldsymbol{Q}} + 2n\boldsymbol{K}^\perp)\cdot\boldsymbol{e} = \mathrm{o}(1)$$

as $\varepsilon \to 0$. Since $\boldsymbol{e}$ is arbitrary, we have

$$\dot{\boldsymbol{Q}} = \text{velocity of vortex} = \frac{2n}{1-\alpha}\boldsymbol{K}^\perp. \tag{4.32}$$

We have already pointed out, this does not mean that we have obtained the vortex velocity because $\boldsymbol{K}$ itself is an unknown quantity.

For a rather particular example of finding an explicit expression of $\boldsymbol{K}$ in the formal asymptotic analysis context, we refer the readers to [Neu 90].

## 2.5 Asymptotic methods for describing vortices in $\mathbb{R}^3$

When we consider co-dimension two-phase transition domains described by a mapping $u$: $\mathbb{R}^3 \to \mathbb{C}$, there is no longer an effective topological tool such as degree theory. The quantity $u^*\nabla u - u\nabla u^*$ loses its effectiveness when compared with the two space dimensional case. Equally, the definition of the global orientation of a vortex losses its sense.

So we will only briefly discuss how to proceed with asymptotic analysis method in 3-space dimensions without getting too much into the technicalities. All the assumptions on vortex spacing, scaling and notation conventions remain the same as in the 2-d case.

### 2.5.1 Steady state case in $\mathbb{R}^3$

The PDE to be discussed is

$$-\Delta u + \frac{1}{\varepsilon^2}(|u|^2 - 1)u = 0. \tag{5.1}$$

We assume that a particular vortex is a non-self-intersecting smooth curve without branching out sections (see Figure 4).

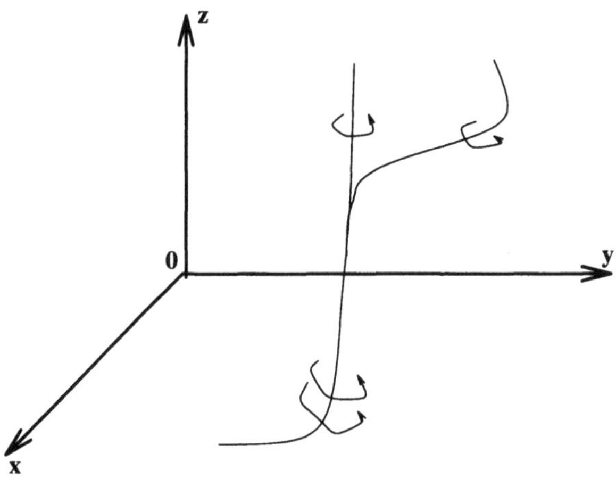


Figure 4: Possible branching out of vortices

We further assume that the vortex curve is described by

$$\boldsymbol{x} = \boldsymbol{x}(s) \tag{5.2}$$

where $s$ is the arc length parameter starting from some arbitrary point on the vortex curve (hence it could assume both positive and negative values depending on the choice of direction and the choice of the point where $s = 0$). In a small neighbourhood of the vortex curve, any point $\boldsymbol{x}$ can be expressed as

$$\boldsymbol{x} = \boldsymbol{x}(s) + r\cos\theta \boldsymbol{n}_1(s) + r\sin\theta \boldsymbol{n}_2(s) \tag{5.3}$$

where $\boldsymbol{n}_1(s)$ and $\boldsymbol{n}_2(s)$ are the principal normal and binormal vectors of the vortex curve. $(r, \theta)$ is the polar co-ordinate in the plane which meets the vortex curve perpendicularly at $\boldsymbol{x}(s)$ with origin fixed at $\boldsymbol{x}(s)$. The choice for the axis where $\theta = 0$ is a technical one. We simply assume that this axis is the principal normal direction of the vortex curve at $\boldsymbol{x}(s)$.

From our explanation, $(s, r, \theta)$ does not form an orthogonal curvilinear co-ordinate system unless $\boldsymbol{x} = \boldsymbol{x}(s)$ represents a straight line and the direction in which $\theta = 0$ is fixed. However, it is very close from being one when we restrict ourselves to a small neighbourhood of the vortex curve.

We look for solutions of the form

$$u = Ue^{i\Theta}. \tag{5.4}$$

Substituting into the equation, we obtain

$$\begin{cases} -\Delta U + |\nabla\Theta|^2 U + \frac{1}{\varepsilon^2}(U^2 - 1)U = 0, \\ U\Delta\Theta + 2\nabla U \cdot \nabla\Theta = 0. \end{cases} \tag{5.5}$$

As the vortex curve can twist around in the space as well as extend to infinity, we can not compute the power expansion of the order parameter directly when $\boldsymbol{x} \to \infty$ as in the 2-d case, so we split our analysis for the core and far fields to get a better picture. The so-called far-field means a finite distance away from the vortex core. The advantage of having this core and far-field separation allows us to reduce the problem into a 2-d one so that the analysis can be carried out.

**Core expansion** Taking into account that the radius of the core region is of the order $\mathcal{O}(\varepsilon)$, we introduce new variables $(s, R, \theta)$ defined by

$$(s, R, \theta) = (s, \frac{r}{\varepsilon}, \theta), \tag{5.6}$$

the PDE system (5.5) can be rewritten in the following form

$$\begin{cases} -\frac{1}{\varepsilon^2 R}\Big(\frac{\partial}{\partial R}(R\frac{\partial U}{\partial R}) + \frac{1}{R}\frac{\partial^2 U}{\partial \theta^2}\Big) + \frac{\kappa}{\varepsilon}(\frac{1}{R}\frac{\partial U}{\partial \theta}\sin(\theta) - \frac{\partial U}{\partial R}\cos\theta) \\ \qquad + \Big(\frac{1}{\varepsilon^2}(\frac{\partial\Theta}{\partial R})^2 + \frac{1}{\varepsilon^2 R^2}(\frac{\partial\Theta}{\partial\theta})^2\Big)U + \frac{1}{\varepsilon^2}(U^2 - 1)U = \mathcal{O}(1), \\ U\Big( - \frac{1}{\varepsilon^2 R}[\frac{\partial}{\partial R}(R\frac{\partial\Theta}{\partial R}) - \frac{1}{R}\frac{\partial^2\Theta}{\partial\theta^2}] + \frac{\kappa}{\varepsilon}(\frac{1}{R}\frac{\partial\Theta}{\partial\theta}\sin\theta - \frac{\partial\Theta}{\partial R}\cos\theta)\Big) \\ \qquad + 2(\frac{1}{\varepsilon^2}\frac{\partial U}{\partial R}\frac{\partial\Theta}{\partial R} + \frac{1}{\varepsilon^2 R^2}\frac{\partial U}{\partial\theta}\frac{\partial\Theta}{\partial\theta}) = \mathcal{O}(1) \end{cases} \tag{5.7}$$

where $\kappa$ is the curvature of the vortex curve.

Taking the first two terms of the core expansion as

$$u = (U_0 + \varepsilon U_1)e^{i(\Theta_0 + \varepsilon\Theta_1)},$$

substituting into the equations in (5.7), we obtain

$$\begin{cases} -\frac{1}{\varepsilon^2 R}\Big(\frac{\partial}{\partial R}(R\frac{\partial(U_0+\varepsilon U_1)}{\partial R}) + \frac{1}{R}\frac{\partial^2(U_0+\varepsilon U_1)}{\partial\theta^2}\Big) \\ +\frac{\kappa}{\varepsilon}(\frac{1}{R}\frac{\partial(U_0+\varepsilon U_1)}{\partial\theta}\sin(\theta) - \frac{\partial(U_0+\varepsilon U_1)}{\partial R}\cos\theta) \\ +\Big(\frac{1}{\varepsilon^2}(\frac{\partial(\Theta_0+\varepsilon\Theta_1)}{\partial R})^2 + \frac{1}{\varepsilon^2 R^2}(\frac{\partial(\Theta_0+\varepsilon\Theta_1)}{\partial\theta})^2)(U_0+\varepsilon U_1) \\ +\frac{1}{\varepsilon^2}((U_0+\varepsilon U_1)^2-1)(U_0+\varepsilon U_1) = \mathcal{O}(1), \\ \\ (U_0+\varepsilon U_1)\Big(-\frac{1}{\varepsilon^2 R}[\frac{\partial}{\partial R}(R\frac{\partial(\Theta_0+\varepsilon\Theta_1)}{\partial R}) - \frac{1}{R}\frac{\partial^2(\Theta_0+\varepsilon\Theta_1)}{\partial\theta^2}] \\ +\frac{\kappa}{\varepsilon}(\frac{1}{R}\frac{\partial(\Theta_0+\varepsilon\Theta_1)}{\partial\theta}\sin\theta - \frac{\partial(\Theta_0+\varepsilon\Theta_1)}{\partial R}\cos\theta)\Big) \\ +2(\frac{1}{\varepsilon^2}\frac{\partial(U_0+\varepsilon U_1)}{\partial R}\frac{\partial(\Theta_0+\varepsilon\Theta_1)}{\partial R} + \frac{1}{\varepsilon^2 R^2}\frac{\partial(U_0+\varepsilon U_1)}{\partial\theta}\frac{\partial(\Theta_0+\Theta_1)}{\partial\theta}) = \mathcal{O}(1) \end{cases} \tag{5.8}$$

The coefficients of the leading order term $1/\varepsilon^2$ give the following equations

$$\begin{cases} -\frac{1}{R}\frac{\partial}{\partial R}(R\frac{\partial U_0}{\partial R}) - \frac{1}{R^2}\frac{\partial^2 U_0}{\partial\theta^2} + ((\frac{\partial\Theta_0}{\partial R})^2 + \frac{1}{R^2}(\frac{\partial\Theta_0}{\partial\theta})^2)U_0 \\ \qquad +(U_0^2-1)U_0 = 0, \\ \\ U_0(-\frac{1}{R}\frac{\partial}{\partial R}(R\frac{\partial\Theta_0}{\partial R}) - \frac{1}{R^2}\frac{\partial^2\Theta_0}{\partial\theta^2}) + 2(\frac{\partial U_0}{\partial R}\frac{\partial\Theta_0}{\partial R} + \frac{1}{R^2}\frac{\partial U_0}{\partial\theta}\frac{\partial\Theta_0}{\partial\theta}) = 0. \end{cases} \tag{5.9}$$

Here the variable $s$ is not present and the equations have a clear polar structure. Hence we make the choice $U_0 = U_0(R)$ and $\Theta_0 = n\theta + \theta_0(s)$ in the equations in (5.9), where $n$ is an arbitrary integer and $\theta_0$ is a constant in the $(R,\theta)$ plane but could depend on $s$, we reduce the system to the following single equation:

$$-\frac{1}{R}(RU_0')' - \frac{n^2}{R^2}U_0 + (U_0^2-1)U_0 = 0. \tag{5.10}$$

This is exactly the same equation as in the 2-d case and the existence of a solution is therefore ensured.

We notice that in the leading order term of the core expansion, the vortex curve arc length parameter $s$ plays no important role. To see the effect of $s$ (indeed, the curvature $\kappa$ of the vortex curve), we have to go on looking at the next order term in the expansion.

The coefficients of $1/\varepsilon$ give the following equations

$$\begin{cases} -\frac{1}{R}\frac{\partial}{\partial R}(R\frac{\partial U_1}{\partial R}) - \frac{1}{R^2}\frac{\partial^2 U_1}{\partial\theta^2} - \kappa U_0'\cos\theta + \frac{2nU_0}{R^2}\frac{\partial\Theta_1}{\partial\theta} \\ +n^2\frac{U_1}{R^2} + 2U_0^2U_1 + (U_0^2-1)U_1 = 0, \\ \\ U_0(-\frac{1}{R}\frac{\partial}{\partial R}(R\frac{\partial\Theta_1}{\partial R}) - \frac{1}{R^2}\frac{\partial^2\Theta_1}{\partial\theta^2}) + 2(\frac{\partial U_0}{\partial R}\frac{\partial\Theta_1}{\partial R} + \frac{1}{R^2}\frac{\partial U_1}{\partial\theta}n) \\ +(U_0+U_1)\frac{n\kappa}{R}\sin\theta = 0. \end{cases} \tag{5.11}$$

We can see now the explicit dependence of $u_1$ and $\Theta_1$ on $s$ through $\kappa$. However, the system is too complicated for us to contemplate a solution here.

**Far field expansion**
It is clear from (5.5) that as $\boldsymbol{x} \to \infty$, the leading order term satisfies

$$U \sim 1 + \mathcal{O}(\varepsilon^2).$$

So the leading order far field solution is of the form

$$u = e^{i\Theta} \tag{5.12}$$

with $\Theta$ satisfying

$$\Delta\Theta = 0. \tag{5.13}$$

**Local structure of the far field solution near a vortex** In order to match the core and far field expansion, it is clear that we have to solve (5.13) with the *topological condition* that

$$\Theta \sim n\theta + \theta_0(s) \text{ as } \inf_{s\in \mathbb{R}} |\boldsymbol{x} - \boldsymbol{x}(s)| \to 0.$$

However, the explicit solution is slightly more complicated than the 2-d case, we stop the discussion here.

### 2.5.2 Evolutionary case in $\mathbb{R}^3$

The PDE is now

$$u_t - \Delta u + \frac{1}{\varepsilon^2}(|u|^2 - 1)u = 0. \tag{5.14}$$

If we write $u = fe^{i\chi}$ with $f = |u|$, the equation (5.14) can be rewritten as the following system:

$$\begin{cases} -\frac{\partial f}{\partial t} + \Delta f = \frac{1}{\varepsilon^2}(f^3 - f) + f|\nabla\chi|^2, \\ f^2\frac{\partial \chi}{\partial t} - \operatorname{div}(f^2\nabla\chi) = 0. \end{cases} \tag{5.15}$$

**Core expansion**
Using similar notation as in the steady state case, we introduce the nearly orthogonal co-ordinate system $(s, r, \theta)$ and then change it to $(s, R, \theta) = (s, r/\varepsilon, \theta)$. Letting the vortex curve equation be $\boldsymbol{x} = \boldsymbol{Q}(s,t)$, we define $\boldsymbol{v} = \dot{\boldsymbol{Q}}$. As the co-ordinate system is not orthogonal, we will use the following notation for derivatives concerned: for any given function $g$,

$$\begin{aligned} \nabla g &= \frac{\partial g}{\partial s}\boldsymbol{e}_s + \frac{1}{\varepsilon}\frac{\partial g}{\partial R}\boldsymbol{e}_R + \frac{1}{\varepsilon R}\frac{\partial g}{\partial \theta}\boldsymbol{e}_\theta + \mathcal{O}(\varepsilon), \\ \Delta g &= \frac{1}{\varepsilon^2 R}\Big(\frac{\partial}{\partial R}(R\frac{\partial g}{\partial r}) + \frac{1}{R}\frac{\partial^2 g}{\partial \theta^2}\Big) + \frac{\kappa}{\varepsilon}(\frac{1}{R}\frac{\partial g}{\partial \theta}\sin\theta - \frac{\partial g}{\partial \theta}\cos\theta) + \mathcal{O}(1) \\ \frac{\partial g}{\partial t} &= -\frac{1}{\varepsilon}\Big((\boldsymbol{v}\cdot\boldsymbol{e}_R)\frac{\partial g}{\partial R} + (\boldsymbol{v}\cdot\boldsymbol{e}_\theta)\frac{1}{R}\frac{\partial g}{\partial \theta}\Big) + \mathcal{O}(1). \end{aligned} \tag{5.16}$$

Hence the system (5.15) is transformed to

$$\begin{aligned} &\varepsilon(\boldsymbol{v}\cdot\boldsymbol{e}_R)\tfrac{\partial f}{\partial R}+(\boldsymbol{v}\cdot\boldsymbol{e}_\theta)\tfrac{\varepsilon}{R}\tfrac{\partial f}{\partial\theta}+\tfrac{1}{R}\Big(\tfrac{\partial}{\partial R}(R\tfrac{\partial f}{\partial r})+\tfrac{1}{R}\tfrac{\partial^2 f}{\partial\theta^2}\Big)\\ &+\varepsilon\kappa(\tfrac{1}{R}\tfrac{\partial f}{\partial\theta}\sin\theta-\tfrac{\partial f}{\partial\theta}\cos\theta)=(f^2-1)f+\Big((\tfrac{\partial\chi}{\partial R}^2+\tfrac{1}{R^2}(\tfrac{\partial\chi}{\partial\theta})^2\Big)f+\mathcal{O}(\varepsilon^2),\end{aligned}\tag{5.17}$$

$$\begin{aligned} &\varepsilon f^2\tfrac{\partial\chi}{\partial t}-\tfrac{1}{R}\Big(\tfrac{\partial}{\partial R}(Rf^2\tfrac{\partial\chi}{\partial R})+\tfrac{\partial}{\partial\theta}(Rf^2\tfrac{\partial\chi}{\partial\theta})\Big)\\ &+\varepsilon\kappa f^2(\tfrac{\partial\chi}{\partial R}\cos\theta-\tfrac{\partial\chi}{\partial\theta}\sin\theta)=\mathcal{O}(\varepsilon^2\nabla\chi,\varepsilon\tfrac{\partial\chi}{\partial s}).\end{aligned}\tag{5.18}$$

Taking inspiration from the steady state solutions, we assume that

$$\begin{aligned} &\nabla\chi\sim\tfrac{1}{\varepsilon}\tfrac{\partial\chi}{\partial\theta}\boldsymbol{e}_\theta+\cdots=\tfrac{1}{\varepsilon}\tfrac{n}{R}\boldsymbol{e}_\theta+\cdots\\ &f\sim f_0(R)+\cdots\end{aligned}\tag{5.19}$$

where $\cdots$ represents lower order terms in the formal expansion. Substituting (5.19) into the equations in (5.15), (5.15-2) is automatically satisfied at the leading order. The leading order equation of (5.15-1) becomes

$$\begin{cases} -f_0''-\frac{f_0'}{R}+\frac{n^2}{R^2}f_0+(f_0^2-1)f_0=0,\\ f_0(0)=0,\qquad f_0(\infty)=1.\end{cases}\tag{5.20}$$

**Far field expansion**

Similar to the 2-d case, we have the leading order terms satisfying

$$\begin{cases} U_0=1,\\ \frac{\partial\chi_0}{\partial t}-\Delta\chi_0=0\end{cases}\tag{5.21}$$

for the modulus of $\psi$ and the phase variable of $\psi$ respectively.

**Local structure of the far field solution near a vortex**

Similar to the steady state case, we need the topological condition

$$\chi_0\sim n\theta+\theta_0+\cdots$$

to be satisfied when solving (5.21-2).

Again, we omit the matching of far field and core expansions for, as in the 2-d case, we will not be able to derive the speed of vortex motion.

## 2.6 Bibliographical remarks

The degree theory has been a useful tool in describing co-dimension 2 phase transition problems in 2-d spaces. For a systematic treatment of degree theory and its recent applications, readers should refer to [FGa 95] and the references therein for more details.

The behaviour of the order parameter near the vortex core has been studied in the finite G-L parameter $\kappa$ regime in [BCh 89] and [CET 94]. In [BBH 94], the behaviour of the order parameter near the vortex core has been established in the case where G-L parameter $\kappa\to\infty$. Related studies can be found in [Kis 95] and [CSW 97].

# Chapter 3
# Asymptotics Involving Magnetic Potential

This is another chapter on formal asymptotic analysis. As a matter of fact, we extend the content of Chapter 2 to include a magnetic field in our discussions. As we will see, the introduction of the magnetic field facilitates the asymptotic analysis a great deal. The presence of the magnetic field allows us to derive an expression for the velocity of vortices that we were unable to obtain in the pure G-L case.

In the following we investigate the formal asymptotic analysis of the vortices in a three-dimensional superconductivity sample. We draw analogies between superconducting vortices and vortices in an inviscid fluid. The formal analysis is done either in the large $\boldsymbol{H}$ large $\kappa$ context or large $\kappa$ finite $\boldsymbol{H}$ context. The large $\kappa$ is achieved through the assumption that $\xi \ll \lambda$ which implies that we are expecting equations on the G-L length scale with diminishing vortex interactions. So the concept is to treat them as vortices in an inviscid fluid are treated. We will make some comments concerning assumptions made from mathematical and/or physical background along the way. In this chapter, we follow mainly the work of [Cha2 95] and [CRi 95].

## 3.1 Basic facts concerning fluid vortices

First, we review briefly how fluid vortices are treated in classical fluid mechanics (see [Bat 67] and [MSa 71] for a complete discussion of the background).

In an irrotational, inviscid fluid, the velocity $\boldsymbol{u}$ of a curvilinear line vortex $\Gamma$ is supposed to satisfy

$$\mathbf{curl}\boldsymbol{u} = \omega_0 \boldsymbol{\delta}_\Gamma(\boldsymbol{x}), \tag{1.1}$$

$$\mathrm{div}\boldsymbol{u} = 0 \tag{1.2}$$

where $\omega_0$ is the circulation of the vortex ($\omega_0 = 2\pi K$ where $K$ is the vortex strength), $\boldsymbol{x} = (x, y, z)$ and

$$\boldsymbol{\delta}_\Gamma(\boldsymbol{x}) = \int_\Gamma \delta(x - x')\delta(y - y')\delta(z - z')d\boldsymbol{x}'$$

where $\delta$ is the Dirac $\delta$-function and $d\boldsymbol{x}' = dx'(s)\boldsymbol{i} + dy'(s)\boldsymbol{j} + dz'(s)\boldsymbol{k}$ is the first differential along the curve $\Gamma = \{(x'(s), y'(s), z'(s)) : \; s \in \text{some interval}\}$.

**Remark 3.1.1** *Because of the $\delta$-function structure, the vortices are independent from each other and different parts of the same vortex do not interact one another.*

Introducing a (standard) vector potential $\boldsymbol{\Omega}$ such that

$$\mathbf{curl}\boldsymbol{\Omega} = \boldsymbol{u}, \tag{1.3}$$

$$\mathrm{div}\boldsymbol{\Omega} = 0. \tag{1.4}$$

Taking **curl** in (1.3) and using (1.1), we have

$$\mathbf{curl}^2\boldsymbol{\Omega} = -\Delta\boldsymbol{\Omega} = \omega_0\boldsymbol{\delta}_\Gamma(\boldsymbol{x}). \tag{1.5}$$

This expression allows us to use the Green's function for the Laplacian in $\mathbb{R}^3$ to obtain

$$\boldsymbol{\Omega} = \frac{\omega_0}{4\pi}\int_\Gamma \frac{1}{s}d\boldsymbol{x}', \;\; \boldsymbol{s} = \boldsymbol{x} - \boldsymbol{x}', \;\; s = |\boldsymbol{s}| \neq 0, \tag{1.6}$$

$$\boldsymbol{u}(\boldsymbol{x}) = \mathbf{curl}\frac{\omega_0}{4\pi}\int_\Gamma \frac{1}{s}d\boldsymbol{x}' = \frac{\omega_0}{4\pi}\int_\Gamma \nabla(\frac{1}{s}) \times d\boldsymbol{x}' = -\frac{\omega_0}{4\pi}\int_\Gamma \frac{\boldsymbol{s}\times d\boldsymbol{x}'}{s^3}. \tag{1.7}$$

This is the well-known Biot-Savart law which gives the equation for the velocity of the fluid in terms of the location of the vortices. In order to obtain the equation of motion of a vortex, we need to take the limit $\boldsymbol{x} \to \boldsymbol{x}'$ in (1.7) and expand the quantities around $\boldsymbol{x}'$.

We consider the induced velocity in the neighbourhood of a point $O$ on the line vortex. We choose rectilinear axes which are parallel to the tangent $\boldsymbol{\tau}$, principal normal $\boldsymbol{n}$, and binormal $\mathbf{b}$ to the line vortex at $O$ with $O$ as the origin, $\boldsymbol{\tau}$, $\boldsymbol{n}$ and $\mathbf{b}$ as the unit direction vectors of the $x_1$, $x_2$ and $x_3$ axes respectively. A position vector of a point in the plane normal to the line vortex at $O$ can be written as

$$\boldsymbol{x} = x_2\boldsymbol{n} + x_3\mathbf{b}.$$

Let $\boldsymbol{x}'$ be another point on the vortex line near $O$, using a Taylor expansion, we have

$$\boldsymbol{x}' \sim l\boldsymbol{\tau} + \frac{1}{2}Cl^2\boldsymbol{n}, \;\; l = |\boldsymbol{x}'|.$$

Here $C$ is the curvature of the vortex line at $O$. Hence, near $O$, we have

$$\begin{aligned} d\boldsymbol{x}' &= (\boldsymbol{\tau} + Cl\boldsymbol{n})dl, \\ \frac{\boldsymbol{s}\times d\boldsymbol{x}'}{s^3} &= \frac{-x_3Cl\boldsymbol{\tau} + x_3\boldsymbol{n} - (x_2 + \frac{1}{2}Cl^2)\mathbf{b}}{x_2^2 + x_3^2 + l^2(1 - x_2C) + \frac{1}{4}C^2l^4}^{3/2} dl. \end{aligned}$$

The contribution to the velocity at the point $(0, x_2, x_3) = (0, r\cos(\theta), r\sin(\theta))$ from the portion of the vortex line $-L < l < L$ is then, letting $m = l/r$,

$$\frac{\omega_0}{4\pi}\int_{-L/r}^{L/r} \frac{(\mathbf{b}\cos\theta - \boldsymbol{n}\sin\theta)/r + \frac{1}{2}Cm^2\mathbf{b}}{(1+m^2(1-Cr\cos\theta)+\frac{1}{4}C^2r^2m^4)^{3/2}}dm.$$

As $r \to 0$, the denominator of the integrand tends to $(1+m^2)^{3/2}$ and we have

$$\begin{aligned} \boldsymbol{u} &= -\frac{\omega_0}{4\pi}\int_\Gamma \frac{\boldsymbol{s}\times\boldsymbol{x}'_s}{s^3}ds \\ &\sim \frac{\omega_0}{2\pi r}(\mathbf{b}\cos\theta - \boldsymbol{n}\sin\theta) - \frac{\omega_0}{4\pi}(\log r)C\mathbf{b} + O(1) \\ &\sim \frac{\omega_0}{2\pi r}\boldsymbol{e}_\theta - \frac{\omega_0}{4\pi}(\log r)C\mathbf{b} + O(1) \end{aligned} \tag{1.8}$$

where $\boldsymbol{e}_\theta$ is defined to be $\mathbf{b}\cos\theta - \boldsymbol{n}\sin\theta$.

The first term in (1.8) corresponds to the rotation of the fluid about the vortex line and doesn't result in any translation. The second term determines the motion of the vortex line and, since it must be evaluated on $r = 0$, it implies that the vortex moves very fast in the binormal direction. This sounds rather unreasonable.

The reason for this is that in deriving (1.8), it is assumed that the fluid has no viscosity. It is the extreme of the physical assumption that the interaction range of the fluid molecules is small. The vortices are represented by $\delta$-functions. This means in fact, that the vortices are perfect mathematical lines. In reality, when we approach the vortex core, the interaction range of the molecules is no longer negligible, so there is a problem of model validity here. If we are not careful, we will draw the wrong conclusion.

A way of avoiding this infinite speed is to look at the problem in a different way. We could assume that there is a vortex core of radius $\varepsilon$, within which the $\delta$-function description is no longer valid. We then study the solutions in this neighbourhood by looking at a core and outer region again. Taking (1.8) as the leading order outer solution, we can then reformulate the second term in the right-hand side of (1.8) as (replacing $r$ by $\varepsilon$)

$$\boldsymbol{v} = -\frac{\omega_0}{4\pi}\log\varepsilon C\mathbf{b} + O(1),$$

where $\varepsilon$ is the radius of the viscous core region.

The use of the potential $\boldsymbol{\Omega}$ is crucial in the study carried above. It allowed the use of the Green's formula and consequently, made the explicit expression of velocity (1.8) possible. In the case of magnetic field, we can play with the magnetic potential and obtain a similar result.

## 3.2 Asymptotic analysis

With the help of a magnetic potential in the G-L model, we can develop a system of equations describing (asymptotically as $\varepsilon \to 0$) the motion of vortices. In this section, we go through the computation for this case. It is not difficult to understand the idea because we have already explained the idea of formal asymptotic analysis using the pure G-L model in the previous chapter.

### 3.2.1 2-D steady state case

We start with the following rescaled G-L system (with $\varepsilon = 1/\kappa$):

$$\begin{cases} (\nabla - i\frac{\boldsymbol{A}}{\varepsilon})^2\psi + \frac{1}{\varepsilon^2}(|\psi|^2 - 1)\psi = 0, \\ \mathbf{curl}^2\boldsymbol{A} + \frac{i\varepsilon}{2}(\psi^*\nabla\psi - \psi\nabla\psi^*) + |\psi|^2\boldsymbol{A} = 0 \end{cases} \tag{2.1}$$

in $\mathbb{R}^2$. We assume that this system exhibits vortex solutions of the following form:

$$\begin{cases} \psi = f(r)\exp(iN\theta), \\ \boldsymbol{A} = A(r)\boldsymbol{e}_\theta \end{cases} \tag{2.2}$$

where $N$ is an integer, $r$ and $\theta$ are the standard polar variables, $\boldsymbol{e}_\theta$ is the unit vector in the azimuthal direction. Substitute (2.2) into (2.1), we conclude that $f$ and $A$ are the solutions of the following system

$$\begin{cases} \frac{1}{r}\frac{d}{dr}(rf)) - (\frac{A}{\varepsilon} - \frac{N}{r})^2 f = \frac{1}{\varepsilon^2}(f^3 - f), \\ \frac{d}{dr}(\frac{1}{r}\frac{d}{dr}(rA)) = f^2(A - \frac{\varepsilon N}{r}), \\ f,\ A \text{ bounded as } r \to \infty,\ \ f \to 1,\ A \to 0 \text{ as } r \to \infty. \end{cases} \tag{2.3}$$

When $\varepsilon = 1$, it was (see [BCh 80]) shown that the system admitted solutions for any integer $N$. Moreover, it was also shown that when $N \neq 0$, $f(0) = 0$ and when $N = 0$, $f(0) \neq 0$.

We now study the formal asymptotic analysis of (2.3) when $\varepsilon \to 0$.

**Far field**

From (2.3), it is straightforward to deduce that

$$\begin{aligned} &f_0^2 = 1 - A_0^2, \\ &\frac{d}{dr}(\frac{1}{r}\frac{d}{dr}(rA_0)) = A_0 - A_0^3, \\ &A_0 \text{ remains bounded as } r \to 0,\ A_0 \to 0 \text{ as } r \to \infty. \end{aligned}$$

The only solution for this problem is obviously $A_0 = 0$, $f_0 = 1$. Therefore, despite the presence of the magnetic potential, the leading order term of the order parameter behaves the same way as that of the Landau case. Another important remark is that in this context, the large $\boldsymbol{H}$ large $\kappa$ case reduces to the finite $\boldsymbol{H}$ large $\kappa$ case.

**Core region**
We rescale lengths by setting $r = \varepsilon R$ to obtain the leading-order core behaviour:

$$\begin{aligned}&\tfrac{1}{R}\tfrac{d}{dR}(R\tfrac{df}{dR}) - (A - \tfrac{N}{R})^2 f = f^3 - f\\ &\tfrac{d}{dR}(\tfrac{1}{R}\tfrac{d}{dR}(RA)) = 0,\\ &f,\ A \text{ bounded as } r \to 0,\\ &f \to 1,\ A \to 0 \text{ as } r \to \infty.\end{aligned} \tag{2.4}$$

This system leads again to $A = 0$ and we go back to the pure G-L case (see [CET 94]).

### 3.2.2 Evolutionary case

The corresponding evolutionary system of (2.1) is

$$\psi_t - i\kappa\phi\psi - (\nabla - i\kappa \boldsymbol{A})^2\psi + \kappa^2(|\psi|^2 - 1)\psi = 0,$$
$$-\mathbf{curl}^2 \boldsymbol{A} = \eta(\boldsymbol{A}_t + \nabla\phi) + \frac{i}{2\kappa}(\psi^*\nabla\psi - \psi\nabla\psi^*) + |\psi|^2 \boldsymbol{A}.$$

Rescale the variables as follows

$$\boldsymbol{A} = \frac{\boldsymbol{A}'}{\kappa},\ \boldsymbol{H} = \frac{\boldsymbol{H}'}{\kappa},\ \Phi = \frac{\Phi'}{\kappa} \tag{2.5}$$

where $\boldsymbol{A}'$, $\boldsymbol{H}'$ and $\Phi'$ are the new variables. Substituting (2.5) into the evolutionary system, dropping the primes, we have

$$\psi_t + i\phi\psi - (\nabla - i\boldsymbol{A})^2\psi + \frac{1}{\varepsilon^2}(|\psi|^2 - 1)\psi = 0, \tag{2.6}$$
$$-\mathbf{curl}^2 \boldsymbol{A} = \eta\varepsilon^2(\boldsymbol{A}_t + \nabla\phi) + \frac{i}{2}(\psi^*\nabla\psi - \psi\nabla\psi^*) + |\psi|^2 \boldsymbol{A} \tag{2.7}$$

where $\varepsilon = 1/\kappa$.

**Remark 3.2.1** *It is noticed that this rescaled form of the G-L system is different from the formula we obtained in Section 4, Chapter 1. The reason is that we didn't rescale time t here as in Section 4, Chapter 1. So t remains on the original length scale.*

*This scaling assumes that the effect of the time derivative on the magnetic field is small in the large $\kappa$ limit. In physics terms, the time scale for the relaxation of the magnetic field is much shorter than the time scale for the motion of vortices and therefore, the magnetic field is taken as quasi-static.*

Because of the gauge invariance, we can introduce the following gauge invariant variables:

$$\boldsymbol{Q} = \boldsymbol{A} - \nabla\chi, \tag{2.8}$$
$$\Theta = \phi - \chi_t \tag{2.9}$$

where $\psi = f\exp(i\chi)$ with $f$ nonnegative and $\chi$ real. We then obtain the following equations for $f$, $\boldsymbol{Q}$ and $\Theta$:

$$-f_t + \Delta f = \frac{1}{\varepsilon^2}(f^3 - f) + f|\boldsymbol{Q}|^2, \tag{2.10}$$

$$f^2\Theta + \operatorname{div}(f^2\boldsymbol{Q}) = 0, \tag{2.11}$$

$$-\mathbf{curl}^2\boldsymbol{Q} = f^2\boldsymbol{Q} + \eta\varepsilon^2(\boldsymbol{Q}_t + \nabla\Theta) \tag{2.12}$$

with $\boldsymbol{H} = \mathbf{curl}\boldsymbol{Q}$. Taking the divergence of (2.12), we have

$$\operatorname{div}(f^2\boldsymbol{Q}) + \eta\varepsilon^2((\operatorname{div}\boldsymbol{Q})_t + \Delta\Theta) = 0. \tag{2.13}$$

### 3.2.3 Far field

Equation (2.10) again implies that $f_0 = 1 + O(\varepsilon^2)$. The leading order terms of $\boldsymbol{Q}$, $\boldsymbol{H}$ and $\Theta$ are consequently determined by

$$\mathbf{curl}^2\boldsymbol{Q}_0 + \boldsymbol{Q}_0 = \mathbf{curl}\boldsymbol{H}_0 + \boldsymbol{Q}_0 = 0, \tag{2.14}$$

$$\Theta_0 + \operatorname{div}\boldsymbol{Q}_0 = 0, \tag{2.15}$$

$$\mathbf{curl}^2\boldsymbol{H}_0 + \boldsymbol{H}_0 = -\Delta\boldsymbol{H}_0 + \boldsymbol{H}_0 = 0. \tag{2.16}$$

To solve these equations, we need to know the behaviour of $\boldsymbol{H}$ as a vortex is approached. In order to have an idea on this matter, we go to a neighbourhood of the core field to investigate.

Let a vortex lie along the curve $\Gamma$, given by $\boldsymbol{x} = (x, y, z) = \boldsymbol{Q}(s, t)$. We define a local co-ordinate system $(s, \rho, \theta)$ as in Chapter 1, Section 5.1. We define the core co-ordinate by introducing $R$ defined by $\rho = \varepsilon R$. The Co-ordinate system $(s, R, \theta)$ thus defined is not orthogonal away from the line $\boldsymbol{Q}(s, t)$. However, it is approximately orthogonal in a small neighbourhood of the vortex curve. Now we compute the various derivatives in the core region ( Superscripts are used to denote the components of the quantities concerned in the corresponding direction. E.g., $B^R$ is the component of $\boldsymbol{B}$ in the $R$ direction and so on, subscripts means partial derivatives, e.g., $B_R = \partial_R B$, the vectors $\boldsymbol{e}_R$, $\boldsymbol{e}_\theta$ and $\boldsymbol{e}_s$ are the unit vectors in the $R$, $\theta$ and $s$ directions respectively) using our new co-ordinates:

$$\nabla g = g_s\boldsymbol{e}_s + \frac{1}{\varepsilon}g_R\boldsymbol{e}_R + \frac{1}{\varepsilon R}g_\theta\boldsymbol{e}_\theta + O(\varepsilon), \tag{2.17}$$

$$\begin{aligned}\operatorname{div}\boldsymbol{B} &= \frac{1}{\varepsilon R}\Big((RB^R)_R + B^\theta_\theta\Big) - C(B^R\cos\theta - B^\theta\sin\theta)\\ &\quad + B^s_s + \tau B^s_\theta + O(\varepsilon),\end{aligned} \tag{2.18}$$

$$\begin{aligned}\Delta g &= \frac{1}{\varepsilon^2 R}((Rg_R)_R + \frac{1}{R}g_{\theta\theta})\\ &\quad + \frac{C}{\varepsilon}(\frac{1}{R}g_\theta\sin\theta - g_R\cos\theta) + O(1),\end{aligned} \tag{2.19}$$

$$\begin{aligned}
\mathbf{curl}\boldsymbol{B} &= \frac{1}{\varepsilon}[\frac{1}{R}(RB^\theta)_R - B^R_\theta)\boldsymbol{e}_s + \frac{1}{R}B^s_\theta\boldsymbol{e}_R - B^s_R\boldsymbol{e}_\theta] \\
&\quad +[\tau(B^R_\theta - (RB^\theta)_R) + C\cos\theta B^s + B^R_s]\boldsymbol{e}_\theta \\
&\quad +\tau RB^s_R\boldsymbol{e}_s + (C\sin\theta B^s - B^\theta_s)\boldsymbol{e}_R + O(\varepsilon), \\
\frac{\partial}{\partial t} &= -\frac{1}{\varepsilon}((\boldsymbol{v}\cdot\boldsymbol{e}_R)\frac{\partial}{\partial R} + (\boldsymbol{v}\cdot\boldsymbol{e}_\theta)\frac{1}{R}\frac{\partial}{\partial\theta}) + O(1) \qquad (2.20)
\end{aligned}$$

where $C$ is the curvature, $-\boldsymbol{v} = \dot{\boldsymbol{Q}}(s,t)$ is the velocity of the vortex and $\tau$ is the torsion of the vortex line.

In term of the core variables, the equations (2.12) and $\boldsymbol{H} = \mathbf{curl}\boldsymbol{Q}$ become

$$\varepsilon H^s = \frac{1}{R}((RQ^\theta)_R - Q^R_\theta) + \varepsilon\tau RQ^s_R + O(\varepsilon^2\boldsymbol{Q}), \quad (2.21)$$

$$\varepsilon H^R = \frac{1}{R}Q^s_\theta + \varepsilon C\sin\theta Q^s - \varepsilon Q^\theta_s + O(\varepsilon^2\boldsymbol{Q}), \quad (2.22)$$

$$\varepsilon H^\theta = -Q^s_R + \varepsilon\tau(Q^R_\theta - (RQ^\theta)_R) + \varepsilon C\cos\theta Q^s + \varepsilon Q^R_s + O(\varepsilon^2\boldsymbol{Q}), \quad (2.23)$$

$$-\varepsilon f^2 Q^s = \frac{1}{R}((RH^\theta)_R - H^R_\theta) + \varepsilon\tau RH^s_R + O(\varepsilon^2\Theta), \quad (2.24)$$

$$-\varepsilon f^2 Q^R = \frac{1}{R}H^s_\theta + \varepsilon C\sin\theta H^s - \varepsilon H^\theta_s + O(\varepsilon^2\Theta), \quad (2.25)$$

$$-\varepsilon f^2 Q^\theta = -H^s_R + \varepsilon\tau(H^R_\theta - (RH^\theta)_R) + \varepsilon C\cos\theta H^s + \varepsilon H^R_s + O(\varepsilon^2\Theta), \quad (2.26)$$

and equations (2.10), (2.11) and (2.13) become

$$\begin{aligned}
&\varepsilon(\boldsymbol{v}\cdot\boldsymbol{e}_R)f_R + (\boldsymbol{v}\cdot\boldsymbol{e}_\theta)\tfrac{\varepsilon}{R}f_\theta + \tfrac{1}{R}((Rf_R)_R + \tfrac{1}{R}f_{\theta\theta}) \\
&+\varepsilon C(\tfrac{1}{R}f_\theta\sin\theta - f_R\cos\theta) = f^3 - f + \varepsilon^2 f|Q|^2 + O(\varepsilon^2),
\end{aligned} \quad (2.27)$$

$$\begin{aligned}
&\varepsilon f^2\Theta + \tfrac{1}{R}((Rf^2Q^R)_R + (f^2Q^\theta)_\theta) \\
&-\varepsilon Cf^2(Q^R\cos\theta - Q^\theta\sin\theta) = O(\varepsilon^2\boldsymbol{Q}, \varepsilon Q^s),
\end{aligned} \quad (2.28)$$

$$\begin{aligned}
&\tfrac{1}{R}((Rf^2Q^R)_R + (f^2Q^\theta)_\theta) - \varepsilon Cf^2(Q^R\cos\theta - Q^\theta\sin\theta) \\
&-\varepsilon\eta[(\boldsymbol{v}\cdot\boldsymbol{e}_R)\tfrac{\partial}{\partial R} + (\boldsymbol{v}\cdot\boldsymbol{e}_\theta)\tfrac{1}{R}\tfrac{\partial}{\partial\theta}][\tfrac{1}{R}(RQ^R)_R + \tfrac{1}{R}Q^\theta_\theta] \\
&+\varepsilon\eta(\tfrac{1}{R}(R\Theta_R)_R + \tfrac{1}{R^2}\Theta_{\theta\theta}) = O(\varepsilon^2\Theta, \varepsilon Q^s).
\end{aligned} \quad (2.29)$$

We impose the following expected conditions at the origin:

$$f \to 0,\ \boldsymbol{Q} \to -\frac{N_k}{\varepsilon R}\boldsymbol{e}_\theta \ \text{ as } \ R \to 0. \quad (2.30)$$

The core solutions then look like:

$$\boldsymbol{Q} \sim \frac{1}{\varepsilon}Q^\theta_0(R)\boldsymbol{e}_\theta + \cdots, \quad (2.31)$$

$$f \sim f_0(R) + \cdots. \quad (2.32)$$

Substituting (2.31) and (2.32) into (2.21)–(2.29), the leading order behaviour of the core field is given by

$$\boldsymbol{Q}_0 = -\frac{N_k}{R}\boldsymbol{e}_\theta, \tag{2.33}$$

$$f_0'' + \frac{1}{R}f_0' - \frac{N_k^2}{R^2}f_0 = f_0^3 - f_0, \tag{2.34}$$

$$f_0(0) = 0, \tag{2.35}$$

$$f_0(\infty) = 1 \tag{2.36}$$

where the final boundary condition (2.36) comes from the need of matching with the outer solution. This system is the same as the leading-order core behaviour of a stationary rectilinear vortex. Expression (2.33) tells us what is the behaviour of $\boldsymbol{Q}$ when a vortex core is approached which in turn, decides the behaviour of $\boldsymbol{H}$ when a vortex core is approached.

From (2.33), we know that

$$\boldsymbol{Q}_0 \sim -\frac{N_k}{\rho}\boldsymbol{e}_\theta \quad \text{as} \quad \rho \to 0. \tag{2.37}$$

Solving (2.14) and (2.37), we obtain

$$H_0 \sim -N_k \log \rho \boldsymbol{e}_s \quad \text{as} \quad \rho \to 0. \tag{2.38}$$

Equations (2.14)–(2.16) with (2.37)–(2.38) imply that for the leading order terms,

$$\Delta H_0 - H_0 = -2\pi \sum_m \int_{\Gamma_m} N_m \delta(x - x_m(s))\delta(y - y_m(s))\delta(z - z_m(s))d\boldsymbol{s} \tag{2.39}$$

where $\boldsymbol{Q}_m(s) = (x_m(s),\ y_m(s),\ z_m(s))$ is the location of the vortex, $\delta$ is the Dirac $\delta$-function, $d\boldsymbol{s} = (dx(s),\ dy(s),\ dz(s))$. Since the solution of the equation

$$\Delta G - G = -\delta(x)\delta(y)\delta(z) \quad \text{in} \quad I\!R^3 \tag{2.40}$$

is

$$G(\boldsymbol{x}) = \frac{1}{4\pi}\frac{e^{-r}}{r} \tag{2.41}$$

where $r = |\boldsymbol{x}|$, the solution of (2.39) is

$$\boldsymbol{H}_0 = \frac{1}{2}\sum_m N_m \int_{\Gamma_m} \frac{e^{-Z_m}}{Z_m} d\boldsymbol{s} \tag{2.42}$$

where $\boldsymbol{Z}_m = \boldsymbol{x} - \boldsymbol{Q}_m$, $Z_m = |\boldsymbol{Z}_m|$. $\Gamma_m$ is the curve representing the position of the $m$-th vortex. Applying (2.14) to (2.42), we have

$$\boldsymbol{Q}_0 = -\frac{1}{2}\sum_m N_m \int_{\Gamma_m} \nabla\left(\frac{e^{-Z_m}}{Z_m}\right) \times d\boldsymbol{s}. \tag{2.43}$$

Now we assume that the vortex line $\Gamma_k$ is approached so that the distance $\rho$ from $\boldsymbol{x}$ to the nearest point on the vortex curve $\boldsymbol{Q}_k(S)$ becomes of order $\varepsilon$ and consider the asymptotic behaviour of (2.44). This will give motivation for the form of the core expansion. We write $\boldsymbol{Q}_0$ in the following form:

$$\boldsymbol{Q}_0 = \frac{N_k}{2}\Big[\int_{S-L}^{S+L} + \int_{S+L}^{\infty} + \int_{-\infty}^{S-L}\Big] \exp(-Z_k)\Big(\frac{1}{Z_k^2}+\frac{1}{Z_k^3}\Big)\boldsymbol{Z}_k \times d\boldsymbol{s}$$
$$-\frac{1}{2}\sum_{m\neq k} N_m \int_{\Gamma_m} \nabla\Big(\frac{\exp(-Z_m)}{Z_m}\Big) \times d\boldsymbol{s}. \qquad (2.44)$$

If we assume that $1/\sqrt{|\log\varepsilon|} \ll L \ll 1$, the major contribution to $\boldsymbol{Q}_0$ comes obviously from the first integral in (2.44). We formally set $s = S + \varepsilon l$ and expand the quantities (to facilitate the computation of the integral)

$$\boldsymbol{Q}_k(S+\varepsilon l) - \boldsymbol{Q}_k(S) = \varepsilon l\boldsymbol{t}(S) + \frac{\varepsilon^2 l^2}{2} C\boldsymbol{n}(S) + \cdots, \qquad (2.45)$$

$$\boldsymbol{Z}_k(S+\varepsilon l) = (\varepsilon R\cos\theta - \frac{\varepsilon^2 l^2}{2}C)\boldsymbol{n}(S) + \varepsilon R\sin\theta\mathbf{b}(S) - \varepsilon l\boldsymbol{t}(S) + \cdots, \qquad (2.46)$$

$$\boldsymbol{t}(S+\varepsilon l) = \boldsymbol{t}(S) + \varepsilon lC\boldsymbol{n}(S) + \cdots, \qquad (2.47)$$

for $l \ll 1/\varepsilon$. Then

$$\int_{S-L}^{S+L} \exp(-Z_k)\Big(\frac{1}{Z_k^2}+\frac{1}{Z_k^3}\Big)\boldsymbol{Z}_k \times d\boldsymbol{s} = \boldsymbol{e}_\theta \frac{1}{\varepsilon R}\int_{-L/\varepsilon R}^{L/\varepsilon R} \frac{1}{(1+u^2)^{3/2}}du$$
$$-\mathbf{b}(S)\int_{-L/\varepsilon R}^{L/\varepsilon R} \frac{Cu^2}{2(1+u^2)^{3/2}}du - \boldsymbol{e}_\theta \frac{3C\cos\theta}{2}\int_{-L/\varepsilon R}^{L/\varepsilon R} \frac{u^2}{(1+u^2)^{5/2}}du + \cdots \qquad (2.48)$$

where $u = l/R$. Evaluating (2.48), we find that $\boldsymbol{Q}_0$ has the following asymptotic behaviour as the vortex line is approached:

$$\boldsymbol{Q}_0 = -\frac{N_k}{\varepsilon R}\boldsymbol{e}_\theta(S) + \frac{N_k C}{2}\log(\varepsilon R)\mathbf{b}(S) - \frac{N_k C}{2}(\log 2 - 1)\mathbf{b}(s)$$
$$-\frac{N_k C}{2}\cos\theta\boldsymbol{e}_\theta(S) - \frac{1}{2}\sum_{m\neq k} N_m \int_{\Gamma_m}\nabla\Big(\frac{\exp(-Z_m)}{Z_m}\Big) \times d\boldsymbol{s}$$
$$+\frac{N_k}{2}\lim_{L\to 0}\Big(\int_{|s-S|>L} \exp(-Z_k)\Big(\frac{1}{Z_k^2}+\frac{1}{Z_k^3}\Big)\boldsymbol{Z}_k \times d\boldsymbol{s} - \frac{1}{2}\log L\mathbf{b}(s)\Big) + \cdots. \qquad (2.49)$$

Using the same method of deriving (1.8), we show that the limit term in the above expression gives a term of order $O(1)$ and is therefore not very important in our expansion.

### 3.2.4 Core region

Now we have obtained the solution in the outer region near a given vortex, we can study the core region to obtain the moving speed of the vortex core. First, we further expand the solutions of (2.21)–(2.29) as follows

$$f = f_0(R) + \varepsilon \log \varepsilon f_1 + \varepsilon f_2 + \cdots , \tag{2.50}$$

$$\boldsymbol{Q} = \frac{N_k}{\varepsilon R}\boldsymbol{e}_\theta + \log \varepsilon \boldsymbol{Q}_1 + \boldsymbol{Q}_2 + \cdots , \tag{2.51}$$

$$\boldsymbol{H} = \log \varepsilon \boldsymbol{H}_0 + \cdots , \tag{2.52}$$

$$\Theta = \frac{\log \varepsilon}{\varepsilon}\Theta_0 + \frac{1}{\varepsilon}\Theta_1 + \cdots , \tag{2.53}$$

$$\boldsymbol{v} = \boldsymbol{v}_0 \log \varepsilon + \boldsymbol{v}_1 + \cdots . \tag{2.54}$$

Substituting these expressions into (2.21)–(2.29), equating the coefficients of $\log \varepsilon$ to zero, we obtain

$$(RQ_1^\theta)_R - Q_{1,\theta}^R = 0, \tag{2.55}$$

$$Q_{1,\theta}^s = Q_{1,R}^s = 0, \tag{2.56}$$

$$(RH_0^\theta)_R - H_{0,\theta}^R = 0, \tag{2.57}$$

$$H_{0,\theta}^s = H_{0,R}^s = 0, \tag{2.58}$$

$$\begin{aligned}&(\boldsymbol{v}_0 \cdot \boldsymbol{e}_R) f_0'(R) + \frac{1}{R}(Rf_{1,R})_R + \frac{1}{R^2} f_{1,\theta\theta} \\ &= 3f_0^2 f_1 - f_1 - \frac{2N_k f_0}{R}\boldsymbol{e}_\theta \cdot \boldsymbol{Q}_1 + \frac{N_k^2 f_1}{R^2},\end{aligned} \tag{2.59}$$

$$f_0^2\Theta_0 - \frac{1}{R}(\frac{2N_k f_0 f_1}{R})_\theta + \frac{1}{R}(Rf_0^2 Q_1^R)_R + \frac{1}{R}(f_0^2 Q_1^\theta)_\theta = 0, \tag{2.60}$$

$$-(\frac{2N_k f_0 f_1}{R})_\theta + (Rf_0^2 Q_1^R)_R + (f_0^2 Q_1^\theta)_\theta + \eta((R\Theta_{0,R})_R + \frac{1}{R}\Theta_{0,\theta\theta}) = 0. \tag{2.61}$$

By direct observation, we know that the solutions satisfy the following conditions:

$$\boldsymbol{H}_0 = -N_k \boldsymbol{e}_s, \tag{2.62}$$

$$f_0^2\Theta_0 = \eta(\frac{1}{R}(R\Theta_{0,R})_R + \frac{1}{R^2}\Theta_{0,\theta\theta}). \tag{2.63}$$

Taking into account of the singularity of $\Theta_0$ as the vortex line is approached, we have, from (2.33)

$$\Theta_0 \sim -\frac{N_k}{R}(\boldsymbol{v}_0 \cdot \boldsymbol{e}_\theta) \quad \text{as} \quad R \to 0. \tag{2.64}$$

The condition (2.55) is a curl type relation, hence we can write the $\theta$ and $R$ components of $\boldsymbol{Q}_1$ in terms of a scalar potential $\psi_1$:

$$Q_1^\theta = \frac{1}{R}\psi_{1,\theta}, \quad Q_1^R = \psi_{1,R}. \tag{2.65}$$

To match with the asymptotic behaviour of the outer solution (2.49), we require that $\psi_1 \sim \text{constant} \times R \sin\theta$ as $R \to \infty$. Hence, we look for solutions of the following form:

$$\psi_1 = \phi_0(R)\sin\theta, \tag{2.66}$$
$$\Theta_0 = U_0 N_k \sigma(R)\sin\theta, \tag{2.67}$$
$$f_1 = g_0(R)\cos\theta, \tag{2.68}$$
$$\boldsymbol{v}_0 = U_0(\cos\theta \boldsymbol{e}_R - \sin\theta \boldsymbol{e}_\theta) = U_0 \boldsymbol{n}. \tag{2.69}$$

The equations (2.59), (2.61) and (2.63) then reduces to

$$\frac{1}{R}(Rg_0')' - 3f_0^2 g_0 + g_0 = -U_0 f_0' - \frac{2N_k f_0 \phi_0}{R^2} + \frac{g_0(N_k^2+1)}{R^2}, \tag{2.70}$$
$$\frac{1}{R}(Rf_0^2\phi_0')' + \eta U_0 N_k(\sigma' + \frac{\eta}{R})' - \frac{f_0^2\phi_0}{R^2} + \frac{2N_k f_0 g_0}{R^2} = 0, \tag{2.71}$$
$$\sigma'' + \frac{1}{R}\sigma' - \frac{1}{R^2}\sigma - \frac{f_0^2}{\eta}\sigma = 0, \tag{2.72}$$

where $' = d/dR$. The boundary conditions are

$$\sigma \sim \frac{1}{R} \quad \text{as} \quad R \to 0, \tag{2.73}$$
$$R\sigma \to 0 \quad \text{as} \quad R \to \infty. \tag{2.74}$$

The difference here with the pure G-L case (without magnetic potential) is that the curvature terms are no longer present (see (5.11) in Chapter 2 and discussions there). Because of the presence of the applied field, the term $H$ in (4.17) and consequently the term $\boldsymbol{K}$ in (4.19) of Chapter 2 do not appear either. Multiplying (2.70) by $Rf_0'$ and the derivative of (2.34) by $Rg_0$, subtracting the results, we have

$$RU_0(f_0')^2 + (Rg_0'f_0' - Rg_0 f_0'')' = -\frac{2N_k f_0' f_0 \phi_0}{R} + \frac{2N_k^2 f_0 g_0}{R^2}. \tag{2.75}$$

Using (2.71) to eliminate the final $g_0$, integrating over $(0,\infty)$ and using the asymptotic behaviour of the $f_0$ and $\Theta_0$ at 0 and $\infty$ we find

$$\lim_{R\to\infty} (\phi_0' + \phi_0/R) = -U_0\beta/N_k, \tag{2.76}$$
$$\beta = \int_0^\infty R(f_0')^2 dR + N_k^2 \int_0^\infty f_0^2 \sigma dR. \tag{2.77}$$

Hence

$$\lim_{R\to\infty} \boldsymbol{Q}_1 = -\frac{U_0\beta}{2N_k}\mathbf{b}. \tag{2.78}$$

For technical reasons, we now proceed to determine one more term in the core expansion. Equating coefficients at the next order in (2.21)–(2.29) gives

$$(RQ_2^\theta)_R - Q_{2,\theta}^R = 0, \tag{2.79}$$

$$Q_{2,\theta}^s = Q_{2,R}^s = 0, \tag{2.80}$$

$$\begin{aligned}(\boldsymbol{v}_1 \cdot \boldsymbol{e}_R)f_0' &+ \frac{1}{R}(Rf_{2,R})_R + \frac{1}{R^2}f_{2,\theta\theta} - Cf_0'\cos\theta \\ &= 3f_0^2 f_2 - f_2 - 2\frac{N_k f_0}{R^2}\psi_{2,\theta} + \frac{N_k^2 f_2}{R^2},\end{aligned} \tag{2.81}$$

$$\begin{aligned}f_0^2\Theta_1 &- \frac{2N_k f_0}{R^2}f_{2,\theta} + \frac{1}{R}(Rf_0\psi_{2,R})_R \\ &+\frac{f_0^2}{R^2}\psi_{2,\theta\theta} + \frac{CN_k f_0^2\sin\theta}{R} = 0,\end{aligned} \tag{2.82}$$

$$\begin{aligned}&-\frac{2N_k f_0}{R^2}f_{2,\theta} + \frac{1}{R}(Rf_0^2\psi_{2,R})_R + \frac{f_0^2}{R^2}\psi_{2,\theta\theta} \\ &-\frac{CN_k f_0^2 \sin\theta}{R} + \eta(\frac{1}{R}(R\Theta_{1,R})_R + \frac{1}{R^2}\Theta_{1,\theta\theta}) = 0.\end{aligned} \tag{2.83}$$

Similar as for the leading term expansion, we use the curl relation in (2.79) to let

$$\boldsymbol{Q}_2^\theta = \frac{1}{R}\psi_{2,\theta}, \boldsymbol{Q}_2^R = \psi_{2,R}. \tag{2.84}$$

Using (2.82) and (2.83), we conclude that $\Theta_1$ also satisfies (2.63). In order to match with the outer solution, we should have $R\Theta_1 \to 0$ as $R \to \infty$, while (2.33) gives $\Theta_1 \sim -\frac{N_k}{R}(\boldsymbol{v}_1 \cdot \boldsymbol{e}_\theta)$ as $R \to 0$. We then assume that the solutions of (2.81)–(2.83) have the form

$$\psi_2 = \phi_1(R)\sin\theta + \phi_2(R)\sin(\theta - \alpha), \tag{2.85}$$

$$f_2 = g_1(R)\cos\theta + g_2(R)\cos(\theta - \alpha), \tag{2.86}$$

$$\Theta_1 = U_1 N_k \sigma(R)\sin\theta + U_2 N_k \sigma(R)\sin(\theta - \alpha), \tag{2.87}$$

$$\begin{aligned}\boldsymbol{v}_1 &= U_1\boldsymbol{n} + U_2(\cos\alpha\boldsymbol{n} + \sin\alpha\mathbf{b}) \\ &= U_1(\cos\theta\boldsymbol{e}_R - \sin\theta\boldsymbol{e}_\theta) \\ &\quad +U_2(\cos(\theta-\alpha)\boldsymbol{e}_R - \sin(\theta-\alpha)\boldsymbol{e}_\theta).\end{aligned} \tag{2.88}$$

Substituting (2.85)–(2.88) into (2.81)–(2.83), we obtain a set of ordinary differential equations. We notice that the equations for $(\phi_2, g_2, U_2)$ are identical to (2.70) and (2.71) satisfied by $(\phi_0, g_0, U_0)$. Hence,

$$\phi_2 \sim -\frac{U_2\beta R}{2N_k} \quad \text{as} \quad R \to \infty. \tag{2.89}$$

The equations for $(\phi_1, g_1)$ are

$$U_1 f_0' + \frac{1}{R}(Rg_1')' = 3f_0^2 g_1 - g_1 - \frac{2N_k f_0 \phi_1}{R^2} + \frac{(N_k^2+1)g_1}{R^2} + Cf_0', \quad (2.90)$$

$$\frac{1}{R}(Rf_0^2\phi_1')' + \eta U_1 N_k(\sigma' + \frac{\sigma}{R})' - \frac{f_0^2\phi_1}{R^2} + \frac{2N_k f_0 g_1}{R^2} - \frac{CN_k f_0^2}{R} = 0. (2.91)$$

Similar as before, multiplying (2.90) by $Rf_0'$, subtracting the derivative of (2.34) multiplied by $Rg_1$, eliminating $g_1$ by using (2.91), and integrating over $(0, R)$, we obtain

$$\phi_1 \sim \frac{N_k CR}{2}(\log R + \gamma - \frac{1}{2} - \frac{U_1\beta}{CN_k^2}) \text{ as } R \to \infty, \quad (2.92)$$

where

$$\gamma = \int_0^1 \frac{f_0^2}{R}dr + \int_1^\infty \frac{f_0^2-1}{R}dR + \frac{1}{N_k^2}\int_0^\infty Rf_0^2 dR. \quad (2.93)$$

Following the same principle as in the leading order term case, we obtain

$$Q_2 \sim -\frac{CN_k}{2}(\cos\theta \boldsymbol{e}_\theta - (\log R + \gamma + \frac{1}{2} - \frac{U_1\beta}{CN_k^2})\mathbf{b})$$
$$-\frac{U_2\beta}{2N_k}(\cos\alpha\mathbf{b} - \sin\alpha\boldsymbol{n})\boldsymbol{Q}_2^s(s)\boldsymbol{e}_s \text{ as } R \to \infty. \quad (2.94)$$

**Matching**

Our final job is to match the core terms $\boldsymbol{Q}_1$ and $\boldsymbol{Q}_2$ with the outer solution $\boldsymbol{Q}_0$ as in the pure G-L case (see Chap 2, Section 4.6). Using (2.78), (2.94) and (2.49), we find that in order to match the core and outer solutions, we must have, for the core variables

$$\boldsymbol{v}_0 = -\frac{CN_k^2}{\beta}\boldsymbol{n}, \quad (2.95)$$

$$\boldsymbol{v}_1 = \frac{CN_k^2}{\beta}(\gamma + \log 2 - \frac{1}{2})\boldsymbol{n} - \frac{2PN_k}{\beta}(\cos\alpha\boldsymbol{n} + \sin\alpha\mathbf{b}) \quad (2.96)$$

where $P$ is defined through the outer field quantities:

$$P(\cos\alpha\mathbf{b} - \sin\alpha\boldsymbol{n}) + Q_0^s(s)\boldsymbol{\tau}(s)$$
$$= -\frac{1}{2}\sum_{m\neq k} N_m \int_{\Gamma_m} \nabla(\exp(-Z_m)/Z_m) \times d\boldsymbol{s}$$
$$+\frac{N_k}{2}\lim_{L\to 0}(\int_{|s-S|>L} \exp(-Z_k)(\frac{1}{Z_k^2} + \frac{1}{Z_k^3})\boldsymbol{Z}_k \times d\boldsymbol{s} - \frac{1}{2}\log L\mathbf{b}(s)). \quad (2.97)$$

We note that for the far field quantities, from (2.14), we have

$$\boldsymbol{J}_0 = \mathbf{curl}\boldsymbol{H}_0 = -\boldsymbol{Q}_0$$

where $\boldsymbol{J}$ is the electric current. Writing accordingly

$$\boldsymbol{J}_{0,ext} = \frac{1}{2}\sum_{m\neq k} N_m \int_{\Gamma_m} \nabla(\exp(-Z_m)/Z_m) \times d\boldsymbol{s} \tag{2.98}$$

and

$$\begin{aligned} \boldsymbol{J}_{0,fin} &= -\frac{N_k}{2}\lim_{L\to 0}\Big(\int_{|s-S|>L} \exp(-Z_k)\Big(\frac{1}{Z_k^2}+\frac{1}{Z_k^3}\Big)\boldsymbol{Z}_k \times d\boldsymbol{s} \\ &\quad -\frac{1}{2}\log L\mathbf{b}(s)\Big), \end{aligned} \tag{2.99}$$

(2.96) becomes

$$\boldsymbol{v}_1 = \frac{CN_k^2}{\beta}\Big(\gamma + \log 2 - \frac{1}{2}\Big)\boldsymbol{n} + \frac{2N_k}{\beta}(\boldsymbol{J}_{0,ext} + \boldsymbol{J}_{0,fin}) \times \boldsymbol{t}. \tag{2.100}$$

Combining (2.95) and (2.100), we obtain the following expression for the velocity of vortex motion:

$$\begin{aligned} \boldsymbol{v} &= -\frac{\log \varepsilon C N_k^2}{\beta}\boldsymbol{n} + \frac{CN_k^2}{\beta}\Big(\gamma + \log 2 - \frac{1}{2}\Big)\boldsymbol{n} \\ &+ \frac{2N_k}{\beta}(\boldsymbol{J}_{0,ext} + \boldsymbol{J}_{0,fin}) \times \boldsymbol{t}. \end{aligned} \tag{2.101}$$

The first two terms here correspond to the local self-induced motion of the vortex line. The final term corresponds to the effect of the Lorentz force.

**Remark 3.2.2** *As this asymptotic analysis is done in the large $\kappa$ context, the corresponding physical background is to have non-interacting vortices. Therefore in the derivation process, we see no term representing vortex interactions.*

## 3.3 Asymptotic analysis of densely packed vortices

So far, we have considered vortices that have finite distances between them. The resulting induced magnetic field is also of finite order. In the formal expansion, we expected that a large magnetic field is required to maintain the vortex presence because of the effect of large $\kappa$ (small $\varepsilon$). This turns out not to be the case because the $1/\varepsilon$ order term of the magnetic potential vanishes automatically as a solution of the resulting limit system. This gives us firm evidence that even as $\kappa \to \infty$, the induced magnetic field is still only proportional to the number of vortices present. In this section, we generalize this conception to cases where there are "many" vortices (but still significantly less than $1/\varepsilon$).

To this end, we consider the case in which the vortex spacing is $O(\delta)$ where $\varepsilon \ll \delta^2 \ll 1$. From our arguments above, this should lead to an induced magnetic field of the order $1/\delta^2$. When we choose $\delta^2 < 1/\log\varepsilon$, the effect of the induced field is much stronger than the effect of the supercurrent surrounding the vortex. In other words, the vortices will be locked into relatively fixed positions by the presence of neighbouring vortices. They can no longer evolve themselves freely in space. We understand this is not usually the case unless the vortices are well placed and well shaped beforehand. So this model is only valid after an initial transitional period during which the vortices have been rearranged and placed by certain means. We expect that the motion due to this induced field will be the dominant term in the formal expansion. We now show formally that this is indeed the case. In this analysis, there will be three length scales on which to consider the problem, namely, $O(1)$, $O(\delta)$ and $O(\varepsilon)$ length scales.

### 3.3.1 Outer region – a mean field model

As $\kappa \to \infty$, from the G-L system, we derived that

$$\mathbf{curl}^2 \boldsymbol{H} + \boldsymbol{H} = 2\pi \boldsymbol{\delta}_\Gamma(\boldsymbol{x}), \tag{3.1}$$

$$\operatorname{div} \boldsymbol{H} = 0. \tag{3.2}$$

These equations were derived under the assumption that the time scale for the relaxation of the magnetic field was much shorter than the time scale for the motion of vortices, so that the magnetic field is quasi-static.

If there are many vortices, we will have

$$\mathbf{curl}^2 \boldsymbol{H} + \boldsymbol{H} = 2\pi \sum_k \boldsymbol{\delta}_{\Gamma_k}(\boldsymbol{x}), \tag{3.3}$$

$$\operatorname{div} \boldsymbol{H} = 0. \tag{3.4}$$

where the sum is taken over all the vortices. We formally define

$$\boldsymbol{\omega}_\eta = \int_{|\boldsymbol{x}-\boldsymbol{x}'|<\eta} \frac{3}{2\eta^3} \sum_k \boldsymbol{\delta}_{\Gamma_k}(\boldsymbol{x}' - \boldsymbol{x}_k) dx' \tag{3.5}$$

and study the case $\eta \to 0$. At the same time, we assume that the number of vortices tends to infinity. More precisely, we let the vortex separation distance be $\delta$ and assume that $\delta \ll \eta$. Hence, as $\eta \to 0$, there are still more and more vortices (proportional $1/\delta^2$) gathering in any small area $\eta^2$.

We assume that the vortices all have unit strength and are not affected by the large number of them present. This is an extreme type II behaviour. The consequence is that the magnetic field generated by the vortices will have to be added up to the order of $O(1/\delta^2)$. Hence we assume that the vorticity

$$\boldsymbol{\omega} = \lim_{\eta \to 0} \boldsymbol{\omega}_\eta \tag{3.6}$$

is of order $O(1/\delta^2)$.

Since we have assumed that $\boldsymbol{\omega}$ has strength $O(1/\delta^2)$, and because Equation (3.3) is linear in both variables, we can divide both sides of the equation by $1/\delta^2$ and renaming the new variables again as $\boldsymbol{H}$ and $\boldsymbol{\omega}$ (to replace the old $\boldsymbol{H}/(1/\delta^2)$ and $\boldsymbol{\omega}/(1/\delta^2)$). Then all the quantities are of finite order and converge as $\eta$, $\delta \to 0$ respectively. The limit quantities satisfy

$$\boldsymbol{H} - \Delta \boldsymbol{H} = \boldsymbol{\omega}, \tag{3.7}$$

$$\mathrm{div}\boldsymbol{H} = 0. \tag{3.8}$$

To obtain an additional equation on $\boldsymbol{\omega}$, we impose a conservation law (using the idea from fluid mechanics). Assuming that the vortices all move uniformly with a given velocity $\boldsymbol{v}$, then, for any surface $S$ which is also moving with velocity $\boldsymbol{v}$, we have the following conservation law:

$$\frac{d}{dt}\int_S \boldsymbol{\omega}\cdot\boldsymbol{n}dS = 0. \tag{3.9}$$

This is equivalent to the differential form

$$\boldsymbol{\omega}_t + \mathbf{curl}(\boldsymbol{\omega}\times\boldsymbol{v}) = \mathbf{0}. \tag{3.10}$$

Now, a constitutive relation will relate $\boldsymbol{v}$ to $\boldsymbol{H}$ and $\boldsymbol{\omega}$. If there is a single vortex, assuming that there is no self-induced motion, we know that $\boldsymbol{v} \perp \boldsymbol{j}$ and $\boldsymbol{v} \perp \boldsymbol{\tau}$ where $\boldsymbol{\tau}$ is the unit tangent direction along the vortex line. Using suitable physical units and choosing a suitable direction of $\boldsymbol{\tau}$, we have $\boldsymbol{v} = \boldsymbol{j}\times\boldsymbol{\tau}$. When there are many vortices, we take the average of this law and let

$$\boldsymbol{v} = \boldsymbol{j}\times\frac{\boldsymbol{\omega}}{|\boldsymbol{\omega}|} \tag{3.11}$$

$$= \mathbf{curl}\boldsymbol{H}\times\frac{\boldsymbol{\omega}}{|\boldsymbol{\omega}|}. \tag{3.12}$$

The resulting complete system is

$$\begin{cases} \boldsymbol{\omega}_t + \mathbf{curl}(\boldsymbol{\omega}\times\boldsymbol{v}) = 0, \\ \mathbf{curl}^2\boldsymbol{H} + \boldsymbol{H} = \boldsymbol{\omega}, \\ \mathrm{div}\boldsymbol{H} = 0, \\ \boldsymbol{v} = \mathbf{curl}\boldsymbol{H}\times\frac{\boldsymbol{\omega}}{|\boldsymbol{\omega}|}. \end{cases}$$

### 3.3.2 Intermediate region

In the mean field context, we can study the problem on the $\delta$ length scale: the independence of the vortices is maintained, and the distance between the vortices is finite. The trick is to rescale the space variables by $\delta$. After the rescaling, it is easy to understand that we obtain the far field equations as in the standard G-L case. So the leading order equations for $\boldsymbol{H}$ and $\boldsymbol{Q}$ are as follows (the same as the leading terms in far field (2.37) and (2.38))

$$\mathbf{curl}\boldsymbol{H}_0 = 0, \tag{3.13}$$

$$\mathbf{curl}\boldsymbol{Q}_0 = 0, \tag{3.14}$$

$$\operatorname{div}\boldsymbol{H}_0 = 0, \tag{3.15}$$

$$\operatorname{div}\boldsymbol{Q}_0 = 0 \tag{3.16}$$

with the condition

$$\boldsymbol{Q}_0 \sim -\frac{N_k}{\delta r}\boldsymbol{e}_\theta \tag{3.17}$$

as each vortex line is approached where $r = R/\delta$ and $R$ and $\theta$ are the curvilinear co-ordinates previously introduced. Hence in the intermediate region,

$$\boldsymbol{H} = \frac{\boldsymbol{H}_0}{\delta^2} + \cdots, \tag{3.18}$$

$$\boldsymbol{Q} = \frac{\boldsymbol{Q}_0}{\delta^2} - \frac{1}{2\delta}\int_{\Gamma_m} \nabla(1/Z_m) \times d\boldsymbol{s} + \cdots \tag{3.19}$$

$$f = 1 + \cdots \tag{3.20}$$

where $\boldsymbol{H}_0$ and $\boldsymbol{Q}_0$ are constants, $Z_m$ and $\Gamma_m$ are as before and $\cdots$ represents higher order terms in the formal expansion.

### 3.3.3 Core region

In the core region, the problem is slightly more complicated. In addition to the $\log\varepsilon/\varepsilon$, $1/\varepsilon$, $\log\varepsilon$, $1$, $\varepsilon\log\varepsilon$, $\varepsilon$, ... terms, we also need to take into account the $1/\delta^2$, $1/\delta$, $1$, $\delta$, ... terms as well as their combinations. We start from $1/\delta^2$ for the $\delta$ involving terms because the magnetic field is of that magnitude. We only state briefly the results. Interested readers can refer to [CRi 95] for more details. Taking into account the two-scale scaling, we have

$$f = f_0(R) + \frac{\varepsilon}{\delta^2}f_1 + \cdots, \tag{3.21}$$

$$\boldsymbol{Q} = -\frac{N_k\boldsymbol{e}_\theta}{\varepsilon R} + \frac{1}{\delta^2}\boldsymbol{Q}_1 + \log\varepsilon\boldsymbol{Q}_2 + \cdots, \tag{3.22}$$

$$\Theta = \frac{1}{\varepsilon\delta^2}\Theta_0 + \frac{\log\varepsilon}{\varepsilon}\Theta_1 + \cdots, \tag{3.23}$$

$$\boldsymbol{H} = \frac{1}{\delta^2}\boldsymbol{H}_0 + \log\varepsilon\boldsymbol{H}_1 + \cdots, \tag{3.24}$$

$$\boldsymbol{v} = \frac{1}{\delta^2}\boldsymbol{v}_0 + \log\varepsilon\boldsymbol{v}_1 + \cdots. \tag{3.25}$$

As before, a scalar potential may be defined for $\boldsymbol{Q}_1$:

$$\boldsymbol{Q}_1^\theta = \frac{1}{R}\psi_{1,\theta}, \tag{3.26}$$

$$\boldsymbol{Q}_1^R = \psi_{1,R}. \tag{3.27}$$

The order-$\delta^{-2}$ solutions of the core equations then take the form

$$\psi_1 = \phi_0(R)\sin(\theta-\alpha), \tag{3.28}$$

$$\Theta_0 = U_0\sigma(R)\sin(\theta-\alpha), \tag{3.29}$$

$$f_1 = g_0(R)\cos(\theta-\alpha), \tag{3.30}$$

$$\boldsymbol{v}_0 = U_0(\cos(\theta-\alpha)\boldsymbol{e}_R - \sin(\theta-\alpha)\boldsymbol{e}_\theta) = U_0(\cos\alpha\boldsymbol{n} + \sin\alpha\mathbf{b}). \tag{3.31}$$

The same kind of computation as for (2.71)–(2.73) leads to

$$\lim_{R\to\infty}\boldsymbol{Q}_1 = -\frac{U_0\beta}{2N_k}(\cos\alpha\mathbf{b} - \sin\alpha\boldsymbol{n}) + \boldsymbol{Q}_1^s(s)\boldsymbol{e}_s. \tag{3.32}$$

Matching this with the solution in the intermediate region, we obtain

$$\boldsymbol{v}_0 = -\frac{2PN_k}{\beta}(\cos\alpha\boldsymbol{n} + \sin\alpha\mathbf{b}). \tag{3.33}$$

Since in the intermediate region, we have

$$\boldsymbol{Q}_0 = P(\cos\alpha\mathbf{b} - \sin\alpha\boldsymbol{n}) + Q_0^s(s)\boldsymbol{e}_s, \tag{3.34}$$

we conclude that

$$\boldsymbol{v}_0 = -\frac{2N_k}{\beta}\boldsymbol{Q}_0\times\boldsymbol{t} = \frac{2N_k}{\beta}\boldsymbol{J}_0\times\boldsymbol{t} \tag{3.35}$$

where $\boldsymbol{J}_0 = -\boldsymbol{Q}_0 = \mathbf{curl}\boldsymbol{H}_0$ is the leading order electric current in the outer region.

Thus we see that in the mean field model, the motion of vortices is dominated by the mean-field behaviour. The self-induced motion is no longer present in the leading order velocity term.

**Conclusion:** Under the context that $\xi \ll \lambda$ (coherence length $\ll$ penetration depth ), the equations of motion of vortices are determined (at least formally) by the assumptions on their density.

If the vortices are few and far between, they move mainly by mean curvature (self-induced motion) and tend to straighten themselves up as we have seen in Section 2.

If there are many vortices (more precisely, with $\varepsilon^2 \ll \delta^2 \ll 1/|\log\varepsilon|$), it seems that the relative motion could be assumed to be frozen and the vortices could be made to move together – This is the essence of the mean field model we have studied in this section.

Finally, if the vortex number is between the two cases above (more precisely, with $\delta^2 \sim 1/|\log\varepsilon|$), we conjecture that the equation of motion would combine the effect of motion by mean curvature and the mean field motion. But even formally, we have difficulty in justifying this and we regard it as an open problem.

## 3.4 Bibliographical remarks

The formal asymptotics of the vortex motion using the complex G-L system has been an active research area in the last few years. In this chapter, we have largely followed the works of [Cha1 95], [Cha2 95] and [CRi 95]. There are many other works such as [E1 94], [E2 94], [PRu 93], [Rub 95] which are also important contributions.

There are also a number of papers on the mathematical and numerical analysis of the mean field models. The details of these are beyond the scope of this book. We refer to [BCE 99], [ESS 98], [ESt1 99], [ESt2 99], [SSt 99] for more details.

# Chapter 4
# Steady State Solutions

In this chapter, we establish some basic mathematical results such as existence of solutions and the vortex structure exhibited by the solutions. Some relationships between the applied magnetic field and the solutions will also be studied.

## 4.1 Existence of steady state solutions

In this section we briefly discuss the existence of solutions for the steady state problem. The idea is to use minimization of energy. It is therefore a fairly straightforward result in the calculus of variations. The interesting points are the interpretation of the various terms involved. We have mainly followed the work of [DGP 92] and [CDG 96].

### 4.1.1 The outside field is a given function, 2-d case

To discuss the minimization of the standard G-L energy, we look at the following rescaled version

$$E(\psi, \boldsymbol{A}) = \frac{1}{2}\int_\Omega \left(\left|(\frac{i}{\kappa}\nabla + \boldsymbol{A})\psi\right|^2 + \frac{1}{2}(|\psi|^2 - 1)^2 + |\mathbf{curl}\boldsymbol{A} - H|^2\right) \tag{1.1}$$

where $\Omega$ is a bounded, simply connected open subset of $\mathbb{R}^2$. We assume that $\partial\Omega$ is at least piecewise $C^2$. We will also make use of the functional

$$F(\psi, \boldsymbol{A}) = E(\psi, \boldsymbol{A}) + \int_\Omega |\mathrm{div}\boldsymbol{A}|^2. \tag{1.2}$$

In this model, the applied field $H$ is a given function. The case when the outside magnetic field is governed by the Maxwell system will be discussed later in this chapter. We will make use of the following subspaces of $\boldsymbol{H}^1(\Omega)$ later on:

$$\boldsymbol{H}^1_n(\Omega) = \{\boldsymbol{Q} \in \boldsymbol{H}^1(\Omega),\ \ \boldsymbol{Q}\cdot\boldsymbol{n} = 0 \text{ on } \partial\Omega\}, \tag{1.3}$$

$$\boldsymbol{H}^1_n(\mathrm{div}, \Omega) = \{\boldsymbol{Q} \in \boldsymbol{H}^1(\Omega),\ \ \mathrm{div}\boldsymbol{Q} = 0 \text{ in } \Omega \text{ and } \boldsymbol{Q}\cdot\boldsymbol{n} = 0 \text{ on } \partial\Omega\}. \tag{1.4}$$

From the discussion of gauge transformation, we know that any element $(\psi, \boldsymbol{A}) \in \mathcal{H}^1 \times \boldsymbol{H}^1$ is gauge equivalent to some $(\xi, \boldsymbol{Q}) \in \mathcal{H}^1 \times \boldsymbol{H}^1_n(\text{div}, \Omega)$ and due to the gauge invariance property of the G-L energy functional, we have

$$E(\psi, \boldsymbol{A}) = E(\xi, \boldsymbol{Q}). \tag{1.5}$$

This observation will help us in discussing the existence of solutions.

**Theorem 4.1.1** *Let $\Omega$ be a simply connected piecewise smooth domain in $I\!R^2$, then $E(\psi, \boldsymbol{A})$ has at least one minimizer belonging to $\mathcal{H}^1(\Omega)\times\boldsymbol{H}^1_n(\text{div}, \Omega)$ and moreover, we have*

$$\min_{\mathcal{H}^1(\Omega)\times\boldsymbol{H}^1(\Omega)} E(\psi, \boldsymbol{A}) = \min_{\mathcal{H}^1(\Omega)\times\boldsymbol{H}^1_n(\text{div},\Omega)} E(\psi, \boldsymbol{A}) = \min_{\mathcal{H}^1(\Omega)\times\boldsymbol{H}^1(\Omega)} F(\psi, \boldsymbol{A}) \tag{1.6}$$

*Proof.* The first equality in (1.6) is by gauge invariance property of the energy.

For the second equality, we use gauge invariance and the energy expression, it is straightforward to see that

$$\min_{\mathcal{H}^1(\Omega)\times\boldsymbol{H}^1(\Omega)} E(\psi, \boldsymbol{A}) \leq \min_{\mathcal{H}^1(\Omega)\times\boldsymbol{H}^1(\Omega)} F(\psi, \boldsymbol{A}) \leq \min_{\mathcal{H}^1(\Omega)\times\boldsymbol{H}^1_n(\text{div},\Omega)} E(\psi, \boldsymbol{A}),$$

Hence (1.6) holds.

Now we only need to show the existence of a solution for $\min_{\mathcal{H}^1(\Omega)\times\boldsymbol{H}^1_n(\text{div},\Omega)} E(\psi, \boldsymbol{A})$ by using a standard argument from the theory of calculus of variations. Let $(\psi_m, \boldsymbol{A}_m)$ be a minimizing sequence, from the expression of the energy, it is straightforward to derive that there is a constant $C$ independent of $m$ such that

$$\int_\Omega |\nabla\psi_m|^2 + |\psi_m|^4 + |\mathbf{curl}\boldsymbol{A}_m|^2 \leq C.$$

Combining this with the fact that

$$\text{div}\boldsymbol{A}_m = 0 \text{ in } \Omega \text{ and } \boldsymbol{A}_m \cdot \boldsymbol{n} = 0 \text{ on } \partial\Omega,$$

we conclude that $(\psi_m, \boldsymbol{A}_m)$ is a bounded sequence in $\mathcal{H}^1(\Omega)\times\boldsymbol{H}^1(\Omega)$. Up to choosing a subsequence (still denoted by the index $m$), there exists a pair $(\psi_0, \boldsymbol{A}_0) \in \mathcal{H}^1(\Omega) \times \boldsymbol{H}^1(\Omega)$ such that

$$(\psi_m, \boldsymbol{A}_m) \rightharpoonup (\psi_0, \boldsymbol{A}_0) \text{ in } \mathcal{H}^1(\Omega) \times \boldsymbol{H}^1(\Omega) \text{ weakly. .}$$

Consequently, by standard Sobolev embedding, we have

$$(\psi_m, \boldsymbol{A}_m) \to (\psi_0, \boldsymbol{A}_0) \text{ in } \mathcal{L}^4(\Omega) \times \boldsymbol{L}^4(\Omega).$$

By standard convergence properties, we have

$$\begin{aligned}
&\lim_{n\to\infty} \int_\Omega \left|(\frac{i}{\kappa}\nabla + \boldsymbol{A}_m)\psi_m\right|^2 + |\mathbf{curl}\boldsymbol{A}_m - H|^2 \\
\geq \; & \int_\Omega \left|(\frac{i}{\kappa}\nabla + \boldsymbol{A}_0)\psi_0\right|^2 + |\mathbf{curl}\boldsymbol{A}_0 - H|^2
\end{aligned}$$

and

$$\lim_{n\to\infty}\int_\Omega (|\psi_m|^2-1)^2 = \int_\Omega (|\psi_0|^2-1)^2.$$

Hence,

$$\min_{\mathcal{H}^1(\Omega)\times \boldsymbol{H}^1_n(\mathrm{div},\Omega)} E(\psi,\boldsymbol{A}) \geq E(\psi_0,\boldsymbol{A}_0).$$

Since $(\psi_0, \boldsymbol{A}_0)$ is admissible, we have

$$\min_{\mathcal{H}^1(\Omega)\times \boldsymbol{H}^1_n(\mathrm{div},\Omega)} E(\psi,\boldsymbol{A}) \leq E(\psi_0,\boldsymbol{A}_0).$$

Hence the Theorem is proved.

**Remark 4.1.1** *It is interesting to observe that the equalities in (1.6) and the gauge invariance property imply that the energy minimizers* $(\psi, \boldsymbol{A})$ *of* $F$ *satisfy automatically* $\mathrm{div}\boldsymbol{A} = 0$ *in* $\Omega$ *and* $\boldsymbol{A}\cdot\boldsymbol{n} = 0$ *on* $\partial\Omega$.

**Remark 4.1.2** *Because the G-L system is the Euler-Langrange equations for the G-L energy, the energy minimizers solve the G-L system*

$$\begin{cases} (\frac{i}{\kappa}\nabla + \boldsymbol{A})^2\psi + (|\psi|^2-1)^2\psi = 0 \ \ in \ \ \Omega, \\ \mathbf{curl}^2\boldsymbol{A} - \mathbf{curl}H + \frac{i}{2\kappa}(\psi^*\nabla\psi - \psi\nabla\psi^*) + |\psi|^2\boldsymbol{A} = 0 \ \ in \ \ \Omega, \\ (\frac{i}{\kappa}\nabla + \boldsymbol{A})\psi\cdot\boldsymbol{n} = 0 \ \ on \ \ \partial\Omega, \\ \mathbf{curl}\boldsymbol{A} = H \ \ on \ \ \partial\Omega, \\ \mathrm{div}\boldsymbol{A} = 0 \ \ in \ \ \Omega, \\ \boldsymbol{A}\cdot\boldsymbol{n} = 0 \ \ on \ \ \partial\Omega \end{cases} \tag{1.7}$$

*As a matter of fact, computing the variation around the minimizer* $(\psi, \boldsymbol{A})$*, we have, for any* $(\tilde{\psi}, \tilde{\mathbf{A}}) \in \mathcal{H}^1 \times \boldsymbol{H}^1$,

$$Re\int_\Omega \Big[\Big(\frac{i}{\kappa}\nabla\psi + \boldsymbol{A}\psi\Big)\cdot\Big(-\frac{i}{\kappa}\nabla\tilde{\psi}^* + \boldsymbol{A}\tilde{\psi}^*\Big) + (|\psi|^2-1)\psi\tilde{\psi}^*\Big] = 0, \tag{1.8}$$

$$Re\int_\Omega [\mathbf{curl}\boldsymbol{A}\mathbf{curl}\tilde{\boldsymbol{A}} + |\psi|^2\boldsymbol{A}\cdot\tilde{\boldsymbol{A}} + \frac{i}{2\kappa}(\psi^*\nabla\psi - \psi\nabla\psi^*)\cdot\tilde{\boldsymbol{A}}] = \int_\Omega H\mathbf{curl}\tilde{\boldsymbol{A}}. \tag{1.9}$$

*These lead to (1.7). It is important to notice that the problem (1.7) has always the trivial solution* $\psi = 0$ *and* $\mathbf{curl}\boldsymbol{A} = H$.

In the following we discuss some basic properties of the minimizers of the G-L energy. We note that many of the interesting mathematical properties are proved for energy minimizers rather than for general solutions of the G-L system. As the G-L energy has stationary points or local energy minimizers which are not global minimizers but satisfy the G-L system, we need to distinguish them from the energy minimizers. Many of the properties we established may or may not be valid for non-energy minimizing solutions. Examples of non energy minimizing solutions will be given later.

### 4.1.2 The outside field is governed by the Maxwell system, 3-d case

If we take the outside Maxwell system into consideration, then the G-L energy in this case should be given as follows (notice that we use $G$ instead of $E$ and $n$ is 3)

$$\begin{aligned} G(\psi, \boldsymbol{A}) &= \int_\Omega \Big( |\frac{i}{\kappa}\nabla\psi + \boldsymbol{A}\psi|^2 + \frac{1}{2}(1-|\psi|^2)^2 \Big) dx \\ &+ \int_{\mathbb{R}^3} \mu |\frac{1}{\mu}\mathbf{curl}\boldsymbol{A} - \boldsymbol{H}|^2 dx + \frac{\gamma}{\kappa}\int_{\partial\Omega} |\psi|^2 d\Gamma \end{aligned} \tag{1.10}$$

Here we have used the particular rescaling scheme (5.7) in Chapter 1, $\mu$ is the permeability of the material and we assume that (the permeability of the superconductor is assumed to be rescaled to be equal to 1)

$$\mu(x) = \begin{cases} 1 & \text{for } x \in \Omega, \\ \mu_e & \text{for } x \in \mathbb{R}^3 \setminus \overline{\Omega}. \end{cases}$$

$\gamma \geq 0$ is a constant which reflects the retarding effect of the material in the exterior domain on the density $|\psi|^2$ at $\partial\Omega$.

The energy is obviously also gauge invariant and consequently, if we minimize the energy over $\mathcal{H}^1(\Omega) \times \boldsymbol{H}^1(\mathbb{R}^3)$ we obtain the existence of energy minimizers exactly the same way as in the previous subsection and we omit the details here.

Moreover, calculating the first variation of the energy, we find that the energy minimizer $(\psi, \boldsymbol{A})$ satisfies the following (slightly different) G-L system:

$$\begin{cases} (\frac{i}{\kappa}\nabla + \boldsymbol{A})^2\psi - \psi + |\psi|^2\psi = 0, & \text{in } \Omega, \\ \mathbf{curl}\Big(\frac{1}{\mu}\mathbf{curl}\boldsymbol{A}\Big) - \mathbf{curl}\boldsymbol{H} + & \\ \Big(\frac{i}{2\kappa}(\psi^*\nabla\psi - \psi\nabla\psi^*) + \boldsymbol{A}|\psi|^2\Big)\chi_\Omega = 0, & \text{in } \mathbb{R}^3, \\ (\frac{i}{\kappa}\nabla + \boldsymbol{A})\psi \cdot \boldsymbol{n} = i\gamma\psi, & \text{on } \partial\Omega, \\ \frac{1}{\mu}\mathbf{curl}\boldsymbol{A} - \boldsymbol{H} \in L^2(\mathbb{R}^3, \mathbb{R}^3). & \end{cases} \tag{1.11}$$

Here the function $\chi_\Omega$ is defined as follows

$$\chi_\Omega(x) = \begin{cases} 1 & x \in \Omega, \\ 0 & x \in \mathbb{R}^3 \setminus \overline{\Omega}. \end{cases} \tag{1.12}$$

We notice that since $\boldsymbol{A}$ is defined on $\mathbb{R}^3$, we can establish existence theorems even if $\Omega$ is a multi-connected domain. We just give the existence results without proof:

**Theorem 4.1.2** *Let $\Omega$ be any piecewise smooth domain in $\mathbb{R}^n$ with $n = 3$, $G(\psi, \boldsymbol{A})$ has at least one energy minimizer belonging to $\mathcal{H}^1(\Omega) \times \boldsymbol{H}^1_{loc}(\mathrm{div}, \mathbb{R}^n)$ with $\frac{1}{\mu}\mathbf{curl}\boldsymbol{A} - \boldsymbol{H} \in \boldsymbol{L}^2(\mathbb{R}^n)$.*

*Moreover, the order parameter $\psi$ satisfies $|\psi|(x) \leq 1$ almost everywhere in $\Omega$.*

## 4.2 Stability and mapping properties of solutions

In this section, we take 2-d case as an example. All properties deduced extend to the 3-d case without alteration.

### 4.2.1 Non-existence of local maxima

We show that the solutions of the G-L system can not be local maxima of the G-L energy. This issue is first raised in [DGP 92] and is aimed at concluding that the solutions of the G-L system are stable in at least some directions.

**Proposition 4.2.1** *A solution of the G-L system (1.7) can not be a local maximizer of $E$ or $F$.*

*Proof.* If we fix $\psi \in \mathcal{H}^1(\Omega)$, $E$ is a convex functional in $\boldsymbol{A}$. On the other hand, if $(\psi, \boldsymbol{A})$ is a solution, we claim that $E(s\psi, \boldsymbol{A})$ is a locally convex function in the real parameter $s$ when $s$ is close to 1. To verify this, we have

$$\frac{d^2E(s\psi,\boldsymbol{A})}{ds^2}|_{s=1}=\int_\Omega (6|\psi|^4-2|\psi|^2+2|(\frac{i}{\kappa}\nabla+\boldsymbol{A})\psi|^2)=4\int_\Omega (|\psi|^4+|(\frac{i}{\kappa}\nabla+\boldsymbol{A})\psi|^2)\geq 0$$

with equality holding if and only if $\psi = 0$. So $(\psi, \boldsymbol{A})$ can not be a local maximizer. This result does not exclude the possibility that the solution $(\psi, \boldsymbol{A})$ is a saddle point.

### 4.2.2 Boundedness of the order parameter

As we understand from the physical background, $|\psi|^2$ is related to the density of the superconducting pairs in the rescaled form, it is important to show that the solution of the equation gives bounded $|\psi|$.

**Proposition 4.2.2** *If $(\psi, \boldsymbol{A})$ is a solution of the G-L system (1.7), then $|\psi| \leq 1$ almost everywhere in $\Omega$.*

*Proof.* Set $\tilde{\psi} = (|\psi| - 1)_+ f$ where $f = \psi/|\psi|$ and where $q_+ = \max\{q, 0\}$. Let $\Omega^+ = \{x \in \Omega, \ |\psi|(x) > 1\}$, we have

$$\frac{i}{\kappa}\nabla\tilde{\psi}^* - \boldsymbol{A}\tilde{\psi}^* = \frac{i}{\kappa}f^*\nabla|\psi| + (|\psi| - 1)(\frac{i}{\kappa}\nabla f^* - \boldsymbol{A}f^*)$$

and

$$-\frac{i}{\kappa}\nabla\psi - \boldsymbol{A}\psi = -\frac{i}{\kappa}f\nabla|\psi| + |\psi|(-\frac{i}{\kappa}\nabla f - \boldsymbol{A}f),$$

so that

$$\mathrm{Re}\Big\{\Big(-\frac{i}{\kappa}\nabla\psi - \boldsymbol{A}\psi\Big)\cdot\Big(\frac{i}{\kappa}\nabla\tilde{\psi}^* - \boldsymbol{A}\tilde{\psi}^*\Big)\Big\} = (\nabla|\psi|)^2 + |\psi|(|\psi| - 1)|\frac{i}{\kappa}\nabla f + \boldsymbol{A}f|^2.$$

Using (1.8)–(1.9), we know that

$$\mathrm{Re}\Big\{\Big(-\frac{i}{\kappa}\nabla\psi-\boldsymbol{A}\psi\Big)\cdot\Big(\frac{i}{\kappa}\nabla\tilde{\psi}^*-\boldsymbol{A}\tilde{\psi}^*\Big)+(|\psi|^2-1)\psi\tilde{\psi}^*\Big\}=0,$$

we have therefore

$$\int_{\Omega^+}(\nabla|\psi|)^2+|\psi|(|\psi|-1)|\frac{i}{\kappa}\nabla f-\boldsymbol{A}f|^2+|\psi|(|\psi|+1)(|\psi|-1)^2=0.$$

Since the integrand is positive, we must have meas$(\Omega^+)=0$ and therefore, $|\psi|\leq 1$ almost everywhere.

We have given the proof for the bound on $|\psi|$ solving (1.7). As a matter of fact, if we impose any other boundary condition on $\psi$, then as long as the boundary condition does not contradict the fact that

$$\Big|\psi|_{\partial\Omega}\Big|\leq C$$

for any arbitrary nonnegative constant $C$, we always have

$$|\psi(x,t)|\leq\max\{C,\ 1\}.$$

### 4.2.3 Constant solutions and mixed state solutions

If the applied magnetic field is large, then we should expect that the superconducting property of the material is destroyed and that the normal solution

$$\psi=0,\quad \mathbf{curl}^2\boldsymbol{A}=\mathbf{curl}\boldsymbol{H}\ \text{ in }\ \Omega\ \text{ and }\ \mathbf{curl}\boldsymbol{A}=H\ \text{ on }\ \partial\Omega$$

should be the energy minimizer. However, we notice that the pair $(0,\boldsymbol{A})$ so defined is **always** a solution of the G-L system.

If the applied field vanishes, i.e. $H=0$, then the ideal superconducting solution is the minimizer. As the proof is short, we give it here

**Proposition 4.2.3** *If $H=0$ and the domain $\Omega$ is simply connected, then $\{(\psi,\boldsymbol{A})=(e^{i\alpha},0)$ for any real constant $\alpha\}$ is the set of global energy minimizers of $E$ under the gauge choice* div$\boldsymbol{A}=0$ *and* $\boldsymbol{A}\cdot\boldsymbol{n}|_{\partial\Omega}=0$.

*Proof.* If $H=0$, then $E(e^{i\alpha},0)=0$, since $E(\psi,\boldsymbol{A})\geq 0$, $(e^{i\alpha},0)$ is a global energy minimizer.

On the other hand, from $E(\psi,\boldsymbol{A})=0$ (where $H=0$), we deduce that

$$(\frac{i}{\kappa}\nabla+\boldsymbol{A})\psi=0,\quad |\psi|=1,\quad \mathbf{curl}\boldsymbol{A}=0. \tag{2.1}$$

Hence, $(\psi,\boldsymbol{A})=(e^{i\alpha},0)$ for some constant $\alpha$.

Combining this proposition and the fact that $(0,\mathbf{0})$ is also a solution of the G-L system, we have a non energy minimizing solution $(0,\mathbf{0})$.

**Remark 4.2.1** *If the domain is not simply connected, the above result is not correct. For example, if* $\Omega = \{1 < |x| < 2,\ x \in \mathbb{R}^2\}$ *and* $n$ *is any integer, we have* $(\psi, \boldsymbol{A}) = (e^{in\kappa\theta}, n(-y, x)/r^2)$ *satisfying (2.1) and the gauge choice condition, where* $\theta$ *is the standard polar angle variable in* $\mathbb{R}^2$. *See also [JM 96] for solutions on torous like domains. The solutions obtained there are not all global energy minimizers and do not have much link with the applied magnetic field. They are examples of non-energy minimizing solutions of the G-L system.*

**Remark 4.2.2** *It is also well known that when* $H = 0$, *there are other solutions to the G-L system. (See [JT 80] for self-dual solutions, [BC 80] for radially symmetric solutions). These are not our main interests here so we do not explain in detail.*

Finally, we ask the question what happens if $H$ is not too small and not too large? Do we get solutions such that $\psi$ vanishes somewhere? If $\psi$ does vanish, does it vanish on co-dimension 2 domains so that the physics observation of vortex structured phase transition domain can be described by the G-L mathematical model? We repeat that when $|\psi(x_0)| = 0$, the superconducting property is regarded as been destroyed at (near) the location $x_0$. In the following, we try to answer some of these questions.

## 4.3 Co-dimension two vortex domain

In this section we establish the fact that if $(\psi, \boldsymbol{A})$ is a global energy minimizer and if $\Omega \subset \mathbb{R}^2$ is **simply connected**, the set

$$S = \{x \in \Omega,\ \ \psi(x) = 0\} \tag{3.1}$$

consists indeed of co-dimension two structures only. In this section, we mainly follow the work of [EMT 93].

When $\Omega \subset \mathbb{R}^3$, it is observed physically that the superconducting property is destroyed along co-dimension two structures (vortex lines) too. If the G-L system is a good description of the physical problem, we should see that the set $\{x \in \Omega,\ \ \psi(x) = 0\}$ consists of space curves only. But this is still an **open** problem from mathematical point of view.

*Before finally starting the detailed mathematical discussion, we make the remark that from a mathematical point of view, we believe that the rigorous justification of the vortex structure and the confirmation of a relation between the applied magnetic field strength and the number of vortices in a superconductor sample are two major issues for which mathematical analysis could enhance the understanding of the mathematical modelling of the physical theory. The justification of the vortex structure for 2-d model is discussed below but the 3-d model is still open. As to the second problem, to our best knowledge, apart from some special situations where people can study a linearized problem (for example, in the case of nucleation), it is still open. Further down the line, the mathematical understanding of vortex motion and the mathematical description of large number of vortices will also enhance the theory.*

We assume in this section that $\Omega$ is a bounded, simply connected, open $C^1$ subset of $\mathbb{R}^2$ and set

$$E(\psi, \boldsymbol{A}) = \frac{1}{2}\int_\Omega [|(\frac{i}{\kappa}\nabla - i\boldsymbol{A})\psi|^2 + \frac{1}{2}(|\psi|^2-1)^2 + |\mathbf{curl}\boldsymbol{A} - H|^2]dx \tag{3.2}$$

where $H$ is the given applied magnetic field. In the 2-d case, $H$ is a scalar. We assume that $H$ is a constant. The results can easily be extended to the case where $H$ is a real analytic function of $(x_1, x_2)$.

**Remark 4.3.1** *We ignore the possibility that the Maxwell system may govern the magnetic field in the outside domain but the proofs can easily be extended to cover the case.*

As we will use regularity results for the solutions, we need to have a strongly elliptic system. Hence we choose the Coulomb's gauge: $\mathrm{div}\boldsymbol{A} = 0$ and $\boldsymbol{A}\cdot\boldsymbol{n} = 0$ on $\partial\Omega$ and we minimize $E(\psi, \boldsymbol{A})$ over $\mathcal{H}^1(\Omega) \times \boldsymbol{H}^1_n(\mathrm{div}, \Omega)$.

**Remark 4.3.2** *The results of this section also apply to various other boundary value problems as long as the boundary conditions are gauge invariant (so that we can choose gauge to make the resulting equation strongly elliptic). For example, the results apply equally to the sets*

$$K_1 = \{\psi \in \mathcal{H}^1(\Omega),\ \boldsymbol{A} \in \boldsymbol{H}^1(\Omega) :\ |\psi| = 1 \ \ on\ \ \partial\Omega,$$
$$\int_\Omega \mathbf{curl}\boldsymbol{A}dx = 2\pi L \ \ with\ L\ given.\}$$
$$K_2 = \{\psi \in \mathcal{H}^1(\Omega),\ \boldsymbol{A} \in \boldsymbol{H}^1(\Omega) :\ \int_\Omega \mathbf{curl}\boldsymbol{A}dx = 2\pi L.\}$$
$$K_3 = \{\psi \in \mathcal{H}^1(\Omega),\ \boldsymbol{A} \in \boldsymbol{H}^1(\Omega) :\ \psi|_{\partial\Omega} = 0.\}$$

*The method of proof can also be altered to cover the case*

$$K_4 = \{\psi \in \mathcal{H}^1(\Omega),\ \boldsymbol{A} \in \boldsymbol{H}^1(\Omega) :\ \psi|_{\partial\Omega} = g \ \ with\ g \neq 0.\}$$

*But this case is not physically interesting, so we do not elaborate on it.*
*For simplicity, we only state the theorem for our designated case.*

It is known that the energy minimizers satisfy

$$\begin{cases} \mathrm{div}\boldsymbol{A} = 0, \\ \boldsymbol{A}\cdot\boldsymbol{n}|_{\partial\Omega} = 0. \end{cases} \tag{3.3}$$

As a consequence, when discussing energy minimization, we only consider solutions in the particular gauge equivalent class where the magnetic potential $\boldsymbol{A}$ satisfies (3.3).

We will find it useful to notice the following fact: as div$\boldsymbol{A} = 0$, (1.7-2) can also be rewritten as

$$\Delta \boldsymbol{A}_j = \operatorname{Im}[\psi(\partial_j + i\boldsymbol{A}_j)\psi^*] \text{ for } j = 1{,}2. \tag{3.4}$$

Before we establish the vortex structure of the solutions, we need the following preliminary results:

**Lemma 4.3.1** *If $\psi = \psi_1 + i\psi_2$ and $\boldsymbol{A}$ are solutions of (1.7), then $\psi_1$, $\psi_2$ are real analytic functions for $x \in \Omega$.*

*Proof.* We note that the solution $(\psi, \boldsymbol{A})$ is also a critical point of

$$\frac{1}{2}\int_\Omega \{|(\frac{i}{\kappa}\nabla - i\boldsymbol{A})\psi|^2 + \frac{1}{2}(|\psi|^2 - 1)^2 + |\nabla \boldsymbol{A}|^2\}dx. \tag{3.5}$$

The analyticity result is then given in [Mo 66], p32, Theorem 1.10.4, iii and vi.

**Lemma 4.3.2** *Let $(\psi, \boldsymbol{A})$ be a solution pair of (1.7) and $y \in S$. Then in a sufficiently small neighbourhood of the point $y$, $u$ is equal to zero either at the point $y$ only or on finitely many piecewise smooth (globally continuous) curves passing through the point $y$.*

*Proof.* Since the $\psi_j$ are analytic functions for $j = 1, 2$, we know that each of them cannot be zero on an open set unless identically zero (note that $\Omega$ is simply connected). If $y$ is a zero point for both $\psi_j$, then we have

$$\psi_1(x_1, x_2)^2 + \psi_2(x_1, x_2)^2 = \sum_{j=0}^{q}(x_1 - y_1)^{n_j}(x_2 - y_2)^{m_j} h_j(x_1, x_2) \tag{3.6}$$

where $h_j$ are non vanishing analytic functions near $(y_1, y_2)$ and $0 \le n_j < n_{j+1}$, $m_j > m_{j+1} \ge 0$. By the Puiseux expansion (originally discovered by Newton in a simpler form)(cf. [BK 86], Section 8.3), the equation

$$\psi_1(x_1, x_2)^2 + \psi_2(x_1, x_2)^2 = 0$$

defines a number of continuous complex curves passing through the point $(y_1, y_2)$ of the form $(y_1 + t^m, y_2 + h(t))$ or $(h(t) + y_1, t^m + y_2)$, where $m$ is an integer and $h$ is an analytic function for small $t$:

$$h(t) = \sum_{j=1}^{\infty} a_j t^j.$$

If there is one real curve among them, it is easy to verify by differentiating that each $a_j$ is real and we can make the sets $\{m\}$ and $\{j,\ a_j \ne 0\}$ have no common factors.

**Lemma 4.3.3** *Let $\psi$ be as in Lemma 3.2. If $\psi = 0$ on a curve then the curve has no end points within the region $\Omega$ and it cannot be a closed curve in $\Omega$.*

*Proof.* Let $\psi = 0$ on a curve which has an end point $(\tilde{x_1}, \tilde{x_2})$ in the interior of $\Omega$. Without losing generality, we assume that $(\tilde{x_1}, \tilde{x_2}) = (0,0)$. In a small neighbourhood of (0,0), the Puiseux expansion applies. If $t \in [-\delta, 0] \to (t^m, h(t))$ or $(h(t), t^m)$ defines a curve ending at $(0,0)$, then $t \in [0, \delta] \to (t^m, h(t))$ or $(h(t), t^m)$, the natural extension of the curve defines a different curve ending at $(0,0)$ on which $|\psi|^2 = 0$ due to the fact that $m$ and $\{j : a_j \neq 0\}$ have no common factor. As a matter of fact, results in algebraic topology ([BK 86], Section 8.3) imply that if $(t^m,\ h(t))$ defines a curve on which $|\psi|^2 = 0$, then $(t^m \exp(ik\pi t/2m),\ h(\exp(ik\pi t/2m)))$ also defines a curve on which $|\psi|^2 = 0$ for $k = 1, 2, \cdots, 2m-1$. This is the reason we assume that they have no common factor.

If the curve has a closed loop in $\Omega$, by Jordan's Theorem, let $\Omega = D_1 \bigcup D_2 \bigcup C$ where $C$ is the closed loop, $D_1$ is the domain inside the loop and $D_2 = \Omega \backslash (D_1 \bigcup C)$. We let

$$w = \begin{cases} \psi(x_1, x_2) & \text{when } (x_1, x_2) \in D_2 \\ -\psi(x_1, x_2) & \text{when } (x_1, x_2) \in D_1 \\ 0 & \text{when } (x_1, x_2) \in C. \end{cases}$$

Then $w$ is in $H^1(\Omega)$ and is also a minimizer of the energy. It is obvious that the boundary condition is unaffected, so $w$ is also a real analytic function. Therefore, each component of $\psi + w$ is real analytic in $\Omega$; but $\psi + w = 0$ on $D_1$ which is impossible unless $\psi + w \equiv 0$ or equivalently, $\psi \equiv 0$.

**Theorem 4.3.1** *Let $\psi$ be as in Lemma 3.2; then $\psi = 0$ only on isolated points or curves which end on the boundary of $\Omega$.*

*Proof.* From Lemma 3.2, if there is a curve on which $\psi = 0$, the curve is continuous. For any compact subset $K$ of $\Omega$, the set $K \bigcap S$ is a closed set. Each connected component of this set must either be a piece of a continuous curve with end points or be a point. But from Lemma 3.3, in the first situation, the end points must be on the boundary.

If there is a point $y$ on which $\psi = 0$ and there is no curve passing through it such that $\psi = 0$, then expression (3.6) at this point does not allow a sequence of points approaching $y$ with $\psi = 0$ at these points. So this point is isolated.

**Corollary 4.3.1** *Let $\psi$ be as in Lemma 3.2; then the set $S$ consists of isolated points only.*

*Proof.* This is now obvious: for the problem of minimizing (3.5), if there is a curve $C$ ending on the boundary of $\Omega$ and $\psi|_C = 0$, then the domain $\Omega$ is divided into $\Omega_1$ and $\Omega_2$ by this curve. If we define

$$w(x) = \begin{cases} \psi(x) & \text{for } x \in \Omega_1, \\ -\psi(x) & \text{for } x \in \Omega_2, \end{cases}$$

then $w$ is clearly an $H^1(\Omega)$ function and also real analytic because it is still an energy minimizer. We have therefore that $\psi + w$ is analytic and $\psi + w = 0 \in \Omega_2$ which is impossible unless $\psi$ and $w$ are identical.

So far, we have not said anything about the existence of zeros for $\psi$. This will be discussed later on in this chapter.

## 4.4 Breakdown of superconductivity

In this section, we give a proof that strong magnetic field indeed destroys superconductivity properties under the G-L regime. We concentrate on the 2-d model only and we mainly follow the work of [GPh 98] and [BPT 98].

For a full discussion of the 3-d case with the Maxwell system governing the magnetic field in the outside domain, we refer to [GPh 98] for more details.

We discuss the model described by (1.10). By solution, we refer to energy minimizers. The constant $\gamma$ is assumed to be nonnegative.

In this setting, a solution is in the normal phase if $\psi \equiv 0$ in $\Omega$ and the full solution is $(0, \boldsymbol{A}_0)$ where $\boldsymbol{A}_0$ satisfies

$$\begin{aligned} &\mathbf{curl}(\tfrac{1}{\mu}\mathbf{curl}\boldsymbol{A}_0) = \mathbf{curl}\boldsymbol{H} \quad \text{in } I\!R^3, \\ &\tfrac{1}{\mu}\mathbf{curl}\boldsymbol{A}_0 - \boldsymbol{H} \in L^2(I\!R^3, I\!R^3). \end{aligned} \tag{4.1}$$

In this section, we only consider the case of a uniform, infinitely-long cylinder in a parallel field here. For a discussion of the full 3-d problem, we refer to [GPh 98] for more details. We assume that the cylinder occupies $\Omega \times (-\infty, \infty)$ where $\Omega$ is a bounded piecewise $C^1$ domain of $I\!R^2$,

$$\boldsymbol{H} = h\boldsymbol{e} \ \ (\text{ which implies } \mathbf{curl}\boldsymbol{H} = 0) \tag{4.2}$$

where $h$ is a constant, $\boldsymbol{e} = \boldsymbol{e}_3 = (0,0,1)^T$, $\boldsymbol{A} = (A_1, A_2, 0)^T$ and all the functions are functions of $(x_1, x_2)$ only.

Let $\kappa$ be fixed, we denote by $\overline{h}$ the upper critical field for the body:

$$\begin{aligned} \overline{h} \ = \ & \inf\{h' : \text{ normal states are the only solutions} \\ & \text{to (1.11) and (4.2) for all } h > h'\}. \end{aligned}$$

For notational preparation, we first give the weak formulation of (1.11): for any $(\phi, \boldsymbol{B}) \in \mathcal{H}^1(\Omega) \times \boldsymbol{H}^1(I\!R^2, I\!R^2)$ with $\boldsymbol{B}$ having compact support,

$$\begin{aligned} \int_\Omega \Big( (\frac{i}{\kappa}\nabla\psi + \boldsymbol{A}\psi) \cdot (\frac{i}{\kappa}\nabla\phi + \boldsymbol{A}\phi)^* \\ + (|\psi|^2 - 1)\psi\phi^* dx = -\frac{\gamma}{\kappa}\int_{\partial\Omega} \psi\phi^* ds, \end{aligned} \tag{4.3}$$

$$\begin{aligned} \int_{I\!R^2} \frac{1}{\mu}\mathbf{curl}\boldsymbol{A} \cdot \mathbf{curl}\boldsymbol{B} dx + \int_\Omega \Big[\frac{i}{2\kappa}(\psi^*\nabla\psi - \psi\nabla\psi^*) \\ + \boldsymbol{A}|\psi|^2\Big] \cdot \boldsymbol{B} dx = 0. \end{aligned} \tag{4.4}$$

Next, we look at the expression $\frac{i}{\kappa}\nabla\psi + \boldsymbol{A}\psi$ for $(\psi, \boldsymbol{A}) \in \mathcal{H}^1(\Omega) \times \boldsymbol{H}^1_{loc}(I\!R^2; I\!R^2)$, we have

$$\mathrm{Re}\Big[\Big(\frac{i}{\kappa}\nabla\psi + \boldsymbol{A}\psi\Big)\psi^*\Big] = \frac{i}{2\kappa}(\psi^*\nabla\psi - \psi\nabla\psi^*) + \boldsymbol{A}|\psi|^2,$$
$$\mathrm{Im}\Big[\Big(\frac{i}{\kappa}\nabla\psi + \boldsymbol{A}\psi\Big)\psi^*\Big] = \frac{1}{2\kappa}\nabla|\psi|^2.$$

Consequently,

$$\frac{i}{\kappa}\nabla\psi + \boldsymbol{A}\psi = \Big\{\Big[\frac{i}{2\kappa}(\psi^*\nabla\psi - \psi\nabla\psi^*) + \boldsymbol{A}|\psi|^2\Big]\frac{1}{|\psi|} + \frac{i}{\kappa}\nabla|\psi|\Big\}\frac{\psi}{|\psi|} \tag{4.5}$$

for almost every $x$ such that $\psi(x) \neq 0$. As the set $\{\psi = 0\}$ consists of isolated points only, (4.5) holds almost everywhere.

Now, we recall the following result that we have discussed in Section 8, Chapter 1.

**Proposition 4.4.1** *Let $(\psi, \boldsymbol{A})$ satisfy (1.11) and (4.2), then* **curl***$\boldsymbol{A}$ is a constant in each connected component of $I\!R^2 \setminus \overline{\Omega}$. Moreover,*

$$\mathbf{curl}\boldsymbol{A} = \mu_e h \boldsymbol{e}$$

*in the unbounded component. Here $\mu_e$ is the value of $\mu$ in the outside domain.*

**Remark 4.4.1** *The above proposition implies that in the long cylinder case, if the base is a simply connected domain in $I\!R^2$, then we can impose* **curl***$\boldsymbol{A} = h$ in $I\!R^2 \setminus \overline{\Omega}$ rather than using the Maxwell system.*

Next, we decide the existence of a normal state:

**Proposition 4.4.2** *A normal solution exists. Moreover, any normal state $(0, \boldsymbol{A}_0) = (0, h\boldsymbol{e}_N)$ with $\boldsymbol{e}_N$ satisfying* **curl**$\boldsymbol{e}_N = \mu\boldsymbol{e} = \mu(0,0,1)^T$.

*Proof.* The main problem is to find a vector potential satisfying $\mathbf{curl}\,\mathbf{b} = \mu\boldsymbol{e}$, then we can let $\boldsymbol{A} = h\mathbf{b}$ and this is a normal state solution.

Since $\mu$ is not a constant, we have to first let $w = \Gamma_2 * (\mu - \mu_e)$ where $*$ is the standard convolution operation and $\Gamma_2(x) = \frac{1}{2\pi}\ln(|x|)$ is the fundamental solution for the Laplacian in $I\!R^2$. Since the function $\mu - \mu_e$ has bounded support in $\overline{\Omega}$, $w$ is well defined with $w \in H^2_{loc}(I\!R^2)$ and $\Delta w = (\mu - \mu_e)$ in $I\!R^2$. Set $\mathbf{b}_N = (-w_y, w_x) + \frac{\mu_e}{2}(-y, x)$. Then $\mathbf{curl}\,\mathbf{b}_N = \mu\boldsymbol{e}$ in $I\!R^2$ and we finish the proof of the first part.

Suppose that $\mathbf{a}_N$ is another weak solution, taking the difference of the equations for $\mathbf{b}_N$ and $\mathbf{a}_N$, we get

$$\int_{I\!R^2} \frac{1}{\mu}\mathbf{curl}(\mathbf{b}_N - \mathbf{a}_N)\cdot\mathbf{curl}\phi\, dx = 0. \tag{4.6}$$

Let $\mathcal{E}$ be the unbounded component of $I\!R^2 \setminus \Omega$. From Proposition 4.1, we have $\mathbf{curl}\,\mathbf{b}_N = \mathbf{curl}\,\mathbf{a}_N$ in $\mathcal{E}$. In $I\!R^2 \setminus \mathcal{E}$, we choose $\phi = \mathbf{b}_N - \mathbf{a}_N$ in a bounded neighbourhood of $I\!R^2 \setminus \mathcal{E}$ in (4.3) and we establish that $\mathbf{curl}\,\mathbf{b}_N = \mathbf{curl}\,\mathbf{a}_N$.

Third, we establish a Sobolev estimate for the magnetic potential:

**Lemma 4.4.1** *Let* $(\xi, \boldsymbol{B})$ *and* $(0, h\boldsymbol{a}_N)$ *be weak solutions to (1.11) and (4.2). Then there is a function pair* $(\psi, \boldsymbol{A})$ *gauge equivalent to* $(\xi, \boldsymbol{B})$ *such that*

$$\int_\Omega |\boldsymbol{A} - h\boldsymbol{a}_N|^2 dx \le C_0 \int_{I\!R^2} |\mathbf{curl}(\boldsymbol{A} - h\boldsymbol{a}_N)|^2 dx \tag{4.7}$$

*where* $C_0$ *depends only on* $\Omega$.

*Proof.* Set $\mathbf{curl}(\boldsymbol{B} - h\boldsymbol{a}_N) = f\boldsymbol{e}_3$. From Propositions 4.1–4.2, we have that $f = 0$ in $\mathcal{E}$ and consequently, $f \in L^2(I\!R^2)$.

Let $w = \Gamma_2 * f$, then standard estimates on the Newtonian potential give $w \in H^2_{loc}(I\!R^2)$, $\nabla w = \nabla\Gamma_2 * f$, $||\nabla w||_{L^2(I\!R^2\setminus\mathcal{E})} < C_0(\Omega)||f||_{L^2(I\!R^2\setminus\mathcal{E})}$ and $\Delta w = f$ (See [GTr 77], Thm 4.5). Thus setting $\tilde{\boldsymbol{A}} = (-w_y, w_x)$, we have $\tilde{\boldsymbol{A}} \in H^1_{loc}(I\!R^2, I\!R^2)$ and $\mathbf{curl}\tilde{\boldsymbol{A}} = \Delta w\boldsymbol{e} = f\boldsymbol{e} = \mathbf{curl}(\boldsymbol{B} - h\boldsymbol{a}_N)$. Let $\boldsymbol{A} = \tilde{\boldsymbol{A}} + h\boldsymbol{a}_N$. It is easy to verify that $\mathbf{curl}(\boldsymbol{B} - \boldsymbol{A}) = 0$ and hence $\boldsymbol{A} = \boldsymbol{B} + \nabla\eta$ for some scalar function $\eta \in H^2_{loc}(I\!R^2)$. Moreover,

$$\begin{aligned}\int_\Omega |\boldsymbol{A} - h\boldsymbol{a}_N|^2 dx &\le \int_{I\!R^2\setminus\mathcal{E}} |\nabla w|^2 dx \le C_0 \int_{I\!R^2\setminus\mathcal{E}} |f|^2 dx \\ &\le C_0 \int_{I\!R^2} |\mathbf{curl}(\boldsymbol{A} - h\boldsymbol{a}_N)|^2 dx.\end{aligned}$$

The proof is finished.

Fourth, a bound on $(i\nabla + \kappa h\boldsymbol{a}_N)\psi$.

**Lemma 4.4.2** *Let* $(\psi, \boldsymbol{A})$ *and* $(0, h\boldsymbol{a}_N)$ *be weak solutions satisfying (1.11) and (4.2), then there is a constant* $C_1 = C_1(\Omega, \mu_e)$ *such that*

$$\int_\Omega |(i\nabla + \kappa h\boldsymbol{a}_N)\psi|^2 dx \le C_1\kappa^2 \int_\Omega |\psi|^2 dx. \tag{4.8}$$

*Proof.* Choosing $\phi = \psi$ in (4.7), using (4.5), we obtain

$$\begin{aligned}&\textstyle\int_\Omega \left(\left|\frac{1}{\kappa}\nabla|\psi|\right|^2 + \left|\left(\frac{i}{2\kappa}(\psi^*\nabla - \psi\nabla\psi^*) + \boldsymbol{A}|\psi|^2\right]\frac{1}{|\psi|}\right|^2\right)dx \\ &= \textstyle\int_\Omega \left|\left(\frac{i}{\kappa}\nabla + \boldsymbol{A}\right)\psi\right|^2 dx \le \int_\Omega (1 - |\psi|^2)|\psi|^2 dx \le \int_\Omega |\psi|^2 dx.\end{aligned} \tag{4.9}$$

Substuting the solutions $(\psi, \boldsymbol{A})$ and $(0, h\boldsymbol{a}_N)$ in (4.4), taking the difference of the respective resulting equations, we have

$$\begin{aligned}&\int_{I\!R^2} \frac{1}{\mu}\mathbf{curl}(\boldsymbol{A} - h\boldsymbol{a}_N)\cdot\mathbf{curl}\boldsymbol{B}dx \\ &= -\int_\Omega \left[\frac{i}{2\kappa}(\psi^*\nabla\psi - \psi\nabla\psi^*) + \boldsymbol{A}|\psi|^2\right]|\psi|^{-1}\cdot|\psi|\boldsymbol{B}dx. \\ &\le \int_\Omega \left(\frac{1}{\varepsilon}\left|\frac{i}{2\kappa}(\psi^*\nabla\psi - \psi\nabla\psi^*) + \boldsymbol{A}|\psi|^2\right|^2|\psi|^{-2} + \varepsilon|\psi|^2|\boldsymbol{B}|^2\right)dx \\ \text{(by (4.9))} \quad &\le \frac{1}{\varepsilon}\int_\Omega |\psi|^2 dx + \varepsilon\int_\Omega |\psi|^2|\boldsymbol{B}|^2 dx \qquad \text{for any } \varepsilon > 0.\end{aligned}$$

Letting $\boldsymbol{B} = \boldsymbol{A} - h\mathbf{a}_N$ in the above estimates, noticing that $|\psi| \leq 1$, we have

$$\int_{\mathbb{R}^2} \frac{1}{\mu} |\mathbf{curl}(\boldsymbol{A} - h\mathbf{a}_N)|^2 dx \leq \frac{1}{\varepsilon} \int_\Omega |\psi|^2 dx + \varepsilon \int_\Omega |\boldsymbol{A} - h\mathbf{a}_N|^2 dx.$$

Combining this inequality with (4.4), we see that we can take $\varepsilon$ sufficiently small to have

$$\int_\Omega |\boldsymbol{A} - h\mathbf{a}_N|^2 dx \leq M \int_\Omega |\psi|^2 dx \tag{4.10}$$

for some constant $M = M(\text{diam}\Omega, \mu_e)$. Next, we write

$$(\frac{i}{\kappa}\nabla + \boldsymbol{A})\psi = (\frac{i}{\kappa}\nabla + h\mathbf{a}_N)\psi + (\boldsymbol{A} - h\mathbf{a}_N)\psi. \tag{4.11}$$

We easily deduce

$$\begin{aligned} \int_\Omega |(\frac{i}{\kappa}\nabla + h\mathbf{a}_N)\psi|^2 dx &\leq 2\int_\Omega |(\frac{i}{\kappa}\nabla + \boldsymbol{A})\psi|^2 dx + 2\int_\Omega |(\boldsymbol{A} - h\mathbf{a}_N)\psi|^2 \\ (\text{by (4.9) and } |\psi| \leq 1) &\leq 2\int_\Omega |\psi|^2 dx + 2\int_\Omega |\boldsymbol{A} - h\mathbf{a}_N|^2 dx \\ (\text{by (4.10)}) &\leq 2(1+M)\int_\Omega |\psi|^2 dx. \end{aligned} \tag{4.12}$$

We choose $C_1 = 2(1+M)$ and the lemma is proved.

We see that if a superconducting state (i.e., a solution with $\psi \not\equiv 0$) exists, then (4.8) implies that the principal eigenvalue for the operator $(i\nabla + \kappa h\mathbf{a}_N)^2$ on $\Omega$ is bounded by $C_1\kappa^2$.

The idea is then if the applied field $h$ is large, this bound may no longer hold and consequently, the problem has normal solution only. In the following, we will show that if $h > \max(\frac{1}{\kappa}, \phi\kappa)$ for some given constant $\phi$, then the bound on the principal eigenvalue fails.

To achieve this, we first study the properties of the first few eigenfunctions of the Laplace operator near the boundary of its definition domain. For any given $\Omega$, we define the $\tau$-neighbourhood in $\Omega$ of $\partial\Omega$ as

$$\Omega_\tau = \{x \in \Omega : \ \text{dist}(x, \ \partial\Omega) < \tau\}.$$

**Lemma 4.4.3** *Let $\Omega$ be a bounded $C^1$ domain in $\mathbb{R}^n$. For any given $\lambda_0 > 0$, there is a constant $d(\lambda_0, \Omega) > 0$ such that when $\lambda \geq \lambda_0$ and,*

$$\int_\Omega |\nabla f|^2 dx \leq \lambda^2 \int_\Omega |f|^2 dx \tag{4.13}$$

*for some $f \in H^1(\Omega)$, then*

$$\int_\Omega |f|^2 dx \leq 2\int_{\Omega\setminus\Omega_{d/\lambda}} |f|^2 dx \tag{4.14}$$

*or equivalently,*

$$\int_{\Omega_{d/\lambda}} |f|^2 dx \le \int_{\Omega\setminus\Omega_{d/\lambda}} |f|^2 dx.$$

*Proof.* Let $\{F_k\}_{k=0}^N$ be a set of open sets covering $\overline{\Omega}$. We assume that $\overline{F}_0 \subset \Omega$ and for each $k$ between 1 and $N$, we have

$$F_k \cap \Omega = \{(x', x_n) :\ g_k(x') < x_n < g_k(x') + \delta_1,\ |x'| < \delta_2\},$$

where $\delta_1$ and $\delta_2$ are positive constants, $(x', x_n)$ are suitably rotated and translated co-ordinates, and $g_k(x') = x_n$ characterizes $\partial\Omega \cap F_k$. We can further assume without loss of generality that $g_k(\cdot)$ is defined for $|x'| \le 2\delta_2$, $|\nabla g_k| < 1$ and

$$\begin{aligned} &\{(x', x_n) :\ g_k(x') < x_n < g_k(x') + 4\delta_1,\ |x'| < 2\delta_2\} \subset \Omega, \\ &\{(x', x_n) :\ g_k(x') - 4\delta_1 < x_n < g_k(x'),\ |x'| < 2\delta_2\} \subset I\!R^n \setminus \overline{\Omega}. \end{aligned} \tag{4.15}$$

Let $f$ be any given $H^1(\Omega)$ function, $t$, $v$ are numbers $\in (0, \delta_1)$. For a fixed $k \ge 1$, we have

$$\begin{aligned} &\int_{\{x_n - g_k(x') = v\}\cap F_k} |f|^2 ds - \int_{\{x_n - g_k(x') = t\}\cap F_k} |f|^2 ds \\ &\le \int_{\{|x'| \le 2\delta_2\}} |f^2(x', g_k(x') + v) - f^2(x', g_k(x') + t)| \sqrt{1 + |\nabla g_k|^2} dx' \\ &\le \int_{\Omega\cap F_k} |\partial_{x_n} f^2| dx. \end{aligned}$$

Integrating in $t$ from 0 to $\delta_1$ we obtain

$$\delta_1 \int_{\{x_n - g_k(x') = v\}\cap F_k} |f|^2 ds \le \int_\Omega |f|^2 dx + 2\delta_1 \int_{\Omega\cap F_k} |\partial_{x_n} f| f dx.$$

Dividing both sides by $\delta_1$ gives

$$\int_{\{x_n - g_k(x') = v\}\cap F_k} |f|^2 ds \le \frac{1}{\delta_1}\Big(\int_\Omega |f|^2 dx + 2\delta_1 \int_\Omega |f||\nabla f| dx\Big).$$

Integrating the above expression in $v$ from 0 to $2d/\lambda$, where $d$ is to be determined. We derive

$$\int_{\{0 < x_n - g_k(x') < 2d/\lambda\}\cap F_k} |f|^2 dx \le \frac{2d}{\lambda\delta_1}\Big(\int_\Omega |f|^2 dx + 2\delta_1 \int_\Omega |f||\nabla f| dx\Big). \tag{4.16}$$

As $x_n = g_k$ defines $\partial\Omega$ and $|\nabla g_k| \le 1$ (the surface is reasonably smooth), it is easy to see that

$$F_k \cap \Omega_{d/\lambda} \subset F_k \cap \{0 < x_n - g_k(x') < 2d/\lambda\} \tag{4.17}$$

when $d/\lambda$ is sufficiently small.

Taking the sum on $k$ for $1 \le k \le N$ in (4.16), we obtain, for some $M_1 = M_1(\delta_1, N)$,

$$\begin{aligned}\int_{\Omega_{d/\lambda}} |f|^2 dx &\le M_1 \frac{d}{\lambda}\Big(\int_\Omega |f|^2 dx + \lambda \int_\Omega |f|^2 dx + \frac{1}{\lambda}\int_\Omega |\nabla f|^2 dx\Big)\\ \text{(using (4.13))} &\le M_2 d \int_\Omega |f|^2 dx\end{aligned}$$

where $M_2 = M_2(\delta_1, N, \lambda_0)$. Because the constant $d$ is yet to be determined, we choose $d = \min(\frac{1}{2M_2}, \frac{\lambda_0\delta_1}{2}, \frac{\lambda_0\delta_2}{2})$ and obtain

$$\int_{\Omega_{d/\lambda}} |f|^2 dx \le \frac{1}{2}\int_\Omega |f|^2 dx.$$

(4.14) now follows from this inequality.

The following is a lower bound of the smallest eigenvalue of the operator $(i\nabla + h\kappa\mathbf{a}_N)^2$ in terms of $h\kappa$.

**Proposition 4.4.3** *There is a continuous function $\sigma(\cdot)\colon t \in [0,\infty) \to I\!R$ with $\sigma(t) > 0$ for $t > 0$ for which $\lim_{t\to\infty}\sigma(t)$ exists with $0 < \lim_{t\to\infty}\sigma(t) < 1$, and such that*

$$\int_{B_r(0)} \Big|\Big(i\nabla + \frac{\omega^2}{2}(-y,x)\Big)\xi\Big|^2 dx \ge \omega^2\sigma(\omega r)\int_{B_r(0)} |\xi|^2 dx \tag{4.18}$$

*for all $\xi \in \mathcal{H}^1(B_r(0))$ and $\omega \ge 0$.*

*Proof.* As the proof of this proposition is interesting in itself, we give it in the next section.

**Remark 4.4.2** *If* $\mathbf{b}$ *is another vector field such that* $\mathbf{b} \in H^1(B_r(0); I\!R^2)$ *with* $\mathbf{curl b} = \mathbf{e}$, *then (4.18) is also valid with* $\frac{1}{2}(-y,x)$ *replaced by* $\mathbf{b}$.

*As a matter of fact, since* $\mathbf{curl a}_N = \mathbf{curl b}$, *there exists a function* $q \in H^2$ *such that* $\nabla q = \mathbf{b} - \frac{1}{2}(-y,x)$. *Using a gauge transformation* $\xi' = \xi e^{i\omega^2 q}$ *and* $\mathbf{b} = \nabla q + \frac{1}{2}(-y,x)$, *we easily deduce that*

$$\begin{aligned}\int_{B_r(0)} |(i\nabla + \omega^2\mathbf{b})\xi'| &= \int_{B_r(0)} \Big|\Big(i\nabla + \frac{\omega^2}{2}(-y,x)\Big)\xi\Big|\\ &\ge \omega^2\sigma(\omega r)\int_{B_r(0)} |\xi|^2 dx\\ &= \omega^2\sigma(\omega r)\int_{B_r(0)} |\xi'|^2 dx.\end{aligned}$$

In the estimate (4.18), the quantity $\omega^2\sigma(\omega r)$ depends not only on $\omega$ but also on $r$. This is inconvenient. In the following, we derive an estimate similar to (4.18) for general $\Omega$ instead of $B_r(0)$ provided $\omega$ is bounded away from 0. Since we assume that $\Omega$ is smooth, $r$ can also be chosen to be bounded away from 0.

**Lemma 4.4.4** *Given $m > 0$, there is a constant $C_2 = C_2(m, \Omega)$ (this dependence of $C_2$ on $\Omega$ is in fact, on the choice of $r$), $0 < C_2 \leq 1$ such that if $\omega^2 \geq m > 0$ then*

$$C_2\omega^2 \int_\Omega |\xi|^2 dx \leq \int_\Omega |(i\nabla + \omega^2 \mathbf{b})\xi|^2 dx \tag{4.19}$$

*for all $\xi \in \mathcal{H}^1(\Omega)$, $\mathbf{b} \in H^1(\Omega; I\!R^2)$ for which* **curlb** $= \boldsymbol{e}$.

*Proof.* If for any $\xi \in \mathcal{H}^1(\Omega)$ satisfying $\int_\Omega |\xi|^2 dx > 0$, we have

$$\int_\Omega |(i\nabla + \omega^2 \mathbf{b})\xi|^2 dx \geq \omega^2 \int_\Omega |\xi|^2 dx,$$

the lemma is proved with $C_2 = 1$. Otherwise, assume that there is a $\xi$ such that

$$\int_\Omega |(i\nabla + \omega^2 \mathbf{b})\xi|^2 dx \leq \omega^2 \int_\Omega |\xi|^2 dx$$

for some $\omega$ with $\omega^2 > m > 0$, then it is easy to see that

$$|\nabla|\xi|| \leq \sqrt{\left(\left[\frac{i}{2}(\xi^* \nabla \xi - \xi \nabla \xi^*) + \omega^2 \mathbf{b}|\xi|^2\right] \frac{1}{|\xi|}\right)^2 + (\nabla|\xi|)^2} \overset{(\text{using}(4.5))}{=} |i\nabla\psi + \omega^2 \mathbf{b}\psi|.$$

Thus,

$$\int_\Omega |\nabla|\xi||^2 dx \leq \omega^2 \int_\Omega |\xi|^2 dx.$$

Lemma 4.3 implies then

$$\int_\Omega |\xi|^2 dx \leq 2 \int_{\Omega\setminus\Omega_{d/\omega}} |\xi|^2 dx. \tag{4.20}$$

Next, we choose a cover of $\Omega \setminus \Omega_{d/\omega}$ consisting of finite many disks $\{B_{d/\omega}(x_k),\ k = 1, \ldots, N(\omega)\}$, each contained in $\Omega$ such that $\sum_{j=1}^{N(\omega)} \chi_{B_{d/\omega}}(x_k) \leq K_1$ with $K_1$ a given constant. Then

$$\begin{aligned}
\int_\Omega |(i\nabla + \omega^2 \mathbf{b})\xi|^2 dx &\geq \int_{\cup_{k=1}^N B_{d/\omega}(x_k)} |(i\nabla + \omega^2 \mathbf{b})\xi|^2 dx \\
&\geq \frac{1}{K_1} \sum_{k=1}^N \int_{B_{d/\omega}(x_k)} |(i\nabla + \omega^2 \mathbf{b})\xi|^2 dx. \\
(\text{Proposition 4.3}) &\geq \frac{1}{K_1}\omega^2 \sigma(d) \sum_{k=1}^N \int_{B_{d/\omega}(x_k)} |\xi|^2 dx \\
&\geq K_2(d)\omega^2 \int_{\Omega\setminus\Omega_{d/\omega}} |\xi|^2 dx \\
(\text{by (4.21)}) &\geq \frac{K_2\omega^2}{2} \int_\Omega |\xi|^2 dx
\end{aligned}$$

Set $C_2 = K_2/2$ and the lemma is proved.

We now establish the principal result in this section.

**Theorem 4.4.1** *There is a constant $\phi = \phi(\mu_e, \Omega)$ such that if $h > \max(1/\kappa, \phi\kappa)$, any weak solution for (1.11) and (4.2) is normal.*

*Proof.* Let $(0, \mathbf{a}_N)$ be a normal state for (4.1)–(4.2) and $(\psi, \boldsymbol{A})$ be a weak solution for (1.11) and (4.2). Up to a suitable gauge choice, we assume that $(\psi, \boldsymbol{A})$ and $(0, h\mathbf{a}_N)$ satisfy (4.7). Set $\omega^2 = h\kappa$, then $\omega^2 \geq 1$ by hypothesis. Letting $m = 1$ in (4.20), we obtain

$$\begin{aligned} C_2 h\kappa \int_\Omega |\psi|^2 dx &\leq \int_\Omega |(i\nabla + h\kappa\mathbf{a}_N)\psi| dx \\ \textit{(Lemma 4.2)} &\leq C_1 \kappa^2 \int_\Omega |\psi|^2 dx. \end{aligned}$$

If we choose $\phi = C_1/C_2$ in the above inequalities, we have

$$h \int_\Omega |\psi|^2 dx \leq \phi\kappa \int_\Omega |\psi|^2 dx.$$

This is impossible unless $\int_\Omega |\psi|^2 dx = 0$ because otherwise this implies that $h \leq \phi\kappa$ contradicting our assumption that $h > \phi\kappa$. This implies that our solution must be a normal state.

## 4.5 A linearized problem

Finally, we will give the proof of Proposition 4.3. This is essentially a linear eigenvalue problem. We mainly follow the work of [BPT 98]. Set

$$G(\xi) = \int_{B_{r_0}} |(i\nabla + \omega^2 \mathbf{a}_N)\xi|^2 dx$$

where $\mathbf{a}_N = \frac{1}{2}(-y, x)^T$ and

$$\lambda = \lambda(\omega, r_0) = \inf \Big\{ \frac{G(\xi)}{||\xi||^2_{L^2(B_{r_0})}} \ : \ \xi \in \mathcal{H}^1(B_{r_0}) \Big\}. \tag{5.1}$$

For $\xi \in \mathcal{H}^1(B_{r_0})$, we expand $\xi$ by its Fourier series

$$\xi(r, \theta) = \sum_{n=-\infty}^{\infty} \xi_n(r) e^{in\theta}$$

where the $\xi_n$ are real valued functions, then

$$\int_0^{r_0} \Big( \sum_{n=-\infty}^{\infty} (|\xi_n'|^2 + n^2 |\xi_n|^2/r^2) \Big) r dr < \infty,$$

$$G(\xi) = 2\pi \int_0^{r_0} \Big( \sum_{n=-\infty}^{\infty} (|\xi_n'|^2 + (\frac{\omega^2 r}{2} - \frac{n}{r})^2 |\xi_n|^2) \Big) r dr.$$

Setting

$$\lambda_n = \lambda_n(\omega, r_0) = \inf\Big\{ \frac{\int_0^{r_0}(|\xi'(r)|^2 + (\frac{\omega^2 r}{2} - \frac{n}{r})^2|\xi|^2)rdr}{\int_0^{r_0}|\xi|^2 rdr} \qquad \xi \in H^1((0,r_0);rdr) \text{ and } \int_0^{r_0}|\xi|^2 rdr > 0\Big\}, \tag{5.2}$$

we have the following standard Sturm-Liouville theory:

**Lemma 4.5.1** *For each $n$, the infimum in (5.2) is taken on by a real valued function $\phi_n(r)$ such that $\phi_n \in H^1((0,r_0);rdr)$, $\phi_n(r) > 0$ for $0 < r \le r_0$, and $\phi_n$ satisfies*

$$\begin{aligned} &\phi_n'' + \frac{\phi_n'}{r} - (\frac{\omega^2 r}{2} - \frac{n}{r})^2\phi_n = -\lambda_n\phi_n \quad in \quad (0,r_0), \\ &\phi_n \quad is\ bounded\ and \quad \phi_n'(r_0) = 0. \end{aligned} \tag{5.3}$$

*Furthermore, any solution of (5.3) is of the form $\alpha\phi_n$ for some complex constant $\alpha$.*

It follows from (5.1), (5.2) and Lemma 5.1 that $G(\xi) \ge \lambda||\xi||_{L^2(B_{r_0})}$ for all $\xi \in \mathcal{H}^1(B_{r_0})$ where $\lambda$ is defined as in (5.1). In particular, we have

$$\lambda(\omega, r_0) = \inf_{n\in Z} \lambda_n(\omega, r_0) \tag{5.4}$$

where $Z$ is the set of all integers. It is also straightforward to verify (from definition) that $\lim_{|n|\to\infty} \lambda_n(\omega, r_0) = \infty$. Thus for any given $\omega$ and $r_0$, $\lambda(\omega, r_0) = \lambda_n(\omega, r_0)$ for a finite number of $n$ and we have

**Lemma 4.5.2** *For given values of $\omega$ and $r_0$, the set of minimizers of (5.1) spans a nontrivial finite-dimensional subspace with a basis given by*

$$S = \{\phi_{n_1}(r)e^{in_1\theta}, \ldots, \phi_{n_k}(r)e^{in_k\theta}\}$$

*where $n_1 < \cdots < n_k$, $\phi_{n_j}$ has the properties described in Lemma 5.1 and $\lambda(\omega, r_0) = \lambda_{n_j}(\omega, r_0)$ for $j = 1, \ldots, k$.*

We have therefore reduced a multi-dimensional eigenvalue problem (5.1) into a sequence of one dimensional problems (5.2)–(5.3) via the relationship (5.4). After the explanation of the basic facts, we now go on to obtain some relevant estimates which give us the desired results. At this stage, the problem (5.2)–(5.3) depends on too many parameters, so we first rescale the problem in the following way:

$$x = \omega r, \quad x_0 = \omega r_0,$$
$$\xi(x) = \xi_n(x) = \phi_n(\frac{x}{\omega}), \quad \mu = \sqrt{\lambda_n}\omega.$$

The eigenvalue problem (5.3) becomes one depending on two parameters $n$ and $\mu_n$ only:

$$\begin{array}{l}\mathcal{L}_n\xi = \xi'' + \frac{\xi'}{x} - (\frac{x}{2} - \frac{n}{x})^2\xi = -\mu_n^2\xi \text{ for } 0 < x < x_0 \\ \text{where } \xi \text{ is bounded, } \xi \geq 0,\ \xi'(x_0) = 0,\ \xi \not\equiv 0.\end{array} \tag{5.5}$$

By the standard Frobenius theory, the solution $\xi_n$ is regular near $x = 0$ and

$$\xi_n(x) = a_n x^{|n|} + o(|x|^{|n|}) \text{ as } x \to 0 \tag{5.6}$$

for some $a_n > 0$. Moreover, $\xi_n > 0$ for $0 < x \leq x_0$. The minimization problem (5.2) becomes

$$\begin{array}{l}\inf \frac{\mathcal{F}(\xi,n,x_0)}{\int_0^{x_0}\xi^2 x dx} = \inf \frac{\int_0^{x_0}[(\xi')^2 + (\frac{x}{2} - \frac{n}{x})^2\xi^2]x dx}{\int_0^{x_0}\xi^2 x dx} \\ \text{in the class } \{\xi \in H^1((0,x_0); x dx) : \int_0^{x_0}\xi^2 x dx > 0\}.\end{array} \tag{5.7}$$

We have also the following standard Sturm-Liouville theory:

**Lemma 4.5.3** *For each $x_0 > 0$, there exists a unique $\mu = \mu(x_0, n) > 0$ for which (5.5) is solvable. The solution $\xi_n(x) = \xi_n(x, x_0)$ is unique under the normalization $\xi_n(x_0) = 1$. Moreover, $\xi_n$ is nonnegative and $\mu^2(x_0, n) = \mathcal{F}(\xi_n, n, x_0)/\int_0^{x_0}\xi_n^2 x dx$.*

**Remark 4.5.1** *We can now compare the constant $\mu^2$ with the constant $\sigma(\omega r)$ discussed in Proposition 4.3. We understand that the largest possible value of $\sigma(\omega r)$ is $\inf_n \mu^2(r, n)$. So we will study the property of $\inf_n \mu^2(r, n)$ in the following.*

**Lemma 4.5.4** $0 < \mu^2(x_0, 0) < \min(1, \frac{1}{4}x_0^2)$ *for* $x_0 > 0$.

*Proof.* Let $v(x) = e^{-x^2/4}$, then $v$ satisfies

$$\mathcal{L}_0 v = v'' + v'/x - (x/2)^2 v = -v.$$

Multiplying this equation by $-xv$ and integrating the product from 0 to $x_0$ gives

$$\int_0^{x_0} v^2 x dx + x_0 v'(x_0)v(x_0) = \mathcal{F}(v) \geq \mu^2(x_0, 0)\int_0^{x_0} v^2 x dx.$$

Since $v' < 0$, we deduce that $\mu^2(x_0, 0) < 1$. On the other hand, since $\int_0^{x_0} v^2 x dx = 1 - e^{-x^2/2}$, we get

$$\mu^2(x_0, 0) \leq \frac{1 - e^{-x_0^2/2} - \frac{1}{2}x_0^2 e^{-x_0^2/2}}{1 - e^{-x_0^2/2}} < \frac{1}{4}x_0^2.$$

Given $r_0$, $\kappa$ and $h$, we wish to identify those $n$ for which

$$\lambda_n((\kappa h)^{1/2}, r_0) = \inf_{n' \in Z} \lambda_{n'}((\kappa h)^{1/2}, r_0) = \lambda((\kappa h)^{1/2}, r_0).$$

This is equivalent to being given $x_0$, $\kappa$ and $h$ and finding those $n$ such that $\mu^2(x_0, n) = \inf_{n' \in Z} \mu^2(x_0, n')$.

**Remark 4.5.2** *When $n < 0$, we have $(\frac{x}{2} - \frac{n}{x})^2 = (\frac{x}{2} + \frac{|n|}{x})^2 \geq 2|n|$. So*

$$\mu^2(x_0, n) \int_0^{x_0} \xi_n^2 x dx = \mathcal{F}(\xi_n) \geq 2|n| \int_0^{x_0} \xi_n x dx.$$

*This implies that $\mu^2(x_0, n) \geq 2|n| > 1$ for $n < 0$. Since $\mu^2(x_0, 0) < 1$ by Lemma 5.4, the cases where $n < 0$ are not interesting and we restrict our investigation to the $n \geq 0$ cases.*

Now we investigate the behaviour of the eigenfunctions $\xi_n$ for $n \geq 0$.

**Lemma 4.5.5** *1) When $n = 0$, then $\xi_n' < 0$ on $(0, x_0)$ and $\xi_n > 0$ on $[0, x_0]$.*

*2) When $n > 0$, let $\mu = \mu(x_0, n)$. If $x_0 \leq \mu + \sqrt{\mu^2 + 2n}$, then $\xi_n' > 0$ on $(0, x_0)$. If $x_0 > \mu + \sqrt{\mu^2 + 2n}$, then there exists $x_{1n} \in (\sqrt{\mu^2 + 2n} - \mu, \sqrt{\mu^2 + 2n} + \mu)$ such that $\xi_n' > 0$ on $(0, x_{1n})$ and $\xi_n' < 0$ on $(x_{1n}, x_0)$.*

*Proof.* 1) We argue for the derivative function first. From (5.6), we know that $\xi_n > 0$ in a neighbourhood of $r = 0$. From (5.5) with $n = 0$, we deduce easily that $\xi_0'(0) = 0$ and $\xi_0''(0) < 0$. Hence, $\xi_0'(r) < 0$ when $r$ is near 0. If it stays this way, 1) is proved.

Otherwise, let $x_1$ be the first positive value such that $\xi_0'(x_1) = 0$. Then from (5.5), $\frac{1}{4}x_1^2 < \mu^2$ is impossible because this implies that $\xi_0''(x_1) \leq 0$, that $x_1$ is a local maximum, contradicting the fact that $\xi_0$ is decreasing on $(0, x_1)$.

If $\frac{1}{4}x_1^2 \geq \mu^2$, we have $\xi_0''(x_1) \geq 0$. Thus for $x \geq x_1$,

$$\frac{1}{x}\frac{d}{dx}(x\xi_0'(x)) = (\frac{x^2}{4} - \mu^2)\xi_0 \geq (\frac{x_1^2}{4} - \mu^2)\xi_0 \geq 0.$$

This implies that $x\xi_0'(x)$ is an increasing function on $(x_1, x_0)$. Since $\xi_0'(x_0) = 0 = \xi_0'(x_1)$, we must have $x_1 = x_0$. Thus, $\xi_0' < 0$ on $(0, x_0)$.

For the function itself, we know that $\xi_0 > 0$ for $x \in [0, x_0)$. If $\xi_0(x_0) = 0$, from the known condition that $\xi_0'(x_0) = 0$, we deduce then $\xi_0''(x_0) = 0$ from (5.5). This implies that $\xi_0 = 0$ in a neighbourhood of $x_0$. This is impossible.

2) From (5.5), we have $\xi_n' > 0$ on a maximum interval $(0, x_{1n})$ with $\xi_n'(x_{1n}) = 0$ and $\xi_n''(x_{1n}) \leq 0$. This implies that $((\frac{x_{1n}}{2} - \frac{n}{x_{1n}})^2 - \mu^2) \leq 0$ which is equivalent to $x_{1n} \in [\sqrt{\mu^2 + 2n} - \mu, \sqrt{\mu^2 + 2n} + \mu]$. Now from (5.5), $x\xi_n'(x)$ is increasing on $[0, \sqrt{\mu^2 + 2n} - \mu]$. Thus in fact, $x_{1n} > \sqrt{\mu^2 + 2n} - \mu$.

If $x_{1n} < x_0 \leq \sqrt{\mu^2 + 2n} + \mu$, then $x\xi_n'(x)$ decreases on $[x_{1n}, x_0]$ and $\xi_n'(x_0) = 0$ is not possible. Thus if $x_0 \leq \sqrt{\mu^2 + 2n} + \mu$, then $x_{1n} = x_0$ and $\xi_n$ is increasing on $[0, x_0]$.

Assume that $x_0 > \sqrt{\mu^2 + 2n} + \mu$.

We have that $x\xi_n'(x)$ decreases on $[\sqrt{\mu^2 + 2n} - \mu, \sqrt{\mu^2 + 2n} + \mu]$ and increases on the intervals $[0, \sqrt{\mu^2 + 2n} - \mu]$ and $[\sqrt{\mu^2 + 2n} + \mu, x_0]$. Since $\xi_n'(x_0) = 0$, we must have $x_{1n} \in (\sqrt{\mu^2 + 2n} - \mu, \sqrt{\mu^2 + 2n} + \mu)$ and $x_{1n}$ must be the unique critical point for $\xi_n$ in $(0, x_0)$.

**Definition 4.5.1** *Let $n \in \{0, 1, \dots, \}$, $\mu \in \mathbb{R}$. Defining $g_n(x) = g_n(x, \mu)$ to be the solution to*

$$\mathcal{L}_n g = g_n'' + \frac{g_n'}{x} - (\frac{x}{2} - \frac{n}{x})^2 g_n = -\mu^2 g_n \quad \text{for } 0 < x < \infty \tag{5.8}$$

*with $g_n(x) = x^n + o(x^n)$ as $x \to 0^+$.*

By the standard theory of ODEs, $g_n$ has an extension which is a real analytic function of $(x, \mu) \in \mathbb{R}^2$ for each $n$.

**Remark 4.5.3** *The function $v_n(x) = x^n e^{-x^2/4}$ is a solution of (5.8) on $(0, \infty)$ with $\mu = 1$. Thus $g_n(x, 1) = x^n e^{-x^2/4}$. Furthermore, $v_n'(x) = 0$ if and only if $x = 0$ or $x = \sqrt{2n}$.*

**Remark 4.5.4** *Let $n \in \{0, 1, \dots\}$ and $x_0 > 0$, set $\mu = \mu(x_0, n)$. Let $\xi_n(x) \equiv \xi_n(x, x_0)$ be the solution of (5.5) with $\xi_n(x_0) = 1$ and assume without loss of generality that $\xi_n(x)$ has been extended to $(0, \infty)$ and it satisfies (5.8). Then*

$$g_n(x, \mu(x_0, n)) = a_n^{-1} \xi_n(x, x_0).$$

Now we study the eigenvalues $\mu(x_0, n)^2$ in greater detail.

**Lemma 4.5.6** *For each $n \in \{0, 1, \dots\}$, the functions $\mu(\cdot, n)$ are analytic on $(0, \infty)$ and*

$$\partial_x \mu^2(x, n) = \frac{\left[ \left(\frac{x}{2} - \frac{n}{x}\right)^2 - \mu^2(x, n) \right] x g_n^2(x, \mu(x, n))}{\int_0^x g_n^2(s, \mu(x, n)) s ds}. \tag{5.9}$$

*Proof.* To simplify notation, fix $n$ as a constant and set $\mu(x) \equiv \mu(x, n)$, $g(x, \mu) \equiv g_n(x, \mu)$. From Remark 5.4, we see that $\mu(x)$ satisfies the equation $g_x(x_0, \mu(x_0)) = 0$ for $x_0 > 0$. Thus $\mu(x)$ is analytic provided $g_{x\mu}(x_0, \mu(x_0)) \neq 0$ for all $x_0 > 0$. To justify this, we differentiate (5.8) with respect to $\mu$ to obtain

$$g_{\mu xx} + \frac{g_{\mu x}}{x} - (\frac{x}{2} - \frac{n}{x})^2 g_\mu + \mu^2 g_\mu = -2\mu g.$$

Multiplying the both sides of the equation by $-gx$ and integrating the result on $(0, x_0)$, we find

$$\begin{aligned}
2\mu(x_0) \int_0^{x_0} g^2 x dx &= -\int_0^{x_0} gx(\mathcal{L}_n + \mu^2) g_\mu dx \\
&= -\int_0^{x_0} g_\mu x(\mathcal{L}_n + \mu^2) g dx - x_0 g_{\mu x}(x_0, \mu(x_0)) g(x_0, \mu(x_0)) \\
&= -x_0 g_{\mu x}(x_0, \mu(x_0)) g(x_0, \mu(x_0))
\end{aligned}$$

where $g = g(x) = g(x, \mu(x_0))$. Hence

$$g_{x\mu}(x_0, \mu(x_0)) = -\frac{2\mu(x_0)\int_0^{x_0} g^2 x dx}{x_0 g(x_0, \mu(x_0))} < 0.$$

Consequently, $\mu(x)$ is analytic. To show that it satisfies (5.9), we differentiate

$$g_x(x_0, \mu(x_0)) = 0$$

with respect to $x_0$:

$$g_{xx}(x_0, \mu(x_0)) + g_{x\mu}(x_0, \mu(x_0))\mu_x(x_0) = 0.$$

Hence

$$(\mu^2)_x = -\frac{2\mu g_{xx}(x_0, \mu(x_0))}{g_{x\mu}(x_0, \mu(x_0))}.$$

From $g'(x_0, \mu(x_0)) = 0$, we obtain that

$$\begin{aligned} g_{xx}(x_0, \mu(x_0)) &= -g'(x_0, \mu(x_0))/x_0 + ((\frac{x_0}{2} - \frac{n}{x_0})^2 - \mu^2(x_0, n))g_n(x_0, \mu(x_0)) \\ &= ((\frac{x_0}{2} - \frac{n}{x_0})^2 - \mu^2(x_0, n))g_n(x_0, \mu(x_0)), \end{aligned}$$

$$g_{x\mu}(x_0, \mu(x_0)) = -\frac{2\mu(x_0)\int_0^{x_0} g^2 x dx}{x_0 g(x_0, \mu(x_0))}.$$

(5.9) then follows easily.

**Lemma 4.5.7** *For each $n \in \{0, 1, \dots\}$, we have*
*1) if $x < \sqrt{2n}$, $\mu(x, n) > 1$,*
*2) if $x > \sqrt{2n}$, $\mu(x, n) < 1$,*
*3) $\lim_{x\to\infty} \mu(x, n) = 1$.*

*Proof.* Fix $n$ and $x_0 > 0$ as constants, let $\mu(x) = \mu(x, n)$ and $g(x) = g_n(x, \mu(x_0))$. Note that $v(x) = x^n e^{-x^2/4}$ satisfies that $\mathcal{L}_n v = -v$. Multiplying (5.8) by $xv$ and integrating on the interval $(0, x_0)$, we obtain, by integration by parts

$$-x_0 g(x_0)v'(x_0) = (1 - \mu^2(x_0)\int_0^{x_0} gvx dx.$$

It is straightforward to verify that $xv' = (n - \frac{1}{2}x^2)v$, we have

$$-g(x_0)v(x_0)(n - \frac{1}{2}x_0^2) = (1 - \mu^2(x_0))\int_0^{x_0} gvx dx.$$

Thus 1) and 2) follow this relationship.

For 3), we let $k = g/v$. Using (5.6), we have $0 < \lim_{x\to 0} k(x) < \infty$ and

$$vk'' + \frac{vk'}{x} + 2k'v' = (1 - \mu^2(x_0))kv.$$

This can be rewritten as

$$(xv^2k')' = (1 - \mu^2)kxv^2.$$

Multiplying this equation by $k$ and integrating over $(0, x_0)$, we have

$$x_0v^2(x_0)k'(x_0)k(x_0) - \int_0^{x_0} v^2k'^2xdx = (1 - \mu^2(x_0)) \int_0^{x_0} g^2xdx.$$

Since $0 = g'(x_0) = (k'v + kv')(x_0)$, at $x_0$,

$$x_0v^2k'k = -x_0vk^2v' = g^2(x_0)(\frac{1}{2}x_0^2 - n).$$

Therefore

$$(\frac{1}{2}x_0^2 - n)g^2(x_0) \geq (1 - \mu^2(x_0)) \int_0^{x_0} g^2xdx. \tag{5.10}$$

From 2), we have $0 < \mu(x_0) < 1$ for $x_0 > \sqrt{2n}$, it follows from (5.9) then that $\mu'(x_0) > 0$ for all large $x_0$. Therefore, $\lim_{x\to\infty} \mu(x_0)$ exists and $\lim_{j\to\infty} \mu'(x_{0j}) = 0$ for some sequence $x_{0j} \to \infty$. Equation (5.9) implies then

$$\lim_{j\to\infty} x_{0j}^2 g^2(x_{0j}) / \int_0^{x_{0j}} g^2xdx = 0.$$

Hence from (5.10), we have

$$\lim_{x_0\to\infty} \mu(x_0) = \lim_{j\to\infty} \mu(x_{0j}) = 1.$$

Next, we study the behaviour of $\mu'(x, n) = \partial_x\mu(x, n)$.

**Lemma 4.5.8** $\mu'(x, 0) > 0$ *for all* $x > 0$.

*Proof.* From Lemma 5.4, we have $\mu^2(x, 0) < \frac{1}{4}x^2$. Hence, from (5.9), we have $\mu'(x, 0) > 0$.

**Lemma 4.5.9** *For each* $n$ *in* $\{1, 2, \dots\}$, *there exists a* $x_n > \sqrt{2n}$ *such that*

*1) If* $0 < x < x_n$, $\mu'(x, n) < 0$,

*2) If* $x > x_n$, $\mu'(x, n) > 0$,

*3)* $x_n = \mu_n + \sqrt{\mu_n^2 + 2n}$

*where* $\mu_n = \mu(x_n, n)$.

*Proof.* Following Remark 5.3, we have $\mu(\sqrt{2n}, n) = 1$. Then $\mu'(\sqrt{2n}, n) < 0$ follows from (5.9). Choose $a_n \in [0, \sqrt{2n})$ such that $(a_n, \sqrt{2n})$ is the maximal interval on which $\mu'(\cdot, n) < 0$.

Assume that $a_n > 0$, then we must have $\mu'(a_n, n) = 0$. We now show that this leads to contradiction.

(5.9) implies that $(\frac{a_n}{2} - \frac{n}{a_n})^2 - \mu^2(a_n, n) = 0$. Hence we have either

$$a_n = x_n^1 = \sqrt{\mu^2(a_n, n) + 2n} - \mu(a_n, n) \tag{5.11}$$

or

$$a_n = x_n^2 = \sqrt{\mu^2(a_n, n) + 2n} + \mu(a_n, n). \tag{5.12}$$

Because $a_n < \sqrt{2n}$, we have $a_n = x_n^1$. This implies then

$$(\frac{x}{2} - \frac{n}{x})^2 - \mu^2(a_n, n) > 0 \text{ for } 0 < x < a_n.$$

The equation

$$(xg_n')' = x[(\frac{x}{2} - \frac{n}{x})^2 - \mu^2(a_n, n)]g_n$$

implies that $x\partial_x g_n(x, \mu(a_n, n))$ is increasing on $[0, a_n]$. It follows that

$$g_n'(a_n, \mu(a_n, n)) > 0,$$

together with $(\frac{a_n}{2} - \frac{n}{a_n})^2 - \mu^2(a_n, n) = 0$, we obtain (the PDE with values substituted in)

$$a_n g_n''(a_n, \mu(a_n, n)) + g_n'(a_n, \mu(a_n, n)) = 0.$$

Consequently,

$$g_n''(a_n, \mu(a_n, n)) < 0.$$

But from

$$(\mu^2)_x(a_n, n) = -\frac{2\mu g_{nxx}(a_n, \mu(a_n, n))}{g_{nx\mu}(a_n, \mu(a_n, n))},$$

we should have $g_{nxx}(a_n, \mu(a_n, n)) = 0$ which is a contradiction. This proves that $\mu'(x, n) < 0$ for $x < \sqrt{2n}$.

We next show that $\mu(\cdot, n)$ has a unique critical point $x_n$. (5.9) implies that $\mu^2(\cdot, n)$ decreases as long as $p_n(x) = (\frac{x}{2} - \frac{n}{x})^2 - \mu^2(x, n) < 0$. As $(\frac{x}{2} - \frac{n}{x})^2$ increases as a polynomial for $x > \sqrt{2n}$ and $p_n(\sqrt{2n}) = -1$, there exists $b_n > \sqrt{2n}$ such that $p_n(x)$ increases to the value 0 on $[\sqrt{2n}, b_n)$. Since $\mu'(x, n) < 0$ for $x < b_n$, $\mu'(b_n, n) = 0$ and from (5.9), $\mu''(b_n, n) > 0$, there is a maximal interval $(b_n, c_n)$ on which $\mu'(\cdot, n) > 0$.

If $c_n < \infty$, we have $\mu'(c_n, n) = 0$ and (5.9) implies that $p_n(c_n) = 0$. As $c_n > \sqrt{2n}$, we have $p'_n(c_n) > 0$. This implies that $p_n(x) < 0$ for $x < c_n$ with $|x - c_n|$ sufficiently small. Thus $\mu'(x, n) < 0$ for such $x$ and this is impossible. This implies that $c_n = \infty$ and $\mu$ has a unique critical point $x_n = b_n$.

To prove 3), we notice that $\mu'(x_n, n) = 0$ implies that $p_n(x_n) = 0$. Since $x_n > \sqrt{2n}$, we see that 3) gives an implicit formula for $x_n$ as in (5.11)–(5.12).

As we have pointed out at the beginning, our main interest is to investigate the behaviour of

$$\mu_e(x) = \inf_n \mu(x, n).$$

From the results of Lemma 5.5, we already know that $\mu_e(x) < 1$ for any finite $x$. In the following, we show that for large $n$, the $\mu(x, n)$ are approximately the translation of each other and we establish the limit results. First, we fix $n \geq 1$ and let

$$y = x - \sqrt{2n}, \quad y_0 = x_0 - \sqrt{2n}, \quad \nu(y_0, n) = \mu(x_0, n).$$

We are interested in the values of $x_0$ in the range $x_0 \geq \sqrt{2n}$ because we have $\mu(x_0, n) \leq 1$ there. In terms of the new variable, this is the range $y_0 \geq 0$. Lemma 5.5 implies, in this case, $0 < \nu(y_0, n) \leq 1$. Let

$$f_n(y) = f_n(y, y_0) = \xi_n(x, x_0) = \frac{g_n(x, \mu(x_0, n))}{g_n(x_0, \mu(x_0, n))} \quad \text{for} \quad -\sqrt{2n} \leq y \leq y_0,$$

(5.5) and (5.8) become, for $-\sqrt{2n} < y$,

$$f''_n + \frac{f'_n}{y + \sqrt{2n}} - y^2\Big[1 - \frac{1}{2}\Big(\frac{y}{y + \sqrt{2n}}\Big)\Big]^2 f_n = -\nu^2(y_0, n) f_n, \tag{5.13}$$

where $f_n \geq 0$, $f_n$ is bounded for $-\sqrt{2n} \leq y \leq y_0$ and $f'_n(y_0) = 0$. We also impose the following normalization condition

$$f_n(y_0) = 1. \tag{5.14}$$

**Lemma 4.5.10** *Let $f_n$ satisfy (5.13) and (5.14). Given $0 < L < \infty$, there is a constant $c$ independent of $n$ and $y_0$ such that if $0 \leq y_0 \leq L$, then*

$$c^{-1} \leq f_n(y) \leq c \quad \textit{for} \quad -1 \leq y \leq y_0,$$
$$|f'_n(y)| \leq c \quad \textit{for} \quad -1 \leq y \leq y_0.$$

*Proof.* For the proposed range, the coefficients in (5.13) are all regular and are all uniformly bounded in $n$ on $[-1, L]$. Thus the lemma follows from standard estimates and the fact that $f_n \geq 0$ and $f_n(y_0) = 1$.

Next, we estimate the functions $f_n$ over the larger range $-\sqrt{2n} \le y \le y_0$.

**Lemma 4.5.11** *For any given $L > 0$, there is a constant $c = c(L)$ such that if $0 \le y_0 \le L$ and $f_n(y) = f_n(y, y_0)$, then for $-\sqrt{2n} \le y \le y_0$,*
*1) $f_n(y) \le ce^{-y^2/4}$ for $n \ge 1$,*
*2) $f_n(y) \le c\big(1 + \frac{y}{\sqrt{2n}}\big)^4 e^{-y^2/4}$ for $n \ge 4$,*
*3) $|f_n'(y)| \le c\big(1 + \frac{y}{\sqrt{2n}}\big)e^{-y^2/8}$ for $n \ge 4$.*

*Proof.* First, we construct the comparison functions. Let

$$w_n(y) = (1 + \frac{y}{\sqrt{2n}})^2 \exp(-\frac{y^2}{4} - \frac{\sqrt{2n}y}{2}),$$

then $w_n$ solves (5.13) and (5.14) with $y_0 = 0$ and $\nu = 1$. Rewriting $w_n$ as

$$w_n(y) = \exp(n \ln(1 + y/\sqrt{2n}) - y^2/4 - \sqrt{2n}y/2)$$

and using the fact that $\ln(1 + x) \le x$ for $x > -1$, it follows that

$$(n - m) \ln(1 + y/\sqrt{2n}) - y^2/4 - \sqrt{2n}y/2 \le -y^2/4 - my/\sqrt{2n} \le m - y^2/4$$

for $m \le n$ and $-\sqrt{2n} \le y \le 0$. If we let $m = 0$ and 4, we see that

$$\begin{aligned} w_n(y) &\le ce^{-y^2/4} && \text{for } n \ge 1, \\ w_n(y) &\le c\big(1 + y/\sqrt{2n}\big)^4 e^{-y^2/4} && \text{for } n \ge 4, \end{aligned} \tag{5.15}$$

where $c$ is independent of $n$.

We now use the maximum principle and $w_n$ as a comparison function to bound $f_n$. We can rewrite (5.13) as

$$f_n'' + a_1 f_n' + a_2 f_n = 0 \quad \text{for} \quad -\sqrt{2n} \le y \le y_0.$$

Since $0 < \nu(y_0, n) \le 1$, we know that $a_2 \le 0$ for $-\sqrt{2n} \le y \le -1$. In addition, $w_n$ satisfies

$$w_n'' + a_1 w_n' + a_2 w_n = (\nu^2 - 1) w_n \le 0.$$

Thus, $w_n$ is a supersolution of the equation on $(-\sqrt{2n}, -1)$. Applying Lemma 5.10 to both $w_n$ and $f_n$, we see that $f_n(-1) \le c \le c^2 w_n(-1)$. Moreover, $f_n(-\sqrt{2n}) = w_n(-\sqrt{2n}) = 0$. Thus, by the maximum principle,

$$f_n \le c^2 w_n \quad \text{for} \quad -\sqrt{2n} \le y \le -1.$$

Properties 1) and 2) follow from this and (5.15).

Integrating (5.13) from $-\sqrt{2n}$ to $y \le y_0$ and using 1) and 2), 3) follows.

We now show that both $\nu^2(y_0, n)$ and $f_n(y, y_0)$ converge as $n \to \infty$.

**Lemma 4.5.12** *For any given $0 < L, M < \infty$, as $n \to \infty$, the function $y_0 \to \nu^2(y_0, n)$ converges in $C^1([0, L])$ and the function $y_0 \to f_n(y, y_0)$ converges in $C([0, L]; C^2([-M, M]))$.*

*Proof.* We first show that the function relation $y_0 \to \nu^2(y_0, n)$ converges in $C([0, L])$. Since $\nu^2(y_0, n)$ is uniformly bounded, for some sequence $y_{0j} \to y_0$ and $n_j \to \infty$, we can choose a subsequence, still denoted by the same sub-index, such that $a^2 = \lim_{j\to\infty} \nu^2(y_{0j}, n_j)$ exists. For this subsequence indexed by $j$, it follows from the standard ODE estimates that $\{f_n(y, y_{0j})\}$ converge in $C^2([-M, M])$ for each $M < \infty$ to a solution $f_\infty(y) = f_\infty(y, y_0)$ of

$$\begin{aligned} &f_\infty'' - y^2 f_\infty = -a^2 f_\infty \quad \text{for } y \in I\!R, \\ &0 \le f_\infty(y) \le ce^{-y^2/4} \quad \text{for } y \le y_0, \\ &f_\infty(y_0) = 1, \quad f_\infty'(y_0) = 0. \end{aligned} \tag{5.16}$$

We now show that $a^2$ is independent of the subsequences we choose so that $\nu_\infty(y_0) \equiv \sqrt{\nu_\infty^2(y_0)}$ is well defined. To this end, let $k_1$ and $k_2$ be two solutions of (5.16) with $a = a_1$ and $a_2$, respectively. Then

$$-a_1^2 \int_{-\infty}^{y_0} k_2 k_1 dy = \int_{-\infty}^{y_0} k_2(k_1'' - y^2 k_1) dy = -a_2^2 \int_{-\infty}^{y_0} k_2 k_1 dy.$$

Hence $a_1 = a_2$. This proves that $y_0 \to \nu^2(y_0, n)$ converges in $C([0, L])$.

It follows that $y_0 \to f_n(y, y_0)$ converges in $C([0, L]; C^2([-M, M]))$ as $n \to \infty$ since the coefficients in (5.13) converges locally uniformly. Using this and Lemma 5.9, we see that the right-hand side of

$$\frac{d}{dy_0} \nu^2(y_0, n) = \frac{\left[y_0^2\left(1 - \frac{y_0}{2(y_0 + \sqrt{2n})}\right)^2 - \nu^2\right](1 + y_0/\sqrt{2n})}{\int_{-\sqrt{2n}}^{y_0} (1 + y/\sqrt{2n}) f_n^2(y, y_0) dy} \tag{5.17}$$

converges uniformly on $[0, L]$. Hence, $\{\nu^2(y_0, n)\}$ converges in $C^1([0, L])$ as $n \to \infty$.

**Remark 4.5.5** *The limit problem satisfied by*

$$f_\infty(y) \equiv f_\infty(y, y_0) = \lim_{n\to\infty} f_n(y, y_0)$$

*and $\nu_\infty = \nu_\infty(y_0) = \lim_{n\to\infty} \nu(y_0, n)$ is*

$$\begin{aligned} &f_\infty'' - y^2 f_\infty = -\nu_\infty^2(y_0) f_\infty \quad \text{for } y \le y_0, \\ &0 \le f_\infty(y) \le ce^{-y^2/4} \quad \text{for } y \le y_0; \\ &f_\infty(y_0) = 1 \quad \text{and} \quad f_\infty'(y_0) = 0, \end{aligned} \tag{5.18}$$

$$\frac{d}{dy_0} \nu_\infty^2(y_0) = \frac{y_0^2 - \nu_\infty^2(y_0)}{\int_{-\infty}^{y_0} f_\infty^2(s, y_0) ds} \quad \text{where } y_0 \ge 0,\ \nu_\infty(0) = 1. \tag{5.19}$$

Next, we show that the critical points of $\{\nu(y_0, n)\}$ converge to the critical point of $\nu_\infty(y_0)$ as $n \to \infty$.

**Lemma 4.5.13** *Let* $y_n = x_n - \sqrt{2n}$ *where* $x_n$ *is defined by Lemma 5.9. Then* $\lim_{n\to\infty} y_n = y_*$ *exists and* $\nu_\infty(y_*) = \inf_{y\geq 0} \nu_\infty(y) = y_*$.

*Proof.* First from Lemmas 5.7 and 5.9, we have $0 \leq y_n \leq 2$. Next, since $\nu(y, n)$ has minimum value when $y = y_n$, it follows that $\nu_\infty(y)$ achives it minimum over the interval $[\underline{\lim}_{n\to\infty} y_n, \overline{\lim}_{n\to\infty} y_n]$. But (5.19) implies that $\nu_\infty$ cannot be a constant on an interval of positive length. Thus $\lim_{n\to\infty} y_n = y_*$ exists and $\nu_\infty$ achives its minimum value at $y_*$. Finally, (5.19) implies that $y_* = \nu_\infty(y_*)$ because $\nu'_\infty(y_*) = 0$.

Finally the main result:

**Theorem 4.5.1** *Let* $\mu_n = \mu(x_n, n) = \nu(y_n, n)$. *Then* $\lim_{n\to\infty} \mu_n = \mu_*$ *exists and* $y_* = \mu_*$. *Moreover,* $0 < \mu_* < 1$.

*Proof.* From Lemma 5.7, we see that if $\lim_{n\to\infty} y_n$ exists, then $\lim_{n\to\infty} \mu_n$ exists and the two limits agree. Note that $\mu_* = \inf_{y\geq 0} \nu_\infty(y)$ and $0 \leq \nu_\infty \leq 1$.

If $\mu_* = 1$, then $\nu_\infty = 1$ and this is not consistent with (5.19). If $\mu_* = 0$, then $y_* = 0$. We have seen that $\nu_\infty(0) = \lim_{n\to\infty} \nu(0, n) = 1$. This would imply $0 = \nu_\infty(0) = 1$ is also impossible. Hence the Lemma is proved.

**Remark 4.5.6** *In [SGe 63], the constant* $\mu^2$ *is estimated numerically to be approximately 0.59.*

The results of this section have independent merit in the study of nucleation phenomenon on a disk domain. As a matter of fact, Remark 5.6 already hinted that we have proved rigorously a classical conjecture of James and De Gennes. For more details and background, we refer to [BPT 98].

Finally, we point out that Lemma 5.7 and Theorem 5.1 imply that as $h$ increases, the number of vortices also increases. This gives the first clear link between the strength of the applied magnetic field and the number of vortices. But we have to clarify that this relation is not yet established for the fully nonlinear G-L system.

## 4.6 Bibliographical remarks

It was in [DGP 92] that the authors established some interesting results related to the solutions of the steady state G-L system. The model that the steady state G-L system with the steady state Maxwell system imposed in the outside domain is established and discussed in [CDG 96]. Some interesting solutions of the Maxwell system in the outside domain when the domain $\Omega$ is of a particular geometric shape have been discussed in [GPh 98].

In [EMT 93], the fact that on a simply connected domain, the steady state 2-d G-L solutions do exhibit vortex structured phase transitions domains was established. The analogous problem for the 3-d G-L system is still open.

In a recent paper [GPh 98], it was shown that when the applied magnetic field $|\boldsymbol{H}|$ is sufficiently large, the G-L system has only the trivial solution $\psi \equiv 0$ and $\mathbf{curl}\boldsymbol{A} = \boldsymbol{H}$. This result justifies mathematically the nucleation theory rigorously. We know that under very strong applied magnetic field, the superconductor becomes a normal conductor. Nucleation theory investigates the vortex structure under the following circumstances: drop the strength of the magnetic field to just below the point where superconductivity property is completely destroyed, how will the superconductivity property recover. The most important physical phenomenon associated with nucleation theory is the so-called surface superconductivity phenomenon: superconductivity recovers near the boundary of the conductor sample, trapping a number of vortices inside. This has been the subject of a number of studies (see for example [Aft 97], [BHe 96], [BPT 98], [Chal 95], [SGe 62] and [SGe 63]).

# Chapter 5
# Evolutionary Solutions

In this Chapter, we establish the existence, uniqueness and asymptotic behaviour of the solutions of a non-stationary G-L superconductivity model. We look at the case when the applied magnetic field is a given function. We know now that this is mainly a 2-d case when related to superconductivity. However, we also include the 3-d case to highlight some mathematical difficulties in dealing with the 3-d problem.

It is proved that when the outside applied field is a given function, the G-L system admits an $\omega$-limit set and a global attractor, which represents exactly all the long time dynamics of the system. We prove that the global attractor obtained consists of exactly the set of steady state solutions and its unstable manifold. Its Hausdorff and fractal dimensions are estimated in terms of the physically relevant G-L parameter, diffusion parameter and applied magnetic field.

The authors noticed that the existence and uniqueness of solutions for the 2-d evolutionary G-L system have been discussed in various papers such as [Du 94], [CHL 93] and [Tan 95], however, those proofs rely on the assumption that the initial condition is $H^1$ and satisfy $|\psi|(x,0) \leq 1$ almost everywhere in $\Omega$. Although those results cover the interesting physical cases, they did not reveal the full effect of the non-linear terms in the equation from a mathematical point of view. Hence we adopt a proof by [TWa 96]. It was shown that under the assumption that the initial data $\psi(x,0)$ is $\mathcal{H}^1$ but without the $L^\infty$ bound, the G-L system admits a unique strong solution. Then, the case where $\psi(x,0)$ is only in $\mathcal{L}^2$ without the $L^\infty$ bound was also investigated. It was shown that the G-L system admits a weak solution.

The existence of solutions to the 3-d G-L superconductivity model with outside magnetic field given by the Maxwell system is also established. Here, we used the $L^\infty$ bound for the order parameter.

After the discussion on the existence, we discuss the $\omega$-limit, establishing the result that the solutions of the evolutionary G-L system converges, as $t \to \infty$, to the set of steady state solutions (not necessarily to a particularly given steady state solution – hence the name $\omega$-limit). The result is established for the 2-d problem but extends to 3-d problem automatically subject to establishing an $L^\infty(0,\infty; H^2(\Omega))$ regularity result for the solutions.

Then we construct, explicitly, absorbing sets for some abstract semi-groups having a Lyapunov functional and consequently prove the existence of global attractors which is applied to the G-L system. This is also carried out for the 2-d problem but extends to 3-d problem automatically subject to establishing an $L^\infty(0,\infty; H^2(\Omega))$ regularity result for the solutions.

Finally, we include brief note from F.H.Lin showing that the limit as $t \to \infty$ of the solution of a particular G-L superconductivity problem (system of PDEs with given initial boundary data) is a singleton. This is a significant improvement on the $\omega$-limit results presented in Section 3. However, we do not include any details of proofs and discussions.

As in the steady state case, we assume that $\Omega$ is a bounded, simply connected, smooth domain in $\mathbb{R}^n$ with $n = 2$ or 3.

## 5.1 2-d solutions with given external field

We mainly follow the work of [TWa 96] here. We use the following rescaled evolutionary system

$$\begin{cases} \psi_t + i\phi\psi + \kappa^2(|\psi|^2 - 1)\psi + (i\nabla + \boldsymbol{A})^2\psi = 0, \\ \eta(\boldsymbol{A}_t + \nabla\phi) + \frac{i}{2}(\psi^*\nabla\psi - \psi\nabla\psi^*) + |\psi|^2\boldsymbol{A} + \mathbf{curl}^2\boldsymbol{A} - \mathbf{curl}H = 0, \\ \mathrm{div}\boldsymbol{A} = 0 \end{cases} \tag{1.1}$$

with the following corresponding initial and boundary conditions:

$$\begin{cases} (i\nabla\psi + \boldsymbol{A}\psi)\cdot\boldsymbol{n} = 0, \quad \mathbf{curl}\boldsymbol{A} = H, \quad \boldsymbol{A}\cdot n = 0 \quad \text{on } \partial\Omega\times(0,T), \\ \psi(x,0) = \psi_0(x), \quad \boldsymbol{A}(x,0) = \boldsymbol{A}_0(x) \quad \text{on } \Omega. \end{cases} \tag{1.2}$$

Here $\psi_0$ and $\boldsymbol{A}_0$ are given functions.

The corresponding G-L energy is

$$E(\psi,\boldsymbol{A}) = \frac{1}{2}\int_\Omega |(i\nabla + \boldsymbol{A})\psi|^2 + \frac{\kappa^2}{2}(|\psi|^2 - 1)^2 + |\mathbf{curl}\boldsymbol{A} - H|^2 dx. \tag{1.3}$$

In comparison, the steady state G-L equations of superconductivity under our choice of gauge are

$$\begin{cases} (i\nabla + \boldsymbol{A})^2\psi + \kappa^2(|\psi|^2 - 1)\psi = 0, & \text{in } \Omega, \\ \mathbf{curl}^2\boldsymbol{A} - \mathbf{curl}H + \frac{i}{2}(\psi^*\nabla\psi - \psi\nabla\psi^*) + |\psi|^2\boldsymbol{A} = 0, & \text{in } \Omega, \\ \mathrm{div}\boldsymbol{A} = 0, & \text{in } \Omega. \end{cases} \tag{1.4}$$

The boundary conditions of steady state G-L equations (1.4) are given by (1.2-1).

In adopting the boundary condition $\mathbf{curl}\boldsymbol{A} = H$, we are accepting that the outside magnetic field is static.

### 5.1.1 Mathematical setting

It is easy to see that in the evolutionary G-L system, the unknown function $\phi$ is the Lagrange multiplier of the magnetic field due to the gauge restraint $\text{div}\boldsymbol{A} = 0$. Therefore, we introduce the following notation $u$ for the unknowns:

$$u = (\psi, \boldsymbol{A}).$$

As in the study of the Navier-Stokes equations of an incompressible fluid, we set

$$\mathbf{V}_2 = \{\boldsymbol{A} \in C^\infty(\Omega) |\ \boldsymbol{A}\cdot\boldsymbol{n}|_{\partial\Omega} = 0, \text{div}\boldsymbol{A} = 0\}. \tag{1.5}$$

Then we define the function spaces

$$\begin{cases} H = H_1 \times H_2, \\ V = V_1 \times \mathbf{V}_2, \\ H_1 = \mathcal{L}^2(\Omega),\ V_1 = \mathcal{H}^1(\Omega), \\ H_2 = \text{ the closure of } \mathbf{V}_2 \text{ for the } \mathbf{L}^2\text{-norm}, \\ V_2 = \text{ the closure of } \mathbf{V}_2 \text{ for the } \mathbf{H}^1\text{-norm}. \end{cases} \tag{1.6}$$

The $L^2$-norms and inner products in $H_1$, $H_2$ and $H$ are denoted by $|\cdot|$ and $(\cdot,\cdot)$. The $H^1$-norm and inner product of $V_1$ are denoted by $||\cdot||$ and $((\cdot,\cdot))$ respectively. It is easy to see that

$$||\boldsymbol{A}|| = (\int_\Omega |\mathbf{curl}\boldsymbol{A}|^2 d\Omega)^{1/2} \tag{1.7}$$

is an equivalent norm of $V_2$, the corresponding inner product is also denoted by $((\cdot,\cdot))$. We also use the notations $||\cdot||$ and $((\cdot,\cdot))$ for the induced norm and inner product in $V$.

Moreover, it is classical that (see among others, [GRa 86] and [Tem 79])

$$H_2 = \{\boldsymbol{A} \in \boldsymbol{L}^2(\Omega) |\ \text{div}\boldsymbol{A} = 0,\ \boldsymbol{A}\cdot\boldsymbol{n}|_{\partial\Omega} = 0\}, \tag{1.8}$$

$$V_2 = \boldsymbol{H}^1_n(\text{div},\Omega) = \{\boldsymbol{A} \in \boldsymbol{H}^1(\Omega) |\ \text{div}\boldsymbol{A} = 0,\ \boldsymbol{A}\cdot\boldsymbol{n}|_{\partial\Omega} = 0\}. \tag{1.9}$$

Hereafter, we also use

$$P = (P_1, P_2) : \ \mathcal{L}^2(\Omega) \times \boldsymbol{L}^2(\Omega) \to H \tag{1.10}$$

to denote the orthogonal projection.

By identifying $H$ (resp. $H_j$, $j = 1,2$) with its dual space $H'$ (resp. $H'_j$, $j = 1,2$), we have

$$V \subset H = H' \subset V', \qquad V_j \subset H_j = H'_j \subset V'_j,$$

for $j = 1,2$.

We need to specify explicitly the function $\phi$ in the weak formulation of the system. It is easy to see that the function $\phi$ is determined uniquely up to an added constant as follows:

$$\eta\nabla\phi + (I - P_2)\Big(\frac{i}{2}(\psi^*\nabla\psi - \psi\nabla\psi^*) + |\psi|^2\boldsymbol{A}\Big) = 0, \quad \forall u = (\psi, \boldsymbol{A}) \in V, \quad (1.11)$$

the operator I being the identity on $\boldsymbol{L}^2$. Therefore, we define a nonlinear operator $\phi = \Phi(u)$ to be the unique solution $\phi$ of (1.11) satisfying

$$\int_\Omega \phi dx = 0. \quad (1.12)$$

**Lemma 5.1.1** *The mapping $\phi = \Phi(u)$ satisfies that*

$$\begin{cases} |\phi|_{L^2(\Omega)} \le \frac{c}{\eta}[|\psi|_{L^4}\|\psi\| + |\psi|_{L^4}^2|u|^{1/2}\|u\|^{1/2}], \text{ if } u \in V, \\ |\phi|_{L^3} \le \frac{c}{\eta}(\|u\|^2 + \|u\||\psi|_{L^4}^2), \text{ if } u \in V, \\ \|\phi\|_{H^1(\Omega)} \le \frac{c}{\eta}[\|u\|^3 + \|u\|^{3/2} \cdot \|u\|_{H^2(\Omega)}^{1/2}], \text{ if } u \in V \bigcap H^2(\Omega). \end{cases} \quad (1.13)$$

*The constant c here is independent of $\eta$, $\kappa$ and $u$. Consequently, $\Phi$ is a continuous mapping from $V$ into $L^2(\Omega)$ (resp. from $V\bigcap H^2$ into $H^1$).*

*Proof.* The first and third equations of (1.13) follow directly from the following computations:

$$\int_\Omega |\psi\nabla\psi^* \cdot \tilde{\boldsymbol{A}}|dx \le |\psi|_{L^3}|\nabla\psi^*|_{L^2}|\tilde{\boldsymbol{A}}|_{L^6} \le c|\psi|_{L^4}\|\psi\|\|\tilde{\boldsymbol{A}}\|,$$

$$\int_\Omega |\psi|^2|\boldsymbol{A} \cdot \tilde{\boldsymbol{A}}|dx \le |\psi|_{L^4}^2|\boldsymbol{A}|_{L^3}|\tilde{\boldsymbol{A}}|_{L^6} \le C|\psi|_{L^4}^2|u|^{1/2}\|u\|^{1/2}\|\tilde{\boldsymbol{A}}\|,$$

$$\int_\Omega |\psi\nabla\psi^* \cdot \tilde{\boldsymbol{A}}|dx \le |\psi|_{L^6}|\nabla\psi|_{L^3}|\tilde{\boldsymbol{A}}|_{L^2} \le C\|u\||\nabla\psi|_{H^{1/2}}|\tilde{\boldsymbol{A}}| \le C\|u\|^{3/2}\|u\|_{H^2}^{1/2}|\tilde{u}|,$$

$$\int_\Omega |\psi|^2|\boldsymbol{A} \cdot \tilde{\boldsymbol{A}}|dx \le |\psi|_{L^6}^2|\boldsymbol{A}|_{L^6}|\tilde{\boldsymbol{A}}|_{L^2} \le C\|u\|^3|\tilde{\boldsymbol{A}}|.$$

On the other hand, we have

$$\int_\Omega |\psi\nabla\psi^* \cdot \tilde{\boldsymbol{A}}|dx \le c|\psi|_{L^6}|\nabla\psi^*||\tilde{\boldsymbol{A}}|_{L^3} \le c\|\psi\|^2\|\tilde{\boldsymbol{A}}\|_{W^{1,3/2}},$$

$$\int_\Omega |\psi|^2|\boldsymbol{A} \cdot \tilde{\boldsymbol{A}}|dx \le c|\psi|_{L^4}^2|\boldsymbol{A}|_{L^6}|\tilde{\boldsymbol{A}}|_{L^3} \le c|\psi|_{L^4}^2\|u\|\|\tilde{\boldsymbol{A}}\|_{W^{1,3/2}}.$$

Hence

$$\|(I - P_2)[\frac{i}{2}(\psi^*\nabla\psi - \psi\nabla\psi^*) + |\psi|^2\boldsymbol{A}]\|_{W^{-1,3}} \le c(\|u\|^2 + |\psi|_{L^4}^2\|u\|).$$

This proves the second equation in (1.13).

We now define some operators related to different terms in the G-L equations. First of all, for the dissipative terms in (1.1), we define two linear operators $L_j : V_j \to V_j'$ $(j = 1, 2)$ by

$$< L_1\psi, \tilde{\psi} >= \int_\Omega \nabla\psi \cdot \nabla\tilde{\psi}^* dx, \tag{1.14}$$

$$< L_2\boldsymbol{A}, \tilde{\boldsymbol{A}} >= \int_\Omega \mathbf{curl}\boldsymbol{A} \cdot \mathbf{curl}\tilde{\boldsymbol{A}} dx, \tag{1.15}$$

for all $u = (\psi, \boldsymbol{A})$ and $\tilde{u} = (\tilde{\psi}, \tilde{\boldsymbol{A}}) \in V$. Classically, $L_j$ $(j = 1,2)$ can be extended naturally as unbounded linear self-adjoint operators on $H_j$ with domains

$$D(L_j) = V_j \bigcap H^2, \quad j = 1,2. \tag{1.16}$$

$L_2$ is also positive definite with compact inverse $L_2^{-1} : H_2 \to H$ while $L_1$ is semi-positive definite with $(L_1 + I)^{-1}$ compact and the following Gårding inequality:

$$< L_1\psi, \psi > + |\psi|^2 \geq \|\psi\|^2. \tag{1.17}$$

For the remaining terms other than the time derivative terms, we define also two (non-linear) operators $R_1$ and $R_2$ as follows:

$$\begin{aligned}(R_1(u), \tilde{\psi}) = \textstyle\int_\Omega [i\Phi(u)\psi\tilde{\psi}^* + \kappa^2(|\psi|^2 - 1)\psi\tilde{\psi}^* \\ +\tfrac{i}{2}(\tilde{\psi}^*\nabla\psi - \psi^*\nabla\tilde{\psi}) \cdot \boldsymbol{A} + |\boldsymbol{A}|^2\psi\tilde{\psi}^*]dx.\end{aligned} \tag{1.18}$$

$$(R_2(u), \tilde{\boldsymbol{A}}) = \frac{1}{\eta}\int_\Omega [\frac{i}{2}(\tilde{\psi}^*\nabla\psi - \psi^*\nabla\tilde{\psi}) \cdot \tilde{\boldsymbol{A}} + |\psi|^2\boldsymbol{A} \cdot \tilde{\boldsymbol{A}} - H \cdot \mathbf{curl}\tilde{\boldsymbol{A}}]dx \tag{1.19}$$

for some $u$, $\tilde{u} \in V$. These two operators enjoy the following properties:

**Lemma 5.1.2**

$$|R_1(u)| \leq C(\|u\|^4 + \|u\|^{5/2}\|u\|_{H^2}^{1/2}), \tag{1.20}$$

$$|R_2(u)| \leq C(1 + \|u\|^4 + \|u\|^{3/2}\|u\|_{H^2}^{1/2}) \tag{1.21}$$

*where $C$ depends on $\kappa$ and $\eta$ but not on $u$ and $\tilde{u}$.*

*Proof.* We only have to check each term in the definition of $R_j(u)$ given by (1.18)–(1.19):

$$\begin{aligned}|\int_\Omega i\phi\psi\tilde{\psi}^* dx| \quad & \leq |\phi|_{L^6}|\psi|_{L^3}|\tilde{\psi}|_{L^2} \\ & \leq c\|\phi\|_{H^1}\|\psi\|^{1/2}|\psi|^{1/2}|\tilde{\psi}|_{L^2} \\ \text{(by (1.13))} \quad & \leq c[\|u\|^3 + \|u\|^{3/2}\|u\|_{H^2}^{1/2}]\|u\||\tilde{\psi}|_{L^2} \\ & \leq c[\|u\|^4 + \|u\|^{5/2}\|u\|_{H^2}^{1/2}]|\tilde{u}|,\end{aligned}$$

$$\int_\Omega \kappa^2|\psi|^2\psi\tilde{\psi}^* dx \le c|\psi|^3_{L^6}|\tilde{\psi}^*|_{L^2} \le c\|u\|^3|\tilde{u}|,$$

$$\int_\Omega \tilde{\psi}^*\nabla\psi\cdot \boldsymbol{A}dx \le |\boldsymbol{A}|_{L^6}|\nabla\psi|_{L^3}|\tilde{\psi}|_{L^2} \le c\|u\|^{3/2}\|u\|^{1/2}_{H^2}|\tilde{u}|,$$

$$\begin{aligned} \int_\Omega \psi^*\nabla\tilde{\psi}\cdot \boldsymbol{A}dx \quad &= -\int_\Omega \tilde{\psi} div(\boldsymbol{A}\psi^*)dx \\ \text{(as in the above inequality)} \quad &\le c\|u\|^{3/2}\|u\|^{1/2}_{H^2}|\tilde{u}|. \end{aligned}$$

Other terms in (1.18) and (1.19) can be estimated in the same fashion.

We are now in a position to state the weak formulation of the G-L system as follows:

**Problem 1.1** (Weak formulation): *For $u_0 = (\psi_0, \boldsymbol{A}_0) \in H$ given, find a solution $u = (\psi, \boldsymbol{A})$ of the G-L system in the following sense:*

$$u \in L^\infty(0,T;H)\bigcap L^2(0,T;V), \text{ for all } T > 0, \tag{1.22}$$

$$\frac{d}{dt}(\psi,\tilde{\psi}) + (L_1\psi,\tilde{\psi}) + (R_1(u),\tilde{\psi}) = 0, \ \forall\tilde{\psi} \in V_1, \tag{1.23}$$

$$\frac{d}{dt}(\boldsymbol{A},\tilde{\boldsymbol{A}}) + \frac{1}{\eta}(L_2\boldsymbol{A},\tilde{\boldsymbol{A}}) + (R_2(u),\tilde{\boldsymbol{A}}) = 0, \ \forall\tilde{\boldsymbol{A}} \in V_2, \tag{1.24}$$

$$u|_{t=0} = u_0. \tag{1.25}$$

Alternatively, (1.18) and (1.19) can be written as

$$\frac{d}{dt}(u,\tilde{u}) + < Lu,\tilde{u} > +(R(u),\tilde{u}) = 0, \ \forall\tilde{u} \in V, \tag{1.26}$$

where

$$\begin{cases} Lu = (L_1\psi, \ \frac{1}{\eta}L_2\boldsymbol{A}), \\ R(u) = (R_1(u), R_2(u)). \end{cases} \tag{1.27}$$

**Remark 5.1.1** *The interpretation of the above weak formulation can be made in the following fashion. First, for any smooth solution $u = (\psi, \boldsymbol{A})$ and $\phi$ of the G-L system (1.1)–(1.2), $u$ satisfies (1.22)–(1.25). On the other hand, if $u$ given by (1.22) is a solution of (1.23)–(1.25), as in the study for the Navier-Stokes equations of an incompressible fluid (see among others [Tem 79]), there is a function $\phi$ such that $u$ and $\phi$ satisfies (1.1) at least in the distributional sense.*

### 5.1.2 Existence and uniqueness of solutions

We present in this section two main existence theorems. The first theorem, Theorem 1.1, shows the existence of global (in time) weak solutions of the G-L system with arbitrary $L^2$-initial data. In the second one, Theorem 1.2, we obtain the existence and uniqueness of global (in time) strong solutions of the G-L equations for arbitrary $H^1$ initial data.

**Theorem 5.1.1** *When $u_0 \in L^2(\Omega)\times \boldsymbol{L}^2(\Omega)$, there is at least one global weak solution $u$ of Problem 1.1, the weak formulation of the G-L system.*

**Theorem 5.1.2** *For any $u_0 \in V$, there is a unique solution for Problem 1.1, i.e., the initial-boundary value problem (1.25)–(1.26), such that, for any $T > 0$,*

$$\begin{aligned} &u \in L^2(0,T; V \bigcap H^2) \bigcap C([0,T]; V),\\ &\phi \in L^2(0,T; H^1(\Omega)) \bigcap C([0,T]; L^2(\Omega)), \end{aligned} \tag{1.28}$$

*where $\phi = \Phi(u)$ defined by (1.11) and (1.12) is the Lagrange multiplier. Moreover, if $u_0 \in D(L)$, the unique solution $(u, \phi)$ satisfies that*

*1) If $u_0$ satisfies some compatibility conditions, then $u \in C([0,\infty); D(L))$.*

*2) For any $T > 0$,*

$$\begin{cases} u \in L^2(0,T; V \bigcap H^3(\Omega)) \bigcap L^\infty(0,T; D(L)),\\ \phi \in L^2(0,T; H^2(\Omega)) \bigcap L^\infty(0,T; H^1(\Omega)). \end{cases} \tag{1.29}$$

**Remark 5.1.2** *It is easy to see that the evolutionary G-L system and the Navier-Stokes equations of an incompressible fluid possess the same type of non-linear terms, namely, $\psi\nabla\psi^*$ here and $v\cdot\nabla v$ for the Navier-Stokes equations, hence the proofs of existence of solutions bear many similarities.*

**Remark 5.1.3** *(Open question) When the space dimension $n = 3$, the existence of a global and/or local (in time) weak solutions with $L^2$ initial data (i.e. $u_0 \in H$) is still open, even if we use the simplified boundary condition $\mathbf{curl}\boldsymbol{A} \times \boldsymbol{n} = \boldsymbol{H} \times \boldsymbol{n}$. Theorems 1.1 and 1.2, however, provides only a complete solution to this problem in the case where $n = 2$.*

### 5.1.3 Proof of Theorem 1.2

We prove this theorem by using the Galerkin method. As we shall see later, the existence of the G-L energy functional (cf. (1.3)) is essential in proving the global existence of solutions, but it is not available for the truncated system of equations in the Galerkin procedure. Therefore, as we mentioned at the beginning of this chapter, we first use the Galerkin method to obtain the existence of local (in time) solutions of the equations, then we apply the G-L functional to the original G-L equations to prove that the local solution can be extended naturally to a unique global solution.

We divide the proof into 5 steps.

*Step 1: Approximate solutions.* We first observe that the linear operator $L$ possesses an orthonormal family of eigenfunctions $w_0, w_1, \dots, w_m, \dots$ with corresponding eigenvalues $\lambda_0, \lambda_1, \dots, \lambda_m, \dots$ such that

$$\begin{cases} 0 = \lambda_0 < \lambda_1 < \cdots < \lambda_j \to \infty, \\ w_j \in D(L). \end{cases} \tag{1.30}$$

Now, we want to find an approximate solution of Problem 1.1 as follows:

$$u_m = \sum_{j=0}^{m} g_{jm}(t) w_j \in W_m, \; g_{jm}(t) \in R^1, \tag{1.31}$$

$$W_m = \text{span}\{w_0, w_1, \dots, w_m\}, \tag{1.32}$$

$$\frac{d}{dt}(u_m, w_j) + (Lu_m, w_j) + (R(u_m), w_j) = 0, \; j = 1, 2, \dots, m, \tag{1.33}$$

$$u_m(0) = P_m u_0, \tag{1.34}$$

the linear operator $P_m$ being the orthogonal projection into $W_m$.

It is easy then to see that (1.33) and (1.34) are a system of ordinary differential equations of $g_{jm}(t)$, $j = 0, 1, \dots, m$ and there exists a unique local solution.

*Step 2: Energy estimates.* We now need to prove some a priori estimates to pass to the limit $m \to \infty$ to obtain a solution of the original problem. To this end, since $w_j$ is an eigenfunction of $L$ corresponding to the eigenvalue $\lambda_j$, we multiply both sides of (1.33) by $\lambda_j g_{jm}(t)$ and add them up for $j = 0, 1, \dots, m$. It follows then easily that

$$\frac{1}{2}\frac{d}{dt}(Lu_m, u_m) + |Lu_m|^2 + (R(u_m), Lu_m) = 0. \tag{1.35}$$

We infer from (1.33) that

$$\frac{1}{2}\frac{d}{dt}|u_m|^2 + (Lu_m, u_m) + (R(u_m), u_m) = 0. \tag{1.36}$$

Therefore,

$$\begin{aligned} &\frac{1}{2}\frac{d}{dt}\left[|u_m|^2 + (Lu_m, u_m)\right] + (Lu_m, u_m) + |Lu_m|^2 \\ &\leq |R(u_m)||u_m| + |R(u_m)||Lu_m| \\ &\leq |u_m||R(u_m)| + 2|R(u_m)|^2 + \frac{1}{4}|Lu_m|^2. \end{aligned}$$

This yields

$$
\begin{aligned}
&\frac{d}{dt}\Big[|u_m|^2 + (Lu_m, u_m)\Big] + (Lu_m, u_m) + \frac{3}{2}|Lu_m|^2 \\
&\le |R(u_m)|^2 + 2|u_m||R(u_m)| \\
&\text{(by (1.20) and (1.21))} \\
&\le c(1 + \|u_m\|^8 + \|u_m\|^5\|u_m\|_{H^2}) \\
&\Big(\text{by } \|u_m\|_{H^2} \le c(|u_m| + |Lu_m|)\Big) \\
&\le c(1 + \|u_m\|^8 + \|u_m\|^5|Lu_m|) \\
&\le \frac{1}{2}|Lu_m|^2 + c(1 + \|u_m\|^{10}) \\
&\Big(\text{by } \|u_m\|^2 \le c(|u_m|^2 + (Lu_m, u_m))\Big) \\
&\le \frac{1}{2}|Lu_m|^2 + c[1 + (|u_m|^2 + (Lu_m, u_m))^5].
\end{aligned}
$$

Namely,

$$
\begin{aligned}
\frac{d}{dt}[|u_m|^2 + (Lu_m, u_m)] + (Lu_m, u_m) + |Lu_m|^2 \\
\le c\{1 + [|u_m|^2 + (Lu_m, u_m)]^5\}.
\end{aligned} \tag{1.37}
$$

Set

$$
y = 1 + |u_m|^2 + (Lu_m, u_m), \tag{1.38}
$$

then

$$
\frac{dy}{dt} \le cy^5.
$$

Integrating this differential inequality, we find

$$
0 < y(t) \le (y(0)^{-4} - ct)^{-1/4}
$$

for $0 \le t \le T_0 = T(\|u_0\|) = \frac{c}{\|u_0\|^8}$. In other words, we have

$$
\|u_m\|^2 \le C, \text{ for } 0 \le t \le T_0 = T(\|u_0\|). \tag{1.39}
$$

The combination of (1.37) and (1.39) implies finally that

$$
u_m \in \text{ a bounded set of } L^2(0, T_0; D(L)) \bigcap L^\infty(0, T_0; V). \tag{1.40}
$$

Moreover, (1.33) amounts to saying that

$$
(u_m)_t + Lu_m + P_m R(u_m) = 0.
$$

Hence

$$\begin{aligned}|(u_m)_t| &\le |Lu_m| + |P_m R(u_m)| \\ &\le \|u_m\|_{H^2} + c(1 + \|u_m\|^4 + \|u_m\|^{5/2}\|u_m\|_{H^1}^{1/2}).\end{aligned}$$

It follows then that, also by (1.33),

$$|(u_m)_t|^2_{L^2(0,T_0;H)} \le c\Big(1 + \int_0^{T_0} (\|u_m(t)\|^{10} + \|u_m(t)\|^2_{H^2})dt\Big) \le C,$$

$C$ being independent of $m$ and $u_m$. That is,

$$(u_m)_t \in \text{ a bounded set of } L^2(0,T_0;H). \tag{1.41}$$

*Step 3: Passage to the limit $m \to \infty$ and existence of solutions.* By (1.40) and (1.41), there is a subsequence $u_{m'}$ of $u_m$ such that as $m' \to \infty$,

$$\begin{cases} u_{m'} \rightharpoonup u \text{ weakly in } L^2(0,T_0;D(L)), \\ u_{m'} \rightharpoonup u \text{ weakly-star in } L^2(0,T_0;D(L)), \\ (u_{m'})_t \rightharpoonup u_t \text{ weakly in } L^2(0,T_0;H). \end{cases} \tag{1.42}$$

Thanks to the compactness of the embedding of $D(L)$ in $V$, the inclusion

$$\{u \in L^2(0,T_0;D(L)) | u' \in L^2(0,T_0;H)\} \subset L^2(0,T_0;V)$$

is compact. Therefore without loss of generality, we may assume

$$u_{m'} \to u \text{ strongly in } L^2(0,T_0;V). \tag{1.43}$$

Then, it is classical to pass to the limit $m' \to \infty$ in (1.33)–(1.34) to obtain that $u$ is a solution of the G-L system (1.1)–(1.2).

*Step 4: Global solutions.* We only have to prove now the local strong solution obtained can be extended uniquely to a global (in time) strong solution of the G-L system. More precisely, let $T_{\max}$ be the maximum time where the local solution $u$ is obtained, namely, $u$ satisfies (1.25)–(1.26) and

$$\begin{cases} u \in L^2_{loc}(0,T_{\max}; V \bigcap H^2) \bigcap C([0,T_{\max}),V), \\ u_t \in L^2_{loc}(0,T_{\max};H), \end{cases} \tag{1.44}$$

we now prove that $T_{\max} = +\infty$. To achieve this, it is enough to establish some *a priori* estimates for the above local strong solutions.

First of all, we consider the G-L functional $E\colon V \to I\!R$ as follows:

$$E(u)(t) = \int_\Omega \Big\{\frac{1}{2}|(i\nabla + \boldsymbol{A})\psi|^2 + \frac{\kappa^2}{4}(|\psi|^2-1)^2 + \frac{1}{2}|\mathbf{curl}\boldsymbol{A} - H|^2\Big\}dx. \tag{1.45}$$

Introduce the following gauge transformation:

$$\begin{cases} \zeta = \psi e^{i\chi}, \\ \boldsymbol{B} = \boldsymbol{A} + \nabla\chi, \\ \chi = \displaystyle\int_0^t \phi(x,s)ds. \end{cases} \tag{1.46}$$

Noticing that

$$\begin{cases} \zeta_t = (\psi_t + i\phi\psi)e^{i\chi}, \\ \boldsymbol{B}_t = \boldsymbol{A}_t + \nabla\phi, \\ (i\nabla + \boldsymbol{B})\zeta = ((i\nabla + \boldsymbol{A})\psi)e^{i\chi}, \end{cases}$$

multiplying the complex conjugate of equation (1.1-1) by $\psi_t + i\phi\psi$ and take the real part, then multiplying (1.1-2) by $\boldsymbol{A}_t + \nabla\phi$. Adding up the results and using gauge invariance, we obtain, noticing that $H$ is independent of $t$,

$$\begin{aligned} 0 &\geq -|\psi_t + i\phi\psi|^2 - \eta|\boldsymbol{A}_t + \nabla\phi|^2 \\ &= \int_\Omega \Big[ \mathrm{Re}[(\psi_t + i\phi\psi)(-(-i\nabla + \boldsymbol{A})^2\bar{\psi} + \tfrac{\kappa^2}{2}(|\psi|^2 - 1)\bar{\psi})] \\ &+(\boldsymbol{A}_t + \nabla\phi)\cdot\Big((\mathbf{curl}^2\boldsymbol{A} - \mathbf{curl}H) + \tfrac{i}{2}(\psi^*\nabla\psi - \psi\nabla\psi^*) + |\psi|^2\boldsymbol{A}\Big)\Big]\, dx \\ &= \int_\Omega \Big[ \mathrm{Re}[(i\nabla + \boldsymbol{B})\zeta_t(-i\nabla + \boldsymbol{B})\boldsymbol{A}\zeta] + \mathrm{Re}[i\boldsymbol{B}_t\zeta\cdot(-i\nabla\bar{\zeta} + \boldsymbol{B}\bar{\zeta}))\\ &+\mathrm{Re}[\tfrac{\kappa^2}{2}(|\zeta|^2 - 1)\bar{\zeta}\zeta_t] + (\mathbf{curl}\boldsymbol{B} - H)\cdot(\mathbf{curl}\boldsymbol{B} - H)_t\Big]dx \\ &= \tfrac{dE}{dt}(\zeta(t), \boldsymbol{B}(t)) = \tfrac{dE}{dt}(u(t)). \end{aligned} \tag{1.47}$$

In the integration process, it is easy to see that the boundary terms concerning $\boldsymbol{A}$ and $\psi$ vanishes due to the Neumann boundary conditions they satisfy.

It is also easy to verify that $E(u(t)) \to \infty$ implies the $H^1$ norms of $\psi$ and $\boldsymbol{A}$ tend to $\infty$.

From the existence of the Lyapunov functional, we derive that (1.39) holds for $0 < T_0 \leq \infty$. From (1.37), we imply that (1.40) also holds for $0 < T_0 \leq \infty$. Subsequently, all ensuing estimates in the Proof of Theorem 1.2 hold for $0 < T_0 \leq \infty$. The proof of existence then follows along the same line.

*Step 5: Uniqueness.*

Let $(\psi_1, \boldsymbol{A}_1)$ and $(\psi_2, \boldsymbol{A}_2)$ be two solutions of the problem and $(\psi, \boldsymbol{A}) = (\psi_1, \boldsymbol{A}_1) - (\psi_2, \boldsymbol{A}_2)$. Substituting into the equations respectively and subtracting the resulting equations, we obtain, for $\psi$:

$$\begin{aligned} &\psi_t - \Delta\psi + 2(\boldsymbol{A}_1\nabla\psi_1 - \boldsymbol{A}_2\nabla\psi_2) + (|\boldsymbol{A}_1|^2\psi_1 - |\boldsymbol{A}_2|^2\psi_2) \\ &+\Big((|\psi_1|^2 - 1)\psi_1 - (|\psi_2|^2 - 1)\psi_2\Big) = 0. \end{aligned} \tag{1.48}$$

Multiplying both sides by $\psi^*$, taking the real part, integrating over $\Omega$ and rearranging the terms, we obtain

$$\begin{aligned}
\frac{1}{2}\frac{\partial}{\partial t}\int_\Omega |\psi|^2 dx \quad &+ \quad \int_\Omega |\nabla\psi|^2 dx \\
&\le \int_\Omega \Big(|\psi|^2 - 2\mathrm{Re}(\nabla\psi_1 \boldsymbol{A}_1 - \nabla\psi_2 \boldsymbol{A}_2)\psi^* \\
&\quad -(|\boldsymbol{A}_1|^2\psi_1 - |\boldsymbol{A}_2|^2\psi_2)\psi^* \\
&\le \int_\Omega \Big(|\psi|^2 - 2\mathrm{Re}(\boldsymbol{A}\nabla\psi_1 + \boldsymbol{A}_2\nabla\psi)\psi^* \\
&\quad -(|\boldsymbol{A}_1|^2\psi + \psi_2(\boldsymbol{A}_1+\boldsymbol{A}_2)\boldsymbol{A})\psi^*\Big) dx \\
&\le |\psi|_{L^2}^2 + |\boldsymbol{A}|_{L^4}|\psi|_{L^4}|\nabla\psi_1|_{L^2} + |\boldsymbol{A}_2|_{L^4}|\psi|_{L^4}|\nabla\psi|_{L^2} \\
&\quad - \int_\Omega |\boldsymbol{A}_1|^2|\psi|^2 dx + |\psi_2|_{L^4}|\boldsymbol{A}_1+\boldsymbol{A}_2|_{L^4}|\boldsymbol{A}|_{L^4}|\psi|_{L^4} \\
&\le |\psi|_{L^2}^2 + |\boldsymbol{A}|_{L^4}|\psi|_{L^4}|\nabla\psi_1|_{L^2} + |\boldsymbol{A}_2|_{L^4}|\psi|_{L^4}|\nabla\psi|_{L^2} \\
&\quad + |\psi_2|_{L^4}|\boldsymbol{A}_1+\boldsymbol{A}_2|_{L^4}|\boldsymbol{A}|_{L^4}|\psi|_{L^4}.
\end{aligned}$$

We know, from the Lyapunov functional, that for some constant $c$ and for $j = 1, 2$,

$$\begin{aligned}
&|\psi_j|_{L^4} \le c, \\
&|\nabla\psi_j|_{L^2} \le c, \\
&|\boldsymbol{A}_j|_{L^4} \le c.
\end{aligned}$$

We derive then

$$\begin{aligned}
\frac{1}{2}\frac{\partial}{\partial t}\int_\Omega |\psi|^2 dx \quad &+ \quad \int_\Omega |\nabla\psi|^2 dx \\
&\le |\psi|_{L^2}^2 + C|\boldsymbol{A}|_{L^4}|\psi|_{L^4} + C|\psi|_{L^4}|\nabla\psi|_{L^2}.
\end{aligned}$$

Here $C$ is some positive constant depending on the initial data and the domain $\Omega$.

Using the standard inequality

$$|f|_{L^4}^2 \le C|f|_{H^1}|f|_{L^2} \le \delta|\nabla f|_{L^2}^2 + c_\delta|f|_{L^2}^2$$

for any (small) positive number $\delta$ and $C_\delta \sim 1/\delta$, we have

$$\begin{aligned}
\frac{1}{2}\frac{\partial}{\partial t}\int_\Omega |\psi|^2 dx \quad &+ \quad \int_\Omega |\nabla\psi|^2 dx \\
&\le |\psi|_{L^2}^2 + C(\delta|\nabla\boldsymbol{A}|_{L^2}^2 + C_\delta|\boldsymbol{A}|_{L^2}^2) \\
&\quad + C(\delta|\nabla\psi|_{L^2}^2 + C_\delta|\psi|_{L^2}^2) + \frac{1}{4}|\nabla\psi|_{L^2}^2.
\end{aligned}$$

Here $C$ may represent different constants but it is independent of $\delta$.

For $\boldsymbol{A}$, we have

$$\begin{aligned}
&\boldsymbol{A}_t - \nabla\mathrm{div}\boldsymbol{A} + \mathbf{curl}^2\boldsymbol{A} - \mathrm{Im}(\psi^*\nabla\psi_1 + \psi_2^*\nabla\psi) \\
&+|\psi_1|^2\boldsymbol{A} + (|\psi_1|^2 - |\psi_2|^2)\boldsymbol{A}_2 = 0.
\end{aligned}$$

Multiplying both sides by $\boldsymbol{A}$, integrating over $\Omega$ and rearranging the terms, we obtain

$$\begin{aligned}\frac{1}{2}\frac{\partial}{\partial t} & \int_\Omega |\boldsymbol{A}|^2 dx + \int_\Omega |\mathrm{div}\boldsymbol{A}|^2 + |\mathbf{curl}\boldsymbol{A}|^2 dx \\ &= \int_\Omega \mathrm{Im}(\psi^* \nabla \psi_1 + \psi_2^* \nabla \psi)\boldsymbol{A} - |\psi_1|^2 |\boldsymbol{A}|^2 - (|\psi_1|^2 - |\psi_2|^2)\boldsymbol{A}_2 \boldsymbol{A} dx \\ &\le |\psi|_{L^4} |\nabla \psi_1|_{L^2} |\boldsymbol{A}|_{L^4} + |\psi_2|_{L^4} |\nabla \psi|_{L^2} |\boldsymbol{A}|_{L^4} + \\ & \quad |\psi_1 + \psi_2|_{L^4} |\psi|_{L^4} |\boldsymbol{A}_2|_{L^4} |\boldsymbol{A}|_{L^4}.\end{aligned}$$

Using similar arguments as for $\psi$, we have

$$\begin{aligned}\frac{1}{2}\frac{\partial}{\partial t}|\boldsymbol{A}|_{L^2}^2 &+ |\mathrm{div}\boldsymbol{A}|_{L^2}^2 + |\mathbf{curl}\boldsymbol{A}|_{L^2}^2 \\ &\le C\Big(|\psi|_{L^4}|\boldsymbol{A}|_{L^4} + |\nabla\psi|_{L^2}|\boldsymbol{A}|_{L^4} + |\psi|_{L^4}|\boldsymbol{A}|_{L^4}\Big) \\ &\le C\Big(\delta|\nabla\psi|^2 + C_\delta|\psi|^2 + \delta|\nabla \boldsymbol{A}|^2 + C_\delta|\boldsymbol{A}|^2\Big) + \frac{1}{4}|\nabla\psi|_{L^2}^2.\end{aligned}$$

Adding the two estimates together, making $\delta$ small, taking into account that

$$|\nabla \boldsymbol{A}|_{L^2}^2 \le C(|\mathrm{div}\boldsymbol{A}|_{L^2}^2 + |\mathbf{curl}\boldsymbol{A}|_{L^2}^2),$$

we can now easily derive the uniqueness results by using Gronwall inequality.

### 5.1.4 Proof of Theorem 1.1

We also proceed by the Galerkin method. As in the previous subsection, we need first to solve (1.33)–(1.34). However, since now we only have $u_0 \in H$, we can not expect to obtain the energy estimates by using the G-L functional. We divide the proof into 4 steps.

*Step 1: Approximate solutions.* We first observe that the eigenfunctions $w_j$ ($j = 0, 1, \dots$) of the linear operator $L$ can be written as

$$w_j = (w_j^{(1)}, w_j^{(2)}),\ j = 0, 1, \dots$$

with super-index (1) for the order parameter and (2) for the magnetic potential. Therefore, the approximate equations (1.33) are equivalent to the following equations:

$$\frac{d}{dt}(\psi_m, w_j^{(1)}) + (L_1\psi_m, w_j^{(1)}) + (R_1(u_m), w_j^{(1)}) = 0, \tag{1.49}$$

$$\frac{d}{dt}(\boldsymbol{A}_m, w_j^{(2)}) + \frac{1}{\eta}(L_2\boldsymbol{A}_m, w_j^{(2)}) + (R_2(u_m), w_j^{(2)}) = 0, \tag{1.50}$$

for $j = 0, 1, \dots, m$.

*Step 2 (energy estimates).* We infer then from (1.48) that

$$\frac{1}{2}\frac{d}{dt}|\psi_m|^2 + \int_\Omega \left[i\Phi(u_m)|\psi_m|^2 + \kappa^2(|\psi_m|^2 - 1)|\psi_m|^2 + |(i\nabla + \boldsymbol{A}_m)\psi_m|^2\right]dx = 0.$$

Taking the real part of the above equation implies that

$$\frac{d}{dt}|\psi_m|^2 + 2\kappa^2|\psi_m|^4_{L^4} + 2|(i\nabla + \boldsymbol{A}_m)\psi_m|^2_{L^2} \le 2\kappa^2|\psi_m|^2 \le \kappa^2|\psi_m|^4_{L^4} + c.$$

Hence

$$\frac{d}{dt}|\psi_m|^2 + \kappa^2|\psi_m|^4_{L^4} + 2|(i\nabla + \boldsymbol{A}_m)\psi_m|^2_{L^2} \le c. \tag{1.51}$$

Integrating (1.50), we obtain that for any $T > 0$,

$$|\psi_m(t)|^2 \le c, \text{ for } t \in [0, T], \tag{1.52}$$

$$\int_0^T |\psi_m|^4 dt \le c, \tag{1.53}$$

$$\int_0^T |(i\nabla + \boldsymbol{A}_m)\psi_m|^2_{L^2} dt \le c, \tag{1.54}$$

where $c$ is a constant independent of $m$ and $u_m$.

On the other hand, we deduce, from (1.49) that

$$\begin{aligned}
&\frac{1}{2}\frac{d}{dt}|\boldsymbol{A}_m|^2 + |\mathbf{curl}\boldsymbol{A}_m|^2 \\
&= (\mathbf{curl}H, \boldsymbol{A}_m) - \mathrm{Re}\int_\Omega (i\nabla\psi_m + \boldsymbol{A}_m\psi_m)\psi_m^* \cdot \boldsymbol{A}_m dx \\
&\le c|\boldsymbol{A}_m| + |(i\nabla\psi_m + \boldsymbol{A}_m\psi_m)||\psi_m|_{L^4}|\boldsymbol{A}_m|_{L^4} \\
&\le c|\boldsymbol{A}_m| + |(i\nabla\psi_m + \boldsymbol{A}_m\psi_m)|^2 + |\psi_m|^2_{L^4}\|\boldsymbol{A}_m\|^2_{H^{1/2}} \\
&\le c|\boldsymbol{A}_m| + |(i\nabla\psi_m + \boldsymbol{A}_m\psi_m)|^2 + c|\psi_m|^2_{L^4}|\boldsymbol{A}_m||\mathbf{curl}\boldsymbol{A}_m| \\
&\le \frac{1}{2}|\mathbf{curl}\boldsymbol{A}_m|^2 + c(1 + |(i\nabla + \boldsymbol{A}_m)\psi_m|^2) + |\psi_m|^4_{L^4}|\boldsymbol{A}_m|^2.
\end{aligned}$$

Therefore

$$\frac{1}{2}\frac{d}{dt}|\boldsymbol{A}_m|^2 + |\mathbf{curl}\boldsymbol{A}_m|^2 \le c(1 + |(i\nabla + \boldsymbol{A}_m)\psi_m|^2) + c|\psi_m|^4_{L^4}|\boldsymbol{A}_m|^2. \tag{1.55}$$

By the Gronwall inequality, we have for any $T > 0$,

$$\begin{aligned}
|\boldsymbol{A}_m(t)|^2 \;&\le\; |P_m u_0|^2 \exp(c\int_0^t |\psi_m(t)|^4_{L^4} dt) \\
&+\; c\int_0^t (1 + |(i\nabla + \boldsymbol{A}_m(s))\psi_m(s)|^2) \\
&\qquad \exp(-c\int_t^s |\psi_m(\tau)|^4_{L^4} d\tau) ds \\
\text{(by (1.51)–(1.53))} \;&\le\; c(|u_0|^2 + 1), \quad t \in [0, T]
\end{aligned}$$

where $c$ is a constant depending on $T$ but not on $m$. Hence, it is easy to see that for any $T > 0$,

$$|\boldsymbol{A}_m(t)| \leq c, \quad \forall t \in [0, T], \tag{1.56}$$

$$\int_0^T |\mathbf{curl}\boldsymbol{A}_m|^2 dt \leq c. \tag{1.57}$$

Since we have

$$\begin{aligned}&\textstyle\int_0^T |\nabla \psi_m|^2 dt \leq \int_0^T |(i\nabla\psi_m + \boldsymbol{A}_m\psi_m)|^2 dt \\ &+c\textstyle\int_0^T \int_\Omega \left(|(\psi_m^* \nabla\psi_m - \psi_m \nabla \psi_m^*)\boldsymbol{A}_m| + |\boldsymbol{A}_m|^2|\psi_m|^2\right) dxdt,\end{aligned} \tag{1.58}$$

by Hölder inequality, we obtain

$$\begin{aligned}\textstyle\int_\Omega |\psi_m^* \nabla \psi_m \cdot \boldsymbol{A}_m| dx \quad &\leq |\nabla\psi_m||\psi_m|_{L^4}|\boldsymbol{A}_m|_{L^4} \\ &\leq c|\nabla\psi_m|^{2/3}|\psi_m|^{1/2}\|\boldsymbol{A}_m\|^{1/2}|\boldsymbol{A}_m|^{1/2} \\ &\leq c|\nabla\psi_m||\psi_m|\|\boldsymbol{A}_m\|^{1/2}|\boldsymbol{A}_m|^{1/2} \\ &\leq \varepsilon|\nabla\psi_m|^2 + c(\varepsilon)|\psi_m|^2\|\boldsymbol{A}_m\|^2|\boldsymbol{A}_m|^2 \\ &+c(\varepsilon)|\psi_m|^4|\boldsymbol{A}_m|\|\boldsymbol{A}_m\|.\end{aligned}$$

Similar computations apply also to other terms in the right-hand side of (1.57) and therefore,

$$\int_0^T |\nabla\psi_m|^2 dt \leq c + c\int_0^T |\psi_m|^2|\boldsymbol{A}_m|^2\|\boldsymbol{A}_m\|^2 dt \leq c \tag{1.59}$$

with $c$ independent of $m$.

*Step 3 (Estimate of $(u_m)_t$).* As in the proof of Lemma 1.1, it is easy to obtain that for any $\delta > 0$,

$$|\phi_m|_{L^2} \leq c|u_m|_{L^{2+\delta}}(\|\psi_m\| + |\psi_m|^2_{L^q}). \tag{1.60}$$

Then for any $\delta > 0$,

$$|\int_\Omega i\phi_m\psi_m\tilde{\psi}dx| \leq c|\phi_m||\psi_m|_{L^{2+\delta}}\|\tilde{\psi}\|, \tag{1.61}$$

which implies that for any $\delta > 0$,

$$\begin{aligned}\|i\psi_m\phi_m\|_{V_1'} \quad &\leq c|\phi_m||\psi_m|_{L^{2+\delta}} \\ &\leq c|u_m|^2_{L^{2+\delta}}(\|\psi_m\| + |\psi_m|^2_{L^4}) \\ &\leq c\|u_m\|^{2\delta/(2+\delta)}|u_m|^{2/(2+\delta)}(\|\psi_m\| + |\psi_m|^2_{L^4}).\end{aligned} \tag{1.62}$$

Thanks to (1.51)–(1.52), (1.55)–(1.56) and (1.58), we infer from (1.60) that

$$P_1(i\psi_m\phi_m) \in \text{ a bounded set of } L^{2-4\delta/(2+3\delta)}(0,T;V_1') \text{ independent of } m. \tag{1.63}$$

By definition,

$$(u_m)_t + PLu_m + PR(u_m) = 0. \tag{1.64}$$

All terms in $PR(u_m)$ can be handled in the same fashion as in (1.61) for $P_1(i\psi_m\phi_m)$, then we obtain that for any $T > 0$, there is a $\delta_1 > 0$ such that

$$(u_m)_t \in \text{ a bounded set of } L^{2-\delta_1}(0,T;V_1') \text{ independently of } u_m \text{ and } m. \tag{1.65}$$

*Step 4 (Passage to the limit $m \to \infty$).* In summary, for any $T > 0$,

$$\begin{cases} u_m \in \text{ a bounded set of } L^2(0,T;V)\bigcap L^\infty(0,T;H), \\ \psi_m \in \text{ a bounded set of } L^4(0,T;L^4(\Omega)), \\ (u_m)_t \in \text{ a bounded set of } L^{2-\delta_1}(0,T;V') \end{cases} \tag{1.66}$$

independently of $u_m$ and $m$. Here $0 < \delta_1 < 1$. We infer also from (1.64) that

$$u_m \in \text{a pre-compact subset of } L^2(0,T;H) \text{ independently of } u_m \text{ and } m. \tag{1.67}$$

Then it is classical to pass to the limit to obtain the existence of a global weak solution of the G-L system.

As we have pointed out, the existence theory has been established without assuming the boundedness of the initial data. Although this is a better mathematical result, we are interested in knowing if the order parameter, as in the steady state case, will remain smaller than 1 if the initial condition is smaller than 1. In fact, we have the following more general result:

**Theorem 5.1.3** *Let $(\psi, \boldsymbol{A})$ be a solution of the G-L system (1.1)–(1.2), if*

$$|\psi|(x,0) \le C$$

*almost everywhere in $\Omega$ for some constant $C$, then*

$$|\psi|(x,t) \le \max\{1, C\}.$$

*Proof.* Let $f = |\psi|$ and $\psi = fe^{i\phi}$, then $f$ is a solution of the following problem

$$\begin{aligned} &f_t - \Delta f + |\boldsymbol{A} - \nabla\phi|^2 f + \kappa^2(f^2-1)f = 0, \\ &f(x,0) = f_0(x), \\ &\partial_{\boldsymbol{n}} f|_{\partial\Omega}(x,t) = 0 \end{aligned} \tag{1.68}$$

where $0 \le f_0(x) \le C$. Let $a = \max\{1, C\}$. Multiplying the both sides of (1.67-1) by $(f-a)_+$, integrating over $\Omega$, we have

$$\begin{aligned} &\frac{1}{2}\Big(\int_\Omega ((f-a)_+)^2 dx\Big)_t + \int_\Omega |\nabla(f-a)_+|^2 dx + \\ &\int_\Omega |\boldsymbol{A} - \nabla\phi|^2 f(f-a)_+ dx + \kappa^2 \int_\Omega (f-a)_+(f^2-1) f dx = 0. \end{aligned}$$

This implies

$$(\int_\Omega ((f-a)_+)^2 dx)_t \le 0.$$

Integrating with respect to $t$ from 0 to $t$, we have

$$\int_\Omega ((f-a)_+)^2(x,t)dx \le \int_\Omega ((f-a)_+)^2(x,0)dx = 0$$

by assumption. The theorem is established.

This result applies also to the model discussed in the next section. We will therefore not repeat it.

## 5.2 Existence of 3-d evolutionary solutions

We have already pointed out, but would like to point out again, that in the following 3-d problem, we have made a technical assumption that the bounded, smooth set $\Omega_0$ is occupied be the superconductor sample. $\Omega$ is another bounded, smooth subset of $I\!R^3$ such that $\Omega_0 \subset \Omega$. In $\Omega \setminus \Omega_0$, we have the Maxwell system. Further beyond, in the set $I\!R^3 \setminus \Omega$, we suppose that $\mathbf{curl}\boldsymbol{A} = \boldsymbol{H}$. If $\Omega$ is sufficiently large, this is not perfect but is a good approximation of the problem from both physical and mathematical point of view. However the biggest advantage is that it made our mathematical existence proof and numerical approximation a lot easier.

We discuss the following system:

$$\begin{cases} \psi_t + i\psi\phi + |\psi|^2\psi - \psi - (\nabla - i\boldsymbol{A})^2\psi = 0, & \boldsymbol{x} \in \Omega_0, \quad t > 0, \\ \boldsymbol{A}_t + \nabla\phi + \mathbf{curl}^2\boldsymbol{A} + \frac{i}{2}(\psi^*\nabla\psi - \psi\nabla\psi^*) & \\ \qquad + |\psi|^2\boldsymbol{A} = 0, & \boldsymbol{x} \in \Omega_0, \quad t > 0, \\ \boldsymbol{A}_{tt} + \nabla\phi_t + \boldsymbol{A}_t + \nabla\phi + \mathbf{curl}^2\boldsymbol{A} = 0, & \boldsymbol{x} \in \Omega \setminus \overline{\Omega_0}, \quad t > 0, \\ \mathrm{div}\boldsymbol{A} + \phi = 0, & \boldsymbol{x} \in \Omega, \quad t > 0, \\ \frac{\partial\psi}{\partial \boldsymbol{n}_0} - i\boldsymbol{A}\cdot\boldsymbol{n}_0\psi = \eta\psi, & \boldsymbol{x} \in \partial\Omega_0, \quad t > 0, \\ \boldsymbol{A}\cdot\boldsymbol{n} = 0, \;\; \mathbf{curl}\boldsymbol{A}\times\boldsymbol{n} = \boldsymbol{H}\times\boldsymbol{n}, & \boldsymbol{x} \in \partial\Omega, \quad t > 0, \\ \psi(\boldsymbol{x},0) = \psi_0(\boldsymbol{x}), & \boldsymbol{x} \in \Omega_0, \\ \boldsymbol{A}(\boldsymbol{x},0) = \boldsymbol{A}_0(\boldsymbol{x}), & \boldsymbol{x} \in \Omega, \\ \boldsymbol{A}_t(x,0) = \boldsymbol{A}_1(x), \;\; \phi(\boldsymbol{x},0) = \phi_0(\boldsymbol{x}), & \boldsymbol{x} \in \Omega \setminus \overline{\Omega_0}, \\ [\boldsymbol{A}] = 0, \;\; [\mathrm{div}\boldsymbol{A}] = 0, \;\; [\mathbf{curl}\boldsymbol{A}\times\boldsymbol{n}] = 0, & x \in \partial\Omega_0, \quad t > 0. \end{cases} \tag{2.1}$$

We follow mainly the recent work of [HTa3].

This time, we describe the existence proof under the following assumption:

$$|\psi_0|(x) = |\psi|(x,0) \le 1 \text{ almost everywhere in } \Omega. \tag{2.2}$$

We know then that $|\psi(x,t)|$ also satisfies the same estimate. We will use the following notation:

$$\begin{aligned} \mathcal{W}^0 &= \mathcal{L}^2(0,T;\mathcal{H}^1(\Omega_0)) \cap \mathcal{H}^1(0,T;(\mathcal{H}^1(\Omega_0))'), \\ \boldsymbol{W}_n &= \boldsymbol{L}^2(0,T;\boldsymbol{H}^1_n(\Omega) \cap \boldsymbol{H}^1(0,T;\boldsymbol{L}^2(\Omega)) \cap \boldsymbol{H}^2(0,T;(\boldsymbol{H}^1_n(\Omega))'). \end{aligned} \tag{2.3}$$

Here, admitting that the dual space of $L^2$ is $L^2$ (no matter real, complex or vector valued), we use $(\mathcal{H}^1(D))'$ to denote the dual space of $\mathcal{H}^1(D)$, $(\boldsymbol{H}_n^1(D))'$ to denote the dual space of $\boldsymbol{H}_n^1(D)$ for any open set $D$ in $\mathbb{R}^3$.

The weak formulation for the system (2.1) is then to find $(\psi, \boldsymbol{A}) \in \mathcal{W}^0 \times \boldsymbol{W}_n$ such that

$$\psi_0(\boldsymbol{x},0) = \psi_0(\boldsymbol{x}),\ \boldsymbol{A}_0(\boldsymbol{x},0) = \boldsymbol{A}_0(\boldsymbol{x}),\ \boldsymbol{A}_t(x,0) = \boldsymbol{A}_1(x) \tag{2.4}$$

and

$$\int_0^T \int_{\Omega_0} \Big(\psi_t\omega - i\mathrm{div}\boldsymbol{A}\psi\omega + (|\psi|^2\psi - \psi)\omega$$
$$+(\nabla - i\boldsymbol{A})\psi(\nabla + i\boldsymbol{A})\omega\Big)d\boldsymbol{x}dt - \int_0^T \int_{\partial\Omega_0} \eta\psi\omega d\Gamma dt = 0, \tag{2.5}$$

$$\int_0^T \int_{\Omega_0} \Big((\boldsymbol{A}_t - \nabla\mathrm{div}\boldsymbol{A})\boldsymbol{B} + \mathbf{curl}\boldsymbol{A}\mathbf{curl}\boldsymbol{B} +$$
$$i(\psi^*\nabla\psi - \psi\nabla\psi^*)\boldsymbol{B} + |\psi|^2\boldsymbol{A}\boldsymbol{B}\Big)d\boldsymbol{x}dt$$
$$+\int_0^T \int_{\Omega\setminus\overline{\Omega_0}} \Big((\boldsymbol{A}_{tt} - \nabla\mathrm{div}\boldsymbol{A}_t)\boldsymbol{B} + (\boldsymbol{A}_t - \nabla\mathrm{div}\boldsymbol{A})\boldsymbol{B}$$
$$+\mathbf{curl}\boldsymbol{A}\mathbf{curl}\boldsymbol{B}\Big)d\boldsymbol{x}dt + \int_0^T \int_{\partial\Omega} (\boldsymbol{H} \wedge \boldsymbol{n})\boldsymbol{B}d\Gamma dt = 0 \tag{2.6}$$

for all $(\omega, \boldsymbol{B}) \in \mathcal{W}^0 \times \boldsymbol{W}_n$.

Now we can announce the main result of this section.

**Theorem 5.2.1** *Let $(\boldsymbol{H}, \psi_0, \boldsymbol{A}_0, \boldsymbol{A}_1)$ be given $\boldsymbol{H}^2(\Omega) \times \mathcal{H}^4(\Omega_0) \times \boldsymbol{H}^4(\Omega) \times \boldsymbol{H}^2(\Omega\setminus\overline{\Omega_0})$ functions ($\boldsymbol{H}$ is therefore assumed to be independent of t but this is not essential in the proofs), and assume that $|\psi_0|(x) \le 1$ for a.e. $x$, then there exists a unique pair $(\psi, \boldsymbol{A})$ which solves the variational problem (2.4)–(2.6) and in addition, we have*

$$\psi \in \mathcal{L}^\infty(0,T;\mathcal{H}^1(\Omega_0)) \cap \mathcal{H}^1(0,T;\mathcal{L}^2(\Omega_0)), \tag{2.7}$$
$$\boldsymbol{A} \in \boldsymbol{L}^\infty(0,T;\boldsymbol{H}_n^1(\Omega)) \cap \boldsymbol{H}^1(0,T;\boldsymbol{L}^2(\Omega)) \cap \boldsymbol{H}^2(0,T;(\boldsymbol{H}_n^1(\Omega))'), \tag{2.8}$$
$$|\psi| \le 1 \text{ almost everywhere on } \Omega_0 \times (0,T). \tag{2.9}$$

*Moreover, we have the following higher regularity result:*

$$\psi \in \mathcal{L}^\infty(0,T;\mathcal{H}^2(\Omega_0)). \tag{2.10}$$

We divide the proof into three parts to tackle the existence, regularity and uniqueness separately.

**The Proof of Existence**

The system couples a diffusive system and a dissipative system together. We first establish some preliminary results which help us to establish estimates in existence and uniqueness proofs.

**Lemma 5.2.1** *Let $(\psi, \boldsymbol{A})$ be a pair of solutions of (3.3)–(3.5), then we have*

$$|\psi|_{L^\infty} \le 1. \tag{2.11}$$

This lemma bears the same proof as that of Theorem 1.3.

**Lemma 5.2.2** *Define the energy functional*

$$\begin{aligned} &E(\psi, \boldsymbol{A})(t) = \\ &\tfrac{1}{2}\int_{\Omega_0}\Big(|(\nabla - i\boldsymbol{A})\psi|^2 + |\mathbf{curl}\boldsymbol{A} - \boldsymbol{H}|^2 + \tfrac{1}{2}(|\psi|^2-1)^2\Big)dx \\ &+\tfrac{1}{2}\int_{\Omega\setminus\overline{\Omega_0}}\Big(|\boldsymbol{A}_t - \nabla \mathrm{div}\boldsymbol{A}|^2 + |\mathbf{curl}\boldsymbol{A} - \boldsymbol{H}|^2\Big)dx - \tfrac{1}{2}\int_{\partial\Omega_0}\eta|\psi|^2 d\Gamma. \end{aligned} \tag{2.12}$$

*Let $(\psi, \boldsymbol{A})$ be a pair of solutions of (2.4)–(2.6), then we have formally*

$$\frac{dE}{dt} \le C_1(t)\sqrt{E} + C_2(t) \tag{2.13}$$

*where both $C_1(t)$ and $C_2(t)$ are both bounded functions.*

*In particular, when $\boldsymbol{H}$ is independent of $t$, we have $C_1(t) = 0$ and $C_2(t) = c_2$ is a constant for any $T > 0$. Moreover, when $\boldsymbol{H}$ is a constant, $E$ is the Lyapunov functional of the G-L system.*

*Proof.* Differentiating both sides of (2.1-1)–(2.1-2) with respect to $t$, we have

$$\begin{aligned} \frac{dE}{dt} &= \mathrm{Re}\Big[\int_{\Omega_0}\Big((\nabla - i\boldsymbol{A})\psi((\nabla + i\boldsymbol{A})(\psi_t^* + i\mathrm{div}\boldsymbol{A}\psi^*) + i(\boldsymbol{A}_t - \nabla\mathrm{div}\boldsymbol{A})\psi^*) \\ &\quad +(\mathbf{curl}\boldsymbol{A} - \boldsymbol{H})(\mathbf{curl}\boldsymbol{A}_t - \boldsymbol{H}_t) + (|\psi|^2 - 1)\psi(\psi_t^* + i\mathrm{div}\boldsymbol{A}\psi^*)\Big)dx \\ &\quad +\int_{\Omega\setminus\overline{\Omega_0}}\Big((\boldsymbol{A}_t - \nabla\mathrm{div}\boldsymbol{A})(\boldsymbol{A}_{tt} - \nabla\mathrm{div}\boldsymbol{A}_t) \\ &\quad +(\mathbf{curl}\boldsymbol{A} - \boldsymbol{H})(\mathbf{curl}\boldsymbol{A}_t - \boldsymbol{H}_t)\Big)dx - \int_{\partial\Omega_0}\eta\psi(\psi_t^* + i\mathrm{div}\boldsymbol{A}\psi^*)d\Gamma\Big] \\ &= \int_{\Omega_0}(-|\psi_t - i\mathrm{div}\boldsymbol{A}\psi|^2 - |\boldsymbol{A}_t - \nabla\mathrm{div}\boldsymbol{A}|^2)dx \\ &\quad +\int_{\Omega\setminus\Omega_0} -|\boldsymbol{A}_t - \nabla\mathrm{div}\boldsymbol{A}|^2 dx \\ &\quad +\int_{\Omega}(-\boldsymbol{H}\mathbf{curl}(\boldsymbol{A}_t - \nabla\mathrm{div}\boldsymbol{A}) + \boldsymbol{H}\boldsymbol{H}_t - \boldsymbol{H}_t(\mathbf{curl}\boldsymbol{A} - \boldsymbol{H}))dx \\ &\quad +\int_{\partial\Omega}((\mathbf{curl}\boldsymbol{A} - \boldsymbol{H}) \times \boldsymbol{n})\cdot(\boldsymbol{A}_t - \nabla\mathrm{div}\boldsymbol{A}) \\ &= \int_{\Omega_0}(-|\psi_t - i\mathrm{div}\boldsymbol{A}\psi|^2 - |\boldsymbol{A}_t - \nabla\mathrm{div}\boldsymbol{A}|^2)dx \\ &\quad +\int_{\Omega\setminus\Omega_0} -|\boldsymbol{A}_t - \nabla\mathrm{div}\boldsymbol{A}|^2 dx \\ &\quad -\int_{\Omega}(\mathbf{curl}\boldsymbol{H}\cdot(\boldsymbol{A}_t - \nabla\mathrm{div}\boldsymbol{A}) + \boldsymbol{H}\boldsymbol{H}_t - \boldsymbol{H}_t(\mathbf{curl}\boldsymbol{A} - \boldsymbol{H}))dx. \end{aligned}$$

Rearranging the terms, we obtain

$$\begin{aligned}\frac{dE}{dt} &\le \frac{1}{2}\Big[\int_{\Omega_0}(-(\psi_t - i\mathrm{div}\boldsymbol{A}\psi|^2 - |\boldsymbol{A}_t - \nabla\mathrm{div}\boldsymbol{A}|^2)dx \\ &\quad - \int_{\Omega\setminus\Omega_0}|\boldsymbol{A}_t - \nabla\mathrm{div}\boldsymbol{A}|^2 dx\Big] \\ &\quad + C\int_\Omega(|\mathbf{curl}\boldsymbol{H}|^2 + |\boldsymbol{H}\boldsymbol{H}_t|)dx + C|\boldsymbol{H}_t|_{L^2(\Omega)}|\mathbf{curl}\boldsymbol{A} - \boldsymbol{H}|_{L^2(\Omega)} \\ &\le C_1(t)\sqrt{E} + C_2(t)\end{aligned}$$

with

$$C_1(t) = C|\boldsymbol{H}_t|_{L^2(\Omega)},\ C_2(t) = C\int_\Omega(|\mathbf{curl}\boldsymbol{H}|^2 + |\boldsymbol{H}\boldsymbol{H}_t|)dx.$$

The Lemma is now proved.

**Lemma 5.2.3** *We have formally*

$$\begin{aligned}&\int_0^T\int_\Omega(|\boldsymbol{A}_t|^2 + |\nabla\mathrm{div}\boldsymbol{A}|^2)dx - \int_\Omega|\mathrm{div}\boldsymbol{A}_0|^2 dx \\ &\le \int_0^T\int_\Omega(|\boldsymbol{A}_t - \nabla\mathrm{div}\boldsymbol{A}|^2)dx \\ &\le \int_0^T\int_\Omega(|\boldsymbol{A}_t|^2 + |\nabla\mathrm{div}\boldsymbol{A}|^2)dx + \int_\Omega|\mathrm{div}\boldsymbol{A}|^2 dx(T).\end{aligned}\tag{2.14}$$

*Proof.* It is easy to see that

$$\begin{aligned}\int_\Omega|\boldsymbol{A}_t - \nabla\mathrm{div}\boldsymbol{A}|^2 &= \int_\Omega(|\boldsymbol{A}_t|^2 + |\nabla\mathrm{div}\boldsymbol{A}|^2 - 2\boldsymbol{A}_t\cdot\nabla\mathrm{div}\boldsymbol{A})dx \\ &= \int_\Omega(|\boldsymbol{A}_t|^2 + |\nabla\mathrm{div}\boldsymbol{A}|^2 + 2\mathrm{div}\boldsymbol{A}_t\mathrm{div}\boldsymbol{A})dx \\ &= \int_\Omega(|\boldsymbol{A}_t|^2 + |\nabla\mathrm{div}\boldsymbol{A}|^2 + \frac{\partial}{\partial t}|\mathrm{div}\boldsymbol{A}|^2)dx,\end{aligned}$$

Integrating both sides of the equality with respect to $t$, it is easy to see the conclusion of the Lemma.

**Lemma 5.2.4** *For the smooth, simply connected domain $\Omega$, we have $\forall\boldsymbol{B}\in\boldsymbol{H}^1_n(\Omega)$,*

$$||\boldsymbol{B}||^2_{\boldsymbol{H}^1(\Omega)} \le C(|\mathrm{div}\boldsymbol{B}|^2_{L^2(\Omega)} + |\mathbf{curl}\boldsymbol{B}|^2_{L^2(\Omega)}).$$

*Proof.* This is a standard result in the study of the Navier-Stokes equations, by noting that our domain $\Omega$ is a ball in $\mathbb{R}^3$ which is smooth and simply connected. With all the above preparation, the existence proof is now easy.

Let $\{\phi_m\}$ and $\{\boldsymbol{B}_m\}$ be smooth bases of $\mathcal{H}^1(\Omega_0)$ and $\boldsymbol{H}^1_{\boldsymbol{n}}(\Omega)$ respectively which satisfy the following orthogonal conditions

$$\int_{\Omega_0} \frac{1}{2}(\phi_m \phi_j^* + \phi_m^* \phi_j)dx = \begin{cases} 0 & \text{if } m \neq j \\ 1 & \text{if } m = j, \end{cases}$$

$$\int_{\Omega} \boldsymbol{B}_m \cdot \boldsymbol{B}_j dx = \begin{cases} 0 & \text{if } m \neq j \\ 1 & \text{if } m = j. \end{cases}$$

Let

$$\psi_m = \sum_{j=1}^{m} a_j(t)\phi_j, \; \boldsymbol{A}_m = \sum_{j=1}^{m} b_j(t)\boldsymbol{B}_j$$

and solve the following ordinary differential equation system where $j = 1, \ldots, m$:

$$\int_{\Omega_0} \Big(\psi_{mt}\phi_j^* - i\text{div}\boldsymbol{A}_m\psi_m\phi_j^* + (|\psi_m|^2\psi_m - \psi_m)\phi_j^*$$
$$+(\nabla - i\boldsymbol{A}_m)\psi_m(\nabla + i\boldsymbol{A}_m)\phi_j^*\Big)d\boldsymbol{x}dt - \int_{\partial\Omega_0} \eta\psi_m\phi_j^* d\Gamma dt = 0, \quad (2.15)$$

$$\int_{\Omega_0} \Big((\boldsymbol{A}_{mt} - \nabla\text{div}\boldsymbol{A}_m)\boldsymbol{B}_j + \textbf{curl}\boldsymbol{A}_m\textbf{curl}\boldsymbol{B}_j +$$
$$i(\psi_m^*\nabla\psi_m - \psi_m\nabla\psi_m^*)\boldsymbol{B}_j + |\psi_m|^2\boldsymbol{A}_m\boldsymbol{B}_j\Big)d\boldsymbol{x}dt$$
$$+\int_{\Omega\setminus\overline{\Omega_0}} \Big((\boldsymbol{A}_{mtt} - \nabla\text{div}\boldsymbol{A}_{mt})\boldsymbol{B}_j$$
$$+(\boldsymbol{A}_{mt} - \nabla\text{div}\boldsymbol{A}_m)\boldsymbol{B}_j + \textbf{curl}\boldsymbol{A}_m\textbf{curl}\boldsymbol{B}_j\Big)d\boldsymbol{x}dt$$
$$+\int_{\partial\Omega} (\boldsymbol{H}\wedge\boldsymbol{n})\boldsymbol{B}_j d\Gamma dt = 0 \quad (2.16)$$

subjecting to the projected (normal $L^2$ projections) initial conditions

$$\psi_m(\boldsymbol{x}, 0) = \psi_{0m}(x), \; \boldsymbol{A}_m(x, 0) = \boldsymbol{A}_{om}(x), \; \boldsymbol{A}_{mt}(x, 0) = \boldsymbol{A}_{1m}(x) \quad (2.17)$$

where the coefficients of $\boldsymbol{A}_{1m}(x)$ are calculated as if $\boldsymbol{A}_1(x)$ is equal to $-(-\nabla\text{div}\boldsymbol{A}_0 + \textbf{curl}^2\boldsymbol{A}_0 + \frac{\xi i}{2\lambda\gamma}(\psi_0^*\nabla\psi_0 - \psi_0\nabla\psi_0^*) + \frac{1}{\lambda^2}|\psi_0|^2\boldsymbol{A}_0)$ in $\Omega_0$. This will not produce the compatibility condition in the discretized problem but as $m \to \infty$, the compatibility condition will be recovered. This doesn't cause any difficulty in the discretized problem because the possible space discontinuity is not important in deriving the ODE system.

Using the a priori estimates techniques and the computation established in the proofs of Lemmas 2.1 and 2.2, noticing that all the formal computations now become rigorous because the discretized functions are smooth across $\partial\Omega_0$, it is similar as in Section 1 to obtain that there exists a constant $C$ independent of $m$ such that

$$\begin{aligned} &\int_{\Omega_0}(|(\nabla - i\boldsymbol{A}_m)\psi_m|^2 + |\psi_m|^4)dx(t) \leq C, \\ &\int_{\Omega}|\mathbf{curl}\boldsymbol{A}_m|^2 dx(t) \leq C, \\ &\int_0^T\int_\Omega |\boldsymbol{A}_{mt}|^2 + |\nabla \mathrm{div}\boldsymbol{A}_m|^2 dx \leq C. \end{aligned} \tag{2.18}$$

Note that

$$\int_\Omega \mathrm{div}\boldsymbol{A}_m dx = \int_{\partial\Omega} \boldsymbol{A}_m \cdot \boldsymbol{n} d\Gamma = 0,$$

the estimate that

$$\int_0^T \int_\Omega |\nabla \mathrm{div}\boldsymbol{A}_m|^2 dx \leq C$$

implies that

$$\int_0^T \int_\Omega |\mathrm{div}\boldsymbol{A}_m|^2 dx \leq C$$

where $C$ represents different constants independent of $m$. The estimates on $\psi_t$ and $\boldsymbol{A}_{tt}$ follows straightforwardly.

It is now enough for us to take the limit $m \to \infty$ in (2.16)–(2.18) to derive (2.4)–(2.6) and we leave the details. Hence the existence of solutions is proved.

**The proof of higher regularity**

The proof of regularity result carries the standard semilinear PDE feature. From the equation

$$\begin{cases} \psi_t - \Delta\psi = 2i\boldsymbol{A}\cdot\nabla\psi - |\boldsymbol{A}|^2\psi - (|\psi|^2 - 1)\psi, \\ \frac{\partial\psi}{\partial \boldsymbol{n}_0} = i\boldsymbol{A}\cdot\boldsymbol{n}_0\psi + \eta\psi, \\ \psi(x,0) = \psi_0, \\ |\psi|(x,t) \leq 1 \end{cases} \tag{2.19}$$

and the facts that $\boldsymbol{A} \in \boldsymbol{L}^\infty(0,T;\boldsymbol{H}^1(\Omega)) \subset \boldsymbol{L}^\infty(0,T;\boldsymbol{L}^6(\Omega))$, $\psi \in \mathcal{L}^\infty(0,T;\mathcal{H}^1(\Omega_0))$ and (2.12), we obtain that

$$\nabla(\boldsymbol{A}\psi) \in L^\infty(0,T;L^{3/2}(\Omega_0)).$$

Hence

$$\begin{cases} \psi_t - \Delta\psi \in \mathcal{L}^\infty(0,T;\mathcal{L}^{3/2}(\Omega_0)), \\ \frac{\partial\psi}{\partial\boldsymbol{n}_0} \in \mathcal{L}^\infty(0,T;\mathcal{W}^{1/3,3/2}(\partial\Omega_0)), \\ \psi(x,0) \in \mathcal{W}^{2,3/2}(\Omega_0). \end{cases} \tag{2.20}$$

This implies, by standard parabolic theory, that

$$\psi \in \mathcal{L}^\infty(0,T;\mathcal{W}^{2,3/2}(\Omega_0))$$

and

$$\nabla\psi \in L^\infty(0,T;W^{1,3/2}(\Omega_0)) \subset L^\infty(0,T;L^3(\Omega_0)).$$

Placing this information back into the equation (2.20), we obtain

$$\begin{cases} \psi_t - \Delta\psi \in \mathcal{L}^\infty(0,T;\mathcal{L}^2(\Omega_0)), \\ \frac{\partial\psi}{\partial \boldsymbol{n}_0} \in \mathcal{L}^\infty(0,T;\mathcal{W}^{1/2,2}(\partial\Omega_0)), \\ \psi(x,0) \in \mathcal{H}^2(\Omega_0) \end{cases} \tag{2.21}$$

which yields

$$\psi \in \mathcal{L}^\infty(0,T;\mathcal{H}^2(\Omega_0)).$$

**Remark 5.2.1** *If $\boldsymbol{H}$ is independent of time $t$, following the proofs of the next section carefully, we can easily conclude that*

$$\psi \in \mathcal{L}^\infty(0,\infty;\mathcal{H}^2(\Omega_0)).$$

*We also notice that there is no $H^2$ regularity results for $\boldsymbol{A}$ at this stage.*

**The proof of uniqueness**

Let $(\psi_1, \boldsymbol{A}_1)$ and $(\psi_2, \boldsymbol{A}_2)$ be two solutions of the problem and $(\psi, \boldsymbol{A}) = (\psi_1, \boldsymbol{A}_1) - (\psi_2, \boldsymbol{A}_2)$. Substituting into the equations respectively and subtracting the resulting equations, we obtain, for $\psi$:

$$\begin{aligned} &\psi_t - \Delta\psi + 2(\boldsymbol{A}_1\nabla\psi_1 - \boldsymbol{A}_2\nabla\psi_2) + (|\boldsymbol{A}_1|^2\psi_1 - |\boldsymbol{A}_2|^2\psi_2) \\ &+\Big((|\psi_1|^2 - 1)\psi_1 - (|\psi_2|^2 - 1)\psi_2\Big) = 0. \end{aligned} \tag{2.22}$$

Multiplying both sides by $\psi^*$, taking the real part, integrating over $\Omega$ and rearranging the terms, we obtain

$$\begin{aligned} &\frac{1}{2}\frac{\partial}{\partial t}\int_\Omega |\psi|^2 dx + \int_\Omega |\nabla\psi|^2 dx \\ \le\ & \int_\Omega \Big(|\psi|^2 - 2\mathrm{Re}(\nabla\psi_1\boldsymbol{A}_1 - \nabla\psi_2\boldsymbol{A}_2)\psi^* - (|\boldsymbol{A}_1|^2\psi_1 - |\boldsymbol{A}_2|^2\psi_2)\psi^* \\ \le\ & \int_\Omega \Big(|\psi|^2 - 2\mathrm{Re}(\boldsymbol{A}\nabla\psi_1 + \boldsymbol{A}_2\nabla\psi)\psi^* - (|\boldsymbol{A}_1|^2\psi + \psi_2(\boldsymbol{A}_1 + \boldsymbol{A}_2)\boldsymbol{A})\psi^*\Big)dx \\ \le\ & |\psi|_{L^2}^2 + |\boldsymbol{A}|_{L^4}|\psi|_{L^4}|\nabla\psi_1|_{L^2} + |\boldsymbol{A}_2|_{L^4}|\psi|_{L^4}|\nabla\psi|_{L^2} \\ & - \int_\Omega |\boldsymbol{A}_1|^2|\psi|^2 dx + |\psi_2|_{L^4}|\boldsymbol{A}_1 + \boldsymbol{A}_2|_{L^4}|\boldsymbol{A}|_{L^4}|\psi|_{L^4} \\ \le\ & |\psi|_{L^2}^2 + |\boldsymbol{A}|_{L^4}|\psi|_{L^4}|\nabla\psi_1|_{L^2} + |\boldsymbol{A}_2|_{L^4}|\psi|_{L^4}|\nabla\psi|_{L^2} \\ & + |\psi_2|_{L^4}|\boldsymbol{A}_1 + \boldsymbol{A}_2|_{L^4}|\boldsymbol{A}|_{L^4}|\psi|_{L^4}. \end{aligned}$$

We know, from the Lyapunov functional, that for some constant $c$ and for $j = 1, 2$,

$$\begin{aligned} &|\psi_j|_{L^4} \le c, \\ &|\nabla\psi_j|_{L^2} \le c, \\ &|\boldsymbol{A}_j|_{L^4} \le c. \end{aligned}$$

We derive then

$$\frac{1}{2}\frac{\partial}{\partial t}\int_\Omega |\psi|^2 dx + \int_\Omega |\nabla\psi|^2 dx \leq |\psi|_{L^2}^2 + C|\boldsymbol{A}|_{L^4}|\psi|_{L^4} + C|\psi|_{L^4}|\nabla\psi|_{L^2}.$$

Here $C$ is some positive constant depending on the initial data and the domain $\Omega$.

Using the standard inequality

$$|f|_{L^4}^2 \leq C|f|_{H^1}|f|_{L^2} \leq \delta|\nabla f|_{L^2}^2 + c_\delta|f|_{L^2}^2$$

for any (small) positive number $\delta$ and $C_\delta \sim 1/\delta$, we have

$$\frac{1}{2}\frac{\partial}{\partial t}\int_\Omega |\psi|^2 dx + \int_\Omega |\nabla\psi|^2 dx$$
$$\leq |\psi|_{L^2}^2 + C(\delta|\nabla \boldsymbol{A}|_{L^2}^2 + C_\delta|\boldsymbol{A}|_{L^2}^2) + C(\delta|\nabla\psi|_{L^2}^2 + C_\delta|\psi|_{L^2}^2) + \frac{1}{4}|\nabla\psi|_{L^2}^2.$$

Here $C$ may represent different constants but it is independent of $\delta$.

Then we look at the equations in $\boldsymbol{A}$. In $\Omega_0$, we have

$$\boldsymbol{A}_t - \nabla\mathrm{div}\boldsymbol{A} + \mathbf{curl}^2\boldsymbol{A} + \frac{i}{2}(\psi_1^*\nabla\psi_1 - \psi_1\nabla\psi_1^* - \psi_2^2\nabla\psi_2$$
$$+\psi_2\nabla\psi_2^*) + |\psi_1|^2\boldsymbol{A}_1 - |\psi_2|^2\boldsymbol{A}_2 = 0.$$

Rearranging the terms, we have

$$\boldsymbol{A}_t - \nabla\mathrm{div}\boldsymbol{A} + \mathbf{curl}^2\boldsymbol{A} - \mathrm{Im}(\psi^*\nabla\psi_1 + \psi_2^*\nabla\psi)$$
$$+|\psi_1|^2\boldsymbol{A} + (|\psi_1|^2 - |\psi_2|^2)\boldsymbol{A}_2 = 0. \tag{2.23}$$

In $\Omega \setminus \overline{\Omega_0}$, we have

$$\boldsymbol{A}_{tt} - \nabla\mathrm{div}\boldsymbol{A}_t + \boldsymbol{A}_t - \nabla\mathrm{div}\boldsymbol{A} + \mathbf{curl}^2\boldsymbol{A} = 0. \tag{2.24}$$

Multiplying both sides of the equations (2.27) and (2.28) by $\boldsymbol{A}_t - \nabla\mathrm{div}\boldsymbol{A}$, integrating over $\Omega_0$ and $\Omega \setminus \overline{\Omega_0}$, we obtain, with the same notation as in (2.24)

$$\frac{1}{2}\frac{\partial}{\partial t}\int_{\Omega\setminus\overline{\Omega_0}} |\boldsymbol{A}_t - \nabla\mathrm{div}\boldsymbol{A}|^2 dx + \int_\Omega |\boldsymbol{A}_t - \nabla\mathrm{div}\boldsymbol{A}|^2 dx + \frac{1}{2}\frac{\partial}{\partial t}\int_\Omega |\mathbf{curl}\boldsymbol{A}|^2 dx$$
$$= \frac{1}{2}\frac{\partial}{\partial t}\int_{\Omega\setminus\overline{\Omega_0}} |\boldsymbol{A}_t - \nabla\mathrm{div}\boldsymbol{A}|^2 dx + \int_\Omega (|\boldsymbol{A}_t|^2 + |\nabla\mathrm{div}\boldsymbol{A}|^2) dx \tag{2.25}$$
$$+\frac{1}{2}\frac{\partial}{\partial t}\int_\Omega (|\mathrm{div}\boldsymbol{A}|^2 + |\mathbf{curl}\boldsymbol{A}|^2) dx$$
$$\leq \frac{1}{2}|\boldsymbol{A}_t - \nabla\mathrm{div}\boldsymbol{A}|^2 + \frac{1}{2}\Big(|\psi\nabla\psi|_{L^2}^2 + |\psi_1^2\boldsymbol{A}|_{L^2}^2 + |(|\psi_1|^2 - |\psi_2|^2)\boldsymbol{A}_2|_{L^2}^2\Big)$$
$$\leq \frac{1}{2}|\boldsymbol{A}_t - \nabla\mathrm{div}\boldsymbol{A}|_{L^2}^2 + \frac{1}{2}\Big(|\nabla\psi|_{L^2}^2 + |\boldsymbol{A}|_{L^2}^2 + 2|\psi_1 - \psi_2|_{L^4}^2|\boldsymbol{A}_2|_{L^4}^2\Big)$$
$$\leq \frac{1}{2}|\boldsymbol{A}_t - \nabla\mathrm{div}\boldsymbol{A}|_{L^2}^2 + \frac{1}{2}\Big(|\nabla\psi|_{L^2}^2 + |\boldsymbol{A}|_{L^2}^2 + C(C_\delta|\psi|_{L^2}^2 + \delta|\nabla\psi|_{L^2}^2\Big). \tag{2.26}$$

Adding up (2.24) and (2.29), adjusting the value of $\delta$, using the fact that $|\boldsymbol{A}|^2_{H^1(\Omega)} \le C(|\mathrm{div}\boldsymbol{A}|^2_{L^2(\Omega)} + |\mathbf{curl}\boldsymbol{A}|^2_{L^2(\Omega)})$ in our function spaces and using Gronwall's inequality, the uniqueness follows.

## 5.3 The existence of an $\omega$-limit set as $t \to \infty$

In this section, we consider the asymptotic behaviour of the solutions of the G-L system in $\mathbb{R}^2$. The 3-d model with Maxwell system in the outside have the same convergence property subjecting to a higher regularity result. We do not discuss the 3-d problem here. We follow mainly the work of [LTa 95].

The domain $\Omega$ is again assumed to be bounded, smooth and simply connected. We will use the following evolutionary G-L system:

$$\psi_t + i\psi\phi + \kappa^2\psi(|\psi|^2 - 1) + (i\nabla + \boldsymbol{A})^2\psi = 0, \tag{3.1}$$

$$\eta(\boldsymbol{A}_t + \nabla\phi) + \mathbf{curl}^2\boldsymbol{A} - \mathbf{curl}H + \frac{i}{2}(\psi^*\nabla\psi - \psi\nabla\psi^*) + |\psi|^2\boldsymbol{A} = 0. \tag{3.2}$$

For the boundary and initial conditions, we have

$$\begin{cases} \boldsymbol{n}\cdot(\nabla - i\boldsymbol{A})\psi = 0, & \text{on } \partial\Omega, \\ \mathbf{curl}\boldsymbol{A} = H, & \text{on } \partial\Omega, \end{cases} \tag{3.3}$$

$$\psi|_{t=0} = \psi_0, \quad \boldsymbol{A}|_{t=0} = \boldsymbol{A}_0. \tag{3.4}$$

The corresponding stationary superconductivity problem is

$$(i\nabla + \boldsymbol{A})^2\psi + \kappa^2\psi(|\psi|^2 - 1) = 0, \tag{3.5}$$

$$\mathbf{curl}^2\boldsymbol{A} - \mathbf{curl}H + \frac{i}{2}(\psi^*\nabla\psi - \psi\nabla\psi^*) + |\psi|^2\boldsymbol{A} = 0 \tag{3.6}$$

with the boundary conditions in (3.3).

The assumptions on the data are as follows:
(A1) $\psi_0 \in \mathcal{H}^1(\Omega)$, $|\psi_0| \le 1$;
(A2) $\boldsymbol{A}_0 \in \boldsymbol{H}^1_n(\mathrm{div}, \Omega)$;
(A3) $H \in \boldsymbol{H}^1(\mathbf{curl}; \Omega)$ and is independent of $t$.

In order to apply the existing theories of elliptic operators, we need to use the following

**Lemma 5.3.1** *If $H \in \boldsymbol{H}^1(\Omega)$, there exists a potential $\boldsymbol{A}_a$ of $H$ such that $\boldsymbol{A}_a \in \boldsymbol{H}^2(\Omega)$ and*

$$\begin{cases} \mathbf{curl}\boldsymbol{A}_a = H, & \mathrm{div}\boldsymbol{A}_a = 0, \quad \text{in } \Omega, \\ \boldsymbol{A}_a\cdot\boldsymbol{n} = 0, & \text{on } \partial\Omega. \end{cases} \tag{3.7}$$

*Furthermore, if $H$ is independent of time $t$, we can also choose $\boldsymbol{A}_a$ to be independent of $t$.*

*Proof.* Simply solve

$$\begin{cases} -\Delta\xi = H, \\ \xi|_{\partial\Omega} = 0, \end{cases}$$

let $\boldsymbol{A}_a = (\partial_2\xi, -\partial_1\xi)^T$ and the first part of lemma is proved.

The proof of the last statement is trivial and we omit the details.

In order to proceed with proofs, we first transform the problem. Let $\boldsymbol{A}_a$ be defined in $\Omega$ as introduced in Lemma 5.3.1, $\widetilde{\boldsymbol{A}} = \boldsymbol{A} - \boldsymbol{A}_a$, it is obvious that $\widetilde{\boldsymbol{A}}$ satisfies that

$$\psi_t + i\phi\psi + (i\nabla + \boldsymbol{A} + \boldsymbol{A}_a)^2\psi + \kappa^2(|\psi|^2 - 1)\psi = 0, \tag{3.8}$$

$$\widetilde{\boldsymbol{A}}_t + \mathbf{curl}^2\widetilde{\boldsymbol{A}} + \nabla\phi = -\frac{i}{2}(\psi^*\nabla\psi - \psi\nabla\psi^*) - |\psi|^2\widetilde{\boldsymbol{A}} - |\psi|^2\boldsymbol{A}_a, \tag{3.9}$$

$$(i\nabla + \boldsymbol{A} + \boldsymbol{A}_a)\psi \cdot \boldsymbol{n} = 0, \quad \mathbf{curl}\widetilde{\boldsymbol{A}} \times \boldsymbol{n} = 0, \quad \widetilde{\boldsymbol{A}}(x, 0) = \boldsymbol{A}_0 - \boldsymbol{A}_a. \tag{3.10}$$

So, Problem (3.1)–(3.4) becomes the transformed problem consisting of (3.8)–(3.10). Since $\boldsymbol{A}$ and $\widetilde{\boldsymbol{A}}$ only differ by a (smooth) function $\boldsymbol{A}_a$ which is independent of time $t$, we take the transformed problem and study the properties of $\widetilde{\boldsymbol{A}}$ as $t \to \infty$ in this section.

In the following, unless otherwise stated, we fix the choice of gauge so that

$$\mathrm{div}\boldsymbol{A} = 0 \text{ in } \Omega, \quad \boldsymbol{A} \cdot \boldsymbol{n} = 0 \text{ on } \partial\Omega \quad \text{and} \quad \int_\Omega \phi\, dx = g(t) \in L_\infty(0, \infty), \tag{3.11}$$

(note that these conditions are equally satisfied by $\widetilde{\boldsymbol{A}}$).

We now reformulate the transformed problem into an equivalent abstract operator problem,

$$\begin{aligned} &\frac{du}{dt} + Fu = f(u) \qquad \text{for } t > 0, \\ &u(0) = u_0 \end{aligned} \tag{3.12}$$

where

$$\begin{aligned} u &= (\psi, \widetilde{\boldsymbol{A}}), \\ f(u) &= (f_1(u), f_2(u)) \\ &= \Big( - i\psi\phi - \kappa^2\psi(|\psi|^2 - 1) - 2i(\widetilde{\boldsymbol{A}} + \boldsymbol{A}_a) \cdot \nabla\psi - |\widetilde{\boldsymbol{A}} + \boldsymbol{A}_a|^2\psi, \\ &\qquad -\mathbf{P}\left(|\psi^2|(\widetilde{\boldsymbol{A}} + \boldsymbol{A}_a) + \frac{i}{2}(\psi^*\nabla\psi - \psi\nabla\psi^*)\right)\Big), \\ u_0 &= (\psi_0, \widetilde{\boldsymbol{A}}_0) = (\psi_0, \boldsymbol{A}_0 - \boldsymbol{A}_a). \end{aligned}$$

In order to maintain compatibility, we assume in addition that the initial condition $\boldsymbol{A}_0$ for $\boldsymbol{A}$ satisfies

$$\mathrm{div}\boldsymbol{A}_0 = 0, \quad \boldsymbol{A}_0 \cdot \boldsymbol{n}|_{\partial\Omega} = 0 \tag{3.13}$$

(notice that these conditions are also valid for $\widetilde{\boldsymbol{A}}_0$).

So $F$ is a mapping from $[\mathcal{W}^{2,\frac{3}{2}}(\Omega) \cap \mathcal{H}^1(\Omega)] \times \boldsymbol{W} \to \mathcal{L}^{\frac{3}{2}}(\Omega) \times \boldsymbol{L}^2(\Omega)$ defined by

$$Fu = (F_1\psi, F_2\boldsymbol{A}) = (-\triangle\psi, \mathbf{Pcurl}^2\widetilde{\boldsymbol{A}}).$$

$\mathbf{P}$ is the orthogonal projection from $\boldsymbol{L}^2(\Omega)$ to $\boldsymbol{L}_n(\text{div}, \Omega)$. Notice that $\boldsymbol{L}_n(\text{div}, \Omega)$ is the subspace of $\boldsymbol{L}^2(\Omega)$ containing the set of all smooth solenoidal vectors with compact support in $\Omega$ as a dense subset.

**Lemma 5.3.2** *$F$ is a sectorial operator and has compact resolvent.*

*Proof.* We can verify that the operator $\mathbf{Pcurl}^2$ is a densely defined self-adjoint operator in the Hilbert space $\boldsymbol{L}_n(\text{div}, \Omega)$. First, it is well defined on the subset of functions in $\mathbf{C}_0^\infty(\Omega)$ with free divergence. By [Tem 79], this set is dense in $\boldsymbol{L}_n(\text{div}, \Omega)$. To prove that $\mathbf{Pcurl}^2$ is self-adjoint, we notice that by using integration by parts, for any $\boldsymbol{B} \in \boldsymbol{W}$, we have

$$(\mathbf{Pcurl}^2\widetilde{\boldsymbol{A}}, \boldsymbol{B}) = (\mathbf{curl}^2\widetilde{\boldsymbol{A}}, \boldsymbol{B}) = (\widetilde{\boldsymbol{A}}, \mathbf{curl}^2\boldsymbol{B}) = (\widetilde{\boldsymbol{A}}, \mathbf{Pcurl}^2\boldsymbol{B}).$$

So $\mathbf{Pcurl}^2$ is a symmetric operator. By the regularity result of [Gri 85] (Theorem 4.2), if $\boldsymbol{B}$ and $\phi$ are solutions to

$$\begin{cases} \mathbf{curl}^2\boldsymbol{B} + \nabla\phi = f(x) \in \boldsymbol{L}^2(\Omega), \\ \text{div}\boldsymbol{B} = 0, \\ \boldsymbol{B} \cdot \boldsymbol{n}|_{\partial\Omega} = 0, \\ \mathbf{curl}\boldsymbol{B}|_{\partial\Omega} = 0, \end{cases}$$

then $\boldsymbol{B} \in \boldsymbol{H}^2(\Omega)$, $\phi \in \boldsymbol{H}^1(\Omega)$. Since $\mathbf{curl}\boldsymbol{B}|_{\partial\Omega} = 0$, it is easy to verify that $\mathbf{curl}^2 B$ is orthogonal to $\nabla\zeta$ in $\boldsymbol{L}^2(\Omega)$ for any $\zeta \in H^1(\Omega)$. Hence, $\mathbf{Pcurl}^2\boldsymbol{B} \in \boldsymbol{L}^2$ implies that $\boldsymbol{B} \in \boldsymbol{H}^2(\Omega)$ and the definition domain is maximum. Consequently, $(\mathbf{Pcurl}^2)^* = \mathbf{Pcurl}^2$. So it is a self-adjoint operator. $\mathbf{Pcurl}^2$ is therefore, a sectorial operator according to [Hen 81], Chap. 1, §1.3. In addition, $-\triangle$ with homogeneous Neumann boundary condition in $\mathbf{L}^p(\Omega)$ is well known to be a sectorial operator for any $p > 1$. Following the discussions in §1.3 of [Hen 81], $F$ is a sectorial operator.

Similarly, it is obvious that $-\triangle$ with homogeneous Neumann boundary condition has compact resolvent, and the proof of the fact that $(\mathbf{Pcurl}^2 + I)^{-1}$ is a compact operator can be found in Theorem 4.2 in [Hen 81]. It now follows easily that $F$, as the direct product of the two operators, has compact resolvent.

Now let us define the operator $S(t) : \mathbb{B} \to \mathbb{B}$, where $\mathbb{B} = \mathcal{H}_n^1(\Omega) \times \boldsymbol{H}_n(\text{div}, \Omega)$, by (here $\mathcal{H}_n^1(\Omega)$ is the set of all $\mathcal{H}^1$ function whose normal derivative along $\partial\Omega$ is 0)

$$S(t)(\psi_0, \boldsymbol{A}_0)(x) = (S_1(t)(\psi_0), S_2(t)(\boldsymbol{A}_0)) = (\psi(x,t), \boldsymbol{A}(x,t)) \text{ (or } (\psi(x,t), \widetilde{\boldsymbol{A}}(x,t))).$$

The existence and uniqueness results for Problem (3.1)–(3.4) after properly fixing the gauge show that the family of maps $\{S(t) : \ t \geq 0\}$ is a dynamical system.

For the dynamical system $S(t)$, we know already that the standard G-L energy is a Lyapunov functional:

$$E(t) = \int_{\Omega} (|(i\nabla + \boldsymbol{A})\psi|^2 + \kappa^2(|\psi|^2 - 1)^2 + |\mathbf{curl}\boldsymbol{A} - H|^2) dx \tag{3.14}$$

Using this fact, we establish some useful estimates:

**Proposition 5.3.1** *Under Assumptions (A1)–(A3), the solutions $(\psi, \boldsymbol{A}, \phi)$ of Problem (3.1)–(3.4) satisfy the following estimate:*

$$\|\psi_t + i\phi\psi\|_{\mathcal{L}^2(Q_\infty)} + \|\boldsymbol{A}_t + \nabla\phi\|_{\boldsymbol{L}^2(Q_\infty)} \leq C.$$

*Proof.* Using the computation in (1.47), we obtain

$$\int_0^T \|\psi_t + i\phi\psi\|^2_{\mathcal{L}^2(\Omega)} + \|\boldsymbol{A}_t + \nabla\phi\|^2_{\boldsymbol{L}^2(\Omega)} dt \leq E(0) - E(T) \leq E(0).$$

Since this is true for any $T$, our result follows.

Concerning the $H^1$ norm of the solutions, we have

**Proposition 5.3.2** *Let $\mathcal{G}$ be defined as*

$$\begin{aligned} \mathcal{G} \;=\; & \{(\zeta, \boldsymbol{B}, \varphi) \textit{ is the solution of Problem (3.1)–(3.4):} \\ & \mathrm{div}\boldsymbol{B} \in \boldsymbol{L}_2(\Omega) \textit{ and is independent of } t, \\ & \boldsymbol{B}\cdot\boldsymbol{n} \in \boldsymbol{H}^{1/2}(\partial\Omega) \textit{ and is independent of } t\}. \end{aligned}$$

*If assumptions (A1)–(A3) hold, the solutions of Problem (3.1)–(3.4) in the set $\mathcal{G}$ satisfy the estimate*

$$\|\zeta\|_{L^\infty(0,T;\mathcal{H}^1(\Omega))} + \|\boldsymbol{B}\|_{L^\infty(0,T;\boldsymbol{H}^1(\Omega))} \leq C \tag{3.15}$$

*where $C$ depends only on the initial data.*

*Proof.* Since $\boldsymbol{B} \in \mathcal{G}$ and $\|\mathrm{div}\boldsymbol{B}\|_{\boldsymbol{L}^2(\Omega)} < C$, by using Lyapunov function and gauge transformation, we can deduce easily the estimate $\|\mathbf{curl}\boldsymbol{B} - H\|_{\boldsymbol{L}^2(\Omega)}(t) \leq E(0) \leq C$, So, $\|\boldsymbol{B}\|_{\boldsymbol{H}^1(\Omega)}(t) \leq C$ here $C$ is a positive constant independent of $t$.

Again from the Lyapunov function, we have

$$\begin{aligned} \|\nabla\zeta\|^2_{\mathcal{L}^2(\Omega)}(t) &\leq \int_{\Omega} |(i\nabla\zeta + \boldsymbol{B}\zeta)|^2 \, dx + \int_{\Omega} |\boldsymbol{B}\zeta|^2 \, dx \\ &\leq E(0) + \|\boldsymbol{B}\|^2_{\boldsymbol{L}^2(\Omega)}(t) \leq CE(0) + C \leq C \end{aligned}$$

where $C$ represents different constants independent of $t$.

Adding up our estimates, we obtain (3.15).

This proposition implies that the second component of $f(u)$ in abstract equation (3.12) maps $I\!B$ into a bounded set of $\boldsymbol{L}^2(\Omega)$ uniformly with respect to $t$. By Sobolev embedding, we have

**Corollary 5.3.1** *If Assumptions (A1)–(A3) hold, the solutions of Problem (3.1)–(3.4) in the set $\mathcal{G}$ satisfy the estimate*

$$\|\boldsymbol{B}\|_{L^\infty(0,\infty;\boldsymbol{L}^6(\Omega))} \le C. \tag{3.16}$$

**Proposition 5.3.3** *In the abstract equation (3.12), we have*

$$\|f(u)\|_{L^\infty(0,\infty;\mathcal{L}^{\frac{3}{2}}(\Omega))\times L^\infty(0,\infty;\boldsymbol{L}^2(\Omega))} \le C$$

*where $C$ is a constant independent of $t$.*

*Proof.* We only need to prove the boundedness of each of the components of $f(u)$. It is obvious that with the estimate in (3.15), the second component satisfies

$$\|f_2(u)\|_{\boldsymbol{L}^2(\Omega)}(t) \le C,$$

where $C$ is a constant independent of $t$.

For the first component, notice that from (3.2), we have, using the known regularity of solutions and boundary conditions,

$$\int_\Omega \boldsymbol{A}_t \cdot \nabla\phi \, dx = -\int_\Omega (\mathrm{div}\boldsymbol{A})_t \phi \, dx + \int_{\partial\Omega} (\boldsymbol{A}\cdot\boldsymbol{n})_t \phi \, d\Gamma = 0,$$

$$\begin{aligned} &\int_\Omega (\mathbf{curl}^2\boldsymbol{A} - \mathbf{curl}H)\cdot\nabla\phi \, dx \\ = &\int_\Omega (\mathbf{curl}\boldsymbol{A} - H)\cdot\mathbf{curl}\nabla\phi \, dx + \int_{\partial\Omega} (\mathbf{curl}\boldsymbol{A} - H)\partial_{\boldsymbol{T}}\phi \, d\Gamma = 0. \end{aligned}$$

Hence,

$$\|\nabla\phi\|^2_{\boldsymbol{L}^2(\Omega)} + \|\boldsymbol{A}_t + \mathbf{curl}^2\boldsymbol{A} - \mathbf{curl}\boldsymbol{h}_a\|^2_{\boldsymbol{L}^2(\Omega)} = \left\|\frac{i}{2}(\psi^*\nabla\psi - \psi\nabla\psi^*) + |\psi|^2\boldsymbol{A}\right\|^2_{\boldsymbol{L}^2(\Omega)}.$$

So, again by (3.15), we have

$$\|\nabla\phi\|_{\boldsymbol{L}^2(\Omega)}(t) \le C, \qquad \text{for all } t > 0.$$

Here $C$ is a constant independent of $t$. With the assumption that $\int_\Omega \phi \, dx = g(t) \in L^\infty(0,\infty)$, we can easily show that

$$\|\phi\|_{L^\infty(0,\infty;H^1(\Omega))} \le C \tag{3.17}$$

(See [Lad 69], Chap. 2). By (3.15) and Corollary 5.3.1, notice also that for any $t$,

$$\|\boldsymbol{A}\cdot\nabla\psi\|_{L^{\frac{3}{2}}(\Omega)}(t) \le \|\boldsymbol{A}\|_{L^6(\Omega)}(t)\|\nabla\psi\|_{L^2(\Omega)}(t) \le C,$$

then we have

$$\|f_1(u)\|_{\mathcal{L}^{\frac{3}{2}}(\Omega)}(t) \le C,$$

where $C$ represents different constants independent of $t$.

If we take $X^0 = \mathcal{L}^{\frac{3}{2}}(\Omega) \times \boldsymbol{L}^2(\Omega)$, $X^1 = \mathcal{D}(F)$, and let $F$ be the operator defined in Section 3, it is easy to see that

$$u_0 \in \mathcal{H}^1(\Omega) \times \boldsymbol{H}^1(\Omega) \subset \mathcal{W}^{1,\frac{3}{2}}(\Omega) \times \boldsymbol{H}^1(\Omega) \subset X^{\frac{1}{2}}(\Omega).$$

(See [Hen 81], p. 77.) So, by Theorem 3.3.6 in [Hen 81], we have, for $\frac{1}{2} < \beta < 1$,

$$\|u\|_{X^\beta}(t) < C(\|u_0\|_{X^{\frac{1}{2}}}(t) + \|f(u)\|_{X^0}(t)),$$

where $C$ is independent of $t$. Especially when $\alpha > \frac{3}{4} > \frac{1}{2}$, we have $X^\alpha \subset \mathcal{H}^1(\Omega) \times \boldsymbol{H}^1(\Omega)$ again from the result in [Hen 81] p77. Consequently, we choose $\frac{3}{4} < \beta < 1$, and use the results of Theorem 3.3.6 in [Hen 81] (note that $\boldsymbol{A}_a \in \boldsymbol{H}^2(\Omega)$ and is independent of $t$) to obtain

**Lemma 5.3.3** *The solution of the dynamical system $\{S(t)(u_0)\}$ is compact in $\mathcal{H}^1(\Omega) \times \boldsymbol{H}^1(\Omega)$ uniformly with respect to $t$.*

With Lemma 5.3.3, following Theorem 4.3.4 of [Hen 81], we have

**Theorem 5.3.1** *Let $S(t)(\psi_0, \boldsymbol{A}_0) = \{\psi, \boldsymbol{A}\}$. Under the assumptions (A1)–(A3), (3.10) and (3.12), we have*

$$\lim_{t\to\infty} \operatorname{dist}\{S(t)(\psi_0, \boldsymbol{A}_0), \boldsymbol{E}\} = 0,$$

*where*

$$\boldsymbol{E} \;=\; \{(\psi_\infty, \boldsymbol{A}_\infty) \in I\!B,\ (\psi_\infty, \boldsymbol{A}_\infty) \quad \text{is a solution of Problem (3.4)–(3.6)}\}.$$

*That is, the solution of (3.1)–(3.4) $\omega$-converges to the set of steady state solutions as $t \to \infty$.*

*Proof.* We only need to prove that $\boldsymbol{E}$ is the set of solutions of Problem (3.4)–(3.6). In fact, by Theorem 4.3.4 in [Hen 81], we have, as $t$ tends to infinity,

$$-E'(t) = \|\psi_t + i\kappa\psi\phi\|_{\mathcal{L}^2(\Omega)} + \|\boldsymbol{A}_t + \nabla\phi\|_{\boldsymbol{L}^2(\Omega)} \to 0.$$

The result then follows easily.

**Remark 5.3.1** *Careful readers may have noticed that we have designed the proof in a way that if we have the appropriate regularity result for the 3-d problem with the Maxwell system in the outside domain, the proof can be carried over without alteration. The same is also true for the discussions in Sections 5–7 in the following.*

## 5.4 An abstract theorem on global attractors

In this section, we prove an existence theorem of global attractors for some abstract semigroup, whose absorbing sets (or stability) are difficult to obtain directly from the *a priori* estimates of the solutions. In this section and in the following Sections 5 and 6, we follow mainly the work of [TWa 96].

We consider, in this section, a general reflexive Banach space $X$. Let $\{S(t)\}$ be a one parameter semigroup of non-linear operators $S(t) : X \to X$, enjoying the following semigroup properties:

$$\begin{cases} S(0) = I = \text{ identity: } X \to X, \\ S(t+s) = S(t)S(s) : \ X \to X, \ \forall s, t \geq 0. \end{cases} \tag{4.1}$$

First, we recall some definitions about the absorbing set and the global attractor (cf. [Tem 88]):

**Definition 5.4.1** *(absorbing set) A subset $B_0$ of $X$ is an absorbing set in $X$ for the semigroup $\{S(t)\}_{t\geq 0}$, if for every bounded set $B$ of $X$, there exists $t = t(B)$ such that*

$$S(t)B \subset B_0, \ \forall t \geq t(B). \tag{4.2}$$

**Definition 5.4.2** *(global attractor)*

*1. A set $\mathcal{A} \subset X$ is a functional invariant set for the semigroup $S(t)$ if*

$$S(t)\mathcal{A} = \mathcal{A}, \ \forall t \geq 0. \tag{4.3}$$

*2. The global attractor $\mathcal{A}$, if it exists, is the unique compact invariant set which attracts every bounded set in $X$, i.e., for any bounded set $B \subset X$,*

$$\lim_{t\to\infty} dist(S(t)B, \mathcal{A}) = 0 \tag{4.4}$$

*where for two sets $B_1$ and $B_2$ of $X$,*

$$dist(B_1, B_2) = \sup_{x\in B_1} \inf_{y\in B_2} |x - y|.$$

A general existence result (see among many others, [Tem 88]) states that the existence of the global attractor is guaranteed by the following conditions:

1. *There exists a bounded absorbing set $B_0$.*

2. *The operators $S(t)$ are uniformly compact. i.e., $\forall B$ bounded, there exists $t_1 = t_1(B) \geq 0$ such that*

$$\bigcup_{t\geq t_1} S(t)B$$

*is relatively compact in $H$.*

Normally, the existence of a bounded absorbing set is the consequence of the following *a priori* estimates of solutions:

$$\|S(t)u_0\| \leq c, \ \forall u_0 \in B \text{ and } t \geq t_1(B) \tag{4.5}$$

with $c$ independent of $t_1(B)$ and $u_0$; As we indicated before, we are unable to obtain the estimate (4.5) via direct estimates.

The main result in this subsection is the following theorem which provides an explicit way of obtaining an absorbing set and hence the global attractor for the G-L system (or any other system with a Lyapunov functional). Before we give the theorem, we need the following definition:

**Definition 5.4.3** *A Lyapunov functional of the semigroup $S(\cdot)$ on a set $B \subset X$ is a continuous function $E$: $X \to \mathbb{R}$ such that*

*1) for any $u_0 \in B$, the function $t \to E(S(t)u_0)$ is decreasing;*

*2) if $E(S(t+\tau)u_0) = E(S(t)u_0)$ for some $\tau > 0$, then $S(t)u_0$ is a fixed point of the semigroup.*

**Theorem 5.4.1** *Let $S(t)$: $X \to X$ be a continuous semigroup satisfying the following properties:*

*1) the semigroup $S(t)$ admits a continuous Lyapunov functional $E(u)$ which satisfies $E(u) \to \infty$ if and only if $|u|_X \to \infty$,*

*2) the set $\mathcal{S}$ of stationary solutions is bounded in $X$,*

*3) For any bounded subset $B$ of $X$, there exists a $t_B > 0$ such that $S(t_B)B$ is a pre-compact subset of $X$.*

*Then,*

*1) for each $\alpha > 0$, the set*

$$B_\alpha = \{u \in V|\ E(u) < \sup_{v \in \mathcal{S}} E(v) + \alpha\} \tag{4.6}$$

*is an absorbing set in $V$ of the semigroup $S(t)$;*

*2) the semigroup $S(t)$ admits a global attractor which consists of exactly the unstable manifold of the set $\mathcal{S}$ of all steady state solutions, i.e.*

$$\mathcal{A} = \mathcal{M}_+(\mathcal{S}). \tag{4.7}$$

*Proof.* First of all, it is easy to prove that for any $u_0 \in X$, we have

$$\lim_{t \to \infty} dist(S(t)u_0, \mathcal{S}) = 0. \tag{4.8}$$

In other words, $\mathcal{S}$ attracts each orbit of the semigroup $S(t)$.

We now show that $B_\alpha$ is an absorbing set of the semigroup in $X$. First, $B_\alpha$ is invariant because the Lyapunov functional is non-increasing in time $t$.

Let $D$ be any bounded set in $V$ and assume that $t_D = 1$, so $S(1)D$ is precompact in $V$. By (3.8) and the continuity of the semigroup $S(t)$, we know that for any $y \in \overline{S(1)D}$, there exists a neighbourhood $N_y$ of $y$ and a $t_y > 0$ such that

$$S(t_y)N_y \subset B_\alpha.$$

Since all these sets $N_y$ forms a cover of $\overline{S(1)D}$, there exists finite many of them, say $N_1, \dots, N_l$, which also forms a cover for $\overline{S(1)D}$. Let

$$T_D = \max\{t_1, \dots, t_l\},$$

by the invariance of $B_\alpha$, we get

$$S(t)D \subset B_\alpha \qquad \text{for all } t \geq T_D.$$

The existence of a global attractor then follows immediately. The structure of the global attractor given by (4.7) can be obtained using Theorem 4.1 on p. 401 of [Tem 88]. The proof is complete.

## 5.5 Global atractor for the G-L sstem

In this section, we study the long time behaviour of the evolutionary G-L system of superconductivity. We first prove the existence of a global attractor of the system, which represents exactly all the long time dynamics of the system. Then in the next section, we establish an upper bound for the Hausdorff and fractal dimensions of the global attractor in terms of the physically relevant parameters, i.e., the G-L parameter $\kappa$, the diffusion related parameter $\eta$ and the applied field $H$.

As we indicated before, we are not able to obtain directly the *a priori* estimates of (1.25)–(1.26), to ensure the existence of the bounded absorbing set. We apply the abstract theorem developed in the previous section to the evolutionary G-L system.

First of all, by Theorem 1.2, we denote by $S(t)$, the semigroup generated by the G-L system, which is defined as follows: for any $u_0 \in V$,

$$S(t)u_0 = u(t) = (\psi(t), \boldsymbol{A}(t)), \tag{5.1}$$

where $u(t)$ is the unique solution of Problem 1.1. The main result in this section is

**Theorem 5.5.1** *If the applied magnetic field $H$ is independent of time $t$, the dynamical system associated with the G-L system (1.1), supplemented by the boundary condition (1.2), possesses a global attractor $\mathcal{A} \subset V$, which is compact, connected and maximal in $V$. Moreover, $\mathcal{A} \subset V \bigcap H^2$ consists of exactly the unstable manifold of the set of all steady-state solutions, i.e.*

$$\mathcal{A} = \mathcal{M}_+(\mathcal{S}) \tag{5.2}$$

*where*

$$\mathcal{S} \subset V \bigcap H^2 \tag{5.3}$$

*is the set of all solutions of the steady-state problem.*

*Proof.* Now we only have to verify the assumptions of Theorem 4.1.

Take $X = H^1(\Omega) \times \boldsymbol{H}^1(\Omega)$.

1) The existence of a Lyapunov functional is guaranteed by (1.47). The fact that $E(u) \to \infty$ as $u \to \infty$ has been discussed in the existence proof.

2) The fact that the stationary points are universally bounded can be established as follows: scalar multiplying the both sides of the following equations

$$-(\nabla - i\boldsymbol{A})^2\psi + \kappa^2(|\psi|^2 - 1)\psi = 0,$$
$$\mathbf{curl}^2\boldsymbol{A} - \mathbf{curl}H + (i\psi, (\nabla - i\boldsymbol{A})\psi) = 0$$

by $\psi$ and $\boldsymbol{A}$ respectively, integrating over $\Omega$, taking into account of boundary conditions and the fact that $|\psi| \le 1$, we obtain, for some universal constant $C$,

$$||(\nabla - i\boldsymbol{A})\psi|| \le C,$$
$$||\mathbf{curl}\boldsymbol{A} - H|| \le C.$$

Together with $\mathrm{div}\boldsymbol{A} = 0$, $\boldsymbol{A} \cdot \boldsymbol{n} = 0$, we obtain the boundedness of the solutions in $X$.

3) In Section 3, we have already established the fact that $S(t)u_0$ is a pre-compact set in $X$.

Hence the theorem has been proved.

Now, we discuss the assumption 2) in Theorem 4.1 a little further. The reason is that we want to establish an explicit bound in terms of the physical parameter for the set of steady state solutions which will be useful later and also, the definition of the set of steady state solutions is interesting and worth an independent investigation.

Formally, there are two steady state G-L systems, i.e. the steady state G-L equations (1.4) with boundary conditions (1.2-1), and the following steady state G-L equations obtained directly from the time dependent G-L equations (1.1):

$$\begin{cases} i\psi\phi + \kappa^2(|\psi|^2 - 1)\psi + (i\nabla + \boldsymbol{A})^2\psi = 0, \\ \eta\nabla\phi + \frac{i}{2}(\psi^*\nabla\psi - \psi\nabla\psi^*) + |\psi|^2\boldsymbol{A} + \mathbf{curl}^2\boldsymbol{A} - \mathbf{curl}H = 0, \\ \mathrm{div}\boldsymbol{A} = 0 \end{cases} \tag{5.4}$$

with boundary conditions (1.2-1). The following lemma shows that these two steady state problems are exactly the same. That is, $\psi_t + i\phi\psi$ and $\boldsymbol{A}_t + \nabla\phi$ disappear together as $t \to \infty$ is an accurate description of the long time dynamics of the G-L system with respect to terms involving time derivatives.

**Lemma 5.5.1** *1. If $u = (\psi, \boldsymbol{A}) \in V$ and $\phi \in H^1(\Omega)$ is a solution of (5.4) and (1.2-1), then $i\psi\phi = 0$ and $\nabla\phi = 0$.*

*2. There is at least one solution $u = (\psi, \boldsymbol{A}) \in V$ to the steady state problem (1.4) and (1.2-1).*

*3. The steady state solutions $u = (\psi, \boldsymbol{A})$ satisfy the following estimates:*

$$\begin{cases} |\psi|_{L^4} \leq c, \\ \|u\| \leq c(\kappa^2 + |H|^2), \quad \forall u \in \mathcal{S}, \end{cases} \tag{5.5}$$

*where $\mathcal{S}$ is the set of all stationary solutions in $V$ and $c$ is independent of $\kappa$, $H$ and $u$.*

*Proof.* 1. By (1.47), we have

$$\frac{dE(u(t))}{dt} + |\psi_t + i\phi\psi|^2 + \eta|\boldsymbol{A}_t + \nabla\phi|^2 = 0.$$

Since $u$ is a stationary solution, we have

$$|i\phi\psi|^2 + \eta|\nabla\phi|^2 = 0.$$

2. The existence of steady state solutions can be obtained by either using the Galerkin method or looking for the minimizers of the G-L functional; we omit the details. 3. First of all, we infer from (1.4-1) that

$$\kappa^2|\psi|_{L^4}^4 + |(i\nabla + \boldsymbol{A})\psi|^2 = \kappa^2|\psi|^2 \leq \frac{\kappa^2}{2}|\psi|_{L^4}^4 + c\kappa^2$$

which implies that

$$\begin{cases} |(i\nabla + \boldsymbol{A})\psi|^2 \leq c\kappa^2, \\ |\psi|_{L^4}^4 \leq c, \end{cases} \tag{5.6}$$

$c$ being independent of $u = (\psi, \boldsymbol{A})$ and $\kappa^2$.

Then noticing that

$$\frac{i}{2}(\psi^*\nabla\psi - \psi\nabla\psi^*) + |\psi|^2\boldsymbol{A} = \mathrm{Re}(i\nabla\psi + \boldsymbol{A}\psi)\psi^*,$$

we deduce from (1.4-2) that

$$\begin{aligned} |\mathbf{curl}\boldsymbol{A}|^2 \quad & \leq |\mathrm{Re}\int_\Omega (i\nabla\psi + \boldsymbol{A}\psi)\psi^* \cdot \boldsymbol{A}dx| + |\int_\Omega H \cdot \mathbf{curl}\boldsymbol{A}dx| \\ & \leq c|(i\nabla + \boldsymbol{A})\psi||\psi|_{L^4}|\boldsymbol{A}|_{L^4} + |H||\mathbf{curl}\boldsymbol{A}| \\ \text{(by (5.6))} \quad & \leq c\kappa|\boldsymbol{A}|_{L^4} + c|H|^2 + \frac{1}{4}|\mathbf{curl}\boldsymbol{A}|^2 \\ & \leq c(\kappa^2 + |H|^2) + \frac{1}{2}|\mathbf{curl}\boldsymbol{A}|^2, \end{aligned}$$

which implies that

$$|\mathbf{curl}\boldsymbol{A}|^2 \leq c(\kappa^2 + |H|^2). \tag{5.7}$$

Therefore,

$$\begin{aligned} |\nabla\psi|^2 &\leq 2|(i\nabla + \boldsymbol{A})\psi|^2 + 2|\boldsymbol{A}\psi|^2 \\ &\leq c + c|\boldsymbol{A}|^2_{L^4}|\psi|^2_{L^4} \\ &\leq c + c|\boldsymbol{A}|^2_{L^4} \\ &\leq c + c|\mathbf{curl}\boldsymbol{A}|^2 \\ &\leq c(\kappa^2 + |H|^2). \end{aligned}$$

The proof is complete.

## 5.6 Physical bounds on the global attractor

In this section, we shall estimate the dimension (fractal and Hausdorff) of the global attractor in terms of the physical relevant parameters $\kappa$ (the G-L parameter), $\eta$ (the diffusion constant) and the applied field $H$. The definition of Hausdorff dimension of a set is classical, we do not repeat here. For any compact set $C$ in a Banach space $X$, let $N_\varepsilon(C)$ be the minimum number of balls of radius $\varepsilon$ necessary to cover $C$, then the fractal dimension of $C$ is defined by

$$\dim_F(C) = \text{limsup}_{\varepsilon\to 0}\frac{logN_\varepsilon(C)}{log\varepsilon^{-1}}. \tag{6.1}$$

Our main result in this section is the following:

**Theorem 5.6.1** *The global attractor $\mathcal{A}$ of the G-L system (1.1)–(1.2) given by Theorem 5.1 has finite dimension. We have the following estimates of the Hausdorff and Fractal dimension in terms of the G-L parameter $\kappa$ and the relaxation time $\eta$:*

$$dim\mathcal{A} \leq c\left(\kappa^4 + \frac{1}{\eta^2}\right)\left(\kappa^4 + |H|^4\right)\left(1 + \frac{|H|^2}{\kappa^2}\right). \tag{6.2}$$

We use the theory developed by Constantin, Foias and Temam. We divide the proof into 4 steps.

*Step 1 (Linearized problem).* For any $u_0 = (\psi_0, \boldsymbol{A}_0) \in \mathcal{A}$ and $u(t) = S(t)u_0 = (\psi(t), \boldsymbol{A}(t))$, we consider the linearized equations of the G-L system around $u$:

$$\frac{\partial\Psi}{\partial t} + F_1'(u)(\Psi, \boldsymbol{B}) = 0, \tag{6.3}$$

$$\frac{\partial\boldsymbol{B}}{\partial t} + \nabla\Pi + F_2'(u)(\Psi, \boldsymbol{B}) = 0, \tag{6.4}$$

$$\text{div}\boldsymbol{B} = 0 \tag{6.5}$$

together with the linearized boundary conditions

$$\frac{\partial \Psi}{\partial n} = 0, \ \boldsymbol{B} \cdot n = 0, \ \mathbf{curl} \boldsymbol{B} \times n = 0, \quad \text{on } \partial\Omega. \tag{6.6}$$

In (6.3) and (6.4), $F_1'$, $F_2'$ and $\Pi$ are given by

$$\begin{aligned} F_1'(u)(\Psi, \boldsymbol{B}) = & -\Delta\Psi + 2i(\boldsymbol{A}\nabla\Psi + \boldsymbol{B}\nabla\psi) + |\boldsymbol{A}|^2\Psi + 2\boldsymbol{A}\boldsymbol{B}\psi \\ & + i(\Phi(u)\Psi + i\psi\Pi) + \kappa^2[(|\psi|^2 - 1)\Psi + \Psi|\psi|^2 + \Psi^*\psi^2], \end{aligned} \tag{6.7}$$

$$\begin{aligned} F_2'(u)(\Psi, \boldsymbol{B}) = & \frac{1}{\eta}\{\frac{i}{2}(\Psi^*\nabla\psi - \Psi\nabla\psi^* + \psi^*\nabla\Psi - \psi\nabla\Psi^*) \\ & + |\psi|^2\boldsymbol{B} + (\psi\Psi^* + \Psi\psi^*)\boldsymbol{A} + \mathbf{curl}^2\boldsymbol{B}\}, \end{aligned} \tag{6.8}$$

$$\nabla\Pi + (I - P_2)F_2'(u)(\Psi, \boldsymbol{B}) = 0, \tag{6.9}$$

$$\int_\Omega \Pi dx = 0. \tag{6.10}$$

As in Lemma 1.1 (cf. (1.12)–(1.13)), $\Pi$ is uniquely defined by (6.9) and (6.10).

It is easy to see that there is a unique strong solution of (6.3)–(6.6) with given initial data in $V$ and given $u(t) = S(t)u_0$ with $u_0 \in \mathcal{A}$.

*Step 2 (CFT theory).* We proceed now by introducing the Constantin-Foias-Temam theory from [CFT 85]. We set

$$\begin{aligned} T_m(\tau) = & -\textstyle\sum_{j=1}^m (F_1'(u(\tau))(\Psi^j(\tau)\boldsymbol{B}^j(\tau), -\Delta\Psi^j(\tau) + \Psi^j(\tau)) \\ & - \textstyle\sum_{j=1}^m (P_2F_2'(u(\tau))\Psi^j(\tau)\boldsymbol{B}^j(\tau), \mathbf{curl}\boldsymbol{B}^j(\tau)). \end{aligned} \tag{6.11}$$

Then we consider solutions $U_1(t), \ldots, U_m(t)$ of the linearized system (6.3)–(6.6) with initial values $\xi_1, \ldots, \xi_m \in V$ respectively. Then we choose $\{(\Psi^j(t), B^j(t))\}_{j=1}^m$ to be an orthonormal basis of

$$Q_mV = \text{span}\{U_1(t), \ldots, U_m(t)\}, \tag{6.12}$$

$Q_m$ being an orthonormal projector of $V$ into $Q_mV$.

Then it is easy to see that (cf. among others, [Tem 88]),

$$|U_1(t) \wedge \cdots \wedge U_m(t)|_{\wedge^m V} = |\xi_1 \wedge \cdots \xi_m|_{\wedge^m V} \exp \int_0^t \text{Re}T_m(\tau)d\tau. \tag{6.13}$$

Moreover, we introduce

$$q_m(t) = \sup_{\mu_0 \in \mathcal{A}} \sup_{\xi_j \in V, \|\xi_j\| \leq 1} [\frac{1}{t}\int_0^t \text{Re}T_m(\tau)d\tau], \tag{6.14}$$

$$q_m = \lim_{t \to \infty} \sup q_m(t). \tag{6.15}$$

Then we quote without proof the main result of the Constantin-Foias-Temam theory from [CFT 85] which will be used in the remaining part of this section.

**Theorem 5.6.2** *If*

$$q_j \leq -\alpha j^{\theta} + \beta, \ \forall j \geq 1, \tag{6.16}$$

*then the Hausdorff and fractal dimension*

$$d_H(\mathcal{A}) \leq m, \ \ d_F(\mathcal{A}) \leq 2m, \tag{6.17}$$

$$m - 1 < (\frac{2\beta}{\alpha})^{1/\theta} \leq m. \tag{6.18}$$

*Step 3 (Estimates of ReT$_m$).* To obtain (6.16) type estimates, we now estimate Re$T_m$. First of all, we have

$$\begin{aligned} &-\mathrm{Re}\textstyle\sum_{j=1}^m[(-\triangle\Psi^j, -\triangle\Psi^j + \Psi^j) + \frac{1}{\eta}(\mathbf{curl}^2\boldsymbol{B}^j, \mathbf{curl}^2\boldsymbol{B}^j)] \\ &= -\textstyle\sum_{j=1}^m[|(I-\triangle)\Psi^j|^2 - \|\Psi^j\|^2 + \frac{1}{\eta}|\mathbf{curl}^2\boldsymbol{B}^j|^2] \\ &\leq -\textstyle\sum_{j=1}^m[|(I-\triangle)\Psi^j|^2 + \frac{1}{\eta}|\mathbf{curl}^2\boldsymbol{B}^j|^2] + m. \end{aligned} \tag{6.19}$$

$$\begin{aligned} &-\mathrm{Re}\textstyle\sum_{j=1}^m(2i\boldsymbol{A}\nabla\Psi^j, -\triangle\Psi^j + \Psi^j) \\ &\leq c\textstyle\sum_{j=1}^m |\nabla\Psi^j|_{L^3}|(I-\triangle)\Psi^j||\boldsymbol{A}|_{L^6} \\ &\leq c\|\boldsymbol{A}\|\textstyle\sum_{j=1}^m \|\Psi^j\|^{1/2}|(I-\triangle)\Psi^j|^{3/2} \\ &\leq c\|\boldsymbol{A}\|\textstyle\sum_{j=1}^m |(I-\triangle)\Psi^j|^{3/2} \\ &\leq cm\|\boldsymbol{A}\|^4 + \textstyle\sum_{j=1}^m |(I-\triangle)\Psi^j|^2. \end{aligned} \tag{6.20}$$

$$\begin{aligned} &-\mathrm{Re}\textstyle\sum_{j=1}^m(2i\boldsymbol{B}^j\nabla\psi, (I-\triangle)\Psi^j) \\ &\leq c\textstyle\sum_{j=1}^m \int_\Omega |\nabla\psi||\boldsymbol{B}^j||(I-\triangle)\Psi^j|dx \\ &\leq c\textstyle\sum_{j=1}^m |\nabla\psi||\boldsymbol{B}^j|_{L^\infty}|(I-\triangle)\Psi^j| \\ \text{(Agmon's inequality)}\ &\leq c\textstyle\sum_{j=1}^m |\nabla\psi|\|\boldsymbol{B}^j\|^{1/2}\|\boldsymbol{B}^j\|_{H^2}^{1/2}|(I-\triangle)\Psi^j| \\ &\leq c\|\psi\|\textstyle\sum_{j=1}^m |\mathbf{curl}^2\boldsymbol{B}^j|^{1/2}|(I-\triangle)\Psi^j| \\ &\leq cm\|\psi\|^4 + \varepsilon\textstyle\sum_{j=1}^m[|(I-\triangle)\Psi^j|^2 + |\mathbf{curl}^2\boldsymbol{B}^j|^2]. \end{aligned} \tag{6.21}$$

$$\begin{aligned} &-\mathrm{Re}\textstyle\sum_{j=1}^m(|\boldsymbol{A}|^2\Psi^j, (I-\triangle)\Psi^j) \\ &\leq c|\boldsymbol{A}|_{L^6}^2\textstyle\sum_{j=1}^m |(I-\triangle)\Psi^j||\Psi^j|_{L^6} \\ &\leq cm\|\boldsymbol{A}\|^4 + \varepsilon\textstyle\sum_{j=1}^m |(I-\triangle)\Psi^j|^2. \end{aligned} \tag{6.22}$$

$$\begin{aligned} &-\mathrm{Re}\textstyle\sum_{j=1}^m(2\boldsymbol{A}\cdot B^j\psi, (I-\triangle)\Psi^j) \\ &\leq cm\|u\|^4 + \varepsilon\textstyle\sum_{j=1}^m |(I-\triangle)\Psi^j|^2. \end{aligned} \tag{6.23}$$

$$\begin{aligned} &-\mathrm{Re}\textstyle\sum_{j=1}^m \kappa^2((|\psi|^2 - 1)\Psi^j + \Psi^j|\psi|^2 + \Psi^{j*}\psi^2, (I-\triangle)\Psi^j) \\ &\leq c\kappa^2\textstyle\sum_{j=1}^m(1 + \|u\|^2)|(I-\triangle)\Psi^j| \\ &\leq c\kappa^2 m(1 + \|u\|^4) + \varepsilon\textstyle\sum_{j=1}^m |(I-\triangle)\Psi^j|^2. \end{aligned} \tag{6.24}$$

$$\begin{aligned}
&-\mathrm{Re}\textstyle\sum_{j=1}^m \frac{1}{\eta}(P_2\frac{i}{2}(\Psi^{j*}\nabla\psi - \Psi^j\nabla\psi^*), \mathbf{curl}^2\boldsymbol{B}^j)\\
&\leq \tfrac{c}{\eta}\textstyle\int_\Omega \sum_{j=1}^m |\mathbf{curl}^2\boldsymbol{B}^j||\Psi^j||\nabla\psi|dx\\
\text{(as in (6.21))}\quad &\leq \tfrac{c}{\eta}\|\psi\|\textstyle\sum_{j=1}^m |(I-\triangle)\Psi^j|_{L^2}^{1/2}|\mathbf{curl}^2\boldsymbol{B}^j|\\
&\leq \tfrac{cm}{\eta^2}\|\psi\|^4 + \varepsilon\textstyle\sum_{j=1}^m[|(I-\triangle)\Psi^j|^2 + \frac{1}{\eta}|\mathbf{curl}^2\boldsymbol{B}^j|^2].
\end{aligned} \tag{6.25}$$

$$\begin{aligned}
&-\mathrm{Re}\textstyle\sum_{j=1}^m \frac{1}{\eta}(P_2\frac{i}{2}(\psi^*\nabla\Psi^j - \psi\nabla\Psi^{j*}), \mathbf{curl}^2\boldsymbol{B})\\
&\leq \tfrac{c}{\eta}\textstyle\sum_{j=1}^m |\psi|_{L^6}|\mathbf{curl}^2 B^j||\nabla\Psi^j|_{L^3}\\
&\leq \tfrac{c}{\eta}\|\psi\|\textstyle\sum_{j=1}^m \|\Psi^j\|_{H^2}^{1/2}|\mathbf{curl}^2 B^j|\\
&\leq \tfrac{c}{\eta}\|\psi\|^2\textstyle\sum_{j=1}^m \|\Psi^j\|_{H^2} + \frac{\varepsilon}{\eta}\sum_{j=1}^m |\mathbf{curl}^2\boldsymbol{B}^j|^2\\
&\leq \varepsilon\textstyle\sum_{j=1}^m[|(I-\triangle)\Psi^j|^2 + \frac{1}{\eta}|\mathbf{curl}^2\boldsymbol{B}^j|^2] + \frac{c}{\eta^2}m\|\psi\|^4.
\end{aligned} \tag{6.26}$$

By Lemma 6.1 below, we have

$$\begin{aligned}
&-\mathrm{Re}\textstyle\sum_{j=1}^m (i\psi\Pi^j, (I-\triangle)\Psi^j)\\
&\leq c\textstyle\sum_{j=1}^m |(I-\triangle)\Psi^j||\psi|_{L^6}|\Pi^j|_{L^3}\\
&\leq \tfrac{c}{\eta}(\|u\| + |\psi|_{L^4}^2)\|\psi\|\textstyle\sum_{j=1}^m |(I-\triangle)\Psi^j|\\
&\leq \varepsilon\textstyle\sum_{j=1}^m |(I-\triangle)\Psi^j|^2 + \frac{c}{\eta^2}m(\|u\|^4 + \|u\|^2|\psi|_{L^4}^4).
\end{aligned} \tag{6.27}$$

Moreover, using Lemma 1.1, we obtain

$$\begin{aligned}
&-\mathrm{Re}\textstyle\sum_{j=1}^m (i\Phi(u)\Psi^j, (I-\triangle)\Psi^j)\\
&\leq c\textstyle\sum_{j=1}^m |(I-\triangle)\Psi^j||\Phi(u)|_{L^3}\|\Psi^j\|\\
&\leq c\textstyle\sum_{j=1}^m |(I-\triangle)\Psi^j|(\|u\|^2 + \|u\|\cdot|\psi|_{L^4}^2)\\
&\leq \varepsilon\textstyle\sum_{j=1}^m |(I-\triangle)\Psi^j|^2 + cm(\|u\|^4 + \|u\|^2|\psi|_{L^4}^4).
\end{aligned} \tag{6.28}$$

Introducing (6.19)–(6.28) in (6.14) leads to

$$\begin{aligned}
\mathrm{Re}T_m(\tau) \leq\ & (1-10\varepsilon)\sum_{j=1}^m\Big(|(I-\triangle)\Psi^j|^2 + \frac{1}{\eta}|\mathbf{curl}^2 B^j|^2\Big)\\
& + cm(1+\kappa^4+\frac{1}{\eta^2})(\|u\|^4 + \|u\|^2|\psi|_{L^4}^4).
\end{aligned}$$

Choosing $\varepsilon = \frac{1}{20}$, the above inequality yields

$$\mathrm{Re}T_m(\tau) \leq -\frac{1}{2}X_m(\tau) + cm(\kappa^4 + \frac{1}{\eta^2})(\|u\|^4 + \|u\|^2|\psi|_{L^4}^4) \tag{6.29}$$

where

$$X_m(\tau) = \sum_{j=1}^m[|(I-\triangle)\Psi^j|^2 + \frac{1}{\eta}\|\boldsymbol{B}^j\|_{H^2}^2]. \tag{6.30}$$

Then we have

$$\begin{aligned}&\tfrac{1}{t}\int_0^t \mathrm{Re}T_m(\tau)d\tau\\ \leq &-\tfrac{1}{2}\tfrac{1}{t}\int_0^t X_m(\tau)d\tau + c(\kappa^4+\tfrac{1}{\eta^2})\tfrac{m}{t}\int_0^t(\|u\|^4+\|u\|^2|\psi|_{L^4}^4)d\tau\\ &\leq -\tfrac{1}{2}\tfrac{1}{t}\int_0^t X_m(\tau)d\tau + c\gamma m(\kappa^4+\tfrac{1}{\eta^2})\end{aligned} \tag{6.31}$$

where

$$\gamma = \sup_{u\in\mathcal{A}}(\|u\|^4+\|u\|^2|\psi|_{L^4}^4). \tag{6.32}$$

*Step 4 (Completion of the proof).* By Lemma 6.2 below, we have now that

$$\begin{aligned}q_m &\leq -cm^2 + c\gamma m(\kappa^4+\eta^{-2})\\ &\leq -cm^2 + c[\gamma(\kappa^4+\eta^{-2})]^2.\end{aligned} \tag{6.33}$$

Finally by Theorem 6.2, we have

$$\begin{aligned}\dim\mathcal{A} &\leq c[(\gamma(\kappa^4+\eta^{-2}))^2]^{1/2}\\ &\leq c(\gamma(\kappa^4+\eta^{-2}))\\ \text{(by Lemma 6.1 below)}\qquad &\leq c\left(\kappa^4+\frac{1}{\eta^2}\right)(\kappa^4+|H|^4)\left(1+\frac{|H|^2}{\kappa^2}\right).\end{aligned}$$

The proof is complete.

The following estimate for $\Pi^j$ was used in the above proof.

**Lemma 5.6.1** *The function $\Pi^j$ defined by*

$$\nabla\Pi^j + (I-P_2)F_2'(u)(\Psi^j,\boldsymbol{B}^j) = 0, \tag{6.34}$$

$$\int_\Omega \Pi^j dx = 0 \tag{6.35}$$

*satisfies the following estimates:*

$$|\Pi^j|_{L^3} \leq c\|(I-P_2)F_2'(u)(\Psi^j,\boldsymbol{B}^j)\|_{W^{-1,3}(\Omega)} \leq c[\|\psi\|+\|\psi\|_{L^4}^2]. \tag{6.36}$$

*Proof.* We only have to prove the second inequality in (5.36). To this end, we consider each term in (6.8).

$$\begin{aligned}&|\int_\Omega \Psi^{j*}\nabla\psi\tilde{\boldsymbol{A}}dx| \leq |\Psi^j|_{L^6}|\nabla\psi||\tilde{\boldsymbol{A}}|_{L^3} \leq c\|\psi\|\|\tilde{\boldsymbol{A}}\|_{W^{1,3/2}(\Omega)},\\ &|\int_\Omega \psi^*\nabla\Psi^j\tilde{\boldsymbol{A}}dx| \leq c\|\psi\|_{L^6}|\nabla\Psi^j||\tilde{\boldsymbol{A}}|_{L^3} \leq c\|\psi\|\|\tilde{\boldsymbol{A}}\|_{W^{1,3/2}(\Omega)},\\ &|\int_\Omega |\psi|^2\boldsymbol{B}^j\cdot\tilde{\boldsymbol{A}}dx| \leq c|\psi^2|_{L^2}|\boldsymbol{B}^j|_{L^6}\|\tilde{\boldsymbol{A}}\|_{L^3} \leq c|\psi|_{L^4}^2\|\tilde{\boldsymbol{A}}\|_{W^{1,3/2}}.\end{aligned}$$

The other terms can be estimated in the same fashion, noticing that

$$\mathbf{curl}^2\boldsymbol{B}^j - P_2\mathbf{curl}^2\boldsymbol{B}^j = 0.$$

The proof is complete.

**Lemma 5.6.2**

$$\sum_{j=1}^{m}[|(I-\triangle)\Psi^j|^2+\|B^j\|_{H^2}^2]\geq cm^2 \tag{6.37}$$

*with c independent of m.*

*Proof.* By Lemma 2.1, p. 302 of [Tem 88], we have

$$\sum_{j=1}^{m}[|(I-\triangle)\Psi^j|^2+\|\boldsymbol{B}^j\|_{H^2}^2]\geq \lambda_1^{1/2}+\cdots+\lambda_m^{1/2}\geq c\sum_{j=1}^{m}j\geq cm^2$$

where $\lambda_1, \ldots, \lambda_m$ are eigenvalues of the operator $u=(\psi,\boldsymbol{A})\to((I-\triangle)\psi,-\triangle\boldsymbol{A})$.

The following lemma provides an upper bound $\gamma$ for the $H^1$-norm of the global attractor defined by (6.32).

**Theorem 5.6.3** *We have*

$$\begin{cases} \sup_{u\in\mathcal{A}}|\psi|_{L^4}^4\leq c(1+\frac{|H|^2}{\kappa^2}), \\ \sup_{u\in\mathcal{A}}\|u\|\leq c(\kappa+|H|)(1+\sqrt{\frac{|H|}{\kappa}}) \end{cases} \tag{6.38}$$

*where $\mathcal{A}$ is the global attractor of the G-L system obtained by Theorem 5.1.*

*Proof.* First of all, we claim that

$$\sup_{u\in\mathcal{A}}E(u)=\sup_{u\in\mathcal{S}}E(u). \tag{6.39}$$

To see this, for any $u_0\in\mathcal{A}$, by Theorem 5.1, there is a complete orbit $\{u(t): t\in \mathbb{R}\}$ such that $u(0)=u_0$ and

$$d(u(t),S)\to 0 \qquad \text{as } t\to\infty. \tag{6.40}$$

On the other hand, $E(u(t))$ is decreasing. Therefore,

$$\sup_{u\in S}E(u)\geq E(u_0).$$

Namely, (6.39) holds.

Then by definition (1.45) and the estimates (5.5), (6.39), it yields that

$$\sup_{u\in\mathcal{A}}E(u)\leq c(\kappa^2+|H|^2). \tag{6.41}$$

This shows that for any $u\in\mathcal{A}$,

$$\begin{cases} |(i\nabla+\boldsymbol{A})\psi|^2\leq c(\kappa^2+|H|^2), \\ \kappa^2|\psi|_{L^4}^4\leq c(\kappa^2+|H|^2), \\ |\mathbf{curl}\boldsymbol{A}|^2\leq c(\kappa^2+|H|^2). \end{cases} \tag{6.42}$$

Hence

$$\begin{aligned}|\nabla\psi|^2 &\le c(|(i\nabla+\boldsymbol{A})\psi|^2+|\boldsymbol{A}\psi|^2_{L^2})\\ &\le c(\kappa^2+|H|^2)+|\boldsymbol{A}|^2_{L^4}|\psi|^2_{L^4}\\ &\le c(\kappa^2+|H|^2)[1+|\psi|^2_{L^4}]\\ &\le c(\kappa^2+|H|^2)(1+\frac{|H|}{\kappa}).\end{aligned}$$

The proof is complete.

## 5.7 The uniqueness of the long time limit of the evolutionary G-L solutions

In this section, we give a brief presentation of the recently achieved results on the uniqueness of the long time limit of the evolutionary G-L solutions. This is a very important result for concluding this chapter, however, the technicalities of the proofs require extensive study of some specific mathematical subjects and are beyond the scope of this monograph. Interested readers are encouraged to read [LDu 97], [Lin 96] plus the references given there for more details.

In this section, we use the following evolutionary G-L system:

$$\begin{cases}\eta\ \dfrac{\partial\psi}{\partial t} &+\ \ i\eta\kappa\Phi\psi+\left(\frac{i\nabla}{\kappa}+\boldsymbol{A}\right)^2\psi-\psi+|\psi|^2\psi=0,\\ \dfrac{\partial\boldsymbol{A}}{\partial t} &+\ \ \nabla\Phi+\mathbf{curl}^2\boldsymbol{A}=-\frac{i}{2\kappa}\left(\psi^*\nabla\psi-\psi\nabla\psi^*\right)-\boldsymbol{A}|\psi|^2,\end{cases}\tag{7.1}$$

in $\Omega\times R_+$ where $\Omega$ is a subset of $\mathbb{R}^3$. The boundary and initial conditions are

$$\psi(x,0)=\psi_0(x),\qquad \boldsymbol{A}(x,0)=\boldsymbol{A}_0(x),\qquad x\in\Omega;\tag{7.2}$$

$$\begin{cases}\left(\frac{1}{\kappa}\nabla+\boldsymbol{A}\right)\psi\cdot\boldsymbol{n} &=& 0,\\ \mathbf{curl}\boldsymbol{A} &=& 0,\ \ (H_0=0),\\ \left(\dfrac{\partial\boldsymbol{A}}{\partial t}+\nabla\Phi\right)\cdot\boldsymbol{n} &=& \boldsymbol{E}\cdot\boldsymbol{n}\ =\ 0,\quad \text{on}\quad \partial\Omega\,.\end{cases}\tag{7.3}$$

We make the following gauge choice:

$$\frac{\partial\chi}{\partial t}=\Phi,\quad\text{and}\quad\text{at } t=0,\tag{7.4}$$

$$\Delta\chi=-\mathrm{div}\boldsymbol{A}\text{ on }\Omega\text{ with }\nabla\chi\cdot n=-\boldsymbol{A}\cdot n\text{ on }\partial\Omega.\tag{7.5}$$

Then, in this gauge, $\Phi\equiv0$, and the system (7.1) reduces to the gradient flow of the energy-functional.

$$E(\psi,\,\boldsymbol{A})=\frac{1}{2}\int_\Omega\left[\left|\left(\frac{i}{\kappa}\nabla+\boldsymbol{A}\right)\psi\right|^2+\frac{1}{2}\left(|u|^2-1\right)^2+\left|\mathbf{curl}\boldsymbol{A}\right|^2\right]dx.\tag{7.6}$$

We know that the flow

$$\frac{dv}{dt} = - \text{grad} \quad E(v), \qquad v(0) = v_0 \,, \tag{7.7}$$

$v = (u,\, \boldsymbol{A})$ has a global classical solution. The main result is the following

**Theorem 5.7.1** *The limit*

$$V_\infty = \lim_{t \to \infty} \, v(t)$$

*exists in* $V = \mathcal{H}^1(\Omega) \times \boldsymbol{H}^1(\Omega)$.

This tells us the for any solution $(\psi(\cdot, t), \boldsymbol{A}(\cdot, t))$, there exists a $(\psi(\cdot), \boldsymbol{A}(\cdot))$, which is a solution of the steady state problem, such that

$$\lim_{t \to \infty} (\psi(\cdot, t), \boldsymbol{A}(\cdot, t)) = (\psi(\cdot), \boldsymbol{A}(\cdot))$$

in the stated function spaces. This is a much better result than the $\omega$-limit we have proved in Section 3.

## 5.8 Bibliographical remarks

The evolutionary system was first derived in [GEl 68]. In [CHO 92], some mathematical observations in relation to the Maxwell system, phase transition theory, scaling and temperature field have been made. Then the existence, uniqueness results were established in a series of papers (see [Du 94], [CHL 93] and [Tan 95]).

The long time behaviour and the fact that the G-L energy is a Lyapunov functional was first established in [LTa 95] and [TWa 96]. In [LDu 97], a much stronger long time convergence property of the solutions was also established. We give the announcement of the results in Section 7 but could not include all the details of proof.

The case with the Maxwell system in the outside domain is first given in [HTa3].

The main concern now is that there are no results for the existence of solutions of the 3-d G-L system when the initial data lie only in $L^2$. From a mathematical point of view, this is a matter of well posedness.

# Chapter 6
# Complex G-L Type Phase Transition Theory

Having dealt with the formal analysis of co-dimension two-phase transition problems in Chapters 2 and 3, we now discuss the rigorous analysis in two space dimensions. The formal asymptotics gave us an indication of the phase transition structures. In this chapter, we give rigorous proofs. However, the problem considered here is not equivalent to that of Chapters 2 and 3. In formal asymptotic analysis, we have the freedom to assume that a vortex exists and that we can expand the solution in a neighbourhood of such a vortex. In rigorous analysis, this is not the case. The boundary condition we impose here is to force the presence of vortices. These conditions therefore are not physical. However, we believe that these results still give us some interesting insight into the nature of the problem and are of great mathematics interest itself, we give it an extended discussion here. We mainly follow the work of [BRi 95] and [BBH 94].

The results imply that the configuration of the vortices should minimize a renormalized energy that involves, among others, boundary conditions. This implies that there is an interaction of vortices through a distant force. This is mainly due to the boundary conditions we imposed out of mathematical considerations.

In the following, $\Omega$ denotes a smooth (at least $C^1$) simply connected bounded domain in $\mathbb{R}^2$. $\varepsilon > 0$ is a small parameter and $\psi_0$: $\partial\Omega \to S^1$ is a smooth map.

In this chapter, we use the equivalent operator $(\nabla - i\boldsymbol{A})$ to replace $(i\nabla + \boldsymbol{A})$ and deal with the $\varepsilon \to 0$ limit of the following 2-d G-L functional

$$G_\varepsilon(\psi, \boldsymbol{A}) = \frac{1}{2}\int_\Omega \left(|(\nabla - i\boldsymbol{A})\psi|^2 + \frac{1}{4\varepsilon^2}(|\psi|^2 - 1)^2 + \frac{1}{2}|\mathbf{curl}\boldsymbol{A} - \boldsymbol{H}|^2\right) dx. \quad (0.1)$$

**Remark 6.0.1** *We only deal with 2-d model because there is no suitable topological boundary condition for the 3-d model which will maintain the presence of vortices as $\varepsilon \to 0$ while keeping $\boldsymbol{H}$ fixed (within this scaling scheme).*

We assume that we have made the Coulomb gauge choice as discussed in Chapter 1. We then need to give the admissible conditions on the class of functions on which the energy (0.1) is to be minimized. As we know from Chapter 1, the ideal case is not to impose any boundary conditions and to minimize the energy in $\mathcal{H}^1 \times \boldsymbol{H}^1$. However, if we choose to do this, we will not be able to control the number of vortices in our bounded domain. It is also possible that if we simply minimize the energy in the function space $\mathcal{H}^1 \times \boldsymbol{H}^1$, we end up having vortices on the boundary of the domain. This is rather difficult to deal with from the mathematical point of view. Therefore, we will have to impose some mathematical conditions that prevent this situation from occurring.

In order to conform to gauge invariance and to control the number of vortices simultaneously, we minimize the energy in the following function class

$$\begin{aligned} V = \{(\psi, \boldsymbol{A}) \in \mathcal{H}^1 \times \boldsymbol{H}^1,\ |\psi| = 1 \text{ on } \partial\Omega, \\ \boldsymbol{J} \cdot \boldsymbol{\tau} = |\psi|^2(\boldsymbol{A} - \nabla\chi) \cdot \boldsymbol{\tau} = g, \deg(\psi, \partial\Omega) = d.\} \end{aligned} \tag{0.2}$$

Here $\boldsymbol{\tau}$ is the unit tangent vector to $\partial\Omega$, $g$ is a given function and $d$ is a given nonnegative integer (when $d$ is negative, we will have to use $|d|$ instead of $d$ in the estimates in the following of this chapter). $\chi$ is the angle variable for the order parameter: $\psi = |\psi|e^{i\chi}$. The definition of degree has been introduced in Section 3 of Chapter 2. Since degree is measured by the integration of $\nabla\chi \cdot \boldsymbol{\tau}$ along $\partial\Omega$, the boundary condition $\boldsymbol{J} \cdot \boldsymbol{\tau} = |\psi|^2(\boldsymbol{A} - \nabla\chi) \cdot \boldsymbol{\tau} = g$ imposes a link between the total magnetic flux $\int_{\partial\Omega} \boldsymbol{A} \cdot \boldsymbol{\tau} ds = \int_\Omega \mathbf{curl}\boldsymbol{A} dx$ and the given topological degree of the order parameter $d$.

## 6.1 Existence and basic properties of solutions

As the boundary conditions are different from those considered previously, we discuss briefly the existence of solutions.

**Theorem 6.1.1** *The energy minimum*

$$\inf_{(\psi, \boldsymbol{A}) \in V} G_\varepsilon(\psi, \boldsymbol{A}) \tag{1.1}$$

*is achieved and there exists a pair of minimizers $(\psi_\varepsilon, \boldsymbol{A}_\varepsilon)$ satisfying the following gauge condition:*

$$\begin{cases} \operatorname{div}\boldsymbol{A}_\varepsilon = 0 & in\ \Omega, \\ \boldsymbol{A}_\varepsilon \cdot \boldsymbol{n} = 0 & on\ \partial\Omega. \end{cases} \tag{1.2}$$

*Moreover, the energy minimizers satisfy the standard G-L system together with the given boundary conditions and in addition, the order parameter solution $\psi_\varepsilon$ satisfies*

$$|\psi_\varepsilon| \le 1 \ in\ \Omega. \tag{1.3}$$

*Proof.* The major difference from before is in the class of admissible functions. We now need to verify that all the boundary conditions are preserved under weak $H^1$ convergence. Then the proof of existence will follow.

First, if $\psi_n$ converges weakly in $H^1$, we know that the sequence $\psi_n|_{\partial\Omega}$ converges weakly in $H^{1/2}(\partial\Omega)$, hence, it converges strongly in $L^2(\partial\Omega)$. The condition that $|\psi_n| = 1$ on $\partial\Omega$ is therefore preserved as $n \to \infty$.

Second, because the condition

$$|\psi_n|^2(\boldsymbol{A}_n - \nabla\chi_n) \cdot \boldsymbol{\tau} = g$$

is equivalent to

$$(\nabla\psi_n) \cdot \boldsymbol{\tau} = (-g + \boldsymbol{A}_n \cdot \boldsymbol{\tau}) i\psi_n,$$

if $(\psi_n, \boldsymbol{A}_n)$ converges weakly in $H^1(\Omega) \times \boldsymbol{H}^1(\Omega)$ and because $\mathbf{curl}\nabla\psi_n = 0$, using the following integration by parts formula

$$\int_{\partial\Omega} \mathbf{W} \cdot (\nabla\psi_n \boldsymbol{\tau}) = \int_{\Omega} \Big(\mathbf{curl}(\nabla\psi_n) \cdot \mathbf{W} - \nabla\psi_n \cdot \mathbf{curl}\mathbf{W}\Big) dx$$

for any smooth $\Omega$ and $\mathbf{W}$, we conclude that the term $(\nabla\psi_n)\boldsymbol{\tau}$ converges weakly in $H^{-1/2}(\partial\Omega)$ to $(\nabla\psi) \cdot \boldsymbol{\tau}$ because $\mathbf{curl}(\nabla\psi_n) \equiv 0$. The continuity of the term $(g + \boldsymbol{A}_n \cdot \boldsymbol{\tau}) i\psi_n$ as $n \to \infty$ is obvious.

Finally, for the degree, we know that

$$\begin{aligned} d &= \frac{1}{2\pi} \int_{\partial\Omega} \nabla\chi_n \cdot \boldsymbol{\tau} = \frac{1}{2\pi} \int_{\partial\Omega} |\psi_n|^2 \nabla\chi_n \cdot \boldsymbol{\tau} \\ &= \frac{1}{2\pi} \int_{\partial\Omega} (-g + \boldsymbol{A}_n \cdot \boldsymbol{\tau}). \end{aligned}$$

The continuity when $n \to \infty$ is also obvious now.

The theorem can now be proved the same way as in Chapter 4.

**Proposition 6.1.1** *We have at least formally,*

$$\partial_{\boldsymbol{n}} h_\varepsilon = -g \quad on \quad \partial\Omega \tag{1.4}$$

*where* $h_\varepsilon = \mathbf{curl}\boldsymbol{A}_\varepsilon$.

*Proof.* From the G-L system, we have

$$\mathbf{curl} h_\varepsilon = -\frac{i}{2}(\psi_\varepsilon^* \nabla\psi_\varepsilon - \psi_\varepsilon \nabla\psi_\varepsilon^*) - |\psi_\varepsilon|^2 \boldsymbol{A}_\varepsilon = |\psi_\varepsilon|^2(\nabla\chi_\varepsilon - \boldsymbol{A}_\varepsilon).$$

Restricting to the boundary and scalar multiplying both sides by the tangential vector $\boldsymbol{\tau}$, using (0.2) and the fact that $\mathbf{curl} h_\varepsilon \cdot \boldsymbol{\tau} = \nabla h_\varepsilon \cdot \boldsymbol{n}$, we obtain (1.4).

**Proposition 6.1.2** *(Pohozaev type equality): Let $D$ be a sub-domain of $\Omega$ and $(\psi_\varepsilon, \boldsymbol{A}_\varepsilon)$ be the energy minimizers of (1.1). We have*

$$\begin{aligned}
& \int_D \frac{1}{\varepsilon^2}(1-|\psi_\varepsilon|^2)^2 - h_\varepsilon^2 dx \\
= & \int_{\partial D} (x\cdot\boldsymbol{n})[(\boldsymbol{\tau}\cdot(\nabla - i\boldsymbol{A}_\varepsilon)\psi_\varepsilon)^2 - (\boldsymbol{n}\cdot(\nabla - i\boldsymbol{A}_\varepsilon)\psi_\varepsilon)^2] \\
+ & \frac{1}{2}\int_{\partial D} (x\cdot\boldsymbol{n})(\frac{1}{\varepsilon^2}(1-|\psi_\varepsilon|^2)^2 - h_\varepsilon^2) \\
- & \int_{\partial D} (x\cdot\boldsymbol{\tau})(\boldsymbol{\tau}\cdot(\nabla - i\boldsymbol{A}_\varepsilon)\psi_\varepsilon,\ \boldsymbol{n}\cdot(\nabla - i\boldsymbol{A}_\varepsilon)\psi_\varepsilon).
\end{aligned} \tag{1.5}$$

*Proof.* A proof for a more complicated case is given in the Section 1 of Chapter 9. To adapt that proof for our situation, just choose $a(x) \equiv 1$ then carry over the proof of Pohozaev equality there word for word.

## 6.2 BBH type upper bound for energy of minimizers

In this part, we derive that under our boundary conditions, the BBH type energy upper bound also holds (see [BBH 94])

**Proposition 6.2.1** *Let $(\psi_\varepsilon, \boldsymbol{A}_\varepsilon)$ be energy minimizers of (1.1) and let $\varepsilon < 1$, we have the following energy upper bound:*

$$G_\varepsilon(\psi_\varepsilon, \boldsymbol{A}_\varepsilon) \le \pi d \log\frac{1}{\varepsilon} + C \tag{2.1}$$

*where $C$ depends only on $\Omega$, $g$ and $d$.*

*Proof.* The idea is similar to that of [BBH 94]. We construct an explicit comparison function $(v_\varepsilon, \boldsymbol{B}_\varepsilon) \in V$ such that

$$G_\varepsilon(v_\varepsilon, \boldsymbol{B}_\varepsilon) \le \pi d \log\frac{1}{\varepsilon} + C.$$

First, we fix some arbitrary $d$ distinct points $a_1, \ldots, a_d$ in $\Omega$ and a small number $R > 0$ such that

$$B(a_j, R) \subset \Omega,$$

and no two of them intersect. We then set

$$\Omega_p = \Omega \setminus \cup_{j=1}^d B(a_j, R)$$

and let $w_0$ be a smooth map from $\Omega_p$ to $S^1$ verifying the following conditions:

$$w_0 = \exp(i\theta_j) = \frac{z-a_j}{|z-a_j|} \quad \text{on } \partial B(a_j, R),\ j = 1, \ldots, d.$$

It is clear that $w_0$ is a smooth map from $\Omega_p$ to $S^1$ and it maps any sub-region of $\Omega_p$ into a region in $I\!R^2$ with zero area, the standard change of variable concept implies then

$$2\frac{\partial w_0}{\partial x_1} \times \frac{\partial w_0}{\partial x_2} = \frac{\partial}{\partial x_1}\Big(w_0 \times \frac{\partial w_0}{\partial x_2}\Big) - \frac{\partial}{\partial x_2}\Big(w_0 \times \frac{\partial w_0}{\partial x_1}\Big) = 2\det \nabla w_0 = 0.$$

We then set (note here the formal definition of $\times$ for complex valued functions, this will be used throughout this chapter)

$$\begin{aligned} B_1^0 &= w_0 \times \frac{\partial w_0}{\partial x_1} = (iw_0, \frac{\partial w_0}{\partial x_1}) = \text{Re}\ (iw_0\frac{\partial w_0^*}{\partial x_1}), \\ B_2^0 &= w_0 \times \frac{\partial w_0}{\partial x_2} = (iw_0, \frac{\partial w_0}{\partial x_2}) = \text{Re}\ (iw_0\frac{\partial w_0^*}{\partial x_2}) \end{aligned}$$

where (and throughout this chapter) $(z_1, z_2)$ =Re$(z_1 z_2^*)$ and $z^*$ is the complex conjugate of the complex number $z$. Let $\boldsymbol{B}^0 = (B_1^0, B_2^0)^T$, we easily verify that

$$(\nabla - i\boldsymbol{B}^0)w_0 = 0 \text{ in } \Omega_p, \tag{2.2}$$

$$\mathbf{curl}\boldsymbol{B}^0 = 0 \text{ in } \Omega_p. \tag{2.3}$$

Let $\boldsymbol{B}^1$ be the solution of the following Dirichlet problem for the Laplacian

$$\begin{cases} \Delta \boldsymbol{B}^1 = 0 \text{ in } \Omega, \\ \boldsymbol{B}^1 = g\boldsymbol{\tau}, \text{ on } \partial\Omega. \end{cases} \tag{2.4}$$

We have the following standard elliptic estimate:

$$\int_\Omega |\boldsymbol{B}^1|^2 + |\nabla \boldsymbol{B}^1|^2 < C \tag{2.5}$$

where $C$ depends on $\Omega$ and $g$. On $\Omega_p$, we let

$$\boldsymbol{B}_\varepsilon = \boldsymbol{B}^0 + \boldsymbol{B}^1, \text{ and } v_\varepsilon = w_0.$$

It is straightforward to verify that

$$\begin{aligned} &|v_\varepsilon| = 1 \text{ on } \Omega_p, \\ &(\nabla - i\boldsymbol{B}_\varepsilon)v_\varepsilon = -i\boldsymbol{B}^1 v_\varepsilon \text{ on } \Omega_p, \\ &\mathbf{curl}\boldsymbol{B}_\varepsilon = \mathbf{curl}\boldsymbol{B}^1 \text{ on } \Omega_p. \end{aligned}$$

We obtain then

$$\begin{aligned} &\frac{1}{2}\int_{\Omega_p} |(\nabla - i\boldsymbol{B}_\varepsilon)v_\varepsilon)|^2 + |\mathbf{curl}\boldsymbol{B}_\varepsilon|^2 + \frac{1}{2\varepsilon^2}(1-|v_\varepsilon|^2)^2 \\ &\qquad = \frac{1}{2}\int_{\Omega_p} |\boldsymbol{B}^1|^2 + |\nabla \boldsymbol{B}^1|^2 < C. \end{aligned} \tag{2.6}$$

We need to define $(v_\varepsilon, \boldsymbol{B}_\varepsilon)$ on $\cup_{j=1}^d B(a_j, R)$ now to complete our construction. Let $\xi$ be a smooth function from $\mathbb{R}$ to $\mathbb{R}$ such that

$$\xi(t) = \begin{cases} 1 & t \geq 1, \\ 0 & t \leq \frac{1}{2} \end{cases}$$

We assume that $\varepsilon < R$ and set, on $B(a_j, R)$,

$$\begin{aligned} v_\varepsilon(x) &= \frac{x - a_j}{|x - a_j|} \xi(\frac{|x - a_j|}{\varepsilon}), \\ \boldsymbol{B}_\varepsilon(x) &= \boldsymbol{B}_\varepsilon(\frac{x - a_j}{|x - a_j|} R) \xi(\frac{|x - a_j|}{R}). \end{aligned}$$

Substituting these terms into the energy expression and integrating directly, we obtain

$$\frac{1}{2} \int_{B(a_j, R)} |(\nabla - i\boldsymbol{B}_\varepsilon) v_\varepsilon)|^2 + |\mathbf{curl} \boldsymbol{B}_\varepsilon|^2 + \frac{1}{2\varepsilon^2}(1 - |v_\varepsilon|^2)^2 < \pi \log \frac{1}{\varepsilon} + C \quad (2.7)$$

where $C$ is a constant depending only on $R$. Combining (2.6) and (2.7), we obtain

$$G_\varepsilon(v_\varepsilon, \boldsymbol{B}_\varepsilon) \leq \pi d \log \frac{1}{\varepsilon} + C. \quad (2.8)$$

The justification that the boundary conditions are satisfied by $(v_\varepsilon, \boldsymbol{B}_\varepsilon)$ is straightforward. Hence we have justified that $(v_\varepsilon, \boldsymbol{B}_\varepsilon) \in V$. (2.1) now follows from (2.8).

## 6.3 Global estimates

The main purpose of this section is to find an $L^\infty$ bound for $\nabla \psi_\varepsilon$ and $\nabla h_\varepsilon$ on the entire domain $\Omega$.

**Proposition 6.3.1** *On $\Omega$, we have*

$$|(\nabla - i\boldsymbol{A}_\varepsilon)\psi_\varepsilon| \leq C/\varepsilon, \quad (3.1)$$

$$|\nabla h_\varepsilon| \leq |(\nabla - i\boldsymbol{A}_\varepsilon)\psi_\varepsilon| \leq C/\varepsilon \quad (3.2)$$

*where $C$ is a constant depending only on $g$, $\Omega$ and $d$.*

*Proof.* Proposition 2.1 implies that

$$\int_\Omega |h_\varepsilon|^2 \leq C \log \frac{1}{\varepsilon}. \quad (3.3)$$

We rescale the problem before proceeding with the proof: let

$$\overline{x} = \frac{x}{\varepsilon}, \quad \text{and } \overline{\Omega} = \frac{1}{\varepsilon}\Omega, \quad (3.4)$$

$$\overline{\psi}(\overline{x}) = \psi_\varepsilon(\varepsilon \overline{x}) = \psi_\varepsilon(x), \quad (3.5)$$

$$\overline{\boldsymbol{A}}(\overline{x}) = \varepsilon \boldsymbol{A}_\varepsilon(\varepsilon \overline{x}). \quad (3.6)$$

We have then

$$(\nabla - i\overline{\boldsymbol{A}})\overline{\psi} = \varepsilon(\nabla - i\boldsymbol{A}_\varepsilon)\psi_\varepsilon(\varepsilon\overline{x}) = \varepsilon(\nabla - i\boldsymbol{A}_\varepsilon)\psi_\varepsilon(x), \tag{3.7}$$

$$\overline{h}(\overline{x}) = \mathbf{curl}\overline{\boldsymbol{A}}(\overline{x}) = \varepsilon^2 h_\varepsilon(x). \tag{3.8}$$

It follows that $(\overline{\psi}, \overline{\boldsymbol{A}})$ minimizes the rescaled functional

$$\overline{G}(v, \boldsymbol{B}) = \int_{\overline{\Omega}} \frac{1}{2}|(\nabla - i\boldsymbol{B})v|^2 + \frac{1}{2\varepsilon^2}|\mathbf{curl}\boldsymbol{B}|^2 + \frac{1}{4}(1 - |v|^2)^2$$

on the function space

$$V = \{(v, \boldsymbol{B}) \in \mathcal{H}^1 \times \boldsymbol{H}^1,\ |v| = 1 \text{ on } \partial\overline{\Omega},\ \boldsymbol{J} \cdot \boldsymbol{\tau} = \overline{g},\ \deg(v, \partial\overline{\Omega}) = d.\}$$

where $\overline{g}(\overline{x}) = \varepsilon g(\varepsilon x)$.

The Euler-Lagrange equations for $(\overline{\psi}, \overline{\boldsymbol{A}})$ become

$$-(\nabla - i\overline{\boldsymbol{A}})^2\overline{\psi} = \overline{\psi}(1 - |\overline{\psi}|^2), \tag{3.9}$$

$$\frac{1}{\varepsilon^2}\mathbf{curl}^2\overline{\boldsymbol{A}} = (i\overline{\psi}, (\nabla - i\overline{\boldsymbol{A}})\overline{\psi}). \tag{3.10}$$

Using $\mathrm{div}\overline{\boldsymbol{A}} = 0$, we can rewrite (3.10) as

$$-\frac{1}{\varepsilon^2}\Delta\overline{\boldsymbol{A}} + |\overline{\psi}|^2\overline{\boldsymbol{A}} = (-i\overline{\psi}, \nabla\overline{\psi}). \tag{3.11}$$

Let $\overline{x}_0 \in \overline{\Omega}$ and we want to show that

$$|(\nabla - i\overline{\boldsymbol{A}})\overline{\psi}| \le C \ \text{ in } \ B(\overline{x}_0, 1). \tag{3.12}$$

We consider the problem under two possibilities.

**Case I:** $\mathrm{dist}(x_0, \partial\Omega) > 2$. Using gauge invariance, we choose $(\tilde{\psi}, \tilde{\boldsymbol{A}})$ which are gauge equivalent to $(\overline{\psi}, \overline{\boldsymbol{A}})$ but satisfy

$$\begin{cases} \mathrm{div}\tilde{\boldsymbol{A}} = 0 & \text{in} \quad B(\overline{x}_0, 2), \\ \tilde{\boldsymbol{A}} \cdot \boldsymbol{n} = 0 & \text{on} \quad \partial B(\overline{x}_0, 2). \end{cases} \tag{3.13}$$

Hence there is a scalar function $\tilde{\xi} \in H^2(B(\overline{x}_0, 2))$ such that

$$\tilde{\boldsymbol{A}} = \mathbf{curl}\tilde{\xi} = (-\tilde{\xi}_{x_2}, \tilde{\xi}_{x_1})^T \tag{3.14}$$

where $\tilde{\xi}$ satisfies

$$\begin{cases} \mathbf{curl}\tilde{\boldsymbol{A}} = \Delta\tilde{\xi} = \overline{h} & \text{in} \quad B(\overline{x}_0, 2), \\ \tilde{\xi} = 0 & \text{on} \quad \partial B(\overline{x}_0, 2). \end{cases}$$

Using (3.3) and (3.8) as well as standard elliptic estimates, we obtain

$$\begin{aligned} ||H\tilde{\xi}||_{L^2(B(\overline{x}_0,2))} &\leq C||\overline{h}||_{L^2(\overline{\Omega})} \leq c\varepsilon^2||h_\varepsilon||_{L^2(\overline{\Omega})} \\ &\leq c\varepsilon||h_\varepsilon||_{L^2(\Omega)} \leq C\varepsilon\sqrt{|\log\varepsilon|} \end{aligned} \tag{3.15}$$

with $H\tilde{\xi} = (\partial^2_{x_ix_j}\tilde{\xi})$ being the Hessian matrix of $\tilde{\xi}$. It follows that

$$||\tilde{\boldsymbol{A}}||^2_{H^1(B(\overline{x}_0,2))} \leq C\varepsilon^2|\log\varepsilon|. \tag{3.16}$$

Because of the gauge condition $\text{div}\tilde{\boldsymbol{A}} = 0$, (3.9) becomes

$$-\Delta\tilde{\psi} = \tilde{\psi}(1-|\tilde{\psi}|^2) - |\tilde{\boldsymbol{A}}|^2\tilde{\psi} - 2i(\tilde{\boldsymbol{A}}, \nabla\tilde{\psi}). \tag{3.17}$$

We use the decomposition

$$\tilde{\psi} = \tilde{\psi}_0 + \tilde{\psi}_1 \quad \text{on} \quad B(\overline{x}_0, 2) \tag{3.18}$$

with $\tilde{\psi}_0$ satisfying

$$\begin{cases} \Delta\tilde{\psi}_0 = 0 & \text{in} \quad B(\overline{x}_0,2), \\ \tilde{\psi}_0 = \tilde{\psi} & \text{on} \quad \partial B(\overline{x}_0,2) \end{cases} \tag{3.19}$$

and $\tilde{\psi}_1$ satisfying

$$\begin{cases} -\Delta\tilde{\psi}_1 = \tilde{\psi}(1-|\tilde{\psi}|^2) - |\tilde{\boldsymbol{A}}|^2\tilde{\psi} - 2i(\tilde{\boldsymbol{A}}, \nabla\tilde{\psi}) & \text{in} \quad B(\overline{x}_0,2), \\ \tilde{\psi}_1 = 0 & \text{on} \quad \partial B(\overline{x}_0,2). \end{cases} \tag{3.20}$$

Being a solution to a Laplace equation with Dirichlet boundary condition, we have

$$|\nabla\tilde{\psi}_0| < C \quad \text{on} \quad B(\overline{x}_0, 3/2), \tag{3.21}$$

for some universal constant $C$ (note that the boundary data is bounded by 1). For any $p < \infty$, using (3.16), we have

$$|||\tilde{\boldsymbol{A}}|^2\tilde{\psi}||_{L^p} \leq ||\tilde{\boldsymbol{A}}^2||_{L^p} = ||\tilde{\boldsymbol{A}}||^2_{L^{2p}} \leq ||\tilde{\boldsymbol{A}}||^2_{H^1} \leq C\varepsilon^2|\log\varepsilon|. \tag{3.22}$$

For $p < 2$ and $q$ such that $1/q + p/2 = 1$, we have

$$\begin{aligned} ||\tilde{\boldsymbol{A}}\cdot\nabla\tilde{\psi}||_{L^p} &\leq ||\tilde{\boldsymbol{A}}\cdot(\nabla - i\tilde{\boldsymbol{A}})\tilde{\psi}||_{L^p} + |||\tilde{\boldsymbol{A}}|^2\tilde{\psi}||_{L^p} \\ &\leq ||(\nabla - i\tilde{\boldsymbol{A}})\tilde{\psi}||_{L^2}||\tilde{\boldsymbol{A}}||_{L^{pq}} + ||\tilde{\boldsymbol{A}}^2||_{L^p} \qquad (3.23) \\ \text{(energy estimate)} \quad &\leq C|\log\varepsilon|||\tilde{\boldsymbol{A}}||_{L^{pq}} + ||\tilde{\boldsymbol{A}}^2||_{L^p} \\ &\leq C|\log\varepsilon|||\tilde{\boldsymbol{A}}||_{H^1} + ||\tilde{\boldsymbol{A}}^2||_{L^p} \\ \text{(using (3.16))} \quad &\leq C\varepsilon|\log\varepsilon|^{3/2}. \qquad (3.24) \end{aligned}$$

It follows from (3.20), (3.22) and (3.24) that for any $p < 2$,

$$\begin{aligned}&||\tilde{\psi}_1||_{W^{2,p}(B(\overline{x}_0,2))}\\&\leq ||\tilde{\psi}(1-|\tilde{\psi}|^2) - |\tilde{\boldsymbol{A}}|^2\tilde{\psi} - 2i(\tilde{\boldsymbol{A}}, \nabla\tilde{\psi})||_{L^p}\\&\leq C_p\varepsilon(\sqrt{|\log\varepsilon|})^3.\end{aligned}$$

Using standard Sobolev embedding, we deduce that for $1 < p < \infty$,

$$||\nabla\tilde{\psi}_1||_{L^p(B(\overline{x}_0,2))} \leq C_p\varepsilon(\sqrt{|\log\varepsilon|})^3 \leq C_p. \tag{3.25}$$

Here $C_p$ denotes different constants depending on $p$. Consequently, for any $1 < p < \infty$,

$$\begin{aligned}||\nabla\tilde{\psi}||_{L^p(B(\overline{x}_0,3/2))} &\leq ||\nabla\tilde{\psi}_0||_{L^p(B(\overline{x}_0,3/2))} + ||\nabla\tilde{\psi}_1||_{L^p(B(\overline{x}_0,3/2))}\\ \text{(using (3.21) and (3.25))} &\leq C_p.\end{aligned} \tag{3.26}$$

With this better estimate, on the smaller domain $B(\overline{x}_0, 3/2)$, equation (3.20) leads to,

$$||\tilde{\psi}_1||_{W^{2,p}(B(\overline{x}_0,3/2))} \leq C_p \ \ \forall\ 1 < p < \infty.$$

By Sobolev embedding, we have the uniform bound

$$|\nabla\tilde{\psi}_1| \leq C \ \text{ on } B(\overline{x}_0, 3/2).$$

Combining with (3.21), we have

$$|\nabla\tilde{\psi}| \leq C \ \text{ on } \ B(\overline{x}_0, 3/2). \tag{3.27}$$

Using (3.11) on $B(\overline{x}_0, 3/2)$ while reducing the domain further, we obtain

$$|\tilde{\boldsymbol{A}}| \leq C \ \text{ on } \ B(\overline{x}_0, 1) \tag{3.28}$$

Combining our results, we deduce (3.12).

**Case II:** $\overline{x}_0 \in \partial\overline{\Omega}$. Choose $(\tilde{\psi}, \tilde{\boldsymbol{A}})$ gauge equivalent to $(\overline{\psi}, \overline{\boldsymbol{A}})$ such that

$$\begin{cases} \operatorname{div}\tilde{\boldsymbol{A}} = 0 & \text{in} \quad B(\overline{x}_0, 2) \cap \overline{\Omega},\\ \tilde{\boldsymbol{A}} \cdot \boldsymbol{n} = 0 & \text{on} \quad \partial(B(\overline{x}_0, 2) \cap \overline{\Omega}).\end{cases} \tag{3.29}$$

As in Case I, we again use the decomposition

$$\tilde{\psi} = \tilde{\psi}_0 + \tilde{\psi}_1 \tag{3.30}$$

with $\tilde{\psi}_0$ satisfying

$$\begin{cases} \Delta\tilde{\psi}_0 = 0 & \text{in} \quad B(\overline{x}_0, 2) \cap \Omega,\\ \tilde{\psi}_0 = \tilde{\psi} & \text{on} \quad \partial(B(\overline{x}_0, 2) \cap \Omega).\end{cases} \tag{3.31}$$

Again as in Case I, we have $\tilde{\boldsymbol{A}} = (-\tilde{\xi}_{x_2}, \tilde{\xi}_{x_1})$ where $\tilde{\xi}$ satisfies

$$\begin{cases} \Delta\tilde{\xi} = \overline{h} & \text{in} \quad B(\overline{x}_0, 2) \cap \overline{\Omega}, \\ \tilde{\xi} = 0 & \text{on} \quad \partial(B(\overline{x}_0, 2) \cap \overline{\Omega}). \end{cases} \tag{3.32}$$

From these definitions, we have then

$$\tilde{\boldsymbol{A}} \cdot \boldsymbol{\tau} = \partial_{\boldsymbol{n}}\tilde{\xi}, \tag{3.33}$$

$$\partial_{\boldsymbol{\tau}}\tilde{\psi} = i\Big(\partial_{\boldsymbol{n}}\tilde{\xi} + g\Big)\tilde{\psi} \quad \text{on} \quad B(\overline{x}_0, 2) \cap \partial\overline{\Omega}. \tag{3.34}$$

Classical linear estimates imply from (3.32) and (3.34) that

$$||\partial_{\boldsymbol{\tau}}\tilde{\psi}||_{H^{1/2}(\partial\overline{\Omega}\cap B(\overline{x}_0,3/2))} \leq C.$$

This implies, if we write $\tilde{\psi} = e^{i\chi}$ on $\partial\overline{\Omega}$, that

$$||\partial_{\boldsymbol{\tau}}\chi||_{H^{1/2}(\partial\overline{\Omega}\cap B(\overline{x}_0,3/2))} \leq C.$$

Hence, since $\tilde{\psi}_0$ is harmonic, it is $H^2$ up to the part of the boundary $B(\overline{x}_0, 2-\delta)\cap\partial\overline{\Omega}$ for any small $\delta > 0$. Consequently, it follows from the standard $L^p$-theory for elliptic equations, $p = 2$, that

$$||\tilde{\psi}_0||_{H^2(B(\overline{x}_0,3/2)\cap\overline{\Omega})} \leq C. \tag{3.35}$$

Similarly, the parallel of (3.16) becomes (notice that the domain has been slightly reduced in order to avoid the corner points produced by $B(\overline{x}_0, 2) \cap \overline{\Omega}$)

$$||\tilde{\boldsymbol{A}}||_{H^1(B(\overline{x}_0,3/2)\cap\overline{\Omega})} \leq C. \tag{3.36}$$

Arguments parallel to (3.20)–(3.27) yield

$$||\tilde{\psi}||_{H^1(B(\overline{x}_0,3/2)\cap\overline{\Omega})} \leq C \tag{3.37}$$

and

$$||(\nabla - i\tilde{\boldsymbol{A}})\tilde{\psi}||_{H^1(B(\overline{x}_0,3/2)\cap\overline{\Omega})} \leq C. \tag{3.38}$$

Since (3.10) gives,

$$-\mathbf{curl}\overline{h} = -\nabla^{\perp}\overline{h} = \varepsilon^2(i\tilde{\psi}, (\nabla - i\tilde{\boldsymbol{A}})\tilde{\psi}),$$

we have

$$|\nabla\overline{h}|_{L^2(B(\overline{x}_0,3/2)\cap\overline{\Omega})} \leq \varepsilon^2|(\nabla - i\tilde{\boldsymbol{A}})\tilde{\psi}|_{L^2(B(\overline{x}_0,3/2)\cap\Omega)} \leq c\varepsilon^2.$$

Hence, using the estimate (same as (3.15)) $||\overline{h}||_{L^2(\overline{\Omega})} \leq C\varepsilon\log(1/\varepsilon)$, we obtain, by Sobolev embedding, for any $p > 2$

$$|\overline{h}|_{L^p(B(\overline{x}_0,3/2)\cap\overline{\Omega})} \leq C\varepsilon^2\log(1/\varepsilon) \to 0.$$

Substituting back into equation (3.32), we have that $\tilde{\xi} \in W^{2,p}(\Omega)$ which combined with (3.34) implies that $\tilde{\psi} \in W^{2-\frac{1}{p},p}(\partial\Omega)$. Therefore it follows from (3.31) and the standard $L^p$-theory for elliptic equations that

$$||\tilde{\psi}_0||_{W^{2,p}(B(\overline{x}_0,5/4)\cap\overline{\Omega})} \leq C.$$

This implies that on $B(\overline{x}_0, 1) \cap \overline{\Omega}$, we have

$$|\nabla\tilde{\psi}_0| \leq C.$$

Following similar arguments after (3.21) in Case I, we complete then the proof of (3.12) as in case I.

## 6.4 Local estimates

In the previous section, we have shown that both $\nabla\psi_\varepsilon$ and $\nabla h_\varepsilon$ are of the order $O(\varepsilon^{-1})$ over the whole domain. This is equivalent to saying that $\varepsilon$ is the characteristic length of variations of $(\psi_\varepsilon, \boldsymbol{A}_\varepsilon)$. Now we derive some estimates on the length scale $\varepsilon^\alpha$ with $\alpha \in (0,1)$ being a fixed number. These estimates are local in the sense that they are made near some given points in the domain (including vortex locations).

To simplify notation, we drop the dependence of $\psi_\varepsilon$, $h_\varepsilon$ and $\boldsymbol{A}_\varepsilon$ on $\varepsilon$ by writing them simply as $\psi$, $h$ and $\boldsymbol{A}$.

**Proposition 6.4.1** *Let $0 < \alpha < 1$ and let $x_0 \in \Omega$. We have*

$$\int_{B(x_0,\varepsilon^\alpha)\cap\Omega} |h|^2 \leq C\varepsilon^\alpha|\log\varepsilon|, \tag{4.1}$$

$$\frac{1}{\varepsilon^2}\int_{B(x_0,\varepsilon^\alpha)\cap\Omega} (1-|u|^2)^2 \leq C \tag{4.2}$$

*where $C$ depends on $g$, $d$, $\alpha$ and $\Omega$ only (not on $\varepsilon$).*

*Proof.* Step 1. Proof of (4.1). From the G-L system, we have (note that $|\mathbf{curl}h| = |\nabla h|$)

$$|\nabla h| = |\mathbf{curl}^2\boldsymbol{A}| = |(i\psi, (\nabla - i\boldsymbol{A})\psi)| \leq |(\nabla - i\boldsymbol{A})\psi|. \tag{4.3}$$

Proposition 2.1 implies then

$$\int_\Omega |\nabla h|^2 dx \leq C|\log\varepsilon|.$$

Standard energy estimate (2.1) implies that

$$\int_\Omega |h|^2 \leq C|\log\varepsilon|.$$

Using Sobolev embedding, we obtain that

$$(\int_\Omega |h|^p dx)^{1/p} = ||h||_p \le C_p ||h||_{W^{1,2}(\Omega)} \le C_p |\log\varepsilon|^{1/2} \ \forall p > 2. \tag{4.4}$$

It follows then, choosing $p = 4$ in the above,

$$\sqrt{\int_{B(x_0,\varepsilon^\alpha)\cap\Omega} |h|^2 dx} \le (\int_\Omega |h|^4 dx)^{1/4} [\text{meas}(B(x_0,\varepsilon^\alpha))]^{\frac{1}{4}} \le C|\log\varepsilon|^{1/2}\varepsilon^{\alpha/2}. \tag{4.5}$$

This is (4.1).

Step 2. Proof of (4.2). Recall that the standard energy estimate (2.1) implies that

$$G_\varepsilon(\psi, \boldsymbol{A}) \le C|\log\varepsilon|.$$

This leads to

$$\frac{1}{2}\int_{B(x_0,\varepsilon^{\alpha/2})\cap\Omega} |(\nabla - i\boldsymbol{A})\psi|^2 + |\mathbf{curl}\boldsymbol{A}|^2 + \frac{1}{2\varepsilon^2}(1-|\psi|^2)^2 \le C|\log\varepsilon|.$$

Using Fubini's theorem, we rewrite the above inequality as

$$\int_{\varepsilon^\alpha}^{\varepsilon^{\alpha/2}} I_r(\psi, \boldsymbol{A})dr \le C|\log\varepsilon| \tag{4.6}$$

where, for any $r \in (\varepsilon^\alpha, \varepsilon^{\alpha/2})$,

$$I_r(\psi, \boldsymbol{A}) = \frac{1}{2}\int_{\partial B(x_0,r)\cap\Omega} |(\nabla - i\boldsymbol{A})\psi|^2 + |\mathbf{curl}\boldsymbol{A}|^2 + \frac{1}{2\varepsilon^2}(1-|\psi|^2)^2. \tag{4.7}$$

We now show that there is some $\rho \in (\varepsilon^\alpha, \varepsilon^{\alpha/2})$ and a constant $C_\alpha$ depending on $\alpha$, $g$ and $d$ but not $\varepsilon$ such that for all $\varepsilon > 0$

$$I_\rho(\psi, \boldsymbol{A}) \le \frac{C_\alpha}{\rho}. \tag{4.8}$$

Assuming by contradiction that (4.8) does not hold for any $\rho \in (\varepsilon^\alpha, \varepsilon^{\alpha/2})$ and any $C_\alpha$. For any given $C_\alpha$, there is some $\varepsilon$ such that (4.8) does not hold. Integrating the inverse inequality of (4.8) for one of these particular $\varepsilon$, we obtain

$$\int_{\varepsilon^\alpha}^{\varepsilon^{\alpha/2}} I_r(\psi, \boldsymbol{A})dr \ \ge \ \int_{\varepsilon^\alpha}^{\varepsilon^{\alpha/2}} \frac{C_\alpha}{r}dr \ge \frac{1}{2}\alpha C_\alpha|\log\varepsilon|.$$

This contradicts (4.6) if $C_\alpha$ is chosen sufficiently large. Now, we proceed while distinguishing the following two possibilities.

Case I: $\partial B(x_0,\rho)\cap\Omega=\emptyset$. On the ball $B(x_0,\rho)$, we apply Proposition 1.2:

$$\begin{aligned}
&\int_{B(x_0,\rho)}\frac{1}{\varepsilon^2}(1-|\psi|^2)^2-|h|^2\\
=&\int_{\partial B(x_0,\rho)}(x\cdot\boldsymbol{n})[(\boldsymbol{\tau}\cdot(\nabla-i\boldsymbol{A}_\varepsilon)\psi_\varepsilon)^2-(\boldsymbol{n}\cdot(\nabla-i\boldsymbol{A}_\varepsilon)\psi_\varepsilon)^2]\\
&+\frac{1}{2}\int_{\partial B(x_0,\rho)}(x\cdot\boldsymbol{n})(\frac{1}{\varepsilon^2}(1-|\psi_\varepsilon|^2)^2-h_\varepsilon^2)\\
&-\int_{\partial B(x_0,\rho)}(x\cdot\boldsymbol{\tau})(\boldsymbol{\tau}\cdot(\nabla-i\boldsymbol{A}_\varepsilon)\psi_\varepsilon,\ \boldsymbol{n}\cdot(\nabla-i\boldsymbol{A}_\varepsilon)\psi_\varepsilon)\\
\le\ & C\rho I_\rho(\psi,\boldsymbol{A})\le C.
\end{aligned}$$

Combined with (4.1) and we obtain the desired result.

Case 2: $\partial B(x_0,\rho)\cap\Omega\neq\emptyset$. We write

$$\partial[B(x_0,\rho)\cap\Omega]=(\partial B(x_0,\rho)\cap\Omega)\cup\Gamma_\rho$$

where

$$\Gamma_\rho=\partial\Omega\cap B(x_0,\rho).$$

Without losing generality, let the origin be a point in $\Omega$, we apply Proposition 1.2 on $\Omega\cap B(x_0,\rho)$ with respect to the origin to obtain, noting that $|\psi|=1$ on $\Gamma_\rho\subset\partial\Omega$,

$$\begin{aligned}
&\int_{B(x_0,\rho)\cap\Omega}\frac{1}{\varepsilon^2}(1-|\psi|^2)^2-|h|^2\le C(\rho+|x_0|)I_\rho(\psi,\boldsymbol{A})\\
&+\int_{\Gamma_\rho}x\cdot\boldsymbol{n}[(\boldsymbol{\tau}\cdot(\nabla-i\boldsymbol{A})\psi)^2-(\boldsymbol{n}\cdot(\nabla-i\boldsymbol{A})\psi)^2]\\
&-\frac{1}{2}\int_{\Gamma_\rho}x\cdot\boldsymbol{n}h^2+\int_{\Gamma_\rho}x\cdot\boldsymbol{\tau}(\boldsymbol{\tau}\cdot(\nabla-i\boldsymbol{A})\psi,\boldsymbol{n}\cdot(\nabla-i\boldsymbol{A})\psi).
\end{aligned}\tag{4.9}$$

We have then

$$\begin{aligned}
&\int_{\Gamma_\rho}x\cdot\boldsymbol{\tau}(\boldsymbol{\tau}\cdot(\nabla-i\boldsymbol{A})\psi,\boldsymbol{n}\cdot(\nabla-i\boldsymbol{A})\psi)\\
\le\ &\frac{1}{2}\int_{\Gamma_\rho}|x_0\cdot\boldsymbol{\tau}|[|\boldsymbol{\tau}\cdot(\nabla-i\boldsymbol{A})\psi|^2+|\boldsymbol{n}\cdot(\nabla-i\boldsymbol{A})\psi|^2].
\end{aligned}$$

Since $\varepsilon$ is very small, we denote by $\pi x_0$ the unique nearest point projection of $x_0$ on $\partial\Omega$ and translate the origin such that $\mathrm{dist}(0,\pi x_0)=4\rho$ and $x_0$ lies on the line segment $[\pi x_0,0]$, we have then

$$x\cdot\boldsymbol{n}\ge 0\ \forall x\in\Gamma_\rho\quad\text{and}\quad\frac{1}{2}x\cdot\boldsymbol{\tau}\le x\cdot\boldsymbol{n}\ \ \forall x\in\Gamma_\rho.$$

(4.9) then yields

$$\begin{aligned}
&\int_{B(x_0,\rho)\cap\Omega}\frac{1}{\varepsilon^2}(1-|\psi|^2)^2-|h|^2\\
\leq\ & C\rho\Big[I_\rho(\psi,\boldsymbol{A})+\int_{\Gamma_\rho}|\boldsymbol{\tau}\cdot(\nabla-i\boldsymbol{A})\psi|^2\Big]\\
\leq\ & C\rho\Big[I_\rho(\psi,\boldsymbol{A})+\int_{\Gamma_\rho}g^2\Big]\leq C
\end{aligned}$$

and this implies (4.2).

## 6.5 The behaviour of solutions near vortices

The purpose of this section is to give some description of the behaviour of the solutions of the G-L system near the location of the vortices using $\varepsilon$ dependent estimates. As in the previous section, we also drop the index $\varepsilon$ for various quantities when there is no possible confusion. Before we announce our main result, we give some preliminary propositions.

**Proposition 6.5.1** *There exist positive constants $\lambda_0$, $\mu_0$ depending only on $d$, $g$ and $\Omega$ such that if $(\psi,\boldsymbol{A})$ is a pair of minimizers of $G_\varepsilon$ on $V$ satisfying*

$$\int_{B(x_0,2l)\cap\Omega}\frac{1}{\varepsilon^2}(1-|\psi|^2)^2\leq\mu_0\quad \textit{with}\ \ x_0\in\Omega, \tag{5.1}$$

$$\frac{l}{\varepsilon}\geq\lambda_0,\qquad l\leq 1, \tag{5.2}$$

*then*

$$|\psi(x)|\geq\frac{1}{2}\quad \textit{on}\ \ \Omega\cap B(x_0,l). \tag{5.3}$$

*Proof.* Writing $\psi=\rho e^{i\phi}$, we have

$$|\nabla\rho|\leq|(\nabla-i\boldsymbol{A})\psi|\leq C/\varepsilon \tag{5.4}$$

because $(\nabla-i\boldsymbol{A})\psi=(\nabla\rho+i\rho(\nabla\phi-\boldsymbol{A}))e^{i\phi}$. Therefore, we have

$$|\rho(x)-\rho(y)|\leq\frac{C}{\varepsilon}|x-y|,\ \forall x,y\in\Omega\cap B(x_0,l). \tag{5.5}$$

Assuming that (5.3) doesn't hold. Then for any $\lambda_0$ and $\mu_0$, there exists a $x_0\in\Omega\cap B(x_0,l)$ such that $\rho(x_0)<1/2$, this implies that

$$|\rho(x)-\rho(x_0)|\leq\frac{C}{\varepsilon}|x-x_0|$$

and thus

$$\rho(x)\leq\frac{1}{2}+\frac{C}{\varepsilon}|x-x_0|\ \ \forall x\in\Omega\cap B(x_0,l).$$

Consequently,

$$1-\rho(x) \geq \frac{1}{2} - \frac{C}{\varepsilon}|x-x_0| \ \ \forall x \in \Omega \cap B(x_0, l).$$

We choose $l = \varepsilon/4C$ and notice that $|x - x_0| \leq l$, we obtain

$$1-\rho(x) \geq \frac{1}{4} \ \ \forall x \in \Omega \cap B(x_0, \varepsilon/4C)$$

and consequently

$$(\rho^2-1)^2 \geq \frac{1}{16} \ \text{ in } \ \Omega \cap B(x_0, \varepsilon/4C).$$

On the other hand, there is a positive constant (depending on $\Omega$ only) such that

$$\text{meas}(\Omega \cap B(x,l)) \geq \alpha l^2 \ \ \forall x \in \Omega, \ \ \forall l < 1.$$

Hence, we have

$$\int_{\Omega \cap B(x_0,\varepsilon/4C)} (\rho^2-1)^2 \geq \frac{\alpha\varepsilon^2}{(16C)^2}$$

provided $\varepsilon/4C \leq 1$. Note that $B(x_0, \varepsilon/4C) \subset B(x_0, 2l)$ where $\varepsilon/4C \leq l$ (since $x_0 \in B(x_0, l)$).

Therefore,

$$\int_{\Omega \cup B(x_0,2l)} (\rho^2-1)^2 \geq \frac{\alpha\varepsilon^2}{(16C)^2}.$$

If we choose $\lambda_0 = 1/4C$ and $\mu_0 < \alpha/(16C)^2$ in (5.1)–(5.2), the above inequality can not hold and we are led to a contradiction. The proposition is proved.

The previous proposition gives a local criterion where $\psi$ does NOT vanish. The following is a global description of the set on which $\psi$ does not vanish.

**Proposition 6.5.2** *For any given $0 < \alpha < 1$ and $x_0 \in \Omega$, there exist a constant $\overline{N}_\alpha$ depending only on $\alpha$, $g$, $d$ and points $x_1^\varepsilon$, $x_2^\varepsilon$, ..., $x_{l_\varepsilon}^\varepsilon$ in $B(x_0, \varepsilon^\alpha) \cap \Omega$ such that*

$$l_\varepsilon \leq \overline{N}_\alpha \tag{5.6}$$

*and*

$$|\psi_\varepsilon(x)| \geq \frac{1}{2} \ \ \forall x \in B(x_0, \varepsilon^\alpha) \cap [\Omega \setminus \left( \cup_{j=1}^{l_\varepsilon} B(x_j^\varepsilon, \lambda_0\varepsilon) \right)]. \tag{5.7}$$

*Proof.* Let $\lambda_0$ be the constant in Proposition 5.1 such that

$$\begin{aligned} &x_j \in \Omega \cap B(x_0, \varepsilon^\alpha), \\ &B(x_j, \tfrac{\lambda_0\varepsilon}{4}) \cap B(x_k, \tfrac{\lambda_0\varepsilon}{4}) = \emptyset \quad \forall j \neq k, \\ &\Omega \cap B(x_0, \varepsilon^\alpha) \subset \cup_{i\in I} B(x_i, \lambda_0\varepsilon). \end{aligned} \tag{5.8}$$

We consider a family of discs $(B(x_j, \lambda_0\varepsilon))_{i\in I}$. We say that the disc $B(x_j, \lambda_0\varepsilon)$ is bad if

$$\int_{B(x_j,2\lambda_0\varepsilon)\cap\Omega} \frac{1}{\varepsilon^2}(1-|\psi_\varepsilon|^2)^2 \geq \mu_0. \tag{5.9}$$

Hence if $x$ does not belong to a bad disc, we deduce, from Proposition 5.1, that $|\psi_\varepsilon(x)| \geq \frac{1}{2}$.

Let $\{B(x_j, \lambda_0\varepsilon)\}_{j\in J}$ be the collection of all possible bad discs, there is an absolute constant $C$ such that

$$|J|\mu_0 \leq \sum_{j\in J} \int_{B(x_j,2\lambda_0\varepsilon)\cap\Omega} \frac{1}{\varepsilon^2}(1-|\psi_\varepsilon|^2)^2 \leq C\int_{B(x_0,\varepsilon^\alpha)\cap\Omega} \frac{1}{\varepsilon^2}(1-|\psi_\varepsilon|^2)^2 \tag{5.10}$$

where $|J|$ =number of elements of $J$. Therefore the number of bad discs is bounded by

$$\frac{C}{\varepsilon^2}\int_{B(x_0,\varepsilon^\alpha)\cap\Omega} (1-|\psi_\varepsilon|^2)^2.$$

Proposition 4.1 implies that the above quantity is bounded by some $N_\alpha$ and so is the number of elements of $J$.

Next, we want to prove that there is a constant $N$ independent of $\alpha$ and $\varepsilon$ such that $N_\alpha \leq N$. To achieve this, we study the energy contribution of the gradient of the order parameter near a vortex location first. We show that in fact, this is of the order $|\log\varepsilon|$. We start with a preparatory lemma:

**Lemma 6.5.1** *Let $D > 0$, $0 < \beta < 1$, $\gamma > 1$ be given constants such that $\gamma\beta < 1$. Let $\rho < \varepsilon^\alpha$ satisfying $\rho^\gamma > \lambda_0\varepsilon$ and*

$$I_\rho(\psi_\varepsilon, \boldsymbol{A}_\varepsilon) < \frac{D}{\rho}, \tag{5.11}$$

$$|\psi_\varepsilon| \geq \frac{1}{2} \quad on \ \ \partial B(x_0,\rho)\cap\Omega, \tag{5.12}$$

$$\deg(\frac{\psi_\varepsilon}{|\psi_\varepsilon|},\ \partial[B(x_0,\rho)\cap\Omega]) = 0. \tag{5.13}$$

*Then there is a constant $\varepsilon_0$ depending on $D$, $\beta$, $\gamma$, $g$, $d$ and $\Omega$ only such that for all $\varepsilon < \varepsilon_0$, we have*

$$|\psi_\varepsilon| \geq \frac{1}{2} \quad on \ \ B(x_0,\rho^\gamma)\cap\Omega. \tag{5.14}$$

*Proof.* First, we introduce the following notation. For any given $K \subset \Omega$, we define

$$G(\psi, \boldsymbol{A}, K) = \frac{1}{2}\int_K |(\nabla - i\boldsymbol{A})\psi|^2 + |\mathbf{curl}\boldsymbol{A}|^2 + \frac{1}{2\varepsilon^2}(1-|\psi|^2)^2$$

and show that there is a constant $C$ depending only on $g$, $d$, $\Omega$ and $\beta$ such that

$$G(\psi, \boldsymbol{A}, \Omega \cap B(x_0, \rho)) \leq C\rho I_\rho(\psi, \boldsymbol{A}) + o(1) \tag{5.15}$$

where $o(1) \to 0$ uniformly as $\varepsilon \to 0$. In other words, the energy in a ball is controlled by the energy on the boundary.

The idea is to construct a comparison function. First, we change the gauge to a local one by considering $(\tilde{\psi}, \tilde{\boldsymbol{A}})$ gauge equivalent to $(\psi, \boldsymbol{A})$ such that

$$\begin{cases} \operatorname{div}\tilde{\boldsymbol{A}} = 0, & \text{in} \quad U = B(x_0, \rho) \cap \Omega, \\ \tilde{\boldsymbol{A}} \cdot \boldsymbol{\tau} = \frac{1}{|\Gamma|} \int_U h & \text{on} \quad \Gamma = \partial U. \end{cases} \tag{5.16}$$

Consequently, there is a scalar function $\xi$ solving the problem

$$\begin{cases} \Delta\xi = h, & \text{on} \quad U, \\ \nabla\xi \cdot \boldsymbol{n} = \frac{1}{|\Gamma|} \int_U h & \text{on} \quad \Gamma, \\ \int_U \xi = 0. \end{cases} \tag{5.17}$$

so that $\tilde{\boldsymbol{A}} = (-\xi_{x_2}, \xi_{x_1})$ in $U$.

By extending the function $\xi$ smoothly to $\Omega$, we can easily define a smooth extension of $\tilde{\boldsymbol{A}}$ to $\Omega$ by setting $\tilde{\boldsymbol{A}} = (-\xi_{x_2}, \xi_{x_1})$. Using (5.17), we have

$$\int_U |\nabla\xi|^2 \leq \int_U |h|^2.$$

By Proposition 4.1, there is some $\beta > 0$ such that

$$\int_U |\tilde{\boldsymbol{A}}|^2 \leq C\varepsilon^\beta |\log\varepsilon| \to 0 \ \text{ as } \varepsilon \to 0. \tag{5.18}$$

We have, by the fact that $\partial_{\boldsymbol{\tau}}\tilde{\psi} = (\nabla - i\tilde{\boldsymbol{A}})\tilde{\psi} \cdot \boldsymbol{\tau} + i\tilde{\boldsymbol{A}} \cdot \boldsymbol{\tau}\tilde{\psi}$,

$$\int_\Gamma |\partial_{\boldsymbol{\tau}}\tilde{\psi}|^2 \leq 2\int_\Gamma |(\nabla - i\boldsymbol{A})\psi|^2 + \int_\Gamma |\tilde{\boldsymbol{A}} \cdot \boldsymbol{\tau}|^2. \tag{5.19}$$

by (5.16),

$$\begin{aligned} \int_\Gamma |\tilde{\boldsymbol{A}} \cdot \boldsymbol{\tau}|^2 &\leq \frac{1}{|\Gamma|}\Big(\int_U |h|\Big)^2 \\ &\leq \frac{1}{|\Gamma|}\Big(\int_U |h|^2 \text{meas}(U)\Big) \leq C\varepsilon^\beta \int_U |h|^2 \leq C\varepsilon^\beta |\log\varepsilon|. \end{aligned}$$

Hence

$$\int_\Gamma |\partial_{\boldsymbol{\tau}}\tilde{\psi}|^2 \leq \frac{C}{\rho} + o(1) \tag{5.20}$$

where $C$ is a constant depending on $D$. Since the degree of $\tilde{\psi}$ along $\Gamma$ is zero, we may write on $\Gamma$ that $\tilde{\psi} = |\tilde{\psi}| \exp(i\tilde{\phi})$ where $\tilde{\phi}$ is a smooth real valued function.

We are now able to construct the comparison function $(v, \boldsymbol{B})$. Let

$$\boldsymbol{B} = \tilde{\boldsymbol{A}} \text{ in } \Omega$$

and

$$\begin{cases} v = \eta \exp(i\phi) & \text{in} \quad U \\ v = \tilde{\psi} & \text{in} \quad \Omega \setminus U. \end{cases} \tag{5.21}$$

Here $\phi$ is the solution of

$$\begin{cases} \Delta\phi = 0 & \text{in} \quad U, \\ \phi = \tilde{\phi} & \text{on} \quad \Gamma = \partial U \end{cases} \tag{5.22}$$

and $\eta$ is the solution of

$$\begin{cases} -\varepsilon^2 \Delta\eta + \eta = 1 & \text{in} \quad U, \\ \eta = |\tilde{\psi}| & \text{on} \quad \Gamma = \partial U. \end{cases} \tag{5.23}$$

Now we show that

$$\int_U |\nabla\phi|^2 \le C\rho I_\rho(\psi, \boldsymbol{A}) + o(1), \tag{5.24}$$

$$\int_U |\nabla\eta|^2 \le o(1), \tag{5.25}$$

$$\frac{1}{\varepsilon^2} \int_U (1-\eta)^2 \le o(1) \tag{5.26}$$

and consequently

$$\frac{1}{2} \int_U |\nabla v|^2 + \frac{1}{4\varepsilon^2} \int_U (1 - |v|^2)^2 \le C\rho I_\rho(\psi, \boldsymbol{A}) + o(1) \le C. \tag{5.27}$$

Here $o(1)$ tends to 0 uniformly as $\varepsilon \to 0$.

Proof of (5.24): We first assume that $\tilde{\phi} = 0$ somewhere along $\partial U$ because the phase is defined uniquely up to an added constant. Let $\boldsymbol{V}$ be a smooth vector valued function such that

$$\begin{cases} \boldsymbol{V}|_{\partial U} = \boldsymbol{n}, \\ |\nabla \boldsymbol{V}| \le \frac{C}{\rho}. \end{cases}$$

Multiplying both sides of (5.22-1) by $\boldsymbol{V} \cdot \nabla\phi$, we obtain

$$0 = \int_U \Delta\phi \boldsymbol{V} \cdot \nabla\phi = -\int_U |\nabla\phi|^2 \text{div} \boldsymbol{V} + \int_{\partial U} |\partial_{\boldsymbol{n}}\phi|^2.$$

So

$$\int_{\partial U} |\partial_{\boldsymbol{n}} \phi|^2 \leq \frac{C}{\rho} \int_U |\nabla \phi|^2.$$

Now multiplying the both sides of (5.22-1) by $\phi$, we have

$$\int_U |\nabla \phi|^2 = \int_{\partial U} \partial_{\boldsymbol{n}} \phi \phi \leq \sqrt{\int_{\partial U} |\partial_{\boldsymbol{n}} \phi|^2 \int_{\partial U} |\phi|^2} \leq \sqrt{\frac{C}{\rho} \int_U |\nabla \phi|^2 \int_{\partial U} |\phi|^2}.$$

So,

$$\begin{aligned}
\int_U |\nabla \phi|^2 &\leq \frac{C}{\rho} \int_{\partial U} |\tilde{\phi}|^2 \leq \frac{C}{\rho} \int_{\partial U} | \int_{\partial U} \partial_{\boldsymbol{\tau}} \tilde{\phi}|^2 \\
&\leq \frac{C}{\rho} \int_{\partial U} \int_{\partial U} |\partial U| |\partial \tilde{\phi}|^2 = \frac{C}{\rho} |\partial U|^2 \int_{\partial U} |\partial_{\boldsymbol{\tau}} \tilde{\phi}|^2 \\
&\leq \frac{C}{\rho} |\partial U|^2 \Big\{ \int_{\partial U} |(\nabla - i\tilde{\boldsymbol{A}})\tilde{\psi}|^2 + \int_{\partial U} |\tilde{\boldsymbol{A}} \cdot \boldsymbol{\tau}|^2 \Big\} \\
&\leq C\rho I_\rho(\tilde{\psi}, \tilde{\boldsymbol{A}}) + o(1)
\end{aligned}$$

and (5.24) is established.

Proof of (5.25)–(5.26): For simplicity, we only give the proof for (5.25)–(5.26) when $\overline{B(x_0, \rho) \subset \Omega}$ so that $U = B(x_0, \rho)$. When not, the notation is more complicated but the proof remains in the same spirit. We choose $\eta$ to be the minimizer of

$$\int_U |\nabla \eta|^2 + \frac{1}{\varepsilon^2} (\eta - 1)^2$$

subject to the Dirichlet boundary condition in (5.23-2). Consider the test function (it is here that $U = B(x_0, \rho)$ simplifies the argument)

$$\overline{\eta} = (\tilde{\psi}(\theta) - 1) f(\frac{r}{\rho}) + 1$$

where $f$ is a smooth function such that $f(0) = 0$, $f(1) = 1$ and $0 \leq f' \leq C$ for some constant $C$, $(r, \theta)$ is the standard polar variables in $B(x_0, \rho)$ with centre at $x_0$. We then establish easily that

$$\int_U |\nabla \overline{\eta}|^2 + \frac{1}{\varepsilon^2} (\overline{\eta} - 1)^2 \leq C$$

which implies that

$$\int_U |\nabla \eta|^2 + \frac{1}{\varepsilon^2} (\eta - 1)^2 \leq C.$$

Let $\boldsymbol{V}$ be defined as previously, multiplying both sides of (5.23-1) by $\boldsymbol{V}\cdot\nabla\eta = \boldsymbol{V}\cdot\nabla(\eta-1)$, we have

$$\begin{aligned} 0 &= \int_U -\Delta\eta\boldsymbol{V}\cdot\nabla\eta + \frac{1}{\varepsilon^2}(\eta-1)\boldsymbol{V}\cdot\nabla(\eta-1) \\ &= \int_U \mathrm{div}\boldsymbol{V}(|\nabla\eta|^2 - \frac{1}{\varepsilon^2}(\eta-1)^2) + \int_{\partial U} -|\partial_{\boldsymbol{n}}\eta|^2 + \frac{1}{\varepsilon^2}(\eta-1)^2. \end{aligned}$$

Hence

$$\int_{\partial U}|\partial_{\boldsymbol{n}}\eta|^2 \le \frac{C}{\rho}\int_U |\nabla\eta|^2 + \frac{1}{\varepsilon^2}(\eta-1)^2 + \frac{1}{\varepsilon^2}\int_{\partial U}(\eta-1)^2.$$

Now, multiplying both sides of (5.23-1) by $\eta-1$, we obtain

$$\begin{aligned} 0 &= \int_U -\Delta\eta(\eta-1) + \frac{1}{\varepsilon^2}(\eta-1)^2 \\ &= \int_U |\nabla\eta|^2 + \frac{1}{\varepsilon^2}(\eta-1)^2 - \int_{\partial U}\partial_{\boldsymbol{n}}\eta(\eta-1). \end{aligned}$$

Hence

$$\begin{aligned} &\int_U |\nabla\eta|^2 + \frac{1}{\varepsilon^2}(\eta-1)^2 \\ \le\ & \int_{\partial U}|\partial_{\boldsymbol{n}}\eta||(\eta-1)| \le \sqrt{\int_{\partial U}|\partial_{\boldsymbol{n}}\eta|^2\int_{\partial U}(\eta-1)^2} \\ \le\ & \sqrt{\left[\frac{C}{\rho}\int_U(|\nabla\eta|^2 + \frac{1}{\varepsilon^2}(\eta-1)^2) + \frac{1}{\varepsilon^2}\int_{\partial U}(\eta-1)^2\right]\int_{\partial U}(\eta-1)^2}. \end{aligned}$$

Setting $a = \int_U |\nabla\eta|^2 + \frac{1}{\varepsilon^2}(\eta-1)^2$ and $b = \int_{\partial U}(\eta-1)^2$, the above estimate becomes

$$a^2 \le \frac{C}{\rho}ab + \frac{1}{\varepsilon^2}b^2.$$

Noting that $\rho$ is of the order $\varepsilon^s$ for some $s<1$, we easily obtain

$$a \le \frac{C}{\varepsilon}b.$$

Hence

$$\int_U |\nabla\eta|^2 + \frac{1}{\varepsilon^2}(\eta-1)^2 \le \frac{C}{\varepsilon}\int_{\partial U}(\eta-1)^2 \le \frac{C\varepsilon}{\rho}\Big(\rho I_\rho(\eta)\Big) = o(1).$$

(5.25)–(5.26) are proved.

It is easy to see that (5.27) follows straightforwardly from (5.24)–(5.26).

Combining (5.27) with (5.18) and Proposition 4.1, we deduce, from the definition of $(v, \boldsymbol{B})$ that

$$G(\psi, \boldsymbol{A}, U) \leq G(v, \boldsymbol{B}, U) \leq C.$$

This completes the proof of (5.15).

Using (5.15), we have

$$\int_{\rho^\gamma}^{\rho} I_r(\psi, \boldsymbol{A}) dr \leq C.$$

Just as in the proof of Proposition 4.1, there is some $r_0 \in [\rho^\gamma, \rho]$ and a constant $C$ independent of $\varepsilon$ such that

$$I_{r_0}(\psi, \boldsymbol{A}) \leq \frac{C}{r_0 \sqrt{|\log r_0|}}.$$

Similar to the proof of (5.15), we deduce

$$\begin{aligned} G(\psi, \boldsymbol{A}, B(x_0, r_0) \cap \Omega) &\leq C r_0 I_{r_0}(\psi, \boldsymbol{A}) + o(1) \\ &\leq \frac{C}{\sqrt{|\log r_0|}} + o(1) \leq \frac{C}{\sqrt{|\log \varepsilon|}} + o(1). \end{aligned}$$

In particular, as $\varepsilon \to 0$,

$$\frac{1}{\varepsilon^2} \int_{\Omega \cap B(x_0, \rho^\gamma)} (1 - |\psi|^2)^2 \leq \frac{C}{\sqrt{|\log \varepsilon|}} + o(1) \to 0 \tag{5.28}$$

and the conclusion of our lemma follows from Proposition 5.1.

Now we are ready to specify that the energy contribution of a bad disc is indeed, of the order $|\log \varepsilon|$. We can also specify that the main contributor to this large energy is the gradient of the order parameter. As a consequence, the number of vortices is bounded by a constant independent of $\alpha$ and $\varepsilon$ because the total energy is bounded $c|\log \varepsilon|$ with $C$ being a universal constant.

**Proposition 6.5.3** *Let $0 < \alpha < 1$ and $x_0 \in \Omega$ be given. Assume that*

$$|\psi_\varepsilon(x_0)| < \frac{1}{2}. \tag{5.29}$$

*There exists a constant $C_\alpha$ depending only on $\alpha$, $g$ and $d$ such that*

$$\int_{B(x_0, \varepsilon^\alpha) \cap \Omega} |(\nabla - i\boldsymbol{A}_\varepsilon)\psi_\varepsilon|^2 \geq C_\alpha |\log \varepsilon|. \tag{5.30}$$

*Proof.* Recall that we have an upper bound $N_\alpha$ on the number of vortices for any $\alpha \in (0,1)$, let

$$\delta_\alpha = \frac{\sqrt{\alpha}-\alpha}{2(\overline{N}_\alpha + 1)},$$

for $k = 1, \ldots, 2(\overline{N}_\alpha + 1)$. We define $I_k = (\varepsilon^{\sqrt{\alpha}-k\delta_\alpha}, \varepsilon^{\sqrt{\alpha}-(k+1)\delta_\alpha})$ and let

$$C_k = [B(x_0, \varepsilon^{\sqrt{\alpha}-k\delta_\alpha}) \setminus B(x_0, \varepsilon^{\sqrt{\alpha}-(k+1)\delta_\alpha})] \cap \Omega.$$

Take $\Omega$ in Proposition 5.2 as $B(x_0,\varepsilon^\alpha)\cap\Omega$, let $x_1, \ldots, x_j, \ldots, x_{l_\varepsilon}$ be the points in $B(x_0, \varepsilon^\alpha) \cap \Omega$ as in Proposition 5.2, it is not difficult to verify that if $\varepsilon < \varepsilon_0$ (depending on $\alpha$, $g$, $d$ and $\Omega$), there is some $k_0$ in $\{1, \ldots, 2(\overline{N}_\alpha + 1)\}$ such that

$$C_{k_0} \cap \left( \cup_{j=1}^{l_\varepsilon} B(x_j, \lambda_0\varepsilon) \right) = \emptyset. \tag{5.31}$$

It follows from (5.31) that

$$|u(x)| \geq \frac{1}{2} \quad \forall x \in C_{k_0}. \tag{5.32}$$

As a consequence, for any $\rho \in I_{k_0}$,

$$d_{k_0} = \deg\left(\frac{\psi_\varepsilon}{|\psi_\varepsilon|}, \partial[B(x_0,\rho) \cap \Omega]\right)$$

is well defined and does not depend on $\rho$.

We now argue that

$$d_{k_0} \neq 0. \tag{5.33}$$

If, by contrast, $d_{k_0} = 0$, adapting the argument of the proof of Proposition 4.1, it is straightforward to find some $\rho_0$ in $I_{k_0}$ such that

$$I_{\rho_0}(\psi, \boldsymbol{A}) < \frac{C}{\rho_0},$$

where $C$ depends only on $\alpha$, $g$, $d$ and $\Omega$. Applying Lemma 5.1 to $B(x_0,\rho_0)\cap\Omega$, we conclude that we must have $|\psi(x_0) \geq \frac{1}{2}$. This is a contradiction and (5.33) holds.

Finally, we give the proof of (5.30): Let $(\tilde{\psi}, \tilde{\boldsymbol{A}})$ be gauge equivalent to $(\psi, \boldsymbol{A})$ be defined as follows: introducing the solution $\tilde{\xi}$ of

$$\begin{cases} \Delta\tilde{\xi} = h & \text{in} \quad B(x_0,\varepsilon^\alpha) \cap \Omega, \\ \tilde{\xi} = 0 & \text{on} \quad \partial B(x_0,\varepsilon^\alpha) \cap \Omega, \\ \partial_{\boldsymbol{n}}\tilde{\xi} = 0 & \text{on} \quad B(x_0,\varepsilon^\alpha) \cap \partial\Omega. \end{cases} \tag{5.34}$$

Since the boundary of the domain $B(x_0, \varepsilon^\alpha) \cap \Omega$ has corners, we have to use the results of [Gri 85] to obtain that

$$||\nabla \xi||_{L^2(B(x_0,\varepsilon^\alpha)\cap\Omega)} + ||H\xi||_{L^2(B(x_0,\varepsilon^\alpha)\cap\Omega)} \le C||h||_{L^2(B(x_0,\varepsilon^\alpha)\cap\Omega)}.$$

where $H\xi$ is the Hessian matrix of $\xi$. Set $\tilde{\boldsymbol{A}} = (-\partial_2\tilde{\xi}, \partial_1\tilde{\xi})$.

The quantity $\tilde{\boldsymbol{A}}$ thus defined satisfies obviously

$$\begin{cases} \operatorname{div}\tilde{\boldsymbol{A}} = 0 & \text{in} \quad B(x_0, \varepsilon^\alpha) \cap \Omega, \\ \tilde{\boldsymbol{A}} \cdot \boldsymbol{n} = 0 & \text{on} \quad \partial B(x_0, \varepsilon^\alpha) \cap \Omega, \\ \tilde{\boldsymbol{A}} \cdot \boldsymbol{\tau} = 0 & \text{on} \quad B(x_0, \varepsilon^\alpha) \cap \partial\Omega. \end{cases} \tag{5.35}$$

This implies

$$\int_{B(x_0,\varepsilon^\alpha)\cap\Omega} |\tilde{\boldsymbol{A}}|^2 + |\nabla\tilde{\boldsymbol{A}}|^2 \le \int_{B(x_0,\varepsilon^\alpha)\cap\Omega} |h|^2 \le C. \tag{5.36}$$

From (5.33) and the definition of the degree, we have

$$\forall \rho \in I_{k_0} \quad |d_{k_0}| = \frac{1}{2\pi}\Big| \int_{\partial[B(x_0,\rho)\cap\Omega]} \frac{1}{|\psi|^2}(\tilde{\psi} \times \partial_{\boldsymbol{\tau}}\tilde{\psi})\Big| \ge 1. \tag{5.37}$$

Since $\tilde{\boldsymbol{A}}$ satisfies $\tilde{\boldsymbol{A}} \cdot \boldsymbol{\tau} = 0$ on $B(x_0, \varepsilon^\alpha) \cap \partial\Omega$, the given boundary condition $|\psi|^2(\boldsymbol{A} - \nabla\chi)\cdot\boldsymbol{\tau} = g$ (or rather, its equivalent form $(\nabla\psi_n)\cdot\boldsymbol{\tau} = (-g + \boldsymbol{A}_n\cdot\boldsymbol{\tau})i\psi_n$) reduces to

$$\partial_{\boldsymbol{\tau}}\tilde{\psi} = ig\tilde{\psi} \quad \text{on} \quad B(x_0, \varepsilon^\alpha) \cap \partial\Omega.$$

We have

$$\Big|\int_{B(x_0,\rho)\cap\partial\Omega} \left(\tilde{\psi} \times \partial_{\boldsymbol{\tau}}\tilde{\psi}\right)\Big| = \Big|\int_{B(x_0,\rho)\cap\partial\Omega} g\Big| \le C\varepsilon^\alpha. \tag{5.38}$$

Combining (5.37) and (5.38), we have

$$\Big|\int_{\partial B(x_0,\rho)\cap\Omega} \frac{1}{|\psi|^2}\left(\tilde{\psi} \times \partial_{\boldsymbol{\tau}}\tilde{\psi}\right)\Big| \ge 2\pi(1 - C\varepsilon^\alpha) \ge \pi.$$

Using the Cauchy-Schwarz inequality, we deduce from above that

$$\forall \rho \in I_{k_0}, \quad \int_{\partial B(x_0,\rho)\cap\Omega} |\nabla\tilde{\psi}|^2 \ge \frac{C}{\rho}.$$

Integrating over the interval $I_{k_0}$ we have

$$\int_{C_{k_0}} |\nabla\tilde{\psi}|^2 \ge C|\log\varepsilon|.$$

Combining this inequality with (5.36), we obtain our desired result

$$\int_{C_{k_0}} |(\nabla - i\tilde{\boldsymbol{A}})\tilde{\psi}|^2 = \int_{C_{k_0}} |(\nabla - i\boldsymbol{A})\psi|^2 \geq C_\alpha |\log \varepsilon|.$$

Now we announce the main theorem of this section:

**Theorem 6.5.1** *There exist an integer $N$, a constant $\lambda_0 > 0$ depending only on $g$, $d$, $\Omega$ but not on $\varepsilon$ and points $x_1^\varepsilon, \dots, x_{N_\varepsilon}^\varepsilon$ in $\Omega$ such that*

$$N_\varepsilon \leq N, \tag{5.39}$$

$$|\psi(x)| \geq \frac{1}{2} \ \textit{for} \ x \in \Omega \setminus (\cup_{j=1}^{N_\varepsilon} B(x_j^\varepsilon, \lambda_0 \varepsilon)). \tag{5.40}$$

*Proof.* Let $(x_j)_{j \in J}$ be a finite family of points in $\Omega$, satisfying (see the proof of Proposition 5.1)

$$\begin{array}{l} x_j \in \Omega, \\ B(x_j, \frac{\lambda_0 \varepsilon}{4}) \cap B(x_k, \frac{\lambda_0 \varepsilon}{4}) = \emptyset \ \ \forall j \neq k, \\ \Omega \subset \cup_{j \in J} B(x_j, \lambda_0 \varepsilon). \end{array} \tag{5.41}$$

As in the Proposition 5.2, we say that $B(x_j, \lambda_0 \varepsilon)$ is a bad disc if there is some $y_j \in B(x_j, \lambda_0 \varepsilon)$ such that

$$|\psi(y_j)| \leq \frac{1}{2}. \tag{5.42}$$

We denote by $J_b$ the set of indices for bad discs. All we have to show is

$$\text{The number of elements of the set } J_b = |J_b| < C \tag{5.43}$$

for some constant $C$ depending on $g$, $d$ and $\Omega$ only.

Let $0 < \alpha < 1$ be a given number, consider the balls $B(y_j, \varepsilon^\alpha)$ for $j \in J_b$ where $y_j$ is one of the points satisfying (5.42) in the bad disc $B(x_j, \varepsilon^\alpha)$. By Proposition 5.3 and (5.42), we know that

$$\int_{B(x_0, \varepsilon^\alpha) \cap \Omega} |(\nabla - i\boldsymbol{A})\psi|^2 \geq C_\alpha |\log \varepsilon|. \tag{5.44}$$

Setting $W = \cup_{j \in J_b} B(x_j, 2\varepsilon^\alpha) \cap \Omega$, we are going to show that

$$\begin{aligned} \int_W |(\nabla - i\boldsymbol{A})\psi|^2 &\geq C \sum_{j \in J_b} \int_{B(x_j, 2\varepsilon^\alpha) \cap \Omega} |(\nabla - i\boldsymbol{A})\psi|^2 \\ &\geq C|J_b||\log \varepsilon| \end{aligned} \tag{5.45}$$

where $C$ is a constant depending only on $\alpha$, $g$, $d$ and $\Omega$. This will imply our theorem because from Proposition 2.1, we have

$$\int_W |(\nabla - i\boldsymbol{A})\psi|^2 \leq \int_\Omega |(\nabla - i\boldsymbol{A})\psi|^2 \leq \pi d |\log \varepsilon| + c$$

and this gives

$$\begin{aligned}|J_b| &= \text{ The number of elements of the set } J_b \\ &\leq \frac{\pi d|\log \varepsilon| + c}{C|\log \varepsilon|} \leq \text{some universal constant.}\end{aligned}$$

To achieve this, we only need to bound the number of balls $B(x_j, 2\varepsilon^\alpha)$ to which any point $y$ in $W$ belongs and to show that this bound is independent of $\varepsilon$. Let $y \in W$ and consider the subset $K_y$ of indices of $J_b$ such that

$$K_y = \{j \in J_b,\ y \in B(y_j, \varepsilon^\alpha)\}.$$

From the definition, it is clear that for $j$ in $K_y$, we have

$$x_j \in B(y, 2\varepsilon^\alpha).$$

From Proposition 5.2, we know that the number of such $x_j$ is bounded by a constant depending just on $\alpha$, $g$, $d$ and $\Omega$. Hence

$$\text{The number of elements of the set } K_y < C$$

and this yields (5.45). Our theorem has been proved.

## 6.6 Global $\varepsilon$-independent estimates

Using Theorem 5.1, we can now establish some global estimates which are independent of $\varepsilon$. To begin with, we establish some preliminary results.

In Theorem 5.1, we have found $N_\varepsilon$ points $x_1^\varepsilon, \ldots, x_{N_\varepsilon}^\varepsilon$ in $\Omega$ and showed that

$$\begin{aligned}&N_\varepsilon \leq N, \\ &|\psi_\varepsilon(x) \geq \frac{1}{2} \quad \text{on} \quad \Omega \setminus \cup_{j=1}^{N_\varepsilon} B(x_j^\varepsilon, \lambda_0 \varepsilon).\end{aligned}$$

Of course the limits as $\varepsilon \to 0$ of the points $x_j^\varepsilon$ are good candidates for being the vortices of the limit order parameter $\psi_*$. However, in Theorem 1.2, we did not exclude the possibility that some of the discs $B(x_j^\varepsilon, \lambda_0\varepsilon)$ may intersect. This might lead to limit vortices with winding number greater than 1. To avoid this situation, we will try to replace these $B(x_j^\varepsilon, \lambda_0\varepsilon)$ by slightly larger discs $B(x_j^\varepsilon, \lambda\varepsilon)$ for $j \in J^\varepsilon$, a subset of $\{1, \ldots, l_\varepsilon\}$ with $\lambda > \lambda_0$ is another constant depending on $g$, $\Omega$, $d$ and such that

$$\begin{aligned}&B(x_j^\varepsilon, \lambda\varepsilon) \cap B(x_k^\varepsilon, \lambda\varepsilon) = \emptyset\ \forall k \neq j, \\ &|\psi_\varepsilon(x)| \geq \frac{1}{2} \quad \text{if} \quad x \notin \cup_{j \in J^\varepsilon} B(x_j^\varepsilon, \lambda\varepsilon).\end{aligned}$$

In the following we also drop the subindex $\varepsilon$ when no confusion is possible as in the previous sections. We define

$$\begin{aligned}\tilde{\Omega} &= \Omega \setminus \cup_{j \in J} B(x_j, \lambda\varepsilon), \\ \hat{\Omega} &= \Omega \setminus \cup_{j \in J} B(x_j, 2\lambda\varepsilon).\end{aligned}$$

**Proposition 6.6.1** *On $\tilde{\Omega}$, we have*

$$\mathbf{curl}(i\frac{\psi_\varepsilon}{|\psi_\varepsilon|^2}, (\nabla - i\boldsymbol{A}_\varepsilon)\psi_\varepsilon) = -h_\varepsilon. \tag{6.1}$$

*Proof.* Since $|\psi| \geq 1/2$ on $\tilde{\Omega}$, (6.1) is well defined. We verify (6.1) locally in $\tilde{\Omega}$, that is, for any simply connected open subset of $\tilde{\Omega}$. On each such simply connected open subset of $\tilde{\Omega}$, we can write

$$\psi = |\psi|e^{i\phi}$$

so that $\phi$ is a well-defined, single valued smooth function. Hence

$$\frac{1}{|\psi|^2}(i\psi, (\nabla - i\boldsymbol{A})\psi) = \nabla\phi - \boldsymbol{A}.$$

(6.1) now follows easily by taking **curl** in both sides of the above equality and by taking into account that $\mathbf{curl}\nabla\phi = 0$ locally.

**Proposition 6.6.2** *Let $\rho = |\psi|$. The magnetic field $h$ satisfies*

$$-\mathrm{div}(\frac{1}{\rho^2}\nabla h) + h = 0 \quad in \quad \tilde{\Omega}. \tag{6.2}$$

*Proof.* First, notice that $-\mathbf{curl}(\frac{1}{\rho^2}\mathbf{curl}h) = \mathrm{div}(\frac{1}{\rho^2}\nabla h)$. As in the previous proof, write locally again $\psi = \rho e^{i\phi}$. Using the G-L equation

$$-\mathbf{curl}h = (i\psi, (\nabla - i\boldsymbol{A})\psi),$$

we obtain

$$\begin{aligned} & -\mathbf{curl}(\frac{1}{\rho^2}\mathbf{curl}h) = \mathrm{div}(\frac{1}{\rho^2}\nabla h) \\ = \quad & \mathbf{curl}(\frac{i\psi}{\rho^2}, (\nabla - i\boldsymbol{A})\psi) = \mathbf{curl}(\boldsymbol{A} - \nabla\phi) = h. \end{aligned}$$

This is (6.2). Here we have used the local relation $\mathbf{curl}\nabla\phi = 0$.

Equation (6.2) is completed with the boundary condition

$$\partial_{\boldsymbol{n}}h\Big( = (i\psi, (\nabla - i\boldsymbol{A})\psi)\cdot\boldsymbol{\tau}\Big) = -g \quad \text{on} \quad \partial\Omega. \tag{6.3}$$

Our next objective is to show some $L^p$ bounds on $|\nabla h_\varepsilon|$. Consider a domain $\Omega \subset \mathbb{R}^2$, a positive function $a : \Omega \to \mathbb{R}$ and a constant $0 < \alpha < 1$ such that

$$\alpha < a < \frac{1}{\alpha}. \tag{6.4}$$

Let $\boldsymbol{f} = (f_1, f_2)^T$ be a vector field in $L^q(\Omega)$ for some $q > 2$ and $\xi \in H^1(\Omega)$ be the unique solution of the following linear problem

$$\int_\Omega a\nabla\xi\nabla\phi + \xi\phi = \int_\Omega \boldsymbol{f}\nabla\phi, \ \forall\phi \in H^1(\Omega). \tag{6.5}$$

Choosing the test function $\phi = 1$ in the above expression, we have

$$\int_\Omega \xi = 0. \tag{6.6}$$

Choosing the test function $\phi = \xi$, we have

$$\int_\Omega |\nabla\xi|^2 + \xi^2 \leq C\int_\Omega |\boldsymbol{f}|^2 \tag{6.7}$$

where $C$ depends on $a$ and therefore, on $\alpha$. Our main estimate is

**Proposition 6.6.3** *For any $q > 2$, we have*

$$||\xi||_{L^\infty(\Omega)} \leq C_q||\boldsymbol{f}||_{L^q(\Omega)} \tag{6.8}$$

*where $C_q$ is a constant depending only on $\alpha$, $\Omega$ and $q$.*

This is a rather standard result concerning estimates of solutions of a linear PDE. As we do not discuss general PDE theory in this book, we refer to [BBH 93] and/or [BRi 95] for the proofs of this result.

Finally, we announce our main result of this section:

**Theorem 6.6.1** *Given $1 < p < 2$ and $q \geq 1$, there are constants $C_p$ (resp. $C_q$) depending only on $d$, $g$, $\Omega$ and $p$ (resp. $q$) such that*

$$\int_\Omega |\nabla h|^p \leq C_p \ \forall p < 2, \tag{6.9}$$

$$\int_\Omega |h|^q \leq C_q \ \forall q < +\infty, \tag{6.10}$$

$$\int_\Omega |(\nabla - i\boldsymbol{A})\psi|^p \leq C_p \ \forall p < 2. \tag{6.11}$$

It is interesting to notice that the right-hand side of these estimates are all independent of $\varepsilon$. The price paid for obtaining these estimates is that the power $p$ on various quantities is smaller than 2.

*Proof.* We prove these inequalities one by one.

Proof of (6.9). Let $\eta$ be a smooth function defined on $\Omega$ satisfying

$$\eta \equiv 1 \ \text{ on } \ \hat{\Omega}, \tag{6.12}$$

$$\eta \equiv 0 \ \text{ on } \ \cup_{j\in J} B(x_j, \lambda\varepsilon) \cap \Omega, \tag{6.13}$$

$$|\nabla\eta| \leq \frac{C}{\varepsilon} \ \text{ on } \ \Omega \tag{6.14}$$

where $C$ is a constant depending only on $\Omega$. Let $a$ be defined on $\Omega$ as

$$\begin{cases} a = \frac{1}{\rho^2} & \text{on} \quad \tilde{\Omega} \ (\text{where} \ \ \rho = |\psi|), \\ a = 1 & \text{on} \quad \cup_{j\in J} B(x_j, \lambda\varepsilon) \cap \Omega. \end{cases} \tag{6.15}$$

It is easy to see that $1 < a < 4$. Let $q$ satisfy $1/p + 1/q = 1$ and $\boldsymbol{f}$ be any function in $L^q(\Omega)$ such that

$$\boldsymbol{f} \equiv 0 \ \text{ on } \ \Omega \cap B(x_j, 2\lambda\varepsilon) \tag{6.16}$$

and $\xi$ be the solution to (6.5). In (6.5), we choose the test function $\phi = \eta h$ to obtain

$$\int_\Omega a\nabla\xi\nabla(\eta h) + \xi\eta h = \int_\Omega \boldsymbol{f}\nabla(\eta h) = \int_\Omega \boldsymbol{f}\nabla h, \tag{6.17}$$

where we have used (6.12) and (6.15) about the support of the functions $\eta$ and $\boldsymbol{f}$. Expanding the left-hand side,

$$\int_\Omega a\nabla\xi\nabla(\eta h) = \int_\Omega a\eta\nabla\xi\nabla h + \int_\Omega ah\nabla\xi\nabla\eta \tag{6.18}$$

and using the support relation again,

$$\begin{aligned} \int_\Omega a\eta\nabla\xi\nabla h &= \int_\Omega \frac{1}{\rho^2}\eta\nabla\xi\nabla h \\ &= \int_\Omega \frac{1}{\rho^2}\nabla(\xi\eta)\nabla h - \int_\Omega \frac{1}{\rho^2}\xi\nabla\eta\nabla h. \end{aligned} \tag{6.19}$$

Recall from Proposition 3.1 and (6.14) that

$$|\nabla h| \le \frac{C}{\varepsilon}, \ |\nabla\eta| \le \frac{C}{\varepsilon} \tag{6.20}$$

we have

$$|\int_\Omega \frac{1}{\rho^2}\xi\nabla\eta\nabla h| \le \sum_{j\in J}(\int_{B(x_j,2\lambda\varepsilon)} \frac{1}{\varepsilon^2})||\xi||_{L^\infty(\Omega)} \le C||\xi||_{L^\infty(\Omega)}. \tag{6.21}$$

Using Proposition 6.2, we may rewrite

$$\int_\Omega \frac{1}{\rho^2}\nabla(\xi\eta)\nabla h + \xi\eta h = \int_{\partial\Omega} \frac{1}{\rho^2}\partial_{\boldsymbol{n}}h(\xi\eta) = \int_{\partial\Omega} \frac{1}{\rho^2}g\xi\eta$$

which implies

$$|\int_\Omega \frac{1}{\rho^2}\nabla(\xi\eta)\nabla h + \xi\eta h| \le C||\xi||_{L^\infty(\Omega)}. \tag{6.22}$$

Finally, we estimate the term

$$\int_\Omega ah\nabla\xi\nabla\eta = \sum_{j\in J}\int_{B(x_j,2\lambda\varepsilon)} ah\nabla\xi\nabla\eta.$$

Letting $h_j = h(x_j)$, we have, by (6.20),

$$|h-h_j| \le |\nabla h||x-x_j| \le \frac{C}{\varepsilon}|x-x_j| \le C \text{ on } B(x_j,2\lambda\varepsilon).$$

Hence, using (6.7), we conclude that

$$\begin{aligned} |\int_{B(x_j,2\lambda\varepsilon)} a(h-h_j)\nabla\xi\nabla\eta| &\le C\int_{B(x_j,2\lambda\varepsilon)} |\nabla\xi||\nabla\eta| \\ &\le C(\int_{B(x_j,2\lambda\varepsilon)} |\nabla\xi|^2)^{1/2} \\ &\le C||\boldsymbol{f}||_{L^2(\Omega)}. \end{aligned} \tag{6.23}$$

Notice that

$$\int_{B(x_j,2\lambda\varepsilon)} ah_j\nabla\xi\nabla\eta = \int_\Omega a\nabla\xi\nabla(\eta_j-1)h_j$$

for $\eta_j$ such that

$$\eta_j = \begin{cases} \eta & \text{on} \quad B(x_j,2\lambda\varepsilon), \\ 1 & \text{on} \quad \Omega\setminus B(x_j,2\lambda\varepsilon). \end{cases}$$

By the variational formulation (6.5) for $\xi$, we have

$$\begin{aligned} |\int_\Omega a\nabla\xi\nabla((\eta_j-1)h_j)| &= |-\int_\Omega(\eta_j-1)h_j\xi| = |-\int_{B(x_j,2\lambda\varepsilon)}(\eta_j-1)h_j\xi| \\ \text{(using Proposition 4.1)} \quad &\le ||\xi||_{L^\infty}(\int_{B(x_j,2\lambda\varepsilon)}|h|+C) \le C||\xi||_{\dot{L}^\infty} \end{aligned} \tag{6.24}$$

Finally, combining (6.17)–(6.19) and (6.21)–(6.24), we obtain

$$\begin{aligned} |\int_\Omega \boldsymbol{f}\cdot\nabla h| &\le C(||\xi||_{L^\infty} + ||\boldsymbol{f}||_{L^2}) \\ \text{(using Proposition 6.3)} \quad &\le C||\boldsymbol{f}||_{L^q(\hat\Omega)}. \end{aligned}$$

Since $\boldsymbol{f}$ is arbitrary subject to satisfying (6.16), we deduce by duality that (recall that $\hat\Omega = \Omega\setminus\cup_{j\in J}B(x_j,2\lambda\varepsilon)$).

$$||\nabla h||_{L^p(\hat\Omega)} \le C_p \ \ \forall p<2. \tag{6.25}$$

Also, since $|\nabla h| \leq C/\varepsilon$ (see (3.2)), we have

$$\sum_{j\in J}\int_{B(x_j,2\lambda\varepsilon)} |\nabla h|^2 \leq \sum_{j\in J}\int_{B(x_j,2\lambda\varepsilon)} (C/\varepsilon)^2 \leq C. \tag{6.26}$$

Combining (6.25) and (6.26), we have proved (6.9).

Proof of (6.10). Since

$$|\int_\Omega h| = |\int_{\partial\Omega} \boldsymbol{A}\cdot\boldsymbol{\tau}| = |2\pi d - \int_{\partial\Omega} g| < C, \tag{6.27}$$

if we denote $\bar{h} = \frac{1}{|\Omega|}\int_\Omega h$ and $H = h - \bar{h}$ we have from (6.9) that

$$\int_\Omega |\nabla H|^p \leq C_p, \quad \forall p < 2.$$

Now we also have $\int_\Omega H dx = 0$ and therefore the Poincaré inequality shows that

$$\|H\|_{W^{1,p}(\Omega)} \leq C_p, \quad \forall p < 2.$$

Hence, by standard Sobolev embedding (note that $p < 2$ is arbitrary),

$$(\int_\Omega |H|^q)^{1/q} \leq \tilde{C}_q, \forall q < \infty$$

and this implies (6.10).

Proof of (6.11). From the G-L system and straightforward computation, we have

$$-\mathbf{curl} h = \mathrm{Re}(i\psi(\nabla + i\boldsymbol{A})\psi^*)),$$
$$(\psi, (\nabla - i\boldsymbol{A})\psi) = \frac{1}{2}\nabla|\psi|^2 = \frac{1}{2}\nabla\rho^2$$

respectively. Therefore, when $\rho \neq 0$,

$$|(\nabla - i\boldsymbol{A})\psi|^2 \leq \frac{1}{\rho^2}(|\mathbf{curl} h|^2 + \rho^2|\nabla\rho|^2) = \frac{1}{\rho^2}(|\nabla h|^2 + \rho^2|\nabla\rho|^2).$$

Since $\rho \geq 1/2$ on $\hat{\Omega}$, it follows that

$$|(\nabla - i\boldsymbol{A})\psi| \leq C(|\nabla h| + |\nabla\rho|) \text{ in } \hat{\Omega}. \tag{6.28}$$

In order to prove (6.11), taking into account of (6.2) and the proof of (6.9), it is sufficient to establish

$$\int_\Omega |\nabla\rho|^p \leq C_p \ \forall p < 2 \tag{6.29}$$

where $C_p$ depends only on $p$ and $\Omega$. To achieve this, we let

$$\overline{\rho} = \text{Max}\Big(\rho, \; 1 - \frac{1}{(\log\varepsilon)^2}\Big)$$

and consider the set

$$K = \Big\{x \in \Omega, \;\; \rho \geq 1 - \frac{1}{(\log\varepsilon)^2}\Big\}.$$

It is clear that from the definition, $\nabla\overline{\rho} = \nabla\rho$ on $K$ and $\nabla\overline{\rho} = 0$ on $\Omega \setminus K$. From the standard energy estimate

$$\frac{1}{\varepsilon^2}\int_\Omega (1-\rho)^2 \leq C|\log\varepsilon|,$$

we deduce that

$$\text{meas}(\Omega \setminus K) \leq C\varepsilon^2|\log\varepsilon|^5.$$

From the equation satisfied by $\psi$, we deduce

$$-\Delta\rho^2 = \frac{2}{\varepsilon^2}\rho^2(1-\rho^2) - 2|(\nabla - i\boldsymbol{A})\psi|^2 \;\text{ in }\; \Omega.$$

Multiplying it by $(1-\overline{\rho})$, we obtain

$$\begin{aligned}\int_K 2\rho|\nabla\rho|^2 \;&\leq\; \int_\Omega 2(1-\overline{\rho})|(\nabla - i\boldsymbol{A})\psi|^2 \\ &\leq\; 2||1-\overline{\rho}||_{L^\infty}\int_\Omega |(\nabla - i\boldsymbol{A})\psi|^2 \\ &\leq\; C||1-\overline{\rho}||_{L^\infty}|\log\varepsilon|.\end{aligned}$$

Since $0 \leq 1-\overline{\rho} \leq 1/|\log\varepsilon|^2$, we obtain

$$\int_K |\nabla\rho|^2 \leq C/|\log\varepsilon| \to 0.$$

On the other hand, by Hölder's inequality, we have

$$\begin{aligned}\int_{\Omega\setminus K} |\nabla\rho|^p \;&\leq\; \Big(\int_{\Omega\setminus K}|\nabla\rho|^2\Big)^{p/2}\text{meas}(\Omega\setminus K)^{1-p/2} \\ &\leq\; C|\log\varepsilon|^{p/2}\Big(\varepsilon^2|\log\varepsilon|^5\Big)^{1-p/2}.\end{aligned}$$

Combining the above two inequalities, we obtain (6.29) and our theorem is proved.

## 6.7 Convergence of the solutions as $\varepsilon \to 0$

Since the number of elements of $J_{\varepsilon_n} = N_\varepsilon$ is a constant for each given $\varepsilon$ and $\{N_\varepsilon\}$ is a bounded set, $N_\varepsilon$ can only take finite many possible values. We may therefore, extract a subsequence $\varepsilon_n \to 0$ (as $n \to \infty$) such that

$$\text{The number of elements of } J_{\varepsilon_n} = \text{ constant } = N_1. \tag{7.1}$$

$$x_j^{\varepsilon_n} \to l_j \in \overline{\Omega} \quad \forall j \in J. \tag{7.2}$$

We denote by $a_1, \ldots, a_{N_0}$ the collection of distinct limits $l_j$ with $N_0 \leq N_1$.

To simplify notation, we will denote $h_{\varepsilon_n}$ by $h_n$, $\boldsymbol{A}_{\varepsilon_n}$ by $\boldsymbol{A}_n$ and $\psi_{\varepsilon_n}$ by $\psi_n$ in the following. Recall that $(\psi_n, \boldsymbol{A}_n)$ satisfies the following gauge choice

$$\begin{cases} \operatorname{div}\boldsymbol{A}_n = 0 & \text{in} \quad \Omega, \\ \boldsymbol{A}_n \cdot \boldsymbol{n} = 0 & \text{on} \quad \partial\Omega \end{cases} \tag{7.3}$$

and that $\boldsymbol{A}_n = (-\xi_{nx_2}, \xi_{nx_1})^T$ with $\xi_n$ being a solution of

$$\begin{cases} \Delta\xi_n = h_n & \text{in} \quad \Omega, \\ \xi_n = 0 & \text{on} \quad \partial\Omega. \end{cases} \tag{7.4}$$

Since Theorem 6.1 implies that $\{h_n\}$ is a bounded sequence in $W^{1,p}(\Omega)$, up to choosing a further subsequence if necessary, there is a $h_* \in W^{1,p}(\Omega)$ such that

$$h_n \to h_* \text{ weakly in } W^{1,p}(\Omega)\ \forall p < 2. \tag{7.5}$$

This leads easily to the existence of $\boldsymbol{A}_* \in W^{2,p}(\Omega, \mathbb{R}^2)$ and $\xi_* \in W^{3,p}(\Omega, \mathbb{R})$ such that

$$\boldsymbol{A}_n \to \boldsymbol{A}_* \text{ weakly in } W^{2,p}(\Omega, \mathbb{R}^2)\ \ \forall p < 2, \tag{7.6}$$

$$\xi_n \to \xi_* \text{ weakly in } W^{3,p}(\Omega, \mathbb{R})\ \ \forall p < 2. \tag{7.7}$$

Also, from Theorem 6.1, we have

$$\int_\Omega |(\nabla - i\boldsymbol{A}_n)\psi_n|^p \leq C_p \ \ \forall p < 2, \tag{7.8}$$

$$\int_\Omega |\boldsymbol{A}_n|^q + \int_\Omega |\nabla \boldsymbol{A}_n|^q \leq C_q \ \ \forall q < +\infty. \tag{7.9}$$

In the latter estimate, we have used the well-known fact that $|\boldsymbol{A}|_{W^{1,q}(\Omega)} \leq$ (div-curl theorem) $C_q(|\operatorname{div}\boldsymbol{A}|_{L^q} + |\mathbf{curl}\boldsymbol{A}|_{L^q} + |\boldsymbol{A}\cdot\boldsymbol{n}|_{W^{1-1/q,q}(\partial\Omega)}) =$ (by the gauge condition) $C_q|\mathbf{curl}\boldsymbol{A}|_{L^q} = C_q|h|_{L^q}$ and the last quantity is bounded by Theorem 6.1. It follows then, that

$$\int_\Omega |\nabla\psi_n|^p \leq C_p \ \ \forall p < 2. \tag{7.10}$$

Hence up to choosing a further subsequence, we assume that there is some map $\psi_* \in W^{1,p}(\Omega; \mathbb{R}^2)$ such that

$$\psi_n \to \psi_* \text{ weakly in } W^{1,p}(\Omega, \mathbb{R}^2)\ \forall p < 2. \tag{7.11}$$

From standard energy estimate, we see that

$$\int_\Omega (1 - |\psi_n|^2)^2 \le C\varepsilon_n^2 |\log \varepsilon_n| \to 0 \tag{7.12}$$

which implies that

$$|\psi_*| = 1 \ \text{ almost everywhere.} \tag{7.13}$$

So far, we have seen the $W^{1,p} \times \mathbf{W}^{2,p}$ weak convergence of $(\psi_\varepsilon, \boldsymbol{A}_\varepsilon)$ (up to choosing a subsequence) with $p < 2$. Now we want to investigate the convergence in the energy spaces, that is, in $H^1$-type spaces.

**Proposition 6.7.1** *Let $K$ be a compact subset of $\overline{\Omega} \setminus \cup_{j=1}^{N_0} \{a_j\}$, we have*

$$\int_K |\nabla h_n|^2 + |\nabla \psi_n|^2 < C_K$$

*where $C_K$ is some constant independent of $n$ and moreover,*

$$h_n \to h_* \ \text{ strongly in } \ H^1(K), \tag{7.14}$$

$$\psi_n \to \psi_* \ \text{ strongly in } \ H^1(K), \tag{7.15}$$

$$\frac{1}{\varepsilon_n^2} \int_K (1 - |\psi_n|^2)^2 \to 0. \tag{7.16}$$

*Proof.* Let $\eta$ be a smooth function with compact support in $\Omega \setminus \cup_{j-1}^{N_0} \{a_j\}$ satisfying $\eta \equiv 1$ on $K$. On $\tilde{\Omega}_{\varepsilon_n}$, by Proposition 6.2, $h_n$ verifies

$$-\mathrm{div}(\frac{1}{\rho_n^2} \nabla h_n) + h_n = 0 \ \text{ in } \tilde{\Omega}_{\varepsilon_n}. \tag{7.17}$$

When $n$ is getting larger, by definition, the set $\tilde{\Omega}_{\varepsilon_n}$ is getting closer to the set $\Omega \setminus \cup_{j-1}^{N_0} \{a_j\}$. When $n$ is sufficiently large, the support of $\eta$ becomes contained in $\tilde{\Omega}_{\varepsilon_n}$. Hence we multiply (7.17) by $\eta h_n$ and integrate over $K$ to obtain

$$\begin{aligned}
\int_K |\nabla h_n|^2 + h_n^2 &\le \int_\Omega |\nabla \eta||\nabla h_n||h_n| + \int_{\partial\Omega} |\partial_{\boldsymbol{n}} h_n||h_n| \\
\text{(use (1.4))} \quad &= \int_\Omega |\nabla \eta||\nabla h_n||h_n| + \int_{\partial\Omega} |g||h_n| \\
&\le C(\int_\Omega |\nabla h_n||h_n| + \int_{\partial\Omega} |h_n|) \\
(|h_n|_{W^{1,p}} \le C\ \forall p < 2) \quad &\le C.
\end{aligned}$$

Moreover, since $h_n \rightharpoonup h_*$ in the distributional sense, we deduce that

$$\int_K |\nabla h_*|^2 + h_*^2 \leq C. \tag{7.18}$$

From this we know that $h_n \rightharpoonup h_*$ in $H^1(K)$ weakly and consequently, $h_* \in H^1(K)$. With this information, we multiply (7.17) by $\eta(h_n - h_*)$ to obtain

$$\begin{aligned}&\int_\Omega \frac{\eta}{\rho_n^2}|\nabla h_n|^2 + h_n\eta(h_n - h_*) = \\ &\int_\Omega (h_* - h_n)\frac{1}{\rho_n^2}\nabla h_n \nabla \eta + \int_\Omega \frac{1}{\rho_n^2}\eta \nabla h_n \nabla h_*.\end{aligned} \tag{7.19}$$

Since we know that $h_n \to h_*$ weakly in $H^1(K)$, we have, using Sobolev embedding, that

$$||h_n - h_*||_{L^q(K)} \to 0 \; \forall q < +\infty.$$

$$\int_\Omega \frac{1}{\rho_n^2}\eta|\nabla h_n|^2 \to \int_\Omega \eta|\nabla h_*|^2.$$

Since $\rho \leq 1$, this implies that

$$\int_\Omega \eta|\nabla h_n|^2 \leq \int_\Omega \frac{1}{\rho_n^2}\eta|\nabla h_n|^2 \to \int_\Omega \eta|\nabla h_*|^2.$$

By lower semi-continuity, we should have

$$\underline{\lim}_{n\to\infty} \int_\Omega \eta|\nabla h_n|^2 \geq \int_\Omega \eta|\nabla h_*|^2,$$

hence

$$\lim_{n\to\infty} \int_\Omega \eta|\nabla h_n|^2 = \int_\Omega \eta|\nabla h_*|^2$$

and this implies that

$$\nabla h_n \to \nabla h_* \text{ strongly in } L^2(K).$$

(7.14) is established.

We now consider (7.15) and (7.16). The equation for $\rho_n$ (see Chapter 1) in $\Omega_{\varepsilon_n}$ is

$$-\Delta\rho_n + \frac{1}{\rho_n^3}|\nabla h_n|^2 = \frac{1}{\varepsilon_n^2}\rho_n(1 - \rho_n^2). \tag{7.20}$$

Multiplying (7.20) by $\eta(1 - \rho_n)$ and integrating, we obtain

$$\begin{aligned}&\int_\Omega \eta|\nabla\rho_n|^2 + \frac{\rho_n}{\varepsilon_n^2}(1 - \rho_n)^2\eta(1 + \rho_n) \\ \leq &\int_{\Omega_{\varepsilon_n}} \frac{\eta(1 - \rho_n)}{\rho_n^3}|\nabla h_n|^2 + \int_\Omega |\nabla\eta\nabla\rho_n(1 - \rho_n)|.\end{aligned} \tag{7.21}$$

Since $\rho_n \to 1$, $\rho_n \leq 1$ and is bounded in $W^{1,p}$, we deduce from (7.14) and dominated convergence that the right-hand side of (7.21) goes to zero. Hence

$$\int_K |\nabla \rho_n|^2 + \frac{1}{\varepsilon_n}(1-\rho_n^2)^2 \to 0. \tag{7.22}$$

This proves (7.16). Moreover, we have, on $K$

$$\begin{aligned}\int_K |(\nabla - i\boldsymbol{A}_n)\psi_n|^2 &= \int_K \frac{1}{\rho_n^2}(i\psi_n, (\nabla - i\boldsymbol{A}_n)\psi_n)^2 + |\nabla \rho_n|^2 \\ &= \int_K \frac{1}{\rho_n^2}|\nabla h_n|^2 + |\nabla \rho_n|^2 \\ (\text{as } n\to\infty) &\to \int_K \frac{1}{\rho_*^2}|\nabla h_*|^2 = \int_K |(\nabla - i\boldsymbol{A}_*)\psi_*|^2.\end{aligned}$$

We deduce that

$$(\nabla - i\boldsymbol{A}_n)\psi_n \to (\nabla - i\boldsymbol{A}_*)\psi_* \ \text{ strongly in } \ L^2(K) \tag{7.23}$$

and (7.15) follows from this and the strong convergence of $\boldsymbol{A}_n$ to $\boldsymbol{A}_*$ in $L^\infty$.

Finally, we give an estimate on how fast the order parameter $|\psi_n|$ approaches 1.

**Proposition 6.7.2** : *We have, for some constant $C$ depends only on $g$, $d$ and $\Omega$,*

$$\frac{1}{\varepsilon_n^2}\int_\Omega (1-|\psi_n|^2)^2 \leq C. \tag{7.24}$$

*Proof.* In order to carry out the proof, we need the result of Proposition 8.1 below (which does not rely on the validity of (7.24)). As our proposition is about convergence, we still give the proof here by assuming the validity of Proposition 8.1. Proposition 8.1 concludes that for each $j$, $a_j$ is not on the boundary of $\Omega$. Let $\mu > 0$ be fixed, sufficiently small so that

$$B(a_j,\mu) \cap B(a_k,\mu) = \emptyset \ \text{ when } \ j \neq k \ \text{ and } \ B(a_j,\mu)\cap\partial\Omega = \emptyset \ \forall j.$$

Using Proposition 7.1, taking $K = K_\mu = B(a_j, 2\mu)\setminus B(a_j,\mu)$, we derive that

$$\int_{K_\mu} \frac{1}{\varepsilon_n^2}(1-|\psi_n|^2)^2 + |(\nabla - i\boldsymbol{A}_n)\psi_n|^2 + |h_n|^2 \leq c(\mu).$$

Using Fubini's theorem, there is some $\mu_n \in (\mu, 2\mu)$ such that, for $n$ large enough,

$$\int_{\partial B(a_j,\mu_n)} \frac{1}{\varepsilon_n^2}(1-|\psi_n|^2)^2 + |(\nabla - i\boldsymbol{A}_n)\psi_n|^2 + |h_n|^2 \leq C(\mu) \tag{7.25}$$

where $C(\mu)$ depends only on $\mu$, $g$, $d$ and $\Omega$. Applying the Pohozaev type inequality (Proposition 1.2) to $(\psi_n, \boldsymbol{A}_n)$ on the set $B(a_j, \mu_n) \cap \Omega$, we deduce

$$\begin{aligned}\int_{B(a_j,\mu)\cap\Omega} \frac{1}{\varepsilon_n^2}(1-|\psi_n|^2)^2 &\leq \int_{B(a_j,\mu_n)\cap\Omega} \frac{1}{\varepsilon_n^2}(1-|\psi_n|^2)^2 \\ &\leq C + \int_{B(a_j,\mu_n)\cap\Omega} |h_n|^2 \leq C. \end{aligned} \tag{7.26}$$

(7.24) follows easily from (7.16) and (7.26).

## 6.8 Main results on the limit functions

In this section, we intend to derive the equations for the limit functions. To this aim, we first introduce the notion of the so-called canonical harmonic maps. Then, we can work out how the minimizers of the original problem converge to the limit functions by using the energy estimates of the canonical harmonic map away from the vortices.

**Definition 6.8.1** *Let $\Omega$ be a smooth, bounded and simply connected open domain in $\mathbb{R}^2$. Fix $n$ points $a_1, \dots, a_n$ in $\Omega$ and $n$ integers $d_1, \dots, d_n$. Set $d = \sum_{j=1}^n d_j$ and let $g$ be a map from $\partial\Omega$ into $S^1$, such that* $\deg(g, \partial\Omega) = d$.

*Let $\Phi_0$ be the solution of*

$$\begin{cases} \Delta\Phi_0 = \sum_{j=1}^n 2\pi d_j \delta_{a_j} & \text{in} \quad \Omega, \\ \partial_{\boldsymbol{n}}\Phi_0 = g \times \partial_{\boldsymbol{\tau}} g & \text{on} \quad \partial\Omega, \\ \int_{\partial\Omega} \Phi_0 = 0. \end{cases} \tag{8.1}$$

*Then there exists a unique harmonic map $u_0 : \Omega \setminus \{a_1, \dots, a_n\} \to S^1$ such that*
*1)* $\deg(u_0, a_j) = d_j$ *($\deg(u_0, a_j)$ is the degree of $u_0$ restricted to any small circle centred at $a_j$ not containing other $a_k'$s),*
*2) $u_0$ is continuous up to $\partial\Omega$ and $u_0|_{\partial\Omega} = g$*
*which is associated to $\Phi_0$, namely,*

$$\begin{cases} u_0 \times \partial_1 u_0 = -\partial_2\Phi_0 & \text{in} \quad \Omega \setminus \{a_1, \dots, a_n\}, \\ u_0 \times \partial_2 u_0 = \partial_1\Phi_0 & \text{in} \quad \Omega \setminus \{a_1, \dots, a_n\}. \end{cases} \tag{8.2}$$

*The function $u_0$ is called the* **canonical harmonic map** *associated to the configuration $(g, a, d)$ where $a = (a_1, \dots, a_n)$, $d = (d_1, \dots, d_n)$.*

Then we establish the following

**Theorem 6.8.1** *Let $u_0$ be a canonical harmonic map associated to $(g, a, d)$, then as $\rho \to 0$*

$$\frac{1}{2}\int_{\Omega_\rho} |\nabla u_0|^2 = \pi \sum_{j=1}^n d_j^2 \log(1/\rho) + W + O(\rho^2) \tag{8.3}$$

*where $\Omega_\rho = \Omega \setminus \cup_{j=1}^n \overline{B(a_j, \rho)}$.*

*Proof.* We first prove the claim that $|\nabla u_0| = |\nabla \Phi_0|$. In fact, since $|u_0| = 1$, we have (denote $u_0 = (u_0^1, u_0^2)$)

$$u_0 \cdot u_{0x_i} = 0, \quad i = 1, 2,$$

that is

$$u_0^1 \cdot u_{0x_i}^1 + u_0^2 \cdot u_{0x_i}^2 = 0, \quad i = 1, 2.$$

This implies by squaring above expressions that

$$\begin{aligned} -2u_0^1 u_0^2 u_{0x_1}^1 u_{0x_1}^2 &= (u_0^1)^2 (u_{0x_1}^1)^2 + (u_0^2)^2 (u_{0x_1}^2)^2, \\ -2u_0^1 u_0^2 u_{0x_2}^1 u_{0x_2}^2 &= (u_0^1)^2 (u_{0x_2}^1)^2 + (u_0^2)^2 (u_{0x_2}^2)^2. \end{aligned}$$

Applying these relations to the following

$$\begin{aligned} u_0 \times u_{0x_1} &= u_0^1 u_{0x_1}^2 - u_0^2 u_{0x_1}^1, \\ u_0 \times u_{0x_2} &= u_0^1 u_{0x_2}^2 - u_0^2 u_{0x_2}^1, \end{aligned}$$

we have

$$\begin{aligned} & |u_0 \times u_{0x_1}|^2 + |u_0 \times u_{0x_2}|^2 \\ = & -2u_0^1 u_0^2 u_{0x_1}^1 u_{0x_1}^2 + (u_0^1)^2 (u_{0x_1}^2)^2 + (u_0^2)^2 (u_{0x_1}^1)^2 \\ & -2u_0^1 u_0^2 u_{0x_2}^1 u_{0x_2}^2 + (u_0^1)^2 (u_{0x_2}^2)^2 + (u_0^2)^2 (u_{0x_2}^1)^2 \\ = & (u_0^1)^2 [(u_{0x_1}^1)^2 + (u_{0x_1}^2)^2] + (u_0^2)^2 [(u_{0x_1}^1)^2 + (u_{0x_1}^2)^2] \\ & + (u_0^1)^2 [(u_{0x_2}^1)^2 + (u_{0x_2}^2)^2] + (u_0^2)^2 [(u_{0x_2}^1)^2 + (u_{0x_2}^2)^2] \\ = & [(u_0^1)^2 + (u_0^2)^2][(u_{0x_1}^1)^2 + (u_{0x_1}^2)^2 + (u_{0x_2}^1)^2 + (u_{0x_2}^2)^2] \\ = & (u_{0x_1}^1)^2 + (u_{0x_1}^2)^2 + (u_{0x_2}^1)^2 + (u_{0x_2}^2)^2 \\ = & |\nabla u_0|^2. \end{aligned}$$

This gives the claim.

Therefore

$$\int_{\Omega_\rho} |\nabla u_0|^2 = \int_{\Omega_\rho} |\nabla \Phi_0|^2 = \int_{\partial\Omega} \Phi_0 \partial_{\boldsymbol{n}} \Phi_0 - \sum_{j=1}^{n} \int_{\partial B(a_j,\rho)} \Phi_0 \partial_{\boldsymbol{n}} \Phi_0. \tag{8.4}$$

Using (8.1-2) and define

$$R_0(x) = \Phi_0(x) - \sum_{j=1}^{n} d_j \log |x - a_j|$$

which is a smooth harmonic function on $\Omega$. Set

$$S_j(x) = \Phi_0(x) - d_j \log |x - a_j|. \tag{8.5}$$

Note that $S_j$ is a smooth harmonic function in some neighbourhood of $a_j$ such that

$$\begin{array}{lll} S_j(x) = \Phi_0(x) - d_j \log\rho & \text{on} & \partial B(a_j,\rho), \\ \partial_{\boldsymbol{n}} S_j = \partial_{\boldsymbol{n}}\Phi_0 - \frac{d_j}{\rho} & \text{on} & \partial B(a_j,\rho), \\ S_j(a_j) = R_0(a_j) + \sum_{j\neq k} d_j \log|a_k - a_j|. & & \end{array} \tag{8.6}$$

Thus we have

$$\int_{\partial B(a_j,\rho)} \partial_{\boldsymbol{n}}\Phi_0\Phi_0 = \int_{\partial B(a_j,\rho)} (\partial_{\boldsymbol{n}} S_j + \frac{d_j}{\rho})(S_j + d_j\log\rho)$$

and, using the fact that $S_j$ is harmonic in $B(a_j,\rho)$, we deduce that

$$\int_{\partial B(a_j,\rho)} \partial_{\boldsymbol{n}}\Phi_0\Phi_0 = \int_{B(a_j,\rho)} |\nabla S_j|^2 + 2\pi d_j S_j(a_j) + 2\pi d_j^2 \log\rho. \tag{8.7}$$

Combining (8.4) and (8.7), we are led to

$$\int_{\Omega_\rho} |\nabla u_0|^2 = 2\pi\sum_{j=1}^n d_j^2\log(1/\rho) + 2W - \sum_{j=1}^n \int_{B(a_j,\rho)} |\nabla S_j|^2 \tag{8.8}$$

which yields the conclusion.

Next, we give an energy lower bound with respect to given vortices and given degree. For simplicity, we only give the proof for the single vortex case.

**Theorem 6.8.2** *Let $u$ be a smooth complex valued function defined on $B_R = B(0,R)$, we assume that*

$$|u| \le 1 \quad in \quad B_R, \tag{8.9}$$

$$|u(x)| \ge a \ \ \forall x \in A_{RR_0} = B_R \setminus B_{R_0}, \tag{8.10}$$

$$\frac{1}{R_0^2}\int_{B_R} (|u|^2 - 1)^2 \le K, \tag{8.11}$$

$$\deg(u, S_r) = d \ \ for\ any \ \ r \in (R_0, R) \tag{8.12}$$

*for some positive constants $a$, $R_0$, $K$ and $d$. We write*

$$u = \rho e^{i\phi} \quad in \quad A_{RR_0},$$

*then*

$$\int_{B_R} |\nabla u|^2 \ge \int_{A_{RR_0}} \rho^2|\nabla\phi|^2 \ge d^2[2\pi\log(R/R_0) - C] \tag{8.13}$$

*where $C$ depends only on $a$ and $K$.*

*Proof.* On $A_{RR_0}$, we write $u(x) = |u(x)|e^{i(d\theta+\eta(x))} = \rho e^{i\phi(x)}$ where $\eta$ is a smooth and single-valued function. We have

$$\nabla\phi = \frac{d}{r}V + \nabla\eta$$

where $V(x)$ is the vector field in $I\!R^2 \setminus \{0\}$ defined by $V(r\cos\theta, r\sin\theta) = (-\sin\theta, \cos\theta)$. Thus,

$$|\nabla\phi|^2 = \frac{d^2}{r^2} + \frac{2d}{r}\partial_{\boldsymbol{T}}\eta + |\nabla\eta|^2 \tag{8.14}$$

where $\partial_{\boldsymbol{T}}\eta = V \cdot \nabla\eta$ is the derivative in the direction tangential to $S_r$. We write

$$\begin{aligned} \int_{A_{RR_0}} \rho^2|\nabla\phi|^2 &= \int_{A_{RR_0}} \frac{\rho^2 d^2}{r^2} + 2\int_{A_{RR_0}} \frac{\rho^2 d}{r}\partial_{\boldsymbol{T}}\eta + \int_{A_{RR_0}} \rho^2|\nabla\eta|^2 \\ &= I_1 + I_2 + I_3. \end{aligned} \tag{8.15}$$

We now evaluate each integral separately. First, we have

$$\begin{aligned} I_1 &= \int_{A_{RR_0}} \frac{d^2}{r^2} - \int_{A_{RR_0}} (1-\rho^2)\frac{d^2}{r^2} \\ &= 2\pi d^2 \log(R/R_0) - \int_{A_{RR_0}} (1-\rho^2)\frac{d^2}{r^2}. \end{aligned} \tag{8.16}$$

Using the Cauchy-Schwartz inequality and (8.11), we have

$$\int_{A_{RR_0}} (1-\rho^2)\frac{1}{r^2} \le \sqrt{\pi K} \tag{8.17}$$

and thus

$$I_1 \ge 2\pi d^2 \log(R/R_0) - \sqrt{\pi K}d^2. \tag{8.18}$$

Next, we use the fact that

$$\int_{S_r} \partial_{\boldsymbol{T}}\eta = 0 \ \ \forall r \in (R_0, R)$$

and we write

$$I_2 = 2\int_{A_{RR_0}} (\rho^2 - 1)\frac{d}{r}\partial_{\boldsymbol{T}}\eta.$$

Hence, by the Cauchy-Schwartz inequality and (8.17), we find that

$$I_2 \le 2\sqrt{K}\Big[\int_{A_{RR_0}} |\nabla\eta|^2\Big]^{1/2} \le 2\frac{d^2K}{a^2} + \frac{a^2}{2}\int_{A_{RR_0}} |\nabla\eta|^2. \tag{8.19}$$

Finally, we have by (8.10) that

$$I_3 \geq a^2 \int_{A_{RR_0}} |\nabla \eta|^2. \tag{8.20}$$

Combining (8.18)–(8.20), we are led to

$$\int_{A_{RR_0}} \rho^2 |\nabla \phi|^2 \geq 2\pi d^2 \log(R/R_0) - d^2(\sqrt{\pi K} + 2K/a^2), \tag{8.21}$$

which is the desired conclusion.

We only make the announcement for the multi-vortex location case. For details, we refer to [BMR 94], Theorem 4.

**Theorem 6.8.3** *Let $a_1, \dots, a_m$ be points in $B_R$ such that*

$$|a_j| \leq R/2 \ \ \textit{for all} \ \ j, \tag{8.22}$$

$$|a_j - a_k| \geq 4R_0 \ \ \textit{for all} \ \ j \neq k. \tag{8.23}$$

*Set $\Omega = B_R \setminus \cup_{j=1}^m B(a_j, R)$ with $R_0 \leq R/4$. Let $\psi$ be a complex valued smooth function defined on $\Omega$. We assume that*

$$0 < a \leq |\psi| \leq 1 \ \ \textit{in} \ \ \Omega, \tag{8.24}$$

$$\frac{1}{R_0^2} \int_\Omega (|\psi|^2 - 1)^2 \leq K, \tag{8.25}$$

$$\deg(\psi, \partial B(a_j, R_0)) = d_j \tag{8.26}$$

*for some constant $a$, $K$ and positive integer $d_j$. We define*

$$\psi_0(z) = \Big(\frac{z - a_1}{|z - a_1|}\Big)^{d_1} \cdots \Big(\frac{z - a_m}{|z - a_m|}\Big)^{d_m}.$$

*Then*

$$\int_\Omega |\nabla \psi|^2 \geq \int_\Omega |\nabla \psi_0|^2 - C||d||^2 m^2 \tag{8.27}$$

*where $||d|| = \sum_j |d_j|$ and $C$ depends only on $a$ and $K$.*

**Remark 6.8.1** *1) The results of Theorem 8.3 implies that*

$$\int_\Omega |\nabla \psi|^2 \geq (\sum_j d_j^2) \log(R/R_0) + O(||d||^2 m^2). \tag{8.28}$$

*2) Theorem 8.3 deals with the case $\Omega = B_R \setminus \cup_{j=1}^m B(a_j, R)$. As it deals with bad disc only, for a general domain, $\Omega$ with bad discs $B_j$, we have naturally $\int_\Omega |\nabla \psi|^2 \geq \sum_j \int_{B_j} |\nabla \psi|^2$ and we can apply (8.28).*

Now we begin describing the location and properties of the limiting vortices.

**Proposition 6.8.1** *For the limit mappings $(\psi_*, \boldsymbol{A}_*)$, we have ($N_0$ is the total number of distinct singular points of $\psi_*$)*

$$N_0 = d, \tag{8.29}$$

$$a_j \notin \partial\Omega \ \ \forall j = 1, \dots, d, \tag{8.30}$$

$$d_j = \deg(\psi_*, a_j) = 1 \ \ \forall j = 1, \dots, d. \tag{8.31}$$

*Proof.* Consider a slightly larger domain $\Omega_1$ such that $\Omega \subset\subset \Omega_1$. First, we extend the definition domain of $\psi_n$ to $\Omega_1$. Recall the definition that for any two complex valued functions $\eta_1$ and $\eta_2$, $\eta_1 \times \eta_2 = (i\eta_1, \eta_2) = \mathrm{Re}(i\eta_1\eta_2^*)$. Since

$$\psi_n \times \partial_{\boldsymbol{T}}\psi_n = g + \boldsymbol{A}_n \cdot \boldsymbol{\tau} \ \text{ on } \ \partial\Omega,$$

similar to the proof of Theorem 8.1, we have

$$|\partial_{\boldsymbol{T}}\psi_n| = |\psi_n \times \partial_{\boldsymbol{T}}\psi_n| = |g + \boldsymbol{A}_n \cdot \boldsymbol{\tau}|,$$

(7.6) then implies

$$\int_{\partial\Omega} |\partial_{\boldsymbol{T}}\psi_n|^2 < C.$$

This enables us to find an extension of $\psi_n$ to $\Omega_1$ (still denoted by $\psi_n$) such that $\psi_n \in H^1(\Omega_1 \setminus \Omega, S^1)$ and

$$\int_{\Omega_1 \setminus \Omega} |\nabla\psi_n|^2 \leq C(\text{ independent of } \varepsilon.\ ) \tag{8.32}$$

We extend $u_*$ to $\Omega_1$ so that the extension is the limit of the extended $\psi_n'$s.

First, we show (8.31). Let $\delta > 0$, consider the balls $B(a_j, \delta)$ and the domain $\Omega_\delta = \Omega_1 \setminus \cup_{j=1}^{N_0} B(a_j, \delta)$. By Theorem 8.1, we have

$$\int_{\Omega_\delta} |\nabla\psi_*|^2 \geq 2\pi \sum_{j=1}^{N_0} |d_j|^2 \log(1/\delta) - C$$

where $C$ depends only on the points $a_j$ and $g$. Since $\psi_n \to \psi_*$ strongly in $H^1$ away from the vortices, we have

$$\int_{\Omega_\delta} |\nabla\psi_n|^2 \geq 2\pi \sum_{j=1}^{N_0} |d_j|^2 \log(1/\delta) - C \tag{8.33}$$

for $n$ sufficiently large. Here $C$ may have different values at different places. On the other hand, Theorem 8.2 implies that

$$\int_{B(a_j,\delta)} |\nabla\psi_n|^2 \geq 2\pi|d_j| \log(\delta/\varepsilon_n) - C. \tag{8.34}$$

Combining (8.32)–(8.34), we obtain

$$\int_\Omega |\nabla\psi_n|^2 \geq 2\pi \sum_{j=1}^{N_0} (|d_j|^2 \log(1/\delta) + |d_j| \log(\delta/\varepsilon_n)) - C \tag{8.35}$$

for any $\delta > 0$, and $n$ sufficiently large (depending on $\rho$).

We also have, using (7.6)–(7.8) and standard Sobolev embedding,

$$\begin{aligned} \int_\Omega |\nabla\psi_n|^2 &= \int_\Omega |(\nabla - i\boldsymbol{A}_n)\psi_n + i\boldsymbol{A}_n\psi_n|^2 \\ &\leq \int_\Omega |(\nabla - i\boldsymbol{A}_n)\psi_n|^2 + R_n \end{aligned} \tag{8.36}$$

where $|R_n| \leq C$ is a constant. Combining (8.35), (8.36) and Proposition 2.1, we obtain (8.31).

We prove (8.30) by contradiction. Let's assume that

$$a_1 \in \partial\Omega. \tag{8.37}$$

Let $R > 0$ be such that

$$\overline{B(a_1, R)} \subset \Omega_1 \setminus \{a_2, \dots, a_d\}.$$

Apply Theorem 8.3 (while taking Remark 8.1 into account) in $\Omega_1 \setminus B(a_1, R)$ with $\Omega_2 = (\Omega_1 \setminus B(a_1, R)) \setminus \cup_{j=2}^d B(a_j, \delta)$, we obtain

$$\frac{1}{2}\int_{\Omega_2} |\nabla\psi_*|^2 \geq \pi(d-1)|\log\delta| - C \tag{8.38}$$

where $C$ depends only on $R$, $d$ and $\Omega_1$. By Theorem 8.2, we have

$$\frac{1}{2}\int_{B(a_1,R)\setminus B(a_1,\eta)} |\nabla\psi_*| \geq 2\pi|\log\eta| - C. \tag{8.39}$$

Hence combining (8.38) and (8.39), we see that

$$\frac{1}{2}\int_{\Omega_3} |\nabla\psi_*|^2 \geq \pi(d+1)|\log\eta| - C$$

where $\Omega_3 = \Omega_1 \setminus \cup_{j=1}^d B(a_j, \eta)$ and $C$ depends only on $R$, $d$ and $\Omega_1$.

Using similar argument as for (8.3), we have

$$\int_{\Omega_\delta} |\nabla\psi_n|^2 \leq 2\pi d|\log\eta| + C$$

(notice now $\sum d_j^2 = d$ because $d_j = 1$). If we take the limit $n \to \infty$, we obtain

$$\frac{1}{2}\int_{\Omega_2} |\nabla\psi_*| \leq \pi d|\log\eta| + C \tag{8.40}$$

where $C$ is independent of $\eta$. (8.39) and (8.40) leads to the impossible estimate

$$|\log \eta| \le C \quad \forall \eta \in (0, R).$$

Hence (8.30) must be true.

Finally, in the process, we have also justified (8.29).

The next result is about the limit equation satisfied by the limit magnetic field $h_*$.

**Proposition 6.8.2** *The magnetic field $h_*$ satisfies the equation*

$$\begin{cases} -\Delta h_* + h_* = 2\pi \sum_{j=1}^d \delta_{a_j} & \text{in} \quad \Omega \\ \partial_{\boldsymbol{n}} h_* = -g & \text{on} \quad \partial\Omega. \end{cases} \tag{8.41}$$

*Proof.* Let $K$ be an arbitrary compact subset of $\Omega \setminus \cup_{j=1}^d \{a_j\}$. Passing to the limit in (6.1), using the convergence properties away from the vortices we have obtained so far, we have

$$-\Delta h_* + h_* = 0 \text{ on } K.$$

For the behaviour of the equation near each vortex point, let $\delta > 0$, we have, using G-L equation,

$$\begin{aligned} \int_{\partial B(a_j,\delta)} \frac{1}{\rho_n^2} \partial_{\boldsymbol{n}} h_n &= \int_{\partial B(a_j,\delta)} \frac{1}{\rho_n^2} \mathbf{curl} h_n \cdot \boldsymbol{\tau} \\ &= -\int_{\partial B(a_j,\delta)} \frac{1}{\rho_n^2} (i\psi_n, \boldsymbol{\tau} \cdot (\nabla - i\boldsymbol{A}_n)\psi_n) \\ &\quad - \int_{\partial B(a_j,\delta)} (\frac{1}{\rho_n^2}(i\psi, \boldsymbol{\tau} \cdot \nabla\psi_n) + \boldsymbol{A}_n \cdot \boldsymbol{\tau}) \\ &= -2\pi d_j + \int_{B(a_j,\delta)} h_n. \end{aligned} \tag{8.42}$$

In the sense of distribution, this implies that

$$\operatorname{div}(\frac{1}{\rho_n^2} \nabla h_n) + h_n \to 2\pi \delta_{a_j}$$

in any small neighbourhood of $a_j$.

We can now pass to the distributional limit in (6.2) on the whole of $\Omega$, this implies (8.41).

Finally, we give the limit equation satisfied by the limit order parameter.

**Proposition 6.8.3** *The map $\psi_*$ is a harmonic map on $\Omega \setminus \cup_{j=1}^d \{a_j\}$. More precisely, it satisfies*

$$-\Delta \psi_* = \psi_* |\nabla \psi_*|^2 \quad \text{on} \quad \Omega \setminus \cup_{j=1}^d \{a_j\}. \tag{8.43}$$

*Moreover, we have the following expression for $\psi_*$:*

$$\psi_* = \prod_{j=1}^{d} \frac{z-a_j}{|z-a_j|} e^{i\phi_*} \tag{8.44}$$

*where $\phi_*$ satisfies*

$$\begin{cases} \Delta\phi = 0 & on \quad \Omega, \\ \phi = \phi_0 & on \quad \partial\Omega, \\ \partial_{\boldsymbol{T}}\phi_0 = -(i\prod_{k=1}^{d} \frac{z-a_k}{|z-a_k|}, \partial_{\boldsymbol{T}} \prod_{k=1}^{d} \frac{z-a_k}{|z-a_k|})) + \partial_{\boldsymbol{n}}\xi + g & on \quad \partial\Omega. \end{cases} \tag{8.45}$$

*Proof.* Taking divergence in the both sides of the magnetic field equation in the G-L system, we obtain

$$\text{div}(i\psi_n, (\nabla - i\boldsymbol{A}_n)\psi_n) = \text{div}\,\mathbf{curl} h_n = 0 \ \text{ in } \ \mathcal{D}'(\Omega). \tag{8.46}$$

Passing to the limit in (8.46), we obtain

$$\text{div}(i\psi_*, (\nabla - i\boldsymbol{A}_*)\psi_*) = 0 \ \text{ in } \ \mathcal{D}'(\Omega). \tag{8.47}$$

Since $\text{div}\boldsymbol{A}_* = 0$ and $|\psi_*| = 1$, we deduce that

$$\text{div}(i\psi_*, \nabla\psi_*) = 0 \ \text{ in } \ \mathcal{D}'(\Omega). \tag{8.48}$$

Expand the expression, we obtain (8.43).

Moreover, (8.44)–(8.45) is obviously a solution of (8.43) by simple substitution. More details are given in [BBH 94], Section 1.3 and we omit the details here. Finally, combining the propositions of Sections 7 and 8, we have proved the following

**Theorem 6.8.4** *Let $(\psi_\varepsilon, \boldsymbol{A}_\varepsilon)$ be a pair of minimizers of*

$$Inf_{(\psi,\boldsymbol{A})\in V} G_\varepsilon(\psi, \boldsymbol{A}) \tag{8.49}$$

*where $V$ and $G_\varepsilon$ have been defined in (0.2) and (0.1) respectively. There is a subsequence $\varepsilon_n \to 0$ and exactly $d$ points $a_1, \ldots, a_d$ in $\Omega$ and $(\psi_*, \boldsymbol{A}_*)$ smooth except at the points $a_1, \ldots, a_d$ such that*

$$\begin{cases} \text{div}\boldsymbol{A}_* = 0 & in \quad \Omega, \\ \boldsymbol{A}_* \cdot \boldsymbol{n} = 0 & on \quad \partial\Omega. \end{cases} \tag{8.50}$$

$$\psi_{\varepsilon_n} \to \psi_* \ \ strongly \ in \ \ H^1_{loc}(\Omega \setminus \cup_{j=1}^{d}\{a_j\}) \ \ and \ \ W^{1,p}(\Omega) \ \ for \ \ p < 2. \tag{8.51}$$

*Setting $h_{\varepsilon_n} = \mathbf{curl}\boldsymbol{A}_{\varepsilon_n}$ and $h_* = \mathbf{curl}\boldsymbol{A}_*$, we have*

$$h_{\varepsilon_n} \to h_* \ \ strongly \ in \ \ H^1_{loc}(\Omega \setminus \cup_{j=1}^{d}\{a_j\}) \ \ and \ \ W^{1,p}(\Omega) \ \ for \ \ p < 2 \tag{8.52}$$

*and $h_*$ satisfies the equation*

$$\begin{cases} -\Delta h_* + h_* = 2\pi \sum_{j=1}^{d} \delta_{a_j} & in \quad \Omega, \\ \partial_{\boldsymbol{n}} h_* = -g & on \quad \partial\Omega. \end{cases} \tag{8.53}$$

## 6.9 Renormalized energies

The purpose of this section is to locate the configuration $(a_j)$, the set of singular points of the limit order parameter function $\psi_*$. For this purpose, we introduce the so-called renormalized energy and show that $(a_j)$ must be a minimizer of the renormalized energy. The renormalized energy has been discussed extensively in [BBH 94]. In the case with magnetic field, it is not so different. So we choose to refer to [BBH 94] for some technical details and do not include all the discussions.

In order to give the renormalized energy, we need some preliminary lemmas to describe the energy expressions for the energy-minimizing harmonic maps. The proofs of these lemmas are exactly parallel to those in Chapter I of [BBH 94]. So some of the proofs (when they can be copied word for word) are omitted.

**Lemma 6.9.1** *Let $G$ be a bounded simply connected smooth domain in $R^2$ and $\omega_i$ $(i=1,\dots,n)$ be $n$ sub-domains in $G$ such that $\overline{\omega}_i \cap \overline{\omega}_j = \emptyset$, $i \neq j$ and $\overline{\omega}_i \subset G$, $\Omega = G \setminus \bigcup_{i=1}^{n} \omega_i$. Define $\mathcal{E}_1 = \{\psi \in H^1(\Omega) : \psi \times \partial_{\boldsymbol{T}}\psi = g, \text{ on } \partial G, \deg(\psi, \partial\omega_i) = d_i\}$ where $g$ is a given function. If $v \in \mathcal{E}_1$ such that*

$$E(v) = \int_\Omega |\nabla v|^2 = E_1, \quad E_1 = \inf_{\psi \in \mathcal{E}_1} \int_\Omega |\nabla \psi|^2,$$

*then*

$$E(v) = \int_\Omega |\nabla \Phi_1|^2,$$

*where $\Phi_1$ solves the following problem*

$$\begin{cases} \Delta\Phi_1 = 0, & \text{in } \Omega, \\ \Phi_1 = Const. = C_i, & \text{on } \partial\omega_i, i = 1,\dots,n, \\ \int_{\partial\omega_i} \frac{\partial\Phi_1}{\partial \boldsymbol{n}} = 2\pi d_i, i = 1,\dots,n, \\ \frac{\partial\Phi_1}{\partial \boldsymbol{n}} = g, \text{ on } \partial\Omega. \end{cases}$$

*Proof.* Since $v$ takes its values in $S^1$ we have

$$v_{x_1} \times v_{x_2} = 0, \quad |\nabla v|^2 = |v \times v_{x_1}|^2 + |v \times v_{x_2}|^2,$$

and hence

$$\frac{\partial}{\partial x_1}(v \times v_{x_2}) + \frac{\partial}{\partial x_2}(-v \times v_{x_1}) = 0,$$

Setting $D = (-v \times v_{x_2} + \Phi_{1x_1}, v \times v_{x_1} + \Phi_{1x_2})$, we have

$$\operatorname{div} D = 0, \quad \int_{\omega_i} D \cdot \boldsymbol{n} = 0, i = 1,\dots,n.$$

Then there exists a function $f$ on $\Omega$ such that $D = (\frac{\partial f}{\partial x_2}, -\frac{\partial f}{\partial x_1})$, that is

$$\begin{cases} v \times v_{x_1} = -\frac{\partial f}{\partial x_1} - \frac{\partial \Phi_1}{\partial x_2}, \\ v \times v_{x_2} = -\frac{\partial f}{\partial x_2} + \frac{\partial \Phi_1}{\partial x_1}, \end{cases}$$

and

$$|\nabla v|^2 = |\nabla \Phi_1|^2 + |\nabla f|^2 + 2(\frac{\partial f}{\partial x_1}\frac{\partial \Phi_1}{\partial x_2} - \frac{\partial f}{\partial x_2}\frac{\partial \Phi_1}{\partial x_1}).$$

Noting that

$$\begin{aligned}
&\int_\Omega (\frac{\partial f}{\partial x_1}\frac{\partial \Phi_1}{\partial x_2} - \frac{\partial f}{\partial x_2}\frac{\partial \Phi_1}{\partial x_1}) \\
&= \int_\Omega \operatorname{div}(f\frac{\partial \Phi_1}{\partial x_2}, -f\frac{\partial \Phi_1}{\partial x_1}) \\
&= \int_{\partial G} f\frac{\partial \Phi_1}{\partial \tau} - \sum_{i=1}^{n}\int_{\partial \omega_i} f\frac{\partial \Phi_1}{\partial \tau} \\
&= \int_{\partial G} f\frac{\partial \Phi_1}{\partial \tau} = -\int_{\partial G} \Phi_1 \frac{\partial f}{\partial \tau} \\
&= -\int_{\partial G} \Phi_1(-(v \times v_\tau) + \frac{\partial \Phi_1}{\partial \boldsymbol{n}}) = 0,
\end{aligned}$$

we have

$$\int_\Omega |\nabla v|^2 = \int_\Omega |\nabla \Phi_1|^2 + |\nabla f|^2 \geq \int_\Omega |\nabla \Phi_1|^2, \text{ for any } v \in \mathcal{E}_1.$$

On the other hand, the problem

$$\begin{cases} v \times v_{x_1} = F_1, \\ v \times v_{x_2} = F_2 \end{cases}$$

has a solution $v : \Omega \to S^1$ if and only if

$$\frac{\partial F_1}{\partial x_2} = \frac{\partial F_2}{\partial x_1}$$

and

$$\int_{\Gamma_i} \mathbf{F} \cdot \tau \in 2\pi Z$$

for each connected component $\Gamma_i$ of $\partial\Omega$ where $Z$ denotes the set of integers. Hence it follows from the definition of $\Phi_1$ that the problem

$$\begin{cases} v \times v_{x_1} = -\frac{\partial \Phi_1}{\partial x_2}, \\ v \times v_{x_2} = \frac{\partial \Phi_1}{\partial x_1} \end{cases}$$

admits a solution $v : \Omega \to S^1$ and we have

$$\int_\Omega |\nabla v|^2 = \int_\Omega |\nabla \Phi_1|^2.$$

Moreover, we may see that $v \times v_\tau = g$ on $\partial G$ and

$$2\pi \deg(v, \partial\omega_i) = \int_{\partial\omega_i} v \times v_\tau = \int_{\partial\omega_i} \frac{\partial\Phi_1}{\partial \boldsymbol{n}} = 2\pi d_i$$

which means that $v \in \mathcal{E}_1$ and $E_1$ is attained by $v \in \mathcal{E}_1$ and $E_1 = \int_\Omega |\nabla\Phi_1|^2$. The lemma is proved.

Now we consider a particular case that $\omega_i = B(a_i, \delta)$ and further consider the limit as $\delta \to 0$. Denote $\Omega_\delta = \Omega \setminus \cup_{i=1}^n B(a_i, \delta)$ where $\Omega$ is now a simply connected domain. Define

$$\mathcal{E}_\delta = \{\psi \in H^1(\Omega_\delta) : \psi \times \partial_{\boldsymbol{\tau}}\psi = g, \text{ on } \partial\Omega, \deg(\psi, \partial B(a_i, \delta)) = d_i\}.$$

We deduce from the above lemma that the minimum

$$\nu_\delta = \inf_{\psi \in \mathcal{E}_\delta} \int_{\Omega_\delta} |\nabla\psi|^2$$

is attained by a function in $\mathcal{E}_\delta$ and

$$\nu_\delta = \int_{\Omega_\delta} |\nabla\Phi_\delta|^2$$

where $\Phi_\delta$ is determined by the above lemma with $\omega_i$ replaced by $B(a_i, \delta)$.

The following two lemmas can be found in [BBH 94].

We have for the minimizer $v_\delta$ obtained above and the canonical harmonic map $v_0$,

**Lemma 6.9.2** (Theorem I.6, [BBH 94]) *As $\delta \to 0$, $v_\delta$ converges to $v_0$ in $C^k(K)$ on every compact subset $K$ of $(\Omega \setminus \{a_1, \ldots, a_n\}) \cup \partial\Omega$.*

**Lemma 6.9.3** (Theorem I.7, [BBH 94]) *Let $\Phi_0$ be as in Definition 8.1 and set*

$$R_0(x) = \Phi_0(x) - \sum_{i=1}^n d_i \log|x - a_i|$$

*so that $R_0(x)$ is a smooth harmonic function on $\Omega$. Thus as $\delta \to 0$,*

$$\frac{1}{2}\int_{\Omega_\delta} |\nabla v_\delta|^2 = \pi(\sum_{i=1}^n d_i^2) \log\frac{1}{\delta} + W + O(\delta)$$

*where $W$ is the so-called renormalized energy relative to the configuration $(a, d, g)$ defined by*

$$\begin{aligned} W = W(a,d,g) \quad &= \quad -\pi \sum_{i \neq j}^n d_i d_j \log|a_i - a_j| \\ &\quad + \frac{1}{2}\int_{\partial\Omega} \Phi_0 \partial_{\boldsymbol{n}} \Phi_0 - \pi \sum_{i=1}^n d_i R_0(a_i). \end{aligned}$$

Now we prove the following

**Lemma 6.9.4** *Let $v_\delta$ and $v$ be the minimizers of the minimization problems*

$$\inf_{\psi\in\mathcal{E}_\delta^1}\int_{\Omega_\delta}|\nabla\psi|^2 \quad and \quad \inf_{\psi\in\mathcal{E}_\delta^2}\int_{\Omega_\delta}|\nabla\psi|^2$$

*respectively, where*

$$\mathcal{E}_\delta{}^1=\{\psi\in H^1(\Omega_\delta):\psi\times\partial_{\boldsymbol{\tau}}\psi|_{\partial\Omega}=g+\partial_{\boldsymbol{n}}\xi_\delta,\ \deg(\psi,\partial B(a_i,\delta))=d_i\},$$

$$\mathcal{E}_\delta{}^2=\{\psi\in H^1(\Omega_\delta):\psi\times\partial_{\boldsymbol{\tau}}\psi|_{\partial\Omega}=g+\partial_{\boldsymbol{n}}\xi_*,\ \deg(\psi,\partial B(a_i,\delta))=d_i\}.$$

*If $\xi_\delta$ converges to $\xi_*$ in $W^{2,2}(\Omega)$, then as $\delta\to 0$ we have*

$$\int_{\Omega_\delta}|\nabla v_\delta|^2=\int_{\Omega_\delta}|\nabla v|^2+o(1).$$

*Proof.* Since the energy of $v_\delta$ and $v$ equal to the energy of $\Phi_\delta$ and $\Phi$ where $\Phi_\delta$ and $\Phi$ are determined by Lemma 9.1 with $g$ replaced by $g+\partial_{\boldsymbol{n}}\xi_\delta$ and $g+\partial_{\boldsymbol{n}}\xi_*$ respectively, we only need to prove that

$$\int_{\Omega_\delta}|\nabla\Phi_\delta|^2=\int_{\Omega_\delta}|\nabla\Phi|^2+o(1).$$

Let $U_\delta=(\Phi_\delta-\Phi)-(\xi_\delta-\xi_*)$, then

$$-\Delta U_\delta=f_\delta$$

where $f_\delta=\Delta(\xi_\delta-\xi_*)$ which converges to zero in $L^2(\Omega)$. The boundary conditions read as follows

$$U_\delta=Const.-(\xi_\delta-\xi_*),\quad \text{on } \partial B(a_i,\delta),$$

and

$$\partial_{\boldsymbol{n}}U_\delta=0,\quad \text{on } \partial\Omega.$$

Impose the condition $\int_{\Omega_\delta}U_\delta=0$ so that the solutions $\Phi_\delta$ and $\Phi$ are unique. Multiplying the equation $-\Delta U_\delta=f_\delta$ by $U_\delta$ we have

$$\int_{\Omega_\delta}|\nabla U_\delta|^2-\sum_{i=1}^n\int_{\partial B(a_i,\delta)}U_\delta\cdot\partial_{\boldsymbol{n}}U_\delta=\int_{\Omega_\delta}U_\delta f_\delta,$$

and then the conclusion follows from the Poincaré inequality.

Now we are in a position to derive the renormalized energy for our problem. Consider $d$ points $b_1,\dots,b_d$ in $\Omega$ and $\delta>0$. We choose $\delta$ sufficiently small so that

$$\begin{aligned}&B(b_j,\delta)\subset\Omega\ \ \forall j\in\{1,\dots,d\},\\&B(b_j,\delta)\cap B(b_k,\delta)=\emptyset\ \text{ for }\ j\neq k.\end{aligned}\tag{9.1}$$

As before, we define $\Omega_\delta = \Omega \setminus \cup_{j=1}^d B(a_j, \delta)$ and consider the space $W_\delta$ defined by

$$W_\delta = \left\{ \begin{array}{c} (\psi, \boldsymbol{A}) \in H^1(\Omega_\delta, S^1) \times H^1(\Omega, I\!R^2) \ \text{ such that} \\ (i\psi, \boldsymbol{\tau} \cdot ((\nabla - i\boldsymbol{A})\psi)) = g \ \text{ on } \ \partial\Omega \\ \deg(\psi, \partial B(b_j, \delta)) = 1. \end{array} \right\}$$

and the functional (note that now, $(|\psi|^2 - 1)^2 = 0$)

$$E_\delta(\psi, \boldsymbol{A}) = \frac{1}{2} \int_{\Omega_\delta} |(\nabla - i\boldsymbol{A})\psi|^2 + \frac{1}{2} \int_\Omega |\mathbf{curl}\boldsymbol{A}|^2.$$

We define

$$\mu_\delta(b) = \mathrm{Inf}_{(\psi, \boldsymbol{A}) \in W_\delta} E_\delta(\psi, \boldsymbol{A}). \tag{9.2}$$

Note that the condition

$$(i\psi, \boldsymbol{\tau} \cdot ((\nabla - i\boldsymbol{A})\psi)) = g \ \text{ on } \ \partial\Omega$$

is equivalent to

$$\psi \times \psi_{\boldsymbol{\tau}} = g + \boldsymbol{A} \cdot \boldsymbol{\tau} \text{ on } \partial\Omega.$$

**Proposition 6.9.1** *$\mu_\delta$ is achieved for $\delta$ sufficiently small so that (9.1) holds and moreover,*

$$\mu_\delta \leq \pi d \log \frac{1}{\delta} + C \tag{9.3}$$

*where $C$ depends only on the configuration $(b_j)$ and $\delta_0$.*

*Proof.* The proof is the same as that of Proposition 2.1, we hence omit the details.

**Proposition 6.9.2** *When $\delta$ is sufficiently small so that (9.1) holds, for the minimizers $(v_\delta, \boldsymbol{B}_\delta)$ of (9.2), we have*

$$\int_\Omega |\mathbf{curl}\boldsymbol{B}_\delta|^2 \leq C, \tag{9.4}$$

$$\int_{\Omega_\delta} |(\nabla - i\boldsymbol{B}_\delta)v_\delta|^2 \geq \pi d \log \frac{1}{\delta} - C \tag{9.5}$$

*where $C$ depends only on $(b_j)$ and $\delta_0$.*

*Proof.* In this proof we omit the subscript $\delta$ for simplicity. Let $(v, \boldsymbol{B})$ be a pair of minimizers of (9.2) satisfying the gauge condition

$$\left\{ \begin{array}{lll} \mathrm{div}\boldsymbol{B} = 0 & \text{in} & \Omega, \\ \boldsymbol{B} \cdot \boldsymbol{n} = 0 & \text{on} & \partial\Omega. \end{array} \right. \tag{9.6}$$

We know that

$$\boldsymbol{B} = \mathbf{curl}\xi$$

for the scalar function $\xi$ solving

$$\begin{cases} \Delta\xi = h = \mathbf{curl}\boldsymbol{B} & \text{in} \quad \Omega, \\ \xi = 0 & \text{on} \quad \partial\Omega. \end{cases} \tag{9.7}$$

We have then

$$\int_{\Omega_\delta} |(\nabla - i\boldsymbol{B})v|^2 = \int_{\Omega_\delta} |\nabla v|^2 + |\nabla\xi|^2 + 2(i\nabla v \cdot \mathbf{curl}\xi, v).$$

We notice that

$$\begin{aligned} &\int_{\Omega_\delta} (i\nabla v \cdot \mathbf{curl}\xi, v) \\ = \quad & -\int_{\Omega_\delta} \xi \mathbf{curl}(iv, \nabla v) + \sum_{j=1}^{d} \int_{\partial B(b_j,\delta)} \xi(iv, \partial_{\boldsymbol{\tau}} v) \\ = \quad & \sum_{j=1}^{d} \int_{\partial B(b_j,\delta)} \xi(iv, \partial_{\boldsymbol{\tau}} v) \end{aligned} \tag{9.8}$$

where $\boldsymbol{\tau}$ is the unit tangential vector to $\partial B(b_j, \delta)$, $\boldsymbol{n}$ is the outward unit normal vector to $\partial B(b_j, \delta)$. Here we have used the fact that $v = e^{i\phi}$ for some local function $\phi$ ($|v| = 1$ by assumption), so $\mathbf{curl}(iv, \nabla v) = \mathbf{curl}\nabla\phi = 0$ locally.

From $\deg(v, \partial B(b_j, \delta)) = 1$, we write (let $\theta_j$ be the standard polar angle variable with respect to $b_j$),

$$v = e^{i(\theta_j + \eta)} \quad \text{on} \quad U_j \setminus B(b_j, \delta)$$

in some fixed neighbourhood $U_j$ of $b_j$. Here $\eta$ is some real single valued function on $U_j \setminus B(b_j, \delta)$. We now have, for $x \in U_j \setminus B(b_j, \delta)$,

$$\nabla v = i(\nabla\theta + \nabla\eta)e^{i(\theta+\eta)}$$

which implies

$$\nabla\eta = -\nabla\theta - ie^{-i(\theta+\eta)}\nabla v$$

and finally

$$|\nabla\eta|^2 \le 2(|\nabla v|^2 + \frac{1}{|x - b_j|^2}). \tag{9.9}$$

On the boundary of $B(b_j, \delta)$, we have

$$(iv, \partial_{\boldsymbol{T}} v) = \frac{1}{\delta} + \partial_{\boldsymbol{T}} \eta.$$

We deduce then, by integrating by parts along $\partial B(b_j, \delta)$,

$$\int_{\partial B(b_j,\delta)} \xi (iv, \partial_{\boldsymbol{T}} v) = 2\pi \overline{\xi}_j + \int_{\partial B(b_j,\delta)} \partial_{\boldsymbol{T}} \xi (\eta - \overline{\eta}_j) \tag{9.10}$$

where

$$\overline{\xi}_j = \frac{1}{|\partial B(b_j,\delta)|} \int_{\partial B(b_j,\delta)} \xi, \qquad \overline{\eta}_j = \frac{1}{|\partial B(b_j,\delta)|} \int_{\partial B(b_j,\delta)} \eta.$$

For any $\xi$ solving (9.7), by Sobolev embedding and elliptic estimates, we have for any $\alpha < 1$,

$$|\xi|_{C^{0,\alpha}} \leq C_\alpha |\Delta \xi|_{L^2}. \tag{9.11}$$

From this we have

$$\begin{aligned} &\textstyle\int_{B(b_j,\delta)} \partial_j \xi = \int_{\partial B(b_j,\delta)} \xi \boldsymbol{e}_j \cdot \boldsymbol{n} = \int_{\partial B(b_j,\delta)} (\xi - \xi(b_j)) \boldsymbol{e}_j \cdot \boldsymbol{n}, \\ &\textstyle\frac{1}{|B(b_j,\delta)|} \left| \int_{B(b_j,\delta)} \partial_j \xi \right| \leq 2\pi C_\alpha \delta^{\alpha-1} |\Delta \xi|_{L^2}. \end{aligned} \tag{9.12}$$

By the standard trace theorem, we obtain

$$\int_{\partial B(b_j,\delta)} |\partial_j \xi - \frac{1}{|B(b_j,\delta)|} \int_{B(b_j,\delta)} \partial_j \xi|^2 \leq C\delta \int_\Omega |H\xi|^2 \leq C\delta \int_\Omega |\Delta \xi|^2. \tag{9.13}$$

Combining (9.12)–(9.13), we have

$$\int_{\partial B(b_j,\delta)} |\nabla \xi|^2 \leq C(\delta + \delta^{2\alpha-1}) \int_\Omega |\Delta \xi|^2. \tag{9.14}$$

Similar to $\xi$, we have also

$$\int_{\partial B(b_j,\delta)} |\eta - \overline{\eta}_j|^2 \leq C\delta \int_U |\nabla \eta|^2 \leq C\delta \Big( \int_{\Omega_\delta} |\nabla v|^2 + 2\pi \log(1/\delta) \Big). \tag{9.15}$$

Combining (9.10), (9.14) and (9.15), we conclude

$$\Big| \int_{\partial B(b_j,\delta)} \partial_{\boldsymbol{T}} \xi (\eta - \overline{\eta}_j) \Big| \leq C\delta^\alpha |\Delta \xi|_{L^2(\Omega)} \Big( |\nabla v|_{L^2(\Omega_\delta)} + \sqrt{\log(1/\delta)} \Big). \tag{9.16}$$

Turning to the first term on the right-hand side of (9.9), we have, by Theorem 8.1,

$$\int_{\Omega_\delta} |\nabla v|^2 \geq 2\pi d \log(1/\delta) - C. \tag{9.17}$$

Combining (9.9), (9.10), (9.12), (9.17) and Proposition 9.1, we obtain

$$|\int_{\Omega_\delta}(|\Delta\xi|^2+|\nabla\xi|^2)+4\pi\sum_{j=1}^{d}\xi(b_j)+R(\delta)|\le C \tag{9.18}$$

where

$$R(\delta)=2\sum_{j=1}^{d}\Big[(\overline{\xi}_j-\xi(b_j))+\int_{\partial B(b_j,\delta)}\partial_{\boldsymbol{T}}\xi(\eta-\overline{\eta}_j)\Big]. \tag{9.19}$$

Combining (9.16) and (9.11), we verify that

$$|R(\delta)|\le C|\Delta\xi|_{L^2(\Omega)}\Big(\delta^{\alpha}|\nabla v|_{L^2(\Omega_\delta)}+\delta^{\alpha/2}\Big). \tag{9.20}$$

Using the equality $\nabla v=(\nabla-i\boldsymbol{B})v+iv\boldsymbol{B}$, we obtain

$$\begin{aligned}\int_{\Omega_\delta}|\nabla v|^2 &\le 2\Big[\int_{\Omega_\delta}|(\nabla-i\boldsymbol{B})v|^2+|\boldsymbol{B}|^2\Big]\\ &\le 2\Big[\int_{\Omega_\delta}|(\nabla-i\boldsymbol{B})v|^2+|\nabla\xi|^2\Big]\\ \text{(by Proposition 9.1)} &\le C\log(1/\delta)\end{aligned} \tag{9.21}$$

Going back to (9.20), we deduce that

$$R(\delta)\to 0 \ \text{ as } \delta\to 0. \tag{9.22}$$

By standard elliptic estimates, we have

$$\Big|\sum_{j=1}^{d}\xi(b_j)\Big|\le C|\Delta\xi|_{L^2(\Omega)}. \tag{9.23}$$

(9.22), (9.23) and (9.18) yield

$$|\Delta\xi|_{L^2(\Omega)}\le C.$$

(9.4) and (9.5) then follow.

**Proposition 6.9.3** *Let $\xi_*$ be the solution of*

$$\begin{cases}-\Delta^2\xi_*+\Delta\xi_*=2\pi\sum_{j=1}^{d}\delta_{b_j} & \text{in} \quad \Omega,\\ \xi_*=0 & \text{on} \quad \partial\Omega,\\ \partial_{\boldsymbol{n}}\Delta\xi_*=-g & \text{on} \quad \partial\Omega,\end{cases}$$

*we have $\xi_b\to\xi_*$ in $W^{2,2}(\Omega)$ as $\delta\to 0$.*

*Proof.* On $\Omega_\delta$, we have, by the fact that $(v_\delta, \boldsymbol{B}_\delta)$ are energy minimizers,

$$\begin{cases} -\Delta^2 \xi_\delta + \Delta \xi_\delta = 0 & \text{in} \quad \Omega, \\ \partial_{\boldsymbol{n}} \Delta \xi_\delta = -g & \text{on} \quad \partial\Omega. \end{cases} \tag{9.24}$$

On the other hand,

$$\int_{\partial B(a_j,\delta)} \partial_{\boldsymbol{n}} \Delta \xi_\delta = -\int_{\partial B(a_j,\delta)} (iv_\delta, \boldsymbol{\tau} \cdot (\nabla - i\boldsymbol{B})v_\delta) = -2\pi + o(1). \tag{9.25}$$

From Proposition 9.2, we know that

$$|\Delta \xi_\delta|_{L^2(\Omega)} + |\nabla \xi_\delta|_{L^2(\Omega)} \le |\Delta \xi_\delta|_{L^2(\Omega)} \le C$$

and hence (passing to a subsequence if necessary) we may assert that

$$\xi_\delta \to \bar{\xi} \text{ in } W^{2,2}(\Omega) \text{ weakly.}$$

Passing to the limit in (9.24) and using (9.25), we see that $\bar{\xi}$ verifies the desired equation and hence $\bar{\xi} = \xi_*$. The full sequence therefore converges by uniqueness of the limit. By the fact that $(v_\delta, \boldsymbol{B}_\delta)$ is an energy minimizer, we have

$$\begin{aligned} & \int_{\Omega_\delta} |(\nabla - i\boldsymbol{B}_\delta)v_\delta|^2 + \int_\Omega |\mathbf{curl}\boldsymbol{B}_\delta|^2 \\ \le & \int_{\Omega_\delta} |(\nabla - i\boldsymbol{A}_*)v_\delta|^2 + \int_\Omega |\mathbf{curl}\boldsymbol{A}_*|^2. \end{aligned} \tag{9.26}$$

Expanding the first term on both sides of above inequality as in the expression just after (9.7), we obtain

$$\begin{aligned} & \int_{\Omega_\delta} |\mathbf{curl}\boldsymbol{B}_\delta|^2 + |\nabla \xi_\delta|^2 + 2(i\nabla v_\delta \cdot \mathbf{curl}\xi_\delta, v_\delta) \\ \le & \int_\Omega |\mathbf{curl}\boldsymbol{A}_*|^2 + |\nabla \xi_*|^2 + 2(i\nabla v_* \cdot \mathbf{curl}\xi_*, v_*). \end{aligned}$$

Since $\xi_\delta \to \xi_*$ weakly in $W^{2,2}(\Omega)$, we obtain

$$\int_\Omega |\mathbf{curl}\boldsymbol{B}_\delta|^2 \le \int_\Omega |\mathbf{curl}\boldsymbol{A}_*|^2 + o(1),$$

that is

$$\int_\Omega |\Delta \xi_\delta|^2 \le \int_\Omega |\Delta \xi_*|^2 + o(1).$$

This implies the strong convergence of $\Delta \xi_\delta$ to $\Delta \xi_*$ in $L^2$ in view of the weak lower semi-continuity and then Proposition 9.3 follows.

**Proposition 6.9.4** *Let*

$$\overline{\mu}_\delta = Inf_{\psi \in G_\delta} \frac{1}{2} \int_{\Omega_\delta} |\nabla \psi|^2,$$
$$G_\delta = \{\psi \in H^1(\Omega_\delta, S^1),\ \deg(\psi, \partial B(b_j, \delta)) = 1,\ \partial_{\boldsymbol{T}} \psi = i(g + \partial_{\boldsymbol{n}} \xi_*)\psi \ \ on\ \ \partial\Omega\}.$$

*We have then*

$$\left| \frac{1}{2} \int_{\Omega_\delta} |\nabla v_\delta|^2 - \overline{\mu_\delta} \right| \to 0 \ \ as\ \delta \to 0. \tag{9.27}$$

*Proof.* Since

$$\partial_{\boldsymbol{T}} v_\delta = i(g + \partial_{\boldsymbol{n}} \xi_\delta) v_\delta \ \text{ on } \ \partial\Omega,$$

we see that $v_\delta \in \mathcal{E}_\delta^1$ (see Lemma 9.4) and then

$$\frac{1}{2} \int_{\Omega_\delta} |\nabla v_\delta|^2 \geq \inf_{\psi \in \mathcal{E}_\delta^1} \frac{1}{2} \int_{\Omega_\delta} |\nabla \psi|^2 := \overline{\mu}_\delta^\delta.$$

On the other hand we have from Lemma 9.4 that

$$\overline{\mu}_\delta^\delta = \overline{\mu}_\delta + o(1).$$

This implies

$$\frac{1}{2} \int_{\Omega_\delta} |\nabla v_\delta|^2 \geq \overline{\mu}_\delta + o(1). \tag{9.28}$$

Recall that

$$R(\delta) \to 0 \ \text{ as } \ \delta \to 0,$$

we deduce

$$\begin{aligned} E_\delta(v_\delta, \boldsymbol{B}_\delta) &= \frac{1}{2} \int_{\Omega_\delta} |\nabla v_\delta|^2 + \frac{1}{2} \int_\Omega (|\Delta \xi_*|^2 + |\nabla \xi_*|^2) \\ &\quad + 2\pi \sum_{j=1}^d \xi_*(b_j) + o(1). \end{aligned} \tag{9.29}$$

Hence, in view of (9.29), we have uncoupled $v$ and $\boldsymbol{B}$. Consequently, in order to minimize $E_\delta(v, \boldsymbol{B})$, it suffices to minimize $\overline{\mu}_\delta$, and this yields

$$\frac{1}{2} \int_{\Omega_\delta} |\nabla v_\delta|^2 \leq \overline{\mu}_\delta + o(1).$$

Proposition 9.4 follows.

**Proposition 6.9.5** *Let $1 < p < 2$, the map $v_\delta$ remains bounded in $W^{1,p}$ and $v_\delta$ converges to $v_*$ strongly in $W^{1,p}$, and in $C^k_{loc}(\Omega \setminus \cup_{j=1}^d \{a_j\})$, where $v_*$ is defined on $\Omega$ by*

$$v_* = \prod_{j=1}^{d} \frac{z-a_j}{|z-a_j|} e^{i\phi_*}$$

*and where $\phi_*$ is harmonic and verifies*

$$\partial_{\boldsymbol{\tau}}\phi_* = -(i\prod_{j=1}^{d} \frac{z-a_j}{|z-a_j|}, \partial_{\boldsymbol{\tau}}(\frac{z-a_j}{|z-a_j|})) + \partial_{\boldsymbol{n}}\xi_* + g.$$

*Proof.* Since $v_\delta$ takes its value into $S^1$, we have

$$\mathrm{div}(iv_\delta, \mathbf{curl} v_\delta) = 0 \text{ in } \Omega_\delta.$$

Let $\Phi$ be the solution of

$$\begin{cases} \Delta\Phi = 2\pi \sum_{j=1}^{d} \delta_{b_j}, \\ \partial_{\boldsymbol{n}}\Phi = g + \partial_{\boldsymbol{n}}\xi_\delta. \end{cases} \tag{9.30}$$

We have

$$\mathrm{div}((iv_\delta, \mathbf{curl} v_\delta) + \nabla\Phi) = 0,$$

and

$$\int_{\partial B(b_j,\delta)} ((iv_\delta, \mathbf{curl} v_\delta) + \nabla\Phi)\cdot \boldsymbol{n} = -\int_{\partial B(b_j,\delta)} (iv_\delta, \partial_{\boldsymbol{\tau}} v_\delta) + \int_{\partial B(b_j,\delta)} \partial_{\boldsymbol{n}}\Phi = 0.$$

Hence there is some $f_\delta \in H^1(\Omega_\delta, I\!R)$ such that

$$(iv_\delta, \mathbf{curl} v_\delta) = -\nabla\Phi + \mathbf{curl} f_\delta,$$

that is

$$\begin{cases} v_\delta \times v_{\delta x_1} = -\frac{\partial f_\delta}{\partial x_1} - \frac{\partial \Phi}{\partial x_2}, \\ v_\delta \times v_{\delta x_2} = -\frac{\partial f_\delta}{\partial x_2} + \frac{\partial \Phi}{\partial x_1}. \end{cases}$$

We have from this equality that

$$\int_{\Omega_\delta} |\nabla v_\delta|^2 = \int_{\Omega_\delta} (|v_\delta \times v_{\delta x_1}|^2 + |v_\delta \times v_{\delta x_2}|^2)$$
$$= \int_{\Omega_\delta} |\nabla\Phi|^2 + \int_{\Omega_\delta} |\nabla f_\delta|^2 + 2\int_{\Omega_\delta} (\frac{\partial f_\delta}{\partial x_1}\frac{\partial\Phi}{\partial x_2} - \frac{\partial f_\delta}{\partial x_2}\frac{\partial\Phi}{\partial x_1}).$$

As in the proof of Lemma 9.1, we have

$$\begin{aligned}
&\int_{\Omega_\delta}\left(\frac{\partial f_\delta}{\partial x_1}\frac{\partial\Phi}{\partial x_2}-\frac{\partial f_\delta}{\partial x_2}\frac{\partial\Phi}{\partial x_1}\right)\\
&=-\int_{\partial\Omega}\Phi\frac{\partial f_\delta}{\partial\tau}+\sum_{i=1}^{d}\int_{\partial B(a_i,\delta)}\Phi\frac{\partial f_\delta}{\partial\tau}\\
&=-\int_{\partial\Omega}\Phi(-v_\delta\times v_{\delta\tau}+\frac{\partial\Phi}{\partial\boldsymbol{n}})=o(1),
\end{aligned}$$

here we have used $\xi_\delta\to\xi$ in $W^{2,2}(\Omega)$ and

$$-\int_{\partial B(b_j,\delta)}(iv_\delta,\partial_{\boldsymbol{\tau}}v_\delta)+\int_{\partial B(b_j,\delta)}\partial_{\boldsymbol{n}}\Phi=0.$$

Therefore

$$\int_{\Omega_\delta}|\nabla v_\delta|^2=\int_{\Omega_\delta}|\nabla\Phi|^2+\int_{\Omega_\delta}|\nabla f_\delta|^2+o(1).$$

It follows from Lemma 9.1 that $\int_{\Omega_\delta}|\nabla\Phi|^2$ is the minimum defined in Lemma 9.1 with $g$ replaced by $g+\partial_{\boldsymbol{n}}\xi_\delta$. Therefore it follows from Lemma 9.1 that

$$\left|\frac{1}{2}\int_{\Omega_\delta}|\nabla\Phi|^2-\overline{\mu}_\delta\right|\to 0\quad\text{as}\quad\delta\to 0. \tag{9.31}$$

From Proposition 9.4, we deduce that

$$\int_{\Omega_\delta}|\nabla f_\delta|^2\to 0.$$

Hence

$$||(iv_\delta,\mathbf{curl}v_\delta)+\nabla\Phi||_{L^2(\Omega_\delta)}\to 0\ \text{ as }\delta\to 0$$

and the convergence in $C^k_{\text{loc}}(\Omega\setminus\cup_{j=1}^d\{b_j\})$ of Proposition 9.5 follows from Lemma 9.2 and the convergence in $W^{1,p}$ follows from the fact that

$$|\nabla\Phi|\sim\frac{d_j}{|x-b_j|},\qquad\text{near } b_j.$$

**Proposition 6.9.6** *We have for any configuration* $(b_j)$,

$$\mu_\delta(b)=W(b)+\pi d\log(1/\delta)+o(1)\ \ \textit{as}\ \ \delta\to 0 \tag{9.32}$$

*where* $W$ *is defined by the renormalized energy*

$$\begin{aligned}
W(b)\ =\ &-\pi\sum_{j\neq k}\log|b_j-b_k|-\sum_{j=1}^{d}R(b_j)\\
&+\pi\sum_{j=1}^{d}\xi_*(b_j)+\frac{1}{2}\int_{\partial\Omega}\Phi(g+\partial_{\boldsymbol{n}}\xi_*)+\frac{1}{2}\int_{\partial\Omega}\Delta\xi_*\partial_{\boldsymbol{n}}\xi_*
\end{aligned}\tag{9.33}$$

*where $\xi_*$ is the solution of*

$$\begin{cases} -\Delta^2\xi_* + \Delta\xi_* = 2\pi\sum_{j=1}^d \delta_{b_j} & in \quad \Omega, \\ \xi_* = 0 & on \quad \partial\Omega, \\ \partial_{\boldsymbol{n}}\Delta\xi_* = -g & on \quad \partial\Omega \end{cases} \tag{9.34}$$

*and $R$ is given by*

$$R(x) = \Phi(x) - \sum_{j=1}^d \log|x - b_j|. \tag{9.35}$$

*In addition, $\Phi$ is the solution of*

$$\begin{cases} \Delta\Phi = 2\pi\sum_{j=1}^d \delta_{b_j} & in \quad \Omega, \\ \partial_{\boldsymbol{n}}\Phi = g + \partial_{\boldsymbol{n}}\xi_* & on \quad \partial\Omega. \end{cases} \tag{9.36}$$

*Proof.* We deduce from Propositions 9.3–9.5 and (9.29) that,

$$\mu_\delta(b) \quad = \quad \frac{1}{2}\int_{\Omega_\delta} |\nabla v_\delta|^2 + \int_\Omega (|\Delta\xi_*|^2 + \nabla\xi_*|^2) + 2\pi\sum_{j=1}^d \xi_*(b_j) + o(1). \tag{9.37}$$

Moreover, Equation (9.34), established in Proposition 9.3, easily implies

$$\frac{1}{2}\int_\Omega (|\Delta\xi_*|^2 + |\nabla\xi_*|^2) = -\pi\sum_{j=1}^d \xi_*(b_j) + \frac{1}{2}\int_{\partial\Omega} \Delta\xi_*\partial_{\boldsymbol{n}}\xi_*. \tag{9.38}$$

(9.27) and Lemma 9.3 imply

$$\frac{1}{2}\int_{\Omega_\delta} |\nabla v_\delta|^2 = \pi d\log(1/\delta) + \omega(b,d,g) + O(\delta) \tag{9.39}$$

where

$$\omega(b,d,g) = -\pi\sum_{j\neq k} \log|b_j - b_k| - \pi\sum_{j=1}^d R(b_j) + \frac{1}{2}\int_{\partial\Omega} \partial_{\boldsymbol{n}}\Phi\Phi.$$

Here $\Phi$ is a function verifying (9.30) and $R$ denotes

$$R(x) = \Phi(x) - \sum_{j=1}^d \log|x - b_j|.$$

Combining (9.37)–(9.39), we obtain the desired result.

Finally, we conclude with

**Theorem 6.9.1** *Let $a_j$ be as in Theorem 8.4, the configuration $a = (a_j)$ minimizes $W$.*

*Proof.* Set

$$I(\varepsilon,\delta) = \mathrm{Inf}_{v\in Q}\frac{1}{2}\int_{B(\delta)}|\nabla v|^2 + \frac{1}{4\varepsilon^2}(1-||v|^2)^2,$$
$$Q = \{v \in H^1(B(\delta), I\!R^2),\ v|_{\partial B(\delta)} = e^{i\theta}\}.$$

Let $b = (b_j)$ be a configuration of $d$ points. We construct a comparison function $v_{\varepsilon,\delta}$ in $H^1(\Omega)$ as follows

(1) $v_{\varepsilon,\delta} = v_\delta$ on $\Omega_\delta$.

(2) On $B(b_i,\delta)\setminus B(b_i,\delta/2)$, $v_{\varepsilon,\delta}$ is a harmonic map from $B(b_i,\delta)\setminus B(b_i,\delta/2)$ to $S^1$ with boundary condition $v_{\varepsilon,\delta}|_{\partial B(b_i,\delta)} = v_\delta$ and $v_{\varepsilon,\delta}|_{\partial B(b_i,\delta/2)} = \frac{z-b_i}{|z-b_i|}$.

(3) On $B(b_i,\delta/2)$, $v_{\varepsilon,\delta}$ is the minimizer of $I(\varepsilon,\delta/2)$. Then, we have

$$\frac{1}{2}\int_{B(b_j,\delta)}|\nabla v_{\varepsilon,\delta}|^2 + \frac{1}{2\varepsilon^2}(1-|v_{\varepsilon,\delta}|^2)^2 \le I(\varepsilon,\delta/2) + C(\delta)$$

where $|C(\delta)| \to 0$ as $\delta \to 0$. It follows that

$$\begin{aligned} G_\varepsilon(\psi_\varepsilon,\boldsymbol{A}_\varepsilon) &\le G_\varepsilon(v_{\varepsilon,\delta},\boldsymbol{A}_\varepsilon)\\ &\le E_\varepsilon(v_\delta,\boldsymbol{A}_*) + dI(\varepsilon,\delta/2) + C(\delta)\\ &\le \mu_\delta(b) + dI(\varepsilon,\delta/2) + C(\delta)\\ &\le W(b) + \pi d\log(1/\delta) + dI(\varepsilon,\delta/2) + C(\delta). \end{aligned} \tag{9.40}$$

Here we have used Proposition 9.6 and $|C(\delta)| \to 0$ as $\delta \to 0$. On the other hand, it follows from the definition of the functional $G_\varepsilon(u,\boldsymbol{A})$ that

$$\begin{aligned} &G_\varepsilon(\psi_\varepsilon,\boldsymbol{A}_\varepsilon)\\ =\ &\frac{1}{2}\int_\Omega|(\nabla - i\boldsymbol{A}_\varepsilon)\psi_\varepsilon|^2 + \frac{1}{2}\int_\Omega|\mathbf{curl}\boldsymbol{A}_\varepsilon|^2 + \frac{1}{4\varepsilon^2}\int_\Omega(|\psi_\varepsilon|^2-1)^2 \qquad (9.41)\\ =\ &\frac{1}{2}\int_{\Omega'_\delta}|(\nabla - i\boldsymbol{A}_\varepsilon)\psi_\varepsilon|^2 + \frac{1}{2}\int_\Omega|\mathbf{curl}\boldsymbol{A}_\varepsilon|^2 + \frac{1}{4\varepsilon^2}\int_{\Omega'_\delta}(|\psi_\varepsilon|^2-1)^2\\ &+\sum_{j=1}^d\left[\frac{1}{2}\int_{B(a_j,\delta)}|(\nabla - i\boldsymbol{A}_\varepsilon)\psi_\varepsilon|^2 + \frac{1}{4\varepsilon^2}\int_{B(a_j,\delta)}(|\psi_\varepsilon|^2-1)^2\right]\\ \ge\ &\frac{1}{2}\int_{\Omega'_\delta}|(\nabla - i\boldsymbol{A}_\varepsilon)\psi_\varepsilon|^2 + \frac{1}{2}\int_\Omega|\mathbf{curl}\boldsymbol{A}_\varepsilon|^2\\ &+\sum_{j=1}^d\left[\frac{1}{2}\int_{B(a_j,\delta)}|\nabla\psi_\varepsilon|^2 + \frac{1}{4\varepsilon^2}\int_{B(a_j,\delta)}(|\psi_\varepsilon|^2-1)^2\right] + O(\delta), \qquad (9.42) \end{aligned}$$

here we have used the facts $|\psi_\varepsilon| \le 1$ and $||\boldsymbol{A}_\varepsilon||_{L^\infty} \le C$, $\Omega'_\delta = \Omega\setminus\cup_{j=1}^d B(a_j,\delta)$.

Arguing as in the proof of Lemma VIII.2 of [BBH 94], we have

$$G_\varepsilon(\psi_\varepsilon, \boldsymbol{A}_\varepsilon) \geq W(a) + \pi d \log\frac{1}{\delta} + \sum_{j=1}^{d}[\frac{1}{2}\int_{B(a_j,\delta/2)} |\nabla\psi_\varepsilon|^2 + \frac{1}{4\varepsilon^2}\int_{B(a_j,\delta/2)} (|\psi_\varepsilon|^2 - 1)^2] + O(\delta), \tag{9.43}$$

where $a = (a_j)$. The summation term on the right-hand side of (9.42) can be dealt with exactly by the arguments also as in the proof of Lemma VIII.2 of [BBH 94], we finally have

$$G_\varepsilon(\psi_\varepsilon, \boldsymbol{A}_\varepsilon) \geq dI(\varepsilon, \delta/2) + W(a) + \pi d \log(1/\delta) + C'(\delta) \tag{9.44}$$

where $C'(\delta) \to 0$ as $\delta \to 0$. Combining (9.40) and (9.43), we obtain, letting $\delta \to 0$,

$$W(a) \leq W(b) \qquad \forall\, b = (b_j).$$

The theorem is proved.

**Remark 6.9.1** *Our final remark about the discussion of this Section is that the results seem to indicate that the applied field is of little importance. However, as we pointed out, the analysis has been carried out for a particular scaling with strong boundary conditions. Although possibly not very relevant to physical reality, it is an interesting mathematical result and so we include it here.*

## 6.10 Bibliographical remarks

Following the work of [BBH 94], a number of investigations have been carried out in looking at the vortex location described by the G-L system under the influence of large $\kappa$ limit and applied magnetic field. In this chapter, we mainly followed the work of [BRi 95].

Just before we finished this book, we learnt about some interesting results by Serfaty (see [Ser1 98] and [Ser2 98]), linking the magnetic field and the number of vortices.

# Chapter 7
# The Slow Motion of Vortices

## 7.1 Introduction

In this chapter, we consider the vortex motion for the G-L heat flow

$$\begin{aligned}
&\psi_t = \Delta\psi + \tfrac{1}{\varepsilon^2}(1 - |\psi|^2)\psi, \\
&\psi(x,t) = g(x), \quad \text{for } x \in \partial\Omega,\ t > 0, \\
&\psi(x,0) = \psi_0(x) \quad \text{for } x \in \Omega.
\end{aligned} \tag{1.1}$$

Here $\Omega$ is a smooth, bounded domain in $I\!R^2$, $\varepsilon$ is a positive parameter, $\psi : \Omega \times I\!R_+ \to I\!R^2$, $g : \partial\Omega \to I\!R^2$ is smooth, and $|g|(x) = 1$, $x \in \partial\Omega$. Naturally we also assume the compatibility condition that $\psi_0(x) = g(x)$ on $\partial\Omega$. In fact, from our assumptions, the initial data $\psi_0(x)$ depends on $\varepsilon$, we should write, in our text, the initial data as $\psi_0^\varepsilon(x)$ rather than $\psi_0(x)$. But for the purpose of simplifying notation, we often write, when there is no confusion, $\psi_0$ rather than $\psi_0^\varepsilon$. It is assumed that the boundary data $g$ is independent of $\varepsilon$.

The system is not exactly the superconductivity system we would like to study. However, as we have to use non physical boundary conditions to force the presence of vortices, the topic may be of limited interest in applications to physics. We include it here for the benefit of mathematical studies. We believe that it is for this reason that there is little available literature on this topic incorporating the quantity $\boldsymbol{A}$ in the discussions. We hence limit ourselves to the discussion of the heat flow.

It has to be said, of course, that if we do impose Neumann type boundary conditions on the order parameter, the problem is more interesting. However, the present methods of proof will no longer be valid for that situation. It is an open problem as to how the equations of vortex motion can be derived with Neumann type boundary conditions being imposed. In the steady state case, there were some attempts of trying to establish the limit $\varepsilon \to 0$ in the Neumann case but no definite conclusions as far as the authors are aware.

In the formal asymptotic analysis of this case, we have seen that we were unable to determine the velocity of the vortex motion. So if we start talking about "rigorous" proof of vortex motion now, the natural question is, what do we expect to get. As a matter of fact, the steady state results conclude that the steady state configuration of the vortices must minimize the renormalized energy. From the long time behaviour of the evolutionary solutions, we understand that the evolutionary solution converges to some steady state ones (in appropriate function spaces). This gives us the motivation to hope that the vortex motion for the evolutionary problem is determined by (firstly discussed systematically, to our knowledge, in [Lin 96])

$$m_j \frac{d}{dt} a_j(t) = -\nabla_{a_j} W_g(a), \;\; j = 1, 2, \ldots, d \tag{1.2}$$

where $\{a_j(t)\}$ are the locations of vortices, $m_j$ are the mobilities of the vortices and, from the steady state energy estimates, we expect that $m_j = \log \frac{1}{\varepsilon}$ where $W_g$ is the renormalized energy as discussed in Chapter 6. The renormalized energy has different expressions in different problems with different boundary conditions involved. We have seen a particular one in Chapter 6. Others can be found in, for example, [BBH 94]. They all have more or less the same feature.

This chapter is about the rigorous results we can expect towards proving these formal formulas. We will establish, as a matter of fact, ample evidence that (1.2) SHOULD be true. But in the end, we fall short of giving an outright proof that (1.2) is correct.

Before we start, we mention the following intermediate result obtained in [RSt 95]:

**Theorem 7.1.1** *Let $\psi_\varepsilon(x,t)$ be a solution of (1.1). Assume that at each time $t > 0$ there exists eaxctly one zero of $\psi_\varepsilon$ located at $q_\varepsilon(t)$, with $\deg(\psi_\varepsilon, \partial B_r, q_\varepsilon(t)) = 1$ for any $0 < r <$dist$(q_\varepsilon(t), \partial\Omega)$. Let $x_0$ be any point in $\Omega \setminus \{0\}$ and denote by $T_\varepsilon$ the infinimum of the set $\{t : \; q_\varepsilon(t) = x_0\}$, assuming that this set is nonempty and that $\psi_0(0) = 0$, i.e., $q_\varepsilon(0) = 0$, then*

$$\liminf_{\varepsilon \to 0} \frac{T_\varepsilon}{|\log \varepsilon|} > 0.$$

This is the first rigorous result towards studying the motion of vortices. In giving this theorem here, we did not mention the conditions on the initial data. As a matter of fact, the initial data were $\varepsilon$-dependent. As we will deal with this issue later, we omit the specifications of the initial data here. We should mention that in this result, no effort was made to claim the exact velocity of the vortices. It simply stated that when $\varepsilon$ is sufficiently small, in order that the vortices of the G-L solutions move a finite distance, we require a time interval of the order $|\log \varepsilon|$. In this chapter, we confirm this result in greater detail. We mainly follow the work of [Lin 96].

We consider the evolution equation of the form

$$\frac{1}{\lambda_\varepsilon}\frac{\partial}{\partial t}\psi_\varepsilon = \Delta\psi_\varepsilon + \frac{1}{\varepsilon^2}\psi_\varepsilon(1-|\psi_\varepsilon|^2) \tag{1.3}$$

in $\Omega \times I\!R_+$ and $\psi_\varepsilon$ satisfies (1.1–2) and (1.1–3).

1) We show that as $\varepsilon \to 0$, if

$$\lim_{\varepsilon\to 0}\frac{\lambda_\varepsilon}{\log(1/\varepsilon)} = 0 \text{ and } \lambda_\varepsilon \to \infty,$$

then, the functions

$$\psi_\varepsilon(x,t) - \prod_{j=1}^{d}\frac{x-b_j}{|x-b_j|}e^{ih(x)}$$

converge to 0 in $L^2_{loc}(\bar{\Omega}\times I\!R_+)$. Here $b_1,\ldots,b_d$ are the vortices for the initial data $\psi_0$. $h$ is a real valued function satisfying $\Delta h(x) = 0$ in $\Omega$ and $h(x) = h_0(x)$ on $\partial\Omega$ so that

$$\prod_{j=1}^{d}\frac{x-b_j}{|x-b_j|}e^{ih_0(x)} = g(x).$$

In plain language, the vortices do not move at all on time length scales which are $\ll |\log\varepsilon|$. 2) We show that as $\varepsilon \to 0$, if

$$\lim_{\varepsilon\to 0}\frac{\lambda_\varepsilon}{\log(1/\varepsilon)} = \infty,$$

then, for any sequence $\varepsilon_n$ monotonically decreasing to 0, there is a subsequence of $\{\psi_{\varepsilon_n}(x,t)\}$ that converges in some generalized sense. Moreover, for a.e. $t > 0$, any limit of any subsequence of $\{\psi_{\varepsilon_n}(x,t)\}$ has the form

$$\prod_{j=1}^{d}\frac{x-a_j}{|x-a_j|}e^{ih_a(x)}.$$

Here $a = (a_1,\ldots,a_d)$ are critical points of $W_g(\cdot)$. In plain language, on a time length scale $\gg |\log\varepsilon|$, the vortices have all settled down to locations determined by the equilibrium minimizers.

The open problem is of course, the middle range where the time length scale is $\sim |\log\varepsilon|$. Unfortunately, there is no result yet as far as we know.

## 7.2 Preliminaries

Let $b_1, \ldots, b_d$ be $d$ distinct points in $\Omega$ where $d = \deg(g, \partial\Omega) > 0$. We know that there exists a constant $\rho_0$ depending on $b = (b_1, \ldots, b_d)$ and $\Omega$ such that for every $0 < \rho \leq \rho_0$ and every $\varepsilon > 0$, there is a $w_\varepsilon \in C^1(\Omega)$ with $w_\varepsilon = g$ on $\partial\Omega$ and

$$\begin{array}{l} E_\varepsilon(w_\varepsilon) \leq dI(\varepsilon, \rho) + W_g(b) + \pi d \log \frac{1}{\rho} + O(\rho), \\ |\nabla w_\varepsilon(x)| \leq c/\varepsilon, \ x \in \Omega, \\ \frac{1}{\varepsilon^2} \int_\Omega (|w_\varepsilon|^2 - 1)^2 dx \leq c \end{array} \tag{2.1}$$

where

$$I(\varepsilon, \rho) = \min\Big\{ \int_{B_\rho(0)} e_\varepsilon(\psi) dx : \ \psi(x) = \frac{x}{|x|} \text{ on } \partial B_\rho(0) \Big\}.$$

We also know that for $\varepsilon$ sufficiently small,

$$\begin{array}{rl} & \min\{E_\varepsilon(\psi) : \ \psi|_{\partial\Omega} = g\} \geq dI(\varepsilon, \rho) + \pi d \log \dfrac{1}{\rho} \\ + & \min\{W_g(a) : a \in \bar{\Omega}^d\} + O(\rho^2). \end{array} \tag{2.2}$$

Now we announce some preliminary results:

**Lemma 7.2.1** *Let $\psi_\varepsilon$ be a minimizer of the functional*

$$E_\varepsilon(\psi) = \frac{1}{2} \int_B \Big[ |\nabla \psi|^2 + \frac{1}{2\varepsilon^2} (|\psi|^2 - 1)^2 \Big] dx$$

*in the unit disc $B$ with $\psi_\varepsilon = g_\varepsilon$ on $\partial B$. Suppose that*

$$\int_{\partial B} \Big[ |\partial_\tau g_\varepsilon|^2 + \frac{1}{2\varepsilon^2} (|g_\varepsilon|^2 - 1)^2 \Big] \leq K, \tag{2.3}$$

*for some constant $K > 0$. Then, for all sufficiently small $\varepsilon > 0$ (depending only on $K$), we have*

$$\text{if } \deg(g_\varepsilon, \partial\Omega) = 0, \quad \text{then } E_\varepsilon(\psi_\varepsilon) \leq C(K), \tag{2.4}$$

$$\text{if } \deg(g_\varepsilon, \partial\Omega) = d \neq 0, \quad \text{then } E_\varepsilon(\psi_\varepsilon) \geq \pi |d| \log \frac{1}{\varepsilon} - C(K). \tag{2.5}$$

*Proof.* It follows from the condition on $g_\varepsilon$ that $g_\varepsilon \in H^1(\partial B)$ embedded in $C^{\frac{1}{2}}(\partial B)$ and

$$|g_\varepsilon| \geq 1 - C\varepsilon^{\frac{1}{4}}.$$

In fact, for any $x \in \partial B$, it follows from the fact that $\{g_\varepsilon\}$ is uniformly bounded in $C^{\frac{1}{2}}(\partial B)$ that

$$\operatorname{osc}_{\partial B \cap B(x, \varepsilon^{\frac{1}{2}})} g_\varepsilon \leq C\varepsilon^{\frac{1}{4}}.$$

On the other hand, we have from $\frac{1}{\varepsilon^2}\int_{\partial B\cap B(x,\varepsilon^{\frac{1}{2}})}(1-|g_\varepsilon|^2)^2 \le K$ that

$$\inf_{\partial B\cap B(x,\varepsilon^{\frac{1}{2}})}(1-|g_\varepsilon|^2) \le C(K)\varepsilon^{\frac{3}{4}}$$

which implies

$$\sup_{\partial B\cap B(x,\varepsilon^{\frac{1}{2}})}|g_\varepsilon| \ge 1-C(K)\varepsilon^{\frac{1}{4}}.$$

Subsituting this into $\operatorname{osc}_{\partial B\cap B(x,\varepsilon^{\frac{1}{2}})} g_\varepsilon \le C\varepsilon^{\frac{1}{4}}$, we have

$$\inf_{\partial B\cap B(x,\sqrt{\varepsilon})}|g_\varepsilon| \ge 1-C\varepsilon^{\frac{1}{4}},$$

which implies that $|g_\varepsilon| \ge 1-C\varepsilon^{\frac{1}{4}}$ on $\partial B$.

Hence, $\deg(g_\varepsilon,\partial B)$ is well defined. If $\deg(g_\varepsilon,\partial B)=0$, we may write $g_\varepsilon = \rho_\varepsilon(\theta)e^{i\phi_\varepsilon(\theta)}$ on $\partial B$. Let $U_\varepsilon^* = \rho_\varepsilon^*(\theta)e^{i\phi_\varepsilon^*(\theta)} \in H^1_{g_\varepsilon}(B)$ where $\phi_\varepsilon^*(\theta)$ and $\rho_\varepsilon^*$ are defined as follows. $\phi_\varepsilon^*(\theta)$ is the harmonic extension of $\phi_\varepsilon$ from $\partial B$ to $B$, that is, $\phi_\varepsilon^*(\theta)$ solves

$$\begin{cases} \Delta\phi_\varepsilon^* = 0, & \text{in } B, \\ \phi_\varepsilon^* = \phi_\varepsilon, & \text{on } \partial B, \end{cases} \tag{2.6}$$

and

$$\rho_\varepsilon^* = \begin{cases} 1, & \text{for } 0\le r\le 1-\varepsilon, \\ -(r-1)\frac{1-\rho_\varepsilon(\theta)}{\varepsilon}+\rho_\varepsilon(\theta), & \text{for } 1-\varepsilon\le r\le 1. \end{cases} \tag{2.7}$$

Using the condition on $g_\varepsilon$, we have from direct computations that

$$E_\varepsilon(\psi_\varepsilon) \le E_\varepsilon(U_\varepsilon^*) \le \frac{1}{2}\int_B |\nabla\phi_\varepsilon^*|^2 + C_1(K)\varepsilon \le C(K). \tag{2.8}$$

The first inequality of Lemma 2.1 is proved.

The second inequality of Lemma 2.1 can be proved using the idea of [BBH 94] Chapter III-Chapter V. The only difference is that we should replace $g:\partial B\to S^1$, a smooth map in [BBH 94], by $g_\varepsilon$ in Lemma 2.1. But this replacement does not affect the proof since we have the condition (2.3).

**Lemma 7.2.2** *Let hypothesis in Lemma 2.1 hold and suppose* $\deg(g_\varepsilon,\partial B)=0$. *Then*

**(i)** $|\psi_\varepsilon(x)|\ge\frac{3}{4}$ *in* $B$ *whenever* $0<\varepsilon\le\varepsilon_K$ *for some* $\varepsilon_K$.

**(ii)** *In general, if* $v\in H^1(B)$ *with* $v=g_\varepsilon$ *on* $\partial B$, $|v(0)|\le 1/2$ *and* $|\nabla v(x)|\le c/\varepsilon$, *then*

$$E_\varepsilon(v) \ge \min\{E_\varepsilon(u):\ u|_{\partial B}=g\}+C_0(K) \tag{2.9}$$

*for some* $C_0(K)>0$.

*Proof.* Proof of (i). We prove this by contradiction. If the conclusion is false, then there is a sequence of minimizer $\{\psi_{\varepsilon_n}\}$ which takes the boundary value $\psi_n = \psi_{\varepsilon_n} = g_{\varepsilon_n}$ where $g_{\varepsilon_n}$ satisfies the condition of Lemma 2.1, but $\inf_B |\psi_n| \leq \frac{3}{4}$. It follows from the condition on $g_{\varepsilon_n}$ that $|g_{\varepsilon_n}|$ converges to 1 uniformly. Without loss of generality we assume $1 - C\varepsilon^{\frac{1}{4}} \leq |g_{\varepsilon_n}| \leq 1$.

Hence if $1 - |x| \leq C\varepsilon_n$, we can claim that $|\psi_n(x)| \geq \frac{4}{5}$. In fact, Let $V_n(x) = \psi_n(\varepsilon_n x)$ which minimizes the energy

$$E_{\varepsilon_n}(V) = \frac{1}{2}\int_{B/\varepsilon_n} [|\nabla V|^2 + \frac{1}{2}(1 - |V|^2)^2].$$

Then $V_n$ solves

$$\begin{cases} -\Delta V_n = V_n(1 - |V_n|^2), & \text{in } B/\varepsilon_n, \\ V_n = g_{\varepsilon_n}(\varepsilon_n x), & \text{in } B/\varepsilon_n. \end{cases}$$

Applying the elliptic estimates to above problem, we have
(a) For $x, y \in \frac{1}{\varepsilon_n}B$ and $|x - y| \leq 1$, $|V_n(x) - V_n(y)| \leq C|x - y|^{\frac{1}{2}}$;
(b) $|\nabla V_n| \leq \frac{C}{R}$, $R \in (0, 1)$, $|x| < \frac{1}{\varepsilon_n} - R$.

Therefore if $|\psi_n(x_n)| \leq \frac{3}{4}$, we have from $|\nabla\psi_n| \leq \frac{C}{\varepsilon_n}$ that for $x \in \{|x - x_n| \leq \eta\varepsilon_n\}$,

$$|\psi_n(x)| \leq \frac{3}{4} + C\eta.$$

This implies that $|\psi_n(x)| \leq \frac{4}{5}$ in $\{|x - x_n| \leq \eta\varepsilon_n\}$ for some $\eta > 0$.

Now we easily see that

$$\begin{aligned} &\frac{1}{\varepsilon_n^2}\int_B (1 - |\psi_n|^2)^2 \geq \frac{1}{\varepsilon_n^2}\int_{B(x_n, \eta\varepsilon_n)} (1 - |\psi_n|^2)^2 \\ &\geq \frac{9}{25}\pi\eta^2 = C(\eta) > 0. \end{aligned}$$

On the other hand it follows from the proof of Lemma 2.1 that

$$E_{\varepsilon_n}(\psi_n) \leq E_{\varepsilon_n}(U^*_{\varepsilon_n}) \leq \frac{1}{2}\int_B |\nabla\phi^*_{\varepsilon_n}|^2 + C_1(K)\varepsilon_n.$$

But since $g_{\varepsilon_n} \to g = e^{i\phi}$ weakly in $H^1(\partial B)$, we know $\int_B |\nabla\phi^*_{\varepsilon_n}|^2 \to \int_B |\nabla\phi^*|^2$ where $\phi^*$ solves (2.6) with boundary data $\phi$. Thus

$$\overline{\lim} E_{\varepsilon_n}(U^*_{\varepsilon_n}) \leq \frac{1}{2}\int_B |\nabla\phi^*|^2. \tag{2.10}$$

We also know that $\psi_n \to \psi$ weakly in $H^1(B)$, $\psi = g$ on $\partial B$, $|\psi| = 1$ a.e. on $B$, and then

$$\begin{aligned} \underline{\lim} E_{\varepsilon_n}(\psi_n) &\geq C(\eta) + \lim \frac{1}{2}\int_B |\nabla\psi_n|^2 \\ &\geq C(\eta) + \frac{1}{2}\int_B |\nabla\phi^*|^2. \end{aligned} \tag{2.11}$$

Combining (2.10) with (2.11), we arrive at a contradiction $0 < C(\eta) \leq 0$. The conclusion (i) of Lemma 2.2 is proved.

We should note that if the condition of (i) holds, then $|v(x)| \leq \frac{3}{4}$ for $|x| \leq \lambda\varepsilon$ for some constant $\lambda$. From this we can derive that the constant $C_0(K) \geq \frac{\lambda^2}{16}$. In fact, it follows from $|v(0)| \leq \frac{1}{2}$ and $|\nabla v| \leq \frac{C}{\varepsilon}$ that $|v(x)| \leq \frac{3}{4}$ when $|x| \leq \lambda\varepsilon$ for some $\lambda > 0$. Therefore $1 - |v|^2 \geq 1 - |v| \geq \frac{1}{4}$ when $|x| \leq \lambda\varepsilon$ for some $\lambda > 0$, we can calculate that

$$\frac{1}{4\varepsilon^2}\int_B (1-|v|^2)^2 \geq \frac{1}{4\varepsilon^2}\int_{B(0,\lambda\varepsilon)} (1-|v|^2)^2 \geq \frac{1}{16}\pi\lambda^2.$$

Now we prove (ii) of Lemma 2.2.

If not, for fixed $\varepsilon > 0$, we may find $v_n \in H^1(B)$ with $v_n = g_\varepsilon$, $|\nabla v_n(x)| \leq \frac{C}{\varepsilon}$ and $|v_n(0)| \leq \frac{1}{2}$, and

$$E_\varepsilon(v_n) \to \min_{H^1_{g_\varepsilon}} E_\varepsilon(u).$$

Since $\deg(g_\varepsilon, \partial B) = 0$, as in the proof of Lemma 2.1, we may write $g_\varepsilon = \rho_\varepsilon(\theta)e^{i\phi_\varepsilon(\theta)}$. Denote the harmonic extension of $\phi_\varepsilon$ to $B$ by $\phi_\varepsilon^*$, and $\rho^*(\theta)$ is defined as in the proof of Lemma 2.1, we construct a function $U_\varepsilon^* = \rho_\varepsilon^*(\theta)e^{i\phi_\varepsilon^*}$. Direct calculations as before give

$$E_\varepsilon(U_\varepsilon^*) \leq \frac{1}{2}\int_B |\nabla\phi_\varepsilon^*|^2 dx + C\varepsilon, \tag{2.12}$$

and since $v_n$ is a minimizing sequence

$$E_\varepsilon(v_n) \leq E_\varepsilon(U_\varepsilon^*).$$

We may assume

$$v_n \to \tilde{v}, \quad \text{in } H^1,$$

where $\tilde{v} = \rho_\varepsilon^{**}e^{i\phi_\varepsilon^{**}}$ is the minimizer of

$$\min_{H^1_{g_\varepsilon}} E_\varepsilon(u).$$

Thus

$$\begin{aligned} \lim_{n\to\infty} E_\varepsilon(v_n) &= \lim_{n\to\infty}\frac{1}{4\varepsilon^2}\int_B (1-|v_n|^2)^2 + \lim_{n\to\infty}\int_B |\nabla v_n|^2 \\ &\geq \frac{\lambda^2}{16} + \int_B (\rho_\varepsilon^{**})^2|\nabla\phi_\varepsilon^{**}|^2. \end{aligned} \tag{2.13}$$

However when $\varepsilon \to 0$, $\rho_\varepsilon^{**} \to 1$ uniformly and $\phi_\varepsilon^* \to \phi^*$, $\phi_\varepsilon^{**} \to \phi^*$ in $H^1$, we are led to a contradiction if we compare (2.12) with (2.13). The conclusion (ii) is proved and in which $C_0(K) = \frac{\lambda^2}{16}$ provided that $|v(x)| \leq \frac{3}{4}$ in $x \in \{x : |x| \leq \lambda\varepsilon\}$.

Now we give the definition of the function class we are going to work on:

**Definition 7.2.1** *Let $\Omega$, $g$ be as in (1.1). A map $\psi : \Omega \to I\!R^2$ belongs to the class $S_g(\lambda, K)$ if*

*1) $\psi \in H^1(\Omega)$, $\psi = g$ on $\partial\Omega$, and $|\psi(x)| \leq 1$ in $\Omega$;*

*2) if $|\psi(x_0)| \leq 1/2$ and $x_0 \in \Omega$, then $|\psi(x)| \leq 3/4$ whenever $x \in \Omega$ and $|x - x_0| \leq \lambda\varepsilon$;*

*3) $E_\varepsilon(\psi) \leq \pi d \log \frac{1}{\varepsilon} + K$.*

If $\psi_\varepsilon$ is a solution of (1.1), then it is easy to verify that for any $t > 0$, $\psi_\varepsilon(\cdot, t) \in S_g(\lambda, K)$ provided that $E_\varepsilon(\psi_0) \leq \pi d \log \frac{1}{\varepsilon} + K$ and $|\nabla \psi_0(x)| \leq c_0/\varepsilon$ where $\lambda$ may depend on $c_0$, $g$ and $\Omega$.

**Theorem 7.2.1** *There are two positive numbers $\varepsilon_0$ and $\alpha_0$ depending only on $\lambda$, $K$, $g$ and $\Omega$ such that, for any $0 < \varepsilon \leq \varepsilon_0$, $\psi \in S_g(\lambda, K)$, there are $N_\varepsilon$ disjoint balls $B_j$ of radius $\varepsilon^{\alpha_j}$, $j = 1, \ldots, N_\varepsilon$, with the following properties:*

*1) $\alpha_0 \leq \alpha_j \leq 1$ for $j = 1, \ldots, N_\varepsilon$ and $N_\varepsilon \leq N_*(\lambda, K)$.*

*2) The set $\{x \in \Omega : \ |\psi(x)| \leq \frac{1}{2}\}$ is contained in $\Omega \cap (\cup_{j=1}^{N_\varepsilon} B_j)$.*

*3) The estimates $\varepsilon^{\alpha_j} \int_{\partial(B_j \cap \Omega)} e_\varepsilon(\psi) \leq c(\alpha_0, \lambda, K)$, $j = 1, \ldots . N_\varepsilon$ are valid. In particular, the degrees $d_j = \deg(\psi, \partial(B_j \cap \Omega))$ are well defined.*

*4) There are exactly $d$ balls, say $B_1, \ldots, B_d$, such that the corresponding degrees $d_j$ are not zero. If we let $x_1, \ldots, x_d$ be the centres of the balls $B_1, \ldots, B_d$, then $min\{|x_j - x_k|, dist(x_j, \partial\Omega) : \ j \neq k, \ j, k = 1, \ldots, d\} \geq \delta(\lambda, K) > 0$. Moreover, each $d_j$ equals 1 for $j = 1, \ldots, d$.*

*5) If $B_j \cap \Omega \neq \emptyset$, then if $B_j \cap \Omega$ is scaled by a factor of size $\varepsilon^{\alpha_j}$, the resulting domain is of diameter 1 and is uniformly Lipschitz (independent of $\varepsilon$ and $j$).*

Before giving the proof, we give the following remark explaining the meaning of the above theorem.

**Remark 7.2.1** *In our discussion, we say that a function $\psi \in S_g(\lambda, K)$ is good at a point $x$ if $|\psi(x)| \geq \frac{1}{2}$ and bad at a point $x$ if $|\psi(x)| < \frac{1}{2}$ and $\deg(\psi, \partial(B_\delta(x) \cap \Omega)) \neq 0$. However, there may be points where $|\psi(x)| < \frac{1}{2}$ but $\deg(\psi, \partial(B_\delta(x) \cap \Omega)) = 0$. These are not really bad points but may cause confusion. This theorem shows that for any $\psi \in S_g(\lambda, K)$, when $\varepsilon$ is sufficiently small, the set $\{x \in \Omega : \ |\psi(x)| \leq \frac{1}{2}\}$ is contained in $\Omega \cap (\cup_{j=1}^{N_\varepsilon} B_j)$ with $B_j$ of radius $\varepsilon^{\alpha_j}$ and $N_\varepsilon$ uniformly bounded.*

*Moreover, there are only $d$ balls ($B_1$, ..., $B_d$) among these $N_\varepsilon$ balls ($B_1$, ..., $B_{N_\varepsilon}$) such that they contain the really bad points, that is,*

$$d_j = \deg(\psi, \partial(B_j(x) \cap \Omega)) \neq 0, \ j = 1, \ldots, d.$$

*The others contain only the not really bad points. That is, $\deg(\psi, \partial(B_j(x) \cap \Omega)) = 0$. Our purpose is to show that there are only $d$ really bad balls.*

*Proof of Theorem 2.1.* Step 1: Starting with a map $\psi \in S_g(\lambda, K)$, we are going to construct a finite sequence of maps such that each map is a simple modification of the preceding one and the final map $v$ has the following properties:

(P1) $v \in S_g(\lambda, K)$;

(P2) the set $\{x \in \Omega : |v(x)| \leq \frac{1}{2}\}$ is contained in $d$ distinct balls $D_j$ centered at $y_j$ and of radius $\varepsilon^{\beta_j}$ for $j = 1, \cdot, d$, where $\beta_j \geq \beta(\lambda, K)$ for $j = 1, \ldots, d$.

(P3) the balls $D_j$, $j = 1, \ldots, d$ satisfy $\varepsilon^{\beta_j} \int_{\partial D_j} e_\varepsilon(V) \leq c(\beta, \lambda, K)$. $\min\{|y_j - y_k|, \operatorname{dist}(y_j, \partial\Omega) : j \neq k,\ j,k = 1, \ldots, d\} \geq \delta(\lambda, K)$, and $d_j = \deg(V, \partial D_j) = 1$ for $j = 1, \ldots, d$.

For this purpose, we assume that $\varepsilon$ is so small that $K \leq \frac{1}{2} \log \frac{1}{\varepsilon}$ and thus $E_\varepsilon(\psi) \leq \pi(d + \frac{1}{2}) \log \frac{1}{\varepsilon}$ and thus $E_\varepsilon(\psi) \leq \pi(d + \frac{1}{2}) \log \frac{1}{\varepsilon}$. Set $\alpha = 2^{-d-k_0}$ where $k_0$ is chosen so that $2^{-k_0+1} \leq \frac{1}{6(d+1)}$. Then it follows from Fubini's theorem that there exists a number $\beta \in [\alpha, 2\alpha]$ such that

$$\begin{aligned}
\pi|d| \log \frac{1}{\varepsilon} + C &\geq E_\varepsilon(\psi) \geq E_\varepsilon(\psi, \Omega \cap B_{\varepsilon^\alpha}(x)) \\
&= \frac{1}{2} \int_{\Omega \cap B_{\varepsilon^\alpha}(x)} (|\nabla\psi|^2 + \frac{1}{2\varepsilon^2}(1 - |\psi|^2)^2) dx \\
&\geq \frac{1}{2} \int_{\varepsilon^{2\alpha}}^{\varepsilon^\alpha} \frac{d\rho}{\rho} \int_{\partial B_\rho(x) \cap \Omega} \rho[|\nabla\psi|^2 + \frac{1}{2\varepsilon^2}(1 - |\psi|^2)^2] ds \\
&= \frac{1}{2} \varepsilon^\beta \log \frac{\varepsilon^\alpha}{\varepsilon^{2\alpha}} \int_{\partial B_{\varepsilon^\beta}(x) \cap \Omega} \rho[|\nabla\psi|^2 + \frac{1}{2\varepsilon^2}(1 - |\psi|^2)^2] ds \\
&= \left( \varepsilon^\beta \int_{\partial B_{\varepsilon^\beta}(x) \cap \Omega} e(\psi) \right) \alpha \log \frac{1}{\varepsilon}.
\end{aligned}$$

This implies, for any $x \in \Omega$, there is $\beta \in [\alpha, 2\alpha]$ such that

$$\varepsilon^\beta \int_{\partial B_{\varepsilon^\beta}(x) \cap \Omega} e_\varepsilon(\psi) \leq c(d)/\alpha. \tag{2.14}$$

In particular, $|\psi(x)| \geq \frac{3}{4}$ for $x \in \partial(B_{\varepsilon^\beta}(x) \cap \Omega)$, and the $\deg(\psi, \partial(B_{\varepsilon^\beta}(x) \cap \Omega))$ is well defined whenever $\varepsilon$ is sufficiently small (depending only on $K$, $d$, $g$ and $\Omega$).

Let $y_0$ be a point in the set $\{x \in \Omega : \ |\psi(x)| \leq \frac{1}{2}\}$. Then we choose a ball $D$ of radius $\varepsilon^\beta$ for some $\beta \in [\alpha, 2\alpha]$ and centred at $y_0$ so that (2.14) is valid (with $y_0$ in place of $x$). If $\deg(\psi, \partial(D \cap \Omega)) = 0$, then we replace $\psi$ inside $D \cap \Omega$ by $\tilde{\psi}$, where $\tilde{\psi}$ minimizes the energy $\int_{D \cap \Omega} e_\varepsilon(v) dx$ with $\tilde{\psi} = \psi$ on $\partial(D \cap \Omega)$. It is easy to see that $\varepsilon^{-\beta}(D \cap \Omega)$ is a Lipschitz domain for which we may apply Lemma 2.2 to conclude that $|\tilde{\psi}| \geq \frac{3}{4}$ in $D \cap \Omega$. This way, we obtain a new map $\psi'(x)$ that equals $\tilde{\psi}$ on $D \cap \Omega$, coincides with $\psi$ on $\Omega \setminus D$ such that

$$E_\varepsilon(\psi'(x)) \leq E_\varepsilon(\psi(x))$$

and that still possesses the property (ii) in the definition for the class $S_g(\lambda, K)$. Indeed, since $|\psi(x)| \geq \frac{3}{4}$ on $\partial(D \cap \Omega)$, for any point $x_0 \in \Omega \setminus D$ with $|\psi(x_0)| \leq \frac{1}{2}$, the ball $\{x: |x-x_0|\le \lambda\varepsilon\}$ will not intersect $\partial(D\cap\Omega)$. In other words, $\psi'(x)\in S_g(\lambda,K)$. We then apply the same modification procedure to $\psi'(x)$ as we did to $\psi(x)$ in the above. We then obtain $\psi_2(x)$. Continue this procedure, say $N_1$ times, we arrive at a new map $\psi_1(x)$, a point

$$y_1\in\{x\in\Omega: |\psi_1(x)|\le\frac{1}{2}\}$$

and a ball $D_1$ centred at $y_1$ of radius $\varepsilon^{\beta_1}$ for which the corresponding estimate (2.14) is valid. Moreover, $\deg(\psi_1,\partial(D_1\cap\Omega))=d_1\neq 0$.

We then keep the ball $D_1$ and let $\Omega_1=\Omega\setminus D_1$. We apply the same arguments as above for $\psi$ on $\Omega$ to $\psi_1$ on $\Omega_1$ with $\beta\in[2\alpha,4\alpha]$. The reason for changing the range of $\beta$ is as follows: for any $x\in\Omega_1$, let $D$ be a ball centered at $x$ and of radius $\varepsilon^\beta$ for some $\beta\in[2\alpha,4\alpha]$, then $D\cap\Omega_1$ is a uniformly Lipschitz domain after the following normalization: since $D$ is a ball centred at $x$ of radius $\varepsilon^\beta$, then $\varepsilon^{-1}D\cap\Omega_1=\{\varepsilon^{-1}x, x\in D\cap\Omega_1\}$ is Lipschitz (its Lipschitz constant is independent of $\beta\in[2\alpha,4\alpha]$, small positive $\varepsilon$ and $x\in\Omega_1$).

Again, we modify $\psi_1$ on $\Omega_1$, a total of $N_2$ times to find another map $\psi_2$, a point $y_2\in\{x\in\Omega_1: |\psi_2(x)|\le\frac{1}{2}\}$, and a ball $D_2$ of radius $\varepsilon^{\beta_2}$ centred at $y_2$ with $\beta_2\in[2\alpha,4\alpha]$ such that $\deg(\psi_2,\partial(D_2\cap\Omega_1))=d_2\neq 0$. We keep the ball $D_2$, let $\Omega_2=\Omega_1\setminus D_2$, then apply the same procedure to $\psi_2$ on $\Omega_2$, then apply the same procedure to $\psi_2$ on $\Omega_2$ with $\beta\in[4\alpha,8\alpha]$, and so on.

We claim that the above procedure has to stop after $d$ iterations of modifying the maps. Indeed, if we find $d+1$ balls, $D_1,\dots,D_{d+1}$, then by applying Lemma 2.1 to $D_j\cap\Omega_{j-1}$, $j=1,\dots,d+1$ and $\Omega_0=\Omega$, since all these domains become uniformly Lipschitz after proper scaling, we have, for $j=1,\dots,d+1$, that

$$\int_{D_j\cap\Omega_{j-1}} e_\varepsilon(\psi_j)\ge\pi|d_j|\log\frac{1}{\varepsilon}(1-2^j\alpha)-c(\alpha,d).$$

Therefore,

$$\begin{aligned}E_\varepsilon(\psi) &\ge E_\varepsilon(\psi_{d+1})\ge\sum_{j=1}^{d+1}\pi|d_j|\log\frac{1}{\varepsilon}(1-2^j\alpha)-C(\alpha,d)\\ &\ge \pi\log\frac{1}{\varepsilon}\sum_{j=1}^{d+1}(1-2^j2^{-d-k_0})\\ &\ge \pi(d+\frac{2}{3})\log\frac{1}{\varepsilon}-c(\alpha,d).\end{aligned}\tag{2.15}$$

This result contradicts the fact $\psi\in S_g(\lambda,K)$ whenever $\varepsilon$ is sufficiently small.

So, after finite many steps, we obtain a new map $\phi$ which satisfies both P1) and P2). Now, we verify that it satisfies P3).

We first show that $d_j = 1, j = 1, \ldots, d$. Since $\sum_{j=1}^d d_j = d$ and $d_j \neq 0$, if $|d_j| > 1$ for some $j$ then $\sum_{j=1}^d |d_j| \geq d+1$. Therefore it follows from the facts

$$2^{-k_0} < \frac{1}{6(d+1)}$$

and

$$1 - 2^j\alpha = 1 - 2^j 2^{-d-k_0} \geq 1 - 2^{-k_0}, \ 1 \leq j \leq d,$$

that

$$\begin{aligned}
\pi d \log\frac{1}{\varepsilon} + C &\geq E_\varepsilon(\psi_\varepsilon) \\
&\geq \sum_{j=1}^d \pi|d_j| \log\frac{1}{\varepsilon}(1 - 2^j\alpha) - C(\alpha, d) \\
&\geq \pi(d+1)\log\frac{1}{\varepsilon}(1 - 2^{-k_0}) - C(\alpha, d) \\
&\geq \pi(d + \frac{5}{6})\log\frac{1}{\varepsilon} - C(\alpha, d)
\end{aligned}$$

which leads to a contradiction. This implies $\sum_{j=1}^d |d_j| \leq d$ and then the fact $d_j = 1$ follows.

To show that the centres of the balls $D_j$ are distinct and lie strictly inside $\Omega$, we consider a new map $\tilde{\nu}$ such that $\tilde{\nu} = \nu$ on $D_j \cap \Omega$, $j = 1, \ldots, d$, and $\tilde{\nu}$ minimizes

$$\int_{\Omega \setminus \cup_{j=1}^d D_j} e_\varepsilon(\psi) dx$$

with $\tilde{\nu} = \nu$ on $\partial(\Omega \setminus \cup_{j=1}^d D_j)$. Then, similar to (9.43) in Chapter 6, using the proof of Lemma VIII.2 of [BBH 94] or Theorem A of [Lin2 95], we have

$$E(\tilde{\nu}) \geq \pi d \log\frac{1}{\rho} + dI(\varepsilon, \rho) + O(\rho) + W_g(\tilde{y}) \tag{2.16}$$

whenever $\varepsilon$ is suitably small and where $\rho \geq \rho(\varepsilon)$, $\tilde{y} = (\tilde{y}_1, \ldots, \tilde{y}_d)$, and $y = (y_1, \ldots, y_d)$ satisfy $|\tilde{y} - y| \leq \rho^2$ since $W_g(\tilde{y}) \to \infty$ whenever two different points $\tilde{y}_j$ and $\tilde{y}_k$ coincide or one of the point $\tilde{y}_j$ tends to $\partial\Omega$. Thus, we conclude that P3) is also true. Note that the value of $\min\{|y_j - y_k|, \mathrm{dist}(y_j, \partial\Omega) : \ j \neq k, \ j, k = 1, \ldots, d\}$ is bounded below by a positive constant depending only on $K$, $g$ and $\Omega$.

Step 2: In above procedure, in order to get the final map $v$, we have first made $N_1$ modifications to $\psi$. Every modification is made when $\deg(\psi, \partial(D \cap \Omega)) = 0$. Since such $D$s are in fact good discs, we can not continue this process forever. After $N_1$ steps, we should arrive at a bad disc $D_1$ on which $\deg(\psi, \partial(D_1 \cap \Omega)) = d_1 \neq 0$. Then continue with the above procedure. In this step, we have to show that the total number of modifications $N = \sum_{j=1}^{d+1} N_j$ that takes place in constructing the map $\nu$ with properties P1), P2) and P3) is uniformly bounded by a constant depending only on $\lambda$, $K$, $g$ and $\Omega$. Here $N_{d+1}$ is the number of modifications that have to cover the set $\{x \in \Omega| \cup_{j=1}^{d} D_j : |U_d(x)| \le \frac{1}{2}\}$ so that the resulting map $\nu$ has the property P2).

To show the above fact, we use (2.2) to obtain that

$$E_\varepsilon(u) - E_\varepsilon(\nu) \le K + C(g, \Omega).$$

On the other hand, all maps in this modification procedure are in class $S_g(\lambda, K)$. We employ the proof of Lemma 2.2 to obtain that each modification decreases the total energy of the map by at least $\frac{\lambda^2}{16}$ whenever $\varepsilon$ is small enough (depending only on $\lambda$, $K$, $g$ and $\Omega$). Thus, $N\frac{\lambda^2}{16} \le K + C(g, \Omega)$, that is,

$$N \le N(\lambda, K).$$

Therefore, the values of the map $\nu$ and $\psi$ can be different only on a union of at most $N$ balls of sizes $\le \varepsilon^\alpha$. In other words, the set $\{x \in \Omega : \ |\psi(x)| \le \frac{1}{2}\}$ is contained in a union of at most $N + d$ balls of sizes $\le \varepsilon^\alpha$.

Step 3: We can now complete the proof of Theorem 2.1. Since the set $\{x \in \Omega : \ |\psi(x)| \le \frac{1}{2}\}$ is contained in a union of at most $N + d$ balls of size $\le \varepsilon^\alpha$, we want to find $N_\varepsilon$ balls $\overline{B}_j$ of radius $\varepsilon^{\overline{\alpha}_j}$, $j = 1, \dots, N_\varepsilon$, with the following properties whenever $\varepsilon \le \varepsilon_0$:

I)$\bar{\alpha}_j \in [\alpha_0, \alpha]$ for $j = 1, \dots, N_\varepsilon$ and $N_\varepsilon \le N + d \le N_*(\lambda, K)$. Here $\alpha_0$ is a positive constant that may depend on $N_*$.

II) The set $\{x \in \Omega : \ |\psi(x)| \le \frac{1}{2}\}$ is contained in $\Omega \cap \bigcup_{j=1}^{N_\varepsilon} \bar{B}_j$, and the ball $\varepsilon^{-\bar{\alpha}_j/3} B_j$ (scale $B_j$ by a factor $\varepsilon^{-\bar{\alpha}_j/3}$ about its centre) are pairwise disjoint for $j = 1, 2, \dots, N_\varepsilon$.

To prove properties I) and II), we need the following

**Lemma 7.2.3** *Let $B_1, \dots, B_N$ be $N$ balls in $\mathbb{R}^2$ with radii not larger than $\varepsilon^\alpha$ for some $\alpha \in (0, 1/4)$ and for $j = 1, \dots, N$. Then there are a positive number $\alpha_0$ (depending only on $\alpha$ and $N$) and balls $\bar{B}_j$ of radius $\varepsilon^{-\bar{\alpha}_j}$ for $j = 1, \dots, N_\varepsilon \le N$ such that I) and II) are valid whenever $\varepsilon$ is sufficiently small.*

*Proof.* Let $A = \cup_{j=1}^{N} B_j$. We are going to prove this lemma on covering by induction on the number of connected components of $A$. If $A$ is connected, then we simply take $\bar{\alpha}_1 = \alpha/3$ and a ball $\bar{B}_1$ of radius $\varepsilon^{-\bar{\alpha}_j} \ge 2N\varepsilon^\alpha$ such that $A \subset \bar{B}_1$. The conclusion of the lemma follows automatically.

Suppose that the conclusions of Lemma 2.3 are true whenever the number of connected components of $A \le k \le N - 1$. Moreover, these $\bar{\alpha}_j$ satisfy $\bar{\alpha}_j \ge \alpha/3^k$ for each $j$. We want to show that Lemma 2.3 is true when the number of the connected components of $A$ is $k + 1 \le N$, and that each $\bar{\alpha}_j$ in the lemma can be chosen to be not less than $\alpha/3^{k+1}$ when $\varepsilon$ is small.

Let $A_1, \dots, A_{k+1}$ be connected components of $A$. Without loss of generality, we may assume that the diameter of $A$ is larger than $3(k+1)\varepsilon^{\frac{2\alpha}{3^{k+1}}}$, otherwise, we simply choose a ball $B$ or radius $\leq \varepsilon^{\frac{2\alpha}{3^{k+1}}}$ that covers $A$ entirely. The lemma then follows.

Now let $x', x'' \in A$ be such that $|x' - x''| = \mathrm{diam} A \geq 3(k+1)\varepsilon^{\frac{2\alpha}{3^{k+1}}}$. We may find a $\rho_0 \in (0, 3(k+1)\varepsilon^{\frac{2\alpha}{3^{k+1}}})$ such that the boundary $\partial B_r(x')$ of the ball $B_r(x')$ will not intersect any of the $A_j$ for $j = 1, \ldots, k+1$ and for any $r \in [\rho_0 - \varepsilon^{\frac{2\alpha}{3^{k+1}}}, \rho_0 + \varepsilon^{\frac{2\alpha}{3^{k+1}}}]$. Then it is obvious that $A \cap B_{\rho_0}(x') = A \sim A'$ contains some of each $A_1, \ldots, A_{k+1}$. We may apply the induction step to both $A'$ and $A''$ to conclude that $A = A' \cup A''$ can be covered by balls $\bar{B}_j$ of radius $\varepsilon^{\bar{\alpha}_j}$, $\bar{\alpha}_j \geq \alpha/3^k$. Now, since $\mathrm{dist}(A', \partial B_{\rho_0}(x')) \geq \varepsilon^{\frac{2\alpha}{3^{k+1}}}$, the conclusions of the covering lemma follow. This completes the induction argument.

Now we can apply the Fubini's theorem to find balls $B_j$ (with the same centre as $\bar{B}_j$) of radius $\varepsilon^{\alpha_j}$, $\alpha_j \in [\bar{\alpha}_j/3, \bar{\alpha}_j]$, such that (i), (ii) and (iii) of Theorem 2.1 are valid. Parts (iv) and (v) follow from the same proof as in Step 1.

**Corollary 7.2.1** *Let $\psi \in S_g(\lambda, K)$ and $B_j$, $j = 1, \ldots, N_\varepsilon$, $0 < \varepsilon \leq \varepsilon_0$ be as in Theorem 2.1. Suppose that $\bar{x}_j$ is the centre of $B_j$ and that $\deg(\psi, \partial B_j) = 1$ for $j = 1, \ldots, d$. Then $N_\varepsilon = d$ whenever*

$$E_\varepsilon(\psi) \leq \pi d \log \frac{1}{\varepsilon} + W_g(\bar{x}) + c_0$$

*where $c_0 = d\gamma + \lambda^2/32$, and $\gamma$ is the value defined in Theorem IX.3 of [BBH 94].*

*Proof.* Suppose $N_\varepsilon > d$. Then we place values of $\psi$ in $B_j$, $j = d+1, \ldots, N_\varepsilon$ by its corresponding values of minimizers of the energy functional on $B_j$ (with the Dirichlet boundary condition given by $\psi$). The resulting map $\nu$ has the similar property (2.16) as for $\tilde{\nu}$. By Theorem IV.3 of [BBH 94],

$$I(\varepsilon, \rho) = \pi \log \frac{\rho}{\varepsilon} + \gamma + o(1)$$

as $\varepsilon \to 0^+$ whenever $\rho \geq \rho(\varepsilon)$. By choosing $\rho$ small, we have

$$E_\varepsilon(\nu) \geq \pi d \log \frac{1}{\varepsilon} + \gamma d + W_g(\bar{x}) + o(1).$$

Since $E_\varepsilon(\psi) \geq E_\varepsilon(\nu) + \lambda^2/16$, we obtain a contradiction from the assumption on the boundedness of $E_\varepsilon(\psi)$:.

$$\frac{\lambda^2}{16} \leq \frac{\lambda^2}{32}.$$

**Corollary 7.2.2** *Let $\psi \in S_g(\lambda, K)$ be as in Corollary 2.1 and*

$$E_\varepsilon(\psi) \geq \pi d \log \frac{1}{\varepsilon} + \gamma d + W_g(\bar{x}) + c_0.$$

*Then*

$$\frac{1}{\varepsilon^2} \int_{\Omega_\varepsilon} (1 - |\psi|^2)^2 dx \leq C \tag{2.17}$$

*for some $C > 0$ depending only on $\lambda$ and $K$. Here $\Omega_\varepsilon = \Omega \setminus \cup_{j=1}^d B_j$.*

*Proof.* We already know that $|\psi(x)| \geq \frac{1}{2}$ on $\Omega_\varepsilon$. We deduce then, using Lemma 2.1 of this chapter,

$$\int_{\cup_{j=1}^d B_j} e_\varepsilon(\psi)dx \geq \sum_{j=1}^d \log\frac{1}{\varepsilon}(1-\alpha_j) - c.$$

On the other hand, by Theorem 8.2 of Chapter 6 for $R_0 = \varepsilon^{\alpha_j}$ and $R = \delta(\lambda, K)$, we obtain

$$\int_{\Omega_\varepsilon} \frac{1}{2}|\nabla\psi|^2 dx \geq \sum_{j=1}^d \pi \log\frac{\delta(\lambda,K)}{\varepsilon^{\alpha_j}} - C.$$

Since $\psi \in S_g(\lambda, K)$, we have

$$\begin{aligned}
\pi d\log\frac{1}{\varepsilon} + K &\geq \int_\Omega e_\varepsilon(\psi) = \int_{\Omega_\varepsilon} e_\varepsilon(\psi) + \int_{\cup_{j=1}^d B_j} e_\varepsilon(\psi) \\
&= \int_{\Omega_\varepsilon} \frac{1}{2}|\nabla\psi|^2 + \frac{1}{4\varepsilon^2}\int_{\Omega_\varepsilon}(1-|\psi|^2)^2 + \int_{\cup_{j=1}^d B_j} e_\varepsilon(\psi) \\
&\geq \frac{1}{4\varepsilon^2}\int_{\Omega_\varepsilon}(1-|\psi|^2)^2 + \sum_{j=1}^d \pi\log\frac{\delta(\lambda,K)}{\varepsilon^{\alpha_j}} + \sum_{j=1}^d \log\frac{1}{\varepsilon}(1-\alpha_j) - C \\
&= \frac{1}{4\varepsilon^2}\int_{\Omega_\varepsilon}(1-|\psi|^2)^2 + \pi d\log\frac{1}{\varepsilon} - C.
\end{aligned}$$

This indicates

$$\frac{1}{\varepsilon^2}\int_{\Omega_\varepsilon}(1-|\psi|^2)^2 dx \leq C$$

where $C$ is a constant depending only on $\lambda$, $K$, $g$ and $\Omega$.

## 7.3 Estimates from below for the mobilities

As in [Lin 96], we assume that the initial data $\psi_0 = \psi_0^\varepsilon$ satisfy the following conditions:

A1) *(i)* $\psi_0$ *is a smooth function and* $|\psi_0(x)| \leq 1$ *in* $\Omega$.

*(ii)* $E_\varepsilon(\psi_0) \leq \pi d\log\frac{1}{\varepsilon} + K_1$ *for a constant* $K_1$.

*(iii)* $\int_\Omega \rho^2(x)e_\varepsilon(\psi_0)dx \leq K_2$ *for some constant* $K_2$ *where* $\rho(x) =$*dist*$(x, \{b_1, \ldots, b_n\})$ *and* $b_1, \ldots, b_d$ *are* $d$ *distinct points in* $\Omega$.

*(iv)* $\psi_0^\varepsilon \to \prod_{j=1}^d \frac{x-b_j}{|x-b_j|}e^{ih_0(x)}$ *in* $H^1_{loc}(\overline{\Omega}\backslash\{b_1, \ldots, b_d\})$ *here* $h_0(x)$ *is a function in* $H^1(\Omega)$.

We now consider the behaviour of the solution over a finite interval. The essence of these assumptions is as follows: if we impose Dirichlet boundary data and assume compatibility, then the initial data will have the same winding number $d$ as the boundary data. A1)(ii) ensures that the initial data has no other vortices (for example, an additional plus-minus pair) except these $d$ vortices. In addition, the initial data is very close to the steady-state solutions and the vortices represented by the initial data are almost radially symmetric.

First, using the gradient form of the equation, we obtain

$$\frac{d}{dt}E_\varepsilon(\psi_\varepsilon(\cdot,t)) = -\int_\Omega |\frac{\partial}{\partial t}\psi_\varepsilon(x,t)|^2 dx.$$

Integrating the both sides of the above equation, we obtain

$$\sup_{t>0}\int_0^t\int_\Omega |\frac{\partial}{\partial t}\psi_\varepsilon(x,\tau)|^2 dx d\tau + E_\varepsilon(\psi_\varepsilon(\cdot,t)) \le E_\varepsilon(\psi_0) \le \pi d\log\frac{1}{\varepsilon} + K. \tag{3.1}$$

By using the same scaling argument and parabolic estimates, we claim that

$$|\nabla\psi_\varepsilon(x,t)|^2 + |\frac{\partial}{\partial t}\psi_\varepsilon(x,t)| \le \frac{c}{\varepsilon^2} \tag{3.2}$$

for some constant $c$ depending only on $g$ and $\Omega$ and for all $t \ge \varepsilon^2$. If the initial data satisfies

$$||\nabla\psi_0^\varepsilon(x)||_{L^\infty} + \sup_{x,y}\frac{|\nabla\psi_0^\varepsilon(x) - \nabla\psi_0^\varepsilon(y)|}{|x-y|^\alpha} \le K_1 \tag{3.3}$$

for some constants $\alpha$ and $K_1$, then (3.2) also holds in $0 < t < \varepsilon^2$. We now prove the claim.

In fact this claim is a consequence of the following lemma

**Lemma 7.3.1** *Suppose $u$ solves the following parabolic equation*

$$u_t = \Delta u + f(u) \qquad on\ \Omega\times(0,\infty),$$

$$u(x,0) = 0 \quad on\ \Omega \qquad u = 0 \quad on\ \partial\Omega.$$

*Then we have for some $C > 0$ independent of $T$ that*

$$\sup_{0\le t\le T}||\nabla u(\cdot,t)||^2_{L^\infty(\Omega)}$$
$$\le C(||f||_{L^\infty(\Omega\times[0,\infty))}||u||_{L^\infty(\Omega\times[0,\infty))} + ||u||^2_{L^\infty(\Omega\times[0,\infty))}). \tag{3.4}$$

*Proof.* In fact, this lemma is a parabolic version of Lemma A.1 and Lemma A.2 of [BBH 93]. Therefore, it can be proved similarly to [BBH 93]. Now we give a sketch for the proof of the interior estimates and omit the proof for the estimates near the boundary.

Assume for simplicity that $0 \in \Omega$ and set $d =$dist$(0,\partial\Omega)$. We shall prove (3.4) at $x = 0$, $t = t_0$ where $t_0 \in [0,T]$.

Let $0 < \gamma \le d$ to be determined later. It is clear that the function $v(y,\tau) = u(\gamma y, t_0 + \gamma^2\tau)$ defined on $B_1 \times [0,1]$ ($B_1 = B(0,1)$) solves

$$v_\tau = \Delta v + \gamma^2 f(\gamma y, t_0 + \gamma^2\tau) \qquad \text{on } \; B_1 \times [0,1].$$

where $f(\cdot,\cdot) = f(u(\cdot,\cdot))$. Then we have from the standard parabolic estimates that

$$|\nabla v(0,\tau)| \le C(\gamma^2 \|f(\gamma y, t_0 + \gamma^2\tau)\|_{L^\infty(B_1\times[0,1])} + \|v\|_{L^\infty(B_1\times[0,1])}) \qquad \forall 0 \le \tau \le 1$$

In particular we get

$$\gamma|\nabla u(0,t_0)| \le C(\gamma^2\|f\|_{L^\infty(\Omega\times[0,\infty))} + \|u\|_{L^\infty(\Omega\times[0,\infty))}). \tag{3.5}$$

If $d \ge (\|u\|_{L^\infty}/\|f\|_{L^\infty})^{1/2}$, we take $\gamma = (\|u\|_{L^\infty}/\|f\|_{L^\infty})^{1/2}$ in (3.5) and obtain

$$|\nabla u(0,t_0)| \le 2C\|f\|_{L^\infty}{}^{1/2}\|u\|_{L^\infty}{}^{1/2}. \tag{3.6}$$

If $d < (\|u\|_{L^\infty}/\|f\|_{L^\infty})^{1/2}$, we take $\gamma = d$ in (3.5) and get

$$|\nabla u(0,t_0)| \le C((\|u\|_{L^\infty}\|f\|_{L^\infty})^{1/2} + \frac{1}{d}\|u\|_{L^\infty}). \tag{3.7}$$

In (3.6) and (3.7), $C$ is independent of $T$. Combining (3.6) with (3.7), we have for $0 \le t \le T$

$$|\nabla u(x,t)|^2 \le C(\|u\|_{L^\infty}\|f\|_{L^\infty} + \frac{1}{dist^2(x,\partial\Omega)}\|u\|^2_{L^\infty})$$

This implies that (3.4) holds on $K \times [0,T]$ for any compact subset $K$ of $\Omega$.

Similarly, we can prove (3.4) near the boundary (see for example, [BBH 93]). The Lemma is proved. The claim (3.2) follows.

We have therefore proved that $\psi_\varepsilon(x,t) \in S_g(\lambda, K)$ for some constant $\lambda$. Also, by using (2.2) on every layer $\Omega \times \{t\}$, we know that

$$E_\varepsilon(\psi_\varepsilon(\cdot,t)) \ge \pi d \log\frac{1}{\varepsilon} - C_1.$$

Combining this with (3.1), we obtain

$$\int_0^\infty \int_\Omega |\frac{\partial}{\partial t}\psi_\varepsilon(x,t)|^2 dxdt \le C_1 + K. \tag{3.8}$$

Using the fact that $|\nabla\rho| \le 1$ in $\Omega$ because it is a distance function, by straightforward differentiation, we obtain

$$\frac{\partial}{\partial t}\int_\Omega \rho^2(x)e_\varepsilon(\psi_\varepsilon(x,t))dx \le \frac{1}{2}\int_\Omega \rho^2(x)|\nabla\psi_\varepsilon|^2dx + 2\int_\Omega |\frac{\partial}{\partial t}\psi_\varepsilon|^2dx. \tag{3.9}$$

Using integrating factor, we have

$$\int_\Omega \rho^2(x)e_\varepsilon(\psi_\varepsilon(x,t))dx \le 2e^t\int_0^t\int_\Omega |\frac{\partial}{\partial t}\psi_\varepsilon|^2dxdt + K_2. \tag{3.10}$$

We define, for any given finite $T > 0$,

$$\delta_0 = \frac{1}{2}\min\{|b_j - b_k|,\ \mathrm{dist}(b_j, \partial\Omega) : \ j \neq k,\ j,k = 1,\dots,d\},$$
$$\Omega_\delta = \Omega \setminus \cup_{j=1}^d B_\delta(b_j)\ \forall\, 0 < \delta < \delta_0,$$
$$Q_{\delta,T} = \Omega_\delta \times [0,T].$$

Our estimates (3.8) and (3.10) can be summarized as: $\psi_\varepsilon \in H^1(Q_{\delta,T})$ and for any $t \in (0,T)$

$$\int_{\Omega_\delta} e_\varepsilon(\psi_\varepsilon(x,t))dx \leq c(\delta, K, K_2)e^T. \tag{3.11}$$

From the estimates we have obtained, we have

**Theorem 7.3.1** *For any subsequence $\varepsilon_n \to 0$, there is a subsequence still denoted by $\varepsilon_n$ so that $\psi_{\varepsilon_n}(x,t) \to \psi_0(x,t)$ weakly in $H^1_{loc}(\overline{\Omega} \setminus \{b_1,\dots,b_d\} \times I\!R_+)$ and strongly in $L^2_{loc}(\overline{\Omega} \times I\!R_+)$. Moreover, the limit function satisfies*

$$\partial_t\psi_0 = \Delta\psi_0 + |\nabla\psi_0|^2\psi_0 \quad in\ \ \Omega \setminus \{b_1,\dots,b_d\} \times I\!R_+,$$
$$\psi_0(x,0) = \prod_{j=1}^d \frac{x - b_j}{|x - b_j|} e^{ih_0(x)}, \tag{3.12}$$
$$\psi_0 = g \quad on \ \ \partial\Omega \times I\!R_+.$$

*Proof.* The local $H^1$ weak convergence is trivial. Noticing that the modulus of the functions are bounded by 1, the $L^2$ convergence is also straightforward. From the energy term $\frac{1}{\varepsilon^2}(|\psi_\varepsilon|^2 - 1)^2$, it is also clear that $|\psi_0(x,t)| = 1$ for a.e. $(x,t)$. Now let us regard $\psi_\varepsilon$ as an $I\!R^2$-valued function and use $\wedge$ to denote the standard exterior product, we have

$$\begin{aligned}\psi_\varepsilon \wedge \partial_t\psi_\varepsilon &= \psi_\varepsilon \wedge \Delta\psi_\varepsilon + \frac{1}{\varepsilon^2}\psi_\varepsilon \wedge (|\psi_\varepsilon|^2 - 1)\psi_\varepsilon \\ &= \psi_\varepsilon \wedge \Delta\psi_\varepsilon \\ &= \mathrm{div}(\psi_{\varepsilon 1}\nabla\psi_{\varepsilon 2}) - \mathrm{div}(\psi_{\varepsilon 2}\nabla\psi_{\varepsilon 1}).\end{aligned}$$

Taking the limit in the above expression, using the fact that $|\psi_0| = 1$ a.e., we obtain the theorem.

Similar to the discussions of Struwe's on the heat flow of harmonic map [Str 94], the function $\psi_0(x,t)$ is smooth in $\overline{\Omega} \setminus \{b_1,\dots,b_d\} \times I\!R_+$. Moreover, for any $t > 0$, $0 < \delta \leq \delta_0$, the degrees $\deg(\psi_0(\cdot,t), \partial B_\rho(b_j)) = 1$ for all $j$ between 1 and $d$ by Assumption A1)(iv).

Consequently,

$$\psi_0(x,t) = \prod_{j=1}^d \frac{x - b_j}{|x - b_j|} e^{i\phi_0(x,t)}. \tag{3.13}$$

Substituting this expression in to (3.12) we know that $\phi_0(x,t)$ satisfies

$$\begin{aligned} \partial_t\phi_0 &= \Delta\phi_0 \text{ in } \Omega\setminus\{b_1,\dots,b_d\}\times I\!R_+, \\ \phi_0(x,0) &= h_0(x), \\ \phi_0(x,t) &= h_0(x) \text{ on } \partial\Omega\times I\!R_+. \end{aligned}$$

Theorem 3.1 is proved.

At this stage, we do not know if the phase function $\phi_0(x,t)$ satisfies the heat equation in $\Omega\times I\!R_+$ rather than in $\Omega\setminus\{b_1,\dots,b_d\}\times I\!R_+$. We also do not know if the phase function $\phi_0(x,t)$ is determined by the limit of the whole family $\psi_\varepsilon(x,t)$ instead of a special sequence $\psi_{\varepsilon_n}(x,t)$. To this question we have

**Theorem 7.3.2** *The phase function $\phi_0(x,t)$ satisfies*

$$\sup_{t>0}[\|\nabla\phi_0(x,t)\|_{L^2}^2]+\int_0^t\int_\Omega|\frac{\partial}{\partial t}\phi_0(x,t)|^2dxdt\le C.$$

*In particular, $\phi_0(x,t)$ satisfies the heat equation in $\Omega\times I\!R_+$. Consequently, we have*

$$\psi_\varepsilon(x,t)\to\prod_{j=1}^d\frac{x-b_j}{|x-b_j|}e^{i\phi_0(x,t)}$$

*in $L^2_{loc}(\overline{\Omega}\times I\!R_+)$ and weakly in $H^1_{loc}(\overline{\Omega}\setminus\{b_1,\dots,b_d\}\times I\!R_+)$ when $\varepsilon\to0$.*

*Proof.* First let us consider a function $u\in S_g(\lambda,K)$, and $B_1,\dots,B_{N_\varepsilon}$ be the balls given in Theorem 2.1 of this chapter. Let $x_1^\varepsilon,\dots,x_d^\varepsilon$ be the centres of balls $B_1,\dots,B_d$. Then, since $\deg(u,\partial B_j)=0$ for $j=d+1,\dots,N_\varepsilon$, one has

$$u(x)=\rho_\varepsilon(x)\prod_{j=1}^d\frac{x-x_j^\varepsilon}{|x-x_j^\varepsilon|}e^{ih_\varepsilon(x)},\quad \text{for all } x\in\Omega\setminus\cup_{j=1}^{N_\varepsilon}B_j:=\Omega_\varepsilon \tag{3.14}$$

where $h_\varepsilon(x)$ is a well-defined function, single-valued in $\Omega_\varepsilon$, and

$$\frac12\le\rho_\varepsilon(x)\le1,\quad x\in\Omega_\varepsilon.$$

Such $h_\varepsilon$ is uniquely determined if $h_\varepsilon(x_0)\in[0,2\pi)$ for some $x_0\in\partial\Omega$.

On the other hand, since $B_j$ is of size $\varepsilon^{\alpha_j}$ and for $u(x)\in S_g(\lambda,K)$,

$$\frac{1}{4\varepsilon^2}\int_{\Omega_\varepsilon}(1-|u|^2)^2\le C$$

(see Corollary 2.2), we have from direct computations that

$$\int_{\Omega_\varepsilon}e_\varepsilon(u)dx\le\pi\sum_{j=1}^d\alpha_j\log\frac1\varepsilon+C(\lambda,K).$$

Moreover, if we denote $e^{i\Theta_\varepsilon} = \prod_{j=1}^d \frac{x-x_j^\varepsilon}{|x-x_j^\varepsilon|}$ with a multi-valued harmonic function $\Theta_\varepsilon$ on $\Omega_\varepsilon$, then

$$\int_{\Omega_\varepsilon} e_\varepsilon(u)dx \geq \int_{\Omega_\varepsilon} \rho_\varepsilon^2|\nabla\Theta_\varepsilon|^2 + \rho_\varepsilon^2|\nabla h_\varepsilon|^2 + 2\rho_\varepsilon^2\nabla\Theta_\varepsilon\cdot\nabla h_\varepsilon.$$

Since $|\nabla\Theta_\varepsilon|(x) \leq \varepsilon^{-\alpha}$ for $x\in\Omega_\varepsilon$ with $\alpha < \frac{1}{2}$ (see step 1 in the proof of Theorem 2.1), we have

$$\begin{aligned}\int_{\Omega_\varepsilon} |\rho_\varepsilon^2 - 1||\nabla\Theta_\varepsilon|^2 &\leq \varepsilon^{-2\alpha}(\int_{\Omega_\varepsilon}|\rho_\varepsilon^2-1|^2dx)^{\frac{1}{2}}|\Omega|^{\frac{1}{2}}\\ &\leq C(K,\Omega)\varepsilon^{-2\alpha+1}(\log\frac{1}{\varepsilon})^{\frac{1}{2}}\end{aligned}$$

and since

$$\int_{\Omega_\varepsilon}|\nabla\Theta_\varepsilon|^2dx \geq \pi\sum_{j=1}^d \alpha_j\log\frac{1}{\varepsilon} - C(\lambda,K),$$

we also have

$$\int_{\Omega_\varepsilon}\rho_\varepsilon^2|\nabla h_\varepsilon|^2 + 2\rho_\varepsilon^2\nabla\Theta_\varepsilon\cdot\nabla h_\varepsilon \leq C(\lambda,K). \tag{3.15}$$

Similarly,

$$\int_{\Omega_\varepsilon}|\rho_\varepsilon^2-1||\nabla\Theta_\varepsilon||\nabla h_\varepsilon| \leq o(1)||\nabla h_\varepsilon||_{L^2(\Omega_\varepsilon)}.$$

In view of (3.15) and $\frac{1}{2}\leq\rho_\varepsilon\leq 1$ for $x\in\Omega_\varepsilon$, to show $\int_{\Omega_\varepsilon}|\nabla h_\varepsilon|^2 \leq C(\lambda,K)$, we only need to verify that

$$\int_{\Omega_\varepsilon}\nabla\Theta_\varepsilon\cdot\nabla h_\varepsilon dx$$

is bounded by a constant. We calculate as follows. Since $\int_{\partial B_j}\frac{\partial\Theta_\varepsilon}{\partial\boldsymbol{n}} = 0$ and $|h_\varepsilon - \overline{h}_\varepsilon| \leq C(\lambda,K)$ where $\overline{h}_\varepsilon = \frac{1}{|\partial B_j|}\int_{\partial B_j} h_\varepsilon dx$, $j=1,\ldots,N_\varepsilon$, we have

$$|\int_{\Omega_\varepsilon}\nabla\Theta_\varepsilon\cdot\nabla h_\varepsilon dx| = |\int_{\partial\Omega} h_\varepsilon\frac{\partial\Theta_\varepsilon}{\partial\boldsymbol{n}} + \sum_{j=1}^{N_\varepsilon}\int_{\partial B_j}(h_\varepsilon-\overline{h}_\varepsilon)\frac{\partial\Theta_\varepsilon}{\partial\boldsymbol{n}}| \leq C(\lambda,K).$$

We now apply the above argument to each $\psi_\varepsilon\in S_g(\lambda,K)$, $t>0$. From (3.11), we see that $x_j^\varepsilon\to b_j$ as $\varepsilon\to 0$ for $j=1,\ldots,d$.

From above discussion we know that for almost every $t$ there holds

$$\psi_{\varepsilon_n}(x,t) \to \prod_{j=1}^d\frac{x-b_j}{|x-b_j|}e^{i\phi_0(x,t)}$$

here $\phi_0(x,t)$ is a weak limit of $h_{\varepsilon_n}(x,t)$ in $H^1(\Omega)$. In order to avoid the trivial ambiguity, we also assume $\phi_0(x_0,t) \in (0,2\pi)$ for some given $x_0 \in \partial\Omega$. Thus $\int_\Omega |\nabla\phi_0(x,t)|^2 dx \le C(\lambda, K)$ for all $t > 0$. From the latter fact one can easily show that $\phi_0(x,t)$ satisfies the heat equation in $\Omega \times \mathbb{R}_+$. This completes the proof of this theorem.

**Remark 7.3.1** *Since $\psi_{\varepsilon_n}(x,t) \to \prod_{j=1}^d \frac{x-b_j}{|x-b_j|} e^{i\phi_0(x,t)}$ in $L^2_{loc}(\overline{\Omega} \times \mathbb{R}_+)$ and weakly in $H^1_{loc}(\overline{\Omega} \setminus \{b_1, \ldots, b_d\} \times \mathbb{R}_+)$, we see that*

$$\int_0^T \int_\Omega |\frac{\partial\phi_0(x,t)}{\partial t}|^2(x,t)dxdt \le \lim_{\varepsilon\to 0} \int_0^T \int_\Omega |\frac{\partial\psi_{\varepsilon_n}(x,t)}{\partial t}|^2(x,t)dxdt \le C(K)$$

**Corollary 7.3.1** *Suppose that the initial data $\psi_0(x)$ in (1.1) satisfies Assumption (A1) of this section with $K_1 \le W_g(b) + C_0$ where $b = (b_1, \ldots, b_d)$ and $C_0$ is as defined in Corollary 2.1. Then for any $T > 0$, the set $G(t) = \{x \in \Omega, |\psi_\varepsilon(x,t)| \le \frac{1}{2}\}$, $t \in (0,T]$, converges uniformly (in $t$) to the set $(b_1, \ldots, b_d)$ in the Hausdorff distance as $\varepsilon \to 0$.*

*Proof.* The proof is a consequence of Corollary 2.1, the estimate (3.11) and Lemma 2.1.

**Remark 7.3.2** *The above Corollary shows that with some suitable initial data, there are no new essential vortices created in a finite time outside any neighbourhood of the initial vortices whenever $\varepsilon$ is small enough. In other words, the interesting point of Theorem 3.1 is that it establishes the fact that under this scaling scheme, the vortices do not move in finite time interval because the limit function has the same vortex locations as the initial data (the initial data have their vortex location defined under the constraints A1)(iv) with respect to $\varepsilon$ given above) and the vortex location no longer depends on time $t$ after we take the limit $\varepsilon \to 0$.*

As a matter of fact, we have a better result:
Under the same assumptions as before, we define

$$\begin{aligned} &\tfrac{1}{\lambda_\varepsilon}\psi_t = \Delta\psi + \tfrac{1}{\varepsilon^2}(1 - |\psi|^2)\psi, \\ &\psi(x,t) = g(x), \quad \text{for } x \in \partial\Omega,\ t > 0, \\ &\psi(x,0) = \psi_0^\varepsilon(x), \quad \text{for } x \in \Omega. \end{aligned} \tag{3.16}$$

Here $\lambda_\varepsilon$ is a positive number for each given $\varepsilon$ and

$$\lim_{\varepsilon\to 0} \lambda_\varepsilon = \infty, \qquad \lim_{\varepsilon\to 0} \frac{\lambda_\varepsilon}{\log\varepsilon} = 0, \tag{3.17}$$

we have

**Theorem 7.3.3** *Suppose that the initial data satisfies Assumption (A1). Then as $\varepsilon \to 0$, the solutions $\psi_\varepsilon(x,t)$ of (3.16) have the following convergence properties:*

$$\psi_\varepsilon(x,t) \to \prod_{j=1}^{d} \frac{x-b_j}{|x-b_j|} e^{ih(x)} \quad in \quad L^2_{loc}(\Omega \times I\!R_+)$$

*where $h$ is a function of $x$ only which satisfies*

$$\Delta h(x) = 0 \quad in \quad \Omega,$$
$$h(x) = h_0(x) \quad on \quad \partial\Omega.$$

Before giving the proof to this theorem, we establish a preliminary result:

**Lemma 7.3.2 (General Convergence Theorem)** *Let $u_\varepsilon(x) \in S_g(\lambda, K)$, then for any sequence of $\varepsilon_n \to 0$, there is a subsequence of $\{u_{\varepsilon_n}(x)\}$ that converges to a mapping of the form*

$$\prod_{j=1}^{d} \frac{x-b_j}{|x-b_j|} e^{ih(x)} \quad in \quad L^2(\Omega) \quad and\ weakly\ in \quad H^1_{loc}(\overline{\Omega} \setminus \{b_1, \ldots, b_d\})$$

*where $\{b_1, \ldots, b_d\}$ are $d$ distinct points in $\Omega$ and $h(x)$ is a $H^1$ function in $\Omega$.*

*Proof.* For $u_\varepsilon \in S_g(\lambda, K)$, we let $B_j$, $j = 1, 2, \ldots, N_\varepsilon$ be balls in Theorem 2.1. replacing $u_\varepsilon$ in each $B_j$ by $U_\varepsilon$ where $U_\varepsilon$ minimizes $\int_{B_j} e_\varepsilon(V)dx$ with $V = u_\varepsilon$ on $\partial B_j$ for $j = d+1, \ldots, N_\varepsilon$. Thus $|U_\varepsilon| \geq \frac{3}{4}$ on each $B_j$, $j = d+1, \ldots, N_\varepsilon$. We still denote the resulting map by $U_\varepsilon$.

Let $\delta \in (\varepsilon^{\alpha_0}, \delta(\lambda, K))$ where $\alpha_0, \delta(\lambda, K)$ are given in Theorem 2.1, and suppose that $\cup_{j=1}^{d} \partial B_\delta(x_j)$ does not intersect with $\cup_{j=d+1}^{N_\varepsilon} B_j$. Except for a set of $\delta \in (\varepsilon^{\alpha_0}, \delta(\lambda, K))$ whose measure$\leq N_*(\lambda, K)\varepsilon^{\alpha_0}$, the latter assumption is valid.

For such a $\delta$, we have

$$\pi d \log \frac{\delta}{\varepsilon} - C(\lambda, K) \leq \sum_{j=1}^{d} \int_{B_\delta(x_j)} e_\varepsilon(U_\varepsilon)dx \leq \int_{A_\varepsilon} e_\varepsilon(U_\varepsilon)dx + C(\lambda, K) \qquad (3.18)$$

where

$$A_\varepsilon = \cup_{j=1}^{d} B_\delta(x_j) \setminus \cup_{j \geq d+1} B_j.$$

Here the first inequality in (3.18) is true because we have from Lemma 2.2 of this chapter that

$$\int_{B_j} e_\varepsilon(U_\varepsilon)dx = \int_{B_j} e_\varepsilon(u_\varepsilon)dx \geq \pi \log \frac{1}{\varepsilon}(1-\alpha_j) - C(\lambda, K)$$

and we also have from Theorem 8.2 of Chapter 6 that

$$\begin{aligned}\int_{B_\delta(x_j)\setminus B_j} e_\varepsilon(U_\varepsilon)dx &\geq \frac{1}{2}\int_{B_\delta(x_j)\setminus B_j} |\nabla U_\varepsilon|^2 dx\\ &\geq \pi \log\frac{\delta}{\varepsilon^{\alpha_j}} - C(\lambda, K), \quad \text{for } j = 1, \dots, d;\end{aligned}$$

and, for the second inequality in (3.18) we have used Lemma 2.1 of this chapter for $U_\varepsilon$ on each $B_j$, $j = d+1, \dots, N_\varepsilon \leq N_*(\lambda, K)$. Since $u_\varepsilon \in S_g(\lambda, K)$, we deduce from (3.18) that

$$\int_{\Omega_\delta \cup (\cup_{j=d+1}^{N_\varepsilon} B_j)} e_\varepsilon(u_\varepsilon)dx \leq C(\lambda, K) + \pi d \log\frac{1}{\delta} \tag{3.19}$$

where $\Omega_\delta = \Omega \setminus \cup_{j=1}^d B_\delta(x_j)$.

Now, for a subsequence $\varepsilon_n \to 0$, we may assume, without loss of generality, that $x_j \to b_j$ as $\varepsilon_n \to 0$. Note that $x_j$ may also depend on $\varepsilon$. We may also assume, by taking subsequence if necessary, that $u_{\varepsilon_n}(x) \to u_*(x)$ weakly in $H^1_{\text{loc}}(\overline{\Omega} \setminus \{b_1, \dots, b_d\})$ and strongly in $L^2(\Omega)$. The conclusion that

$$u_*(x) = \prod_{j=1}^d \frac{x - b_j}{|x - b_j|} e^{ih(x)} \quad \text{with } h(x) \in H^1(\Omega)$$

follows from the proof of Theorem 3.2. *Proof of Theorem 3.3.* Using the same computation as before by taking into account of the quantity $\lambda_\varepsilon$, we have by multiplying the first equation in (3.16) by $\partial_t \psi_\varepsilon$ that

$$\begin{aligned}&\sup_{t>0} \frac{1}{\lambda_\varepsilon}\int_0^t \int_\Omega |\frac{\partial}{\partial t}\psi_\varepsilon|^2 dxdt + E_\varepsilon(\psi_\varepsilon)(t) \leq E_\varepsilon(\psi_0^\varepsilon)\\ &\leq \pi d \log\frac{1}{\varepsilon} + K,\end{aligned} \tag{3.20}$$

and

$$\int_\Omega \rho^2(x) e_\varepsilon(\psi_\varepsilon(x,t))dx \leq 2\lambda_\varepsilon e^t \int_0^t \frac{1}{\lambda_\varepsilon}\int_\Omega |\frac{\partial}{\partial t}\psi_\varepsilon|^2 dxdt + K_2 \leq C(K)\lambda_\varepsilon e^t. \tag{3.21}$$

Note that $\psi_\varepsilon(\cdot, t) \in S_g(\lambda, K)$ for each given $t > 0$ whenever $\varepsilon$ is sufficiently small, we have, by above Lemma 3.2, that

$$\psi_{\varepsilon_n}(x,t) \to \prod_{j=1}^d \frac{x - a_j}{|x - a_j|} e^{ih_a(x)}$$

in $L^2(\Omega)$ and weakly in $H^1_{loc}(\overline{\Omega}\setminus\{a_1, \dots, a_d\})$ where $a_1, \dots, a_d$ are $d$ distinct points in $\Omega$ which might depend on $t$. However, (3.21) implies that up to a rearrangement, $a_j = b_j$.

Since

$$\lambda_{\varepsilon_n}\int_0^\infty\int_\Omega|\Delta\psi_{\varepsilon_n}+\frac{1}{\varepsilon_n^2}\psi_{\varepsilon_n}(1-|\psi_{\varepsilon_n}|^2)|^2dxdt\le C(K),$$

we see that there is a subsequence (still denoted by $\varepsilon_n$) such that

$$\|\Delta\psi_{\varepsilon_n}+\frac{1}{\varepsilon_n^2}\psi_{\varepsilon_n}(1-|\psi_{\varepsilon_n}|^2)\|_{L^2(\Omega)}(t)\to 0,\quad \text{for a.e. } t>0.$$

This together with Lemma 3.2 yields $h_a(x)=h(x)$, $\Delta h(x)=0$.

In summary, we have shown that for any sequence $\varepsilon_n\to 0$, there is a subsequence of $\psi_{\varepsilon_n}$ such that for almost all $t>0$,

$$\psi_{\varepsilon_n}(x,t)\to\prod_{j=1}^d\frac{x-b_j}{|x-b_j|}e^{ih(x)}\quad\text{in } L^2(\Omega),\quad \Delta h(x)=0.$$

In particular,

$$\int_0^T\|\psi_{\varepsilon_n}(x,t)-\prod_{j=1}^d\frac{x-b_j}{|x-b_j|}e^{ih(x)}\|^2_{L^2(\Omega)}dt\to 0$$

as $\varepsilon_n\to 0$, for any $T$. This follows from Egorov's theorem (see [Yos 78] p16) and the fact that

$$\|\psi_{\varepsilon_n}(x,t)-\prod_{j=1}^d\frac{x-b_j}{|x-b_j|}e^{ih(x)}\|_{L^2(\Omega)}\le C(\Omega),\quad \forall t>0.$$

**Remark 7.3.3** *This theorem implies that the vortices will not move on a time scale of $\lambda_\varepsilon$ (which tends to $\infty$ but still small compared with $\log(1/\varepsilon)$) as $\varepsilon\to 0$.*

**Remark 7.3.4** *If we replace $\lambda_\varepsilon$ in (3.16) by $\delta\log\frac{1}{\varepsilon}$ for small positive number $\delta$, then it follows from the proof of above theorem that for any $\varepsilon_n\to 0$, there is a subsequence $\psi_{\varepsilon_n}$ such that*

$$\|\Delta\psi_{\varepsilon_n}+\frac{1}{\varepsilon_n^2}\psi_{\varepsilon_n}(1-|\psi_{\varepsilon_n}|^2)\|_{L^2(\Omega)}(t)\to 0,\quad \textit{for a.e. } t.$$

*Moreover, for any $t>0$, there is a subsequence $\psi_{\varepsilon_n}$ that converges to $\prod_{j=1}^d\frac{x-a_j}{|x-a_j|}e^{ih_a(x)}$ in $L^2(\Omega)$ and weakly in $H^1_{\rm loc}(\Omega\setminus\{a_1,\dots,a_d\})$ for some $a\in\Omega^d$ with $|b-a|\le\eta(K)$ where $\eta(K)\to 0$ when $\delta\to 0$. Finally, for a.e. $t$, $\Delta h(x)=0$ in $\Omega$.*

## 7.4 Estimates from above for the mobilities

For a steady state solution $\psi_\varepsilon(x)$ of (1.1) with $E_\varepsilon(\psi_\varepsilon) \leq \pi d \log\frac{1}{\varepsilon} + K$, it follows from Chapter X of [BBH 94] that:

For any sequence $\varepsilon_n \to 0$, there is a subsequence $\{\psi_{\varepsilon_n}\}$ which converges to a map of the form

$$\prod_{j=1}^{d} \frac{x-a_j}{|x-a_j|} e^{ih(x)} \quad \text{in } L^2(\Omega) \cap H^1_{loc}(\overline{\Omega} \setminus \{a_1, \ldots, a_d\}),$$

for some $d$ distinct points $\{a_1, \ldots, a_d\}$ in $\Omega$. Moreover, $\Delta h(x) = 0$ in $\Omega$, and the point $a = (a_1, \ldots, a_d)$ is a critical point of the renormalized energy $W_g(\cdot)$ in $\overline{\Omega}^d$. If in addition $\psi_\varepsilon$ minimizes $E_\varepsilon(u)$, then in [BCP 93], the following result was proved:

*For a solution $\psi_\varepsilon(x,t)$ of (1.1) with the initial data $\psi_0^\varepsilon$ satisfying Assumption A1), there is a time $T(\varepsilon, \psi_0^\varepsilon, g, \Omega)$ such that $\psi_\varepsilon(x,t)$ has exactly $d$ distinct zeros of degree 1 whenever $t \geq T(\varepsilon, \psi_0^\varepsilon, g, \Omega)$. Moreover, if $a_1^\varepsilon(t), \ldots, a_d^\varepsilon(t)$ are the zeros of $\psi_\varepsilon(x,t)$, then all $a_j^\varepsilon(t) \in C^1(T,\infty)$ and $\lim_{t\to\infty} a_j^\varepsilon(t) = a_j^\varepsilon$ exists for $j = 1, \ldots, d$.*

The aim of this section is to establish various facts similar to the above result for all time $t \geq T_\varepsilon$ where $T_\varepsilon$ is chosen so that

$$\lim_{\varepsilon\to 0} \frac{T_\varepsilon}{\log(1/\varepsilon)} = \infty.$$

For this reason, we look at the following problem (under the same assumption on the boundary and initial data as in the previous section):

$$\begin{array}{l} \frac{1}{T_\varepsilon}\psi_t = \Delta\psi + \frac{1}{\varepsilon^2}(1-|\psi|^2)\psi, \\ \psi(x,t) = g(x), \quad \text{for } x \in \partial\Omega,\ t > 0, \\ \psi(x,0) = \psi_0^\varepsilon(x) \quad \text{for } x \in \Omega. \end{array} \tag{4.1}$$

From the discussions made in the previous section, some straightforward conclusions are as follows: let $\{\varepsilon_n\}$ be a sequence decreasing monotonically to 0, we may find a subsequence of $\{\psi_{\varepsilon_n}\}$ still denoted by the same sequence, such that as $\varepsilon_n \to 0$,

$$||\Delta\psi_{\varepsilon_n} + \frac{1}{\varepsilon_n^2}\psi_{\varepsilon_n}(1-|\psi_{\varepsilon_n}|^2)||_{L^2(\Omega)}(t) \to 0 \text{ a.e. } t, \tag{4.2}$$

$$||\nabla\psi_{\varepsilon_n}||(t)||\Delta\psi_{\varepsilon_n} + \frac{1}{\varepsilon_n^2}\psi_{\varepsilon_n}(1-|\psi_{\varepsilon_n}|^2)||(t) \to 0 \text{ a.e. } t, \tag{4.3}$$

$$\begin{aligned} & C\log(1/\varepsilon_n)\int_0^\infty ||\Delta\psi_{\varepsilon_n} + \frac{1}{\varepsilon_n^2}\psi_{\varepsilon_n}(1-|\psi_{\varepsilon_n}|^2)||(t)dt \\ \leq\ & C\log(1/\varepsilon_n)/T_n \to 0 \text{ a.e. } t. \end{aligned} \tag{4.4}$$

We now define the limit set of the sequence of sets $S_g(\lambda, K)$ (the definition of $S_g(\lambda, K)$ really depends on $\varepsilon$ and the limit we just mentioned is with respect to the limit $\varepsilon \to 0$).

**Definition 7.4.1** *A function* $V(x)$ *of the form* $\prod_{j=1}^{d} \frac{x-a_j}{|x-a_j|} e^{ih_a(x)}$ *belongs to* $S_g(t)$ *for a given* $t$ *if there is a sequence* $\{\psi_n\}$ *with* $\psi_n = \psi_{\varepsilon_n}(x,t) \in S_g(\lambda, K, (\varepsilon_n))$ *such that* $\psi_n$ *converges to* $V(x)$ *in* $L^2(\Omega)$ *and weakly in* $H^1_{loc}(\overline{\Omega} \setminus \{a_1, \dots, a_n\})$*. Here* $\lambda$ *and* $K$ *are fixed constants. As a matter of fact, the definition of* $S_g(t)$ *depends strongly on the chosen sequence* $\{\psi_n\}$.

Our discussions so far (Theorem 2.1, Theorem 3.2, Lemma 3.2) have established the following facts: if $V \in S_g(t)$, then there exist constants $\delta(\lambda, K)$ and $C(\lambda, K)$ independent of $V$ such that

$$\min\{|a_j - a_k|, \operatorname{dist}(a_j, \partial\Omega) : j \neq k, \ j,k = 1, \dots, d\} \geq \delta(\lambda, K) > 0,$$
$$||h_a||_{H^1(\Omega)} \leq C(\lambda, K).$$

Moreover, for a.e. $t$, from Theorem 3.3 we have $\Delta h_a(x) = 0$ for all $V(x) \in S_g(t)$.

To establish the property of the points $a = (a_1, \dots, a_d)$, we first establish the following

**Proposition 7.4.1** *If* $V \in S_g(t)$ *and* $\{\psi_n\}$ *is the associated sequence. Then for some subsequence still denoted by* $n \to \infty$, *and for some* $R \in (0, \delta(\lambda, K))$, *we have*

$$\psi_n \to V \ \text{ strongly in } \ H^1(\partial B_R), \tag{4.5}$$
$$\frac{1}{\varepsilon_n^2} \int_{\partial B_R} (|\psi_n|^2 - 1)^2 \to 0. \tag{4.6}$$

*Proof.* Since $\psi_n \in S_g(\lambda, K)$ and $\psi_n$ converges to

$$V(x) = \prod_{j=1}^{d} \frac{x - a_j}{|x - a_j|} e^{ih_a(x)}.$$

Using Theorem 2.1 and the proof of Lemma 3.2, we deduce that there is an $R \in [\delta/2, \ \delta]$ with $\delta = \delta(\lambda, K)$ and $N$ is defined as $N_*$ in Theorem 2.1 such that the annular domains

$$A_j = \left\{x : \ R - \frac{\delta}{4N} \leq |x - a_j| \leq R + \frac{\delta}{4N}\right\}, \ j = 1, \dots, d,$$

has the property that, when $n$ is sufficiently large,

$$\sum_{j=1}^{d} \int_{A_j} e_{\varepsilon_n}(\psi_n) dx \leq C(\lambda, K), \tag{4.7}$$
$$|\psi_n(x)| \geq \frac{1}{2} \ \text{ on } \ \cup_{j=1}^{d} A_j. \tag{4.8}$$

Then for each point $y$ such that $|y - a_j| = R$, $j = 1, \dots, d$, we let $B = B(y, \rho)$ with $\rho \in [\delta/8N, \ \delta/4N]$ such that

$$\int_B e_{\varepsilon_n}(\psi_n) dx + \int_{\partial B} e_{\varepsilon_n}(\psi_n) d\Gamma \leq C(\lambda, K) \tag{4.9}$$

is valid for a subsequence of $n \to \infty$. The fact that (4.7) implies (4.9) comes from the geometric theory and is beyond the scope of this book, interested readers can refer to [Sim1 83] (p. 57) for more details.

Because the energy is bounded, $\deg(\psi_n, \partial B) = 0$. Estimate (4.9) implies that there is an $x_0 \in \partial B$ such that $|\,|\psi_n| - 1\,|(x_0) \le \varepsilon_n^2/C\rho \le c\varepsilon_n$. Consequently

$$|\,|\psi_n| - 1\,|(x) \le c\sqrt{\int_{\partial B} |\nabla\psi_n|(1 - |\psi_n|)d\Gamma + c\varepsilon_n^2} \le C(\lambda, K)\varepsilon_n^{1/4}.$$

Next, we prove that under (4.9) and (4.2), $\psi_n$ converges strongly in $H^1(B)$ as $n \to \infty$ and

$$\frac{1}{\varepsilon_n^2}\int_B (|\psi_n|^2 - 1)^2 dx \to 0.$$

For notational convenience, we now regard $\psi_n$ as complex valued functions. Hence we write $\psi_n = \rho_n e^{i\theta_n}$ so that

$$\operatorname{div}(\rho_n^2 \nabla\theta_n) = f_n \in L^2(B),\ \theta_n|_{\partial B} \in H^1(\partial B), \tag{4.10}$$

$$\Delta\rho_n + \frac{\rho_n}{\varepsilon_n^2}(1 - \rho_n^2) - |\nabla\theta_n|^2\rho_n = g_n \in L^2(B) \tag{4.11}$$

with

$$\begin{aligned}
&|\rho_n - 1|(x) \le C(\lambda, K)\varepsilon_n^{1/4} \quad \text{for} \quad x \in \partial B,\\
&\int_{\partial B} e_{\varepsilon_n}(\rho_n)d\Gamma \le C(\lambda, K),\\
&|\rho_n(x)| \ge 1/2 \quad \text{on} \quad B,\\
&\|f_n\|_{L^2(B)} + \|g_n\|_{L^2(B)} \to 0 \quad \text{as} \quad n \to \infty.
\end{aligned}$$

We conclude now that $\theta_n \to \theta_0$ in $H^1(B)$ as $n \to \infty$ where $\Delta\theta_0 = 0$ in $B$ and $\theta_0|_{\partial B}$ is the weak limit of $\theta_n|_{\partial B}$ in $H^1(\partial B)$.

To proceed with the proof, we quote the following result of Giaquinta (see [Gia 83], Theorem 2.2 of Chapter V).

*For the system*

$$-\partial_\alpha A_j^\alpha(x, u, \nabla u) + B_j(x, u, \nabla u) = 0 \quad j = 1, 2, \ldots, N$$

*where $x \in \Omega \subset \mathbb{R}^n$ and $u : \Omega \to \mathbb{R}^N$. We assume that*

$$r = \frac{2n}{n-2} \quad \textit{if } n > 2; \quad r \in (0, \infty) \quad \textit{if } n = 2.$$

*Then under the assumptions*

$$\begin{aligned}
&|A_j^\alpha(x,u,\nabla u)| \le C_1(|\nabla u| + |u|^{r/2} + f_{j\alpha}),\\
&|B_j(x,u,\nabla u) \le C_2(|\nabla u|^{2(1-1/r)} + |u|^{r-1} + f_j),\\
&f_{j\alpha} \in L^2(\Omega), \quad f_j \in L^{\frac{r}{r-1}}(\Omega),\\
&A_j^\alpha(x,u,p)p_\alpha^j \ge \lambda |p|^2 - C_2|u|^r - f^2,\\
&f \in L^2(\Omega), \quad \lambda > 0, \quad C_1, C_2 \ge 0,\\
&f,\ f_{j\alpha} \in L^\sigma(\Omega),\\
&f_j \in L^s(\Omega) \text{ with } \sigma > 2 \text{ and } s > \frac{r}{r-1},
\end{aligned}$$

*there exists an exponent $p > 2$ such that, if $u \in H^1(\Omega, I\!R^N)$ is a weak solution of the above system, then $u \in H^{1,p}_{loc}(\Omega, R^N)$. Moreover for $B_{R/2} \subset B_R \subset \Omega$, we have*

$$\begin{aligned}
\Big(\frac{1}{|B_{R/2}|}\int_{B_{R/2}} (|u|^r + |\nabla u|^2)^{p/2} dx\Big)^{1/p} \le\ & C\Big(\frac{1}{|B_R|}\int_{B_R} (|u|^r + |\nabla u|^2) dx]^{1/2} +\\
& [\frac{1}{|B_R|}\int_{B_R} (f^2 + f_{j\alpha}^2)^{p/2} dx]^{1/p} +\\
& R[\frac{1}{|B_R|}\int_{B_R} |f_j|^{\frac{rp}{2(r-1)}} dx]^{\frac{2(r-1)}{pr}}\Big).
\end{aligned}$$

Applying this result to equation (4.10) and noting that

$$\frac{1}{2} \le \rho_n \le 1, \quad f_n \in L^2(B)$$

(this is equivalent to $r = 3$ in the above Giaquinta's result) we have

$$\begin{aligned}
\Big(\frac{1}{|B_{R/2}|}\int_{B_{R/2}} |\nabla \theta_n|^p dx\Big)^{1/p} \le\ & C\Big([\frac{1}{|B_R|}\int_{B_R} |\nabla \theta_n|^2 dx]^{1/2}\\
+\ & R[\frac{1}{|B_R|}\int_{B_R} |f_n|^{\frac{2p}{3}} dx]^3 2p\Big) \le C_0.
\end{aligned}$$

Therefore we have (noting that $g_n \in L^2(B)$, $\frac{1}{2} \le \rho_n \le 1$)

$$g_n + |\nabla \theta_n|^2 \rho_n = \Delta \rho_n + \frac{\rho_n}{\varepsilon_n^2}(1 - \rho_n^2) \in L^p(B) \ \text{ for some } \ p > 1.$$

Combining this with (4.11) and the boundary estimate $1 - \rho_n \le C(\lambda, K)\varepsilon_n^{1/4}$, we have $1 - \rho_n \le C\varepsilon_n^\beta$ for some constants $\beta$ and $C$.

Multiplying (4.11) by $\rho_n$ and integrate it over $B$, using the fact that $|\psi| \geq 1/2$ on $B$, we have

$$\begin{aligned}
\frac{1}{4\varepsilon_n^2}\int_B (1-\rho_n^2)dx &\leq \frac{1}{\varepsilon_n^2}\int_B \rho_n^2(1-\rho_n^2)dx \\
&\leq \int_B |\nabla\rho_n|^2 + |\nabla\theta_n|^2|\rho_n|^2 + |\rho_n g_n| + \int_{\partial B} \rho_n \partial_{\boldsymbol{n}} \rho_n \\
&\leq \int_B |\nabla\rho_n|^2 + |\nabla\theta_n|^2|\rho_n|^2 + |\rho_n g_n| + \frac{1}{2}\int_{\partial B} |\nabla\rho_n|^2 + |\rho_n|^2 \\
&\leq C(\lambda, K). \qquad (4.12)
\end{aligned}$$

Consequently,

$$\frac{1}{\varepsilon_n^2}\int_B (1-\rho_n^2)^2 dx \leq C\varepsilon_n^\beta \to 0 \ \text{ as } \ n\to\infty.$$

We let $V_n = \rho_n^2 - 1$, it satisfies the following equation (multiply the equation (4.11) by $\rho_n$ and rearrange the terms)

$$\begin{cases} -\Delta V_n + \frac{2\rho_n^2}{\varepsilon_n^2} V_n = -2\rho_n^2|\nabla\theta_n|^2 + 2g_n\rho_n - 2|\nabla\rho_n|^2 \ \text{ in } B, \\ |V_n| \leq C(\lambda, K)\varepsilon_n^{1/4} \ \text{ on } \partial B. \end{cases} \qquad (4.13)$$

Multiplying (4.13) by $V_n$ and integrate over $B$, we have

$$\begin{aligned}
\int_B |\nabla V_n|^2 + \frac{1}{2\varepsilon_n^2}V_n^2 &\leq -2\int_B V_n\rho_n^2(-2|\nabla\rho_n|^2 + |\nabla\theta_n|^2 \\
&\qquad +2|g_n|)dx + \int_{\partial B} V_n \partial_{\boldsymbol{n}} V_n \\
&\leq o(1) + C(\lambda, K)\int_{\partial B} V_n^2 \to 0 \ \text{ as } n\to\infty. \qquad (4.14)
\end{aligned}$$

Since $|\nabla V_n|^2 = 2|\rho_n\nabla\rho_n|^2 \geq |\nabla\rho_n|^2$, up to choosing a subsequence, we conclude that $\psi_n \to V$ in $H^1(B)$.

Since the domain

$$A = \cup_{j=1}^d \left\{x: \ R - \frac{\delta}{16N} \leq |x - a_j| \leq R + \frac{\delta}{16N}\right\}$$

is covered by a finite number of our $B$'s defined above, using finite many repeating processes of choosing subsequences, it is easy to find a sequence $\psi_n$ which converges to $V$ on $A$ and

$$\frac{1}{\varepsilon_n^2}\int_A (1-\rho_n^2)^2 dx \to 0 \ \text{ as } n\to\infty.$$

Finally, we just have to choose $R$ so that (4.5) and (4.6) hold. It is straightforward now to use standard geometric measure theory.

Next, we show that

**Theorem 7.4.1** *For a.e. $t > 0$, and for any $V \in S_g(t)$, the corresponding $a = (a_1, \dots, a_d)$ is a critical point of $W_g(a)$ in $\Omega^d$.*

*Proof.* We let $\{\psi_n(x)\} = \{\psi_{\varepsilon_n}(x,t)\}$ where $t$ is fixed and satisfies (4.3). $\{\psi_n(x)\}$ converges to $V \in S_g(t)$. Multiplying the equation $\psi_{nt} = \Delta\psi_n + \frac{1}{\varepsilon_n^2}\psi_n(1-|\psi_n|^2)$ by $\partial_{x_k}\psi_n$ and integrating over $B_R(a_j)$ for some small suitable $R$ and a given $a_j$, we find

$$-\int_{\partial B_R} \partial_{\boldsymbol{n}}\psi_n \partial_k \psi_n + \frac{1}{2}\int_{\partial B_R} |\nabla\psi_n|^2 \boldsymbol{n}_k = -\frac{1}{4\varepsilon_n^2}\int_{\partial B_R}(1-\rho_n^2)^2\boldsymbol{n}_k + o(1) \quad (4.15)$$

where

$$\begin{aligned} |o(1)| &= |\frac{1}{T_{\varepsilon_n}}\int_{B_R} \partial_t\psi_n\partial_k\psi_n dx| \\ &\le \frac{1}{2T_{\varepsilon_n}}\int_{B_R}(|\partial_t\psi_n|^2 + |\nabla\psi_n|^2)dx \to 0 \text{ as } n\to\infty \text{ for a.e. } t. \end{aligned}$$

(The quantity $\frac{1}{2T_{\varepsilon_n}}\int_{B_R}|\partial_t\psi_n|^2dx$ tends to 0 can easily be justified by the scaling $t' = T_{\varepsilon_n}t$ and the assumption that $\log(1/\varepsilon)/T_\varepsilon \to 0$ as $\varepsilon \to 0$. The quantity $\frac{1}{2T_{\varepsilon_n}}\int_{B_R}|\nabla\psi_n|^2dx$ tends to 0 follows from $\log(1/\varepsilon)/T_\varepsilon \to 0$ as $\varepsilon \to 0$.)

Let $e^{i\phi(x)} = (x-a_j)/|x-a_j|$ and write $V(x) = e^{i\phi(x)+iH(x)}$ in a neighbourhood of $a_j$ for $j = 1, \dots, d$, then $\Delta H(x) = 0$ in any neighbourhood of $a_j$ not containing any other $a_k$. Using Proposition 4.1 and (4.15), choosing appropriate $R$, we have

$$-\int_{\partial B_R}\partial_{\boldsymbol{n}}V\partial_k V^* + \frac{1}{2}\int_{\partial B_R}|\nabla V|^2\boldsymbol{n}_k = 0. \quad (4.16)$$

Replacing $V$ by $e^{i\phi(x)+iH(x)}$, we have, for any $k$,

$$-\int_{\partial B_R}\partial_{\boldsymbol{n}}H(\partial_k H + \frac{\boldsymbol{\tau}_k}{R}) + \int_{\partial B_R}(\frac{1}{R}\partial_{\boldsymbol{\tau}}H + \frac{1}{2}|\nabla H|^2)\boldsymbol{n}_k = 0.$$

This and the fact that $\Delta H(x) = 0$ in $B_R$ implies that $\nabla H(a_j) = 0$. By the definition of $W_g$, this is equivalent to say that

$$\nabla W_g(a) = 0.$$

In the above discussion, we have regarded $t$ as a fixed constant. Taking into account that $t$ is a variable, Theorem 4.1 should be reformulated as follows:

**Theorem 7.4.2** *Let $\psi_\varepsilon$ be the solution of (4.1) with initial data $\psi_0^\varepsilon$ satisfying assumption A1). Then for any sequence $\varepsilon_n \to 0$, there is a subsequence of $\{\varepsilon_n\}$ still denoted by $\{\varepsilon_n\}$ such that $\psi_{\varepsilon_n}$ satisfies*

*1) for a.e.* $t > 0$, $dist(\psi_{\varepsilon_n}(x,t), S_g(t)) \to 0$ *as* $n \to \infty$,
*2) for a.e.* $t > 0$ *and any* $V \in S_g(t)$,

$$V(x) = \prod_{j=1}^{d} \frac{x - a_j}{|x - a_j|} e^{ih_a(x)}, \quad \Delta h_a(x) = 0$$

*and* $a = (a_1, \dots, a_d)$ *is a critical point of* $W_g(a)$.
*Here* $dist(W, S_g(t)) = \inf\{\|W - V\|_{L^2(\Omega)} : V \in S(t)\}$.

This theorem tells us that if $T_\varepsilon$ is far greater than $\log(1/\varepsilon)$, then the limit vortices all go to a fixed set: the critical points of $W_g(a)$. There will be no motion of vortices in general unless the set of critical points of $W_g(a)$ consists of some continuum. This means that the migration of vortices has already finished if the time scale is larger than $\log(1/\varepsilon)$.

We can now conclude that the time scale $T_\varepsilon = c\log(1/\varepsilon)$ is the scale on which we expect to see some migration of the vortices under our particular setting. That is, if we look at the following problem as $\varepsilon \to 0$:

$$\begin{aligned} &\tfrac{1}{\log(1/\varepsilon)} \psi_t = \Delta\psi + \tfrac{1}{\varepsilon^2}(1 - |\psi|^2)\psi, \\ &\psi(x,t) = g(x), \quad \text{for } x \in \partial\Omega,\ t > 0, \\ &\psi(x,0) = \psi_0^\varepsilon(x) \quad \text{for } x \in \Omega. \end{aligned} \tag{4.17}$$

From the results of last section, we can reasonably guess that the law of motion is

$$M\frac{d}{dt}a(t) = -\nabla_a W_g(a) \tag{4.18}$$

because the vortices will eventually settle down in the set of critical points of $W_g(a)$.

Unlike for the pinning case, there is no rigorous proof yet for this fact. The difficulty is to show that $a(t)$ depends on $t$ smoothly and the limit of $(a(t+\delta t) - a(t))/\delta t$ as $\delta t \to 0$ is equal to the desired quantity.

## 7.5 Bibliographical remarks

We have already pointed out at the beginning of this chapter that due to the difficulty in dealing with boundary conditions, little material on the rigorous proof of vortex motion equation for the G-L superconductivity model is available. The works on the pure G-L case are given in [RSt 95] and [Lin 96]. Also, in [LDu 97], the issue of steady state vortex distribution and the convergence of the solutions of the evolutionary model to that of the steady state model is discussed.

# Chapter 8
# Thin Plate/Film G-L Models

In this chapter, we take a flat thin 3-d plate/film lying in the $xy$-plane and let the thickness (in the $z$-direction) of the plate/film tend to zero. Theoretically, the limit would yield the G-L superconductivity model for an ideal 2-d plate/film. Naturally, it is very interesting to see if this 2-d plate/film model has anything to do with the 2-d G-L system. The conclusion is, it depends on which 3-d model we start from.

If we start with imposing the G-L system in the thin plate/film and imposing the Maxwell system in the outside domain, then we obtain a 2-d limit model that is simpler than the 2-d G-L system. At the leading order, we have $\mathbf{curl}\boldsymbol{A}_0 = \boldsymbol{H}$ in $\mathbb{R}^3$ with $\boldsymbol{H}$ being the applied magnetic field, and that the order parameter term $\psi_0$ is a function of $(x, y)$ only (no more $z$-dependence) which satisfies

$$\frac{1}{p}(i\nabla + \boldsymbol{A}_0(x,y,0))p(i\nabla + \boldsymbol{A}_0(x,y,0))\psi + \kappa^2(|\psi|^2 - 1)\psi = 0. \tag{0.1}$$

Here $p = p(x, y)$ is the plate thickness parameter to be specified later. When the plate is perfectly flat, $d = 1$. It is interesting to point out that $\mathbf{curl}\boldsymbol{A} = \boldsymbol{H}$ implies total magnetic field penetration yet this total penetration of the applied magnetic field does not cause the total destruction of the superconducting property. It implies that the thin plate/film geometry has physical properties distinct from that of the imposition of 3-d bodies.

For pure mathematical curiosity, we also investigate the imposition of the natural boundary condition of the G-L energy $\mathbf{curl}\boldsymbol{A} \times \boldsymbol{n} = \boldsymbol{H} \times \boldsymbol{n}$ on the boundary of the thin plate/film rather than having the Maxwell system in the exterior domain. To be clear, we steer away temporarily from the terminology "superconductivity". It is interesting to see that the limit model is the standard 2-d G-L system with the 2-d applied field being $H_3$ – the third component of the original applied field. This implies that under this setting, the components of the applied field that are parallel to the thin plate/film have little effect on the behaviour of the order parameter solution. The normal component of the applied field interacts with the plate/film in a manner governed by the 2-d G-L system. This latter model presents an interesting picture for the interaction between a smaller physical dimension and the coupling field. It implies that there must be an "engagement area" for an applied field to influence the order parameter. For example, the applied field component that is perpendicular to the plate/film will have a positive impact on the order parameter solution because it is "caught up" by the surface area of the plate/film. The applied field components that are parallel to the film are not directly facing the sample. Consequently, they have no impact on the leading term of the order parameter. Being a thin plate/film in the $xy$-plane, the sample has also lost its ability to produce a vortex-like phase transition domain lying in the $xy$-plane. Hence, a parallel field can do little to influence the order parameter. We are not aware of any physical experiment justifying these. However, this model does come from an (probably) incorrect initial physical setting and is probably of pure academic interests only at the moment. We include it because we feel that this argument of engagement area in connection with the minimum coupling principle (the way the magnetic potential is involved in the order parameter equation) is very interesting and may find its application in other fields.

First, we define some notations for this chapter. Let $\Omega_0$ be an open bounded subset of $I\!R^2$ with $\partial\Omega_0 \in C^2$. We consider a thin film of variable upper and lower surfaces defined by $\Omega_\varepsilon = \{x = (x_1, x_2, x_3) : x' = (x_1, x_2) \in \Omega_0,\ x_3 \in \varepsilon(-b(x'), a(x'))\}$ where $\varepsilon > 0$ is a small parameter, $a$ and $b$ are assumed to be functions in $C^2(\bar{\Omega}_0)$, and $a(x') \geq c_a > 0$ and $b(x') \geq c_b > 0$ for all $x' \in \bar{\Omega}_0$. We define $p(x,y) = a(x,y) + b(x,y)$.

In the following of this chapter, when we discuss the time dependent problems, we also use the following notation: $Q_\varepsilon = \Omega_\varepsilon \times (0,T)$ for some given $T > 0$, $\Gamma_\varepsilon = \partial\Omega_\varepsilon \times (0,T)$.

## 8.1 The outside Maxwell system – steady state case

We take the steady state case as an example for the problem where the outside field is governed by the Maxwell system. We mainly followed the conclusion of the first part and the layout of the proofs of the 2nd part of [CDG 96].

We consider the G-L energy

$$\begin{aligned} G_\varepsilon(\psi, \underline{\boldsymbol{A}}) &= \int_{\Omega_\varepsilon} \Big( |(i\nabla + \underline{\boldsymbol{A}})\psi|^2 + \frac{\kappa^2}{2}(|\psi|^2 - 1)^2 \Big) dx dy dz \\ &\quad + \int_{I\!R^3} |\mathbf{curl}\underline{\boldsymbol{A}} - \boldsymbol{H}|^2 dx dy dz. \end{aligned} \tag{1.1}$$

Let $\bar{\boldsymbol{A}}$ be a magnetic potential of given applied field $\boldsymbol{H}$ satisfying:

$$\mathbf{curl}\bar{\boldsymbol{A}} = \boldsymbol{H},$$
$$\mathrm{div}\bar{\boldsymbol{A}} = 0 \ \ \text{(by gauge choice)},$$

and let $\boldsymbol{A} = \underline{\boldsymbol{A}} - \bar{\boldsymbol{A}}$, the energy can be rewritten as

$$\begin{aligned} G_\varepsilon(\psi, \boldsymbol{A}) &= \int_{\Omega_\varepsilon} \Big( |(i\nabla + \boldsymbol{A} + \bar{\boldsymbol{A}})\psi|^2 + \frac{\kappa^2}{2}(|\psi|^2 - 1)^2 \Big) dxdydz \\ &\quad + \int_{\mathbb{R}^3} |\mathbf{curl}\boldsymbol{A}|^2 dxdydz. \end{aligned} \tag{1.2}$$

Using a gauge transformation, we consider

$$\min_{(\psi, \boldsymbol{A}) \in \mathcal{H}^1(\Omega_\varepsilon) \times \boldsymbol{H}^1(\mathrm{div}, \mathbb{R}^3)} G_\varepsilon(\psi, \boldsymbol{A}) \tag{1.3}$$

Let $(\psi_\varepsilon, \boldsymbol{A}_\varepsilon)$ be an energy minimizer of $G_\varepsilon(\psi, \boldsymbol{A})$, it is easy to know that $(\psi_\varepsilon, \boldsymbol{A}_\varepsilon + \bar{\boldsymbol{A}})$ is an energy minimizer of $G_\varepsilon(\psi, \underline{\boldsymbol{A}})$. The existence of energy minimizers has been discussed in Chapter 4.

### 8.1.1 The energy bound

Now we establish the energy bound for the solutions $(\psi_\varepsilon, \boldsymbol{A}_\varepsilon)$. First, notice that $(\psi, \boldsymbol{A}) = (0, \mathbf{0})$ is an admissible function for the problem (1.3), it is straightforward to obtain

**Proposition 8.1.1** *For any $\varepsilon > 0$, we have*

$$\min_{(\psi, \boldsymbol{A}) \in \mathcal{H}^1(\Omega_\varepsilon) \times \boldsymbol{H}^1(\mathrm{div}, \mathbb{R}^3)} G_\varepsilon(\psi, \boldsymbol{A}) \le G_\varepsilon(0, \mathbf{0}) = \varepsilon \frac{\kappa^2}{2} \int_{\Omega_0} p(x, y) dxdy \le C\varepsilon. \tag{1.4}$$

This implies

**Corollary 8.1.1** *There is a constant $C > 0$ such that for any $\varepsilon > 0$,*

$$||\nabla \boldsymbol{A}_\varepsilon||_{L^2(\mathbb{R}^3)} \le C\sqrt{\varepsilon}. \tag{1.5}$$

*Proof.* From Proposition 1.1, we know already that

$$||\mathbf{curl}\boldsymbol{A}_\varepsilon||_{L^2(\mathbb{R}^3)} \le C\sqrt{\varepsilon}. \tag{1.6}$$

For any smooth $\boldsymbol{A}$ with compact support in $\mathbb{R}^3$, we have

$$\begin{aligned} & ||\mathbf{curl}\boldsymbol{A}||^2_{L^2(\mathbb{R}^3)} + ||\mathrm{div}\boldsymbol{A}||^2_{L^2(\mathbb{R}^3)} \\ = & \int_{\mathbb{R}^3} \sum_{j \ne k} \Big( (\partial_j A_k)^2 - \partial_j A_k \partial_k A_j \Big) dxdydz + ||\mathrm{div}\boldsymbol{A}_\varepsilon||^2_{L^2(\mathbb{R}^3)} \\ = & \sum_{j \ne k} \int_{\mathbb{R}^3} \Big( (\partial_j A_k)^2 - \partial_j (A_k \partial_k A_j) + A_k \partial_k \partial_j A_j \Big) + (\mathrm{div}\boldsymbol{A})^2 dxdydz \\ = & \sum_{j \ne k} \int_{\mathbb{R}^3} \Big( (\partial_j A_k)^2 - 0 - A_k \partial_k \partial_k A_k + \boldsymbol{A}_k \partial_k \mathrm{div}\boldsymbol{A} \Big) + (\mathrm{div}\boldsymbol{A})^2 dxdydz \end{aligned}$$

$$= \sum_{j\neq k} \int_{I\!R^3} \left((\partial_j A_k)^2 + (\partial_k A_k)^2 - (\mathrm{div}\boldsymbol{A})^2\right) + (\mathrm{div}\boldsymbol{A})^2 dxdydz$$
$$= \sum_{j,k=1}^{3} \int_{I\!R^3} (\partial_j A_k)^2 dxdydz = ||\nabla \boldsymbol{A}||^2_{L^2(I\!R^3)}.$$

But the energy minimizer $\boldsymbol{A}_\varepsilon$ satisfies $\mathrm{div}\boldsymbol{A}_\varepsilon = 0$, so we have, by density argument,

$$||\mathbf{curl}\boldsymbol{A}_\varepsilon||^2_{L^2(I\!R^3)} = ||\nabla \boldsymbol{A}_\varepsilon||^2_{L^2(I\!R^3)}.$$

Combining this with (1.6), we obtain the Corollary.

(1.5) implies that $\nabla \boldsymbol{A}_\varepsilon \to 0$ in $\boldsymbol{L}^2(I\!R^3)$. This means that the magnetic field approaches the given applied magnetic field. That is

$$\mathbf{curl}\underline{\boldsymbol{A}}_\varepsilon \to \boldsymbol{H} \quad \text{as } \varepsilon \to 0.$$

Because these estimates are all small, we need to rescale the variables to obtain further information. Introduce

$$\tilde{\psi}(x,y,\tilde{z}) = \psi(x,y,\varepsilon\tilde{z}),\ \tilde{\boldsymbol{A}}(x,y,\tilde{z}) = \boldsymbol{A}(x,y,\varepsilon\tilde{z}),\ \tilde{\bar{\boldsymbol{A}}}(x,y,\tilde{z}) = \bar{\boldsymbol{A}}(y,y,\varepsilon\tilde{z}) \quad (1.7)$$

on the new domain

$$\tilde{\Omega} = \{(x,y,z) \in I\!R^3 |\ x' = (x,y) \in \Omega_0,\ -b(x,y) < \tilde{z} < a(x,y)\}. \quad (1.8)$$

In the rest of this section, unless otherwise stated clearly, we work on the new variables exclusively. If there is no confusion, we will drop the "$\tilde{\cdot}$" sign. We also introduce the notation that $\nabla' = (\partial_x, \partial_y)^T$.

Changing variables in Corollary 1.1 and using standard Sobolev interpolation theory, we obtain

**Corollary 8.1.2** *There exists a constant $C > 0$ such that for any $\varepsilon > 0$,*

$$||\nabla' \boldsymbol{A}_\varepsilon||_{L^2(I\!R^3)} \le C,$$
$$||\partial_z \boldsymbol{A}_\varepsilon||_{L^2(I\!R^3)} \le C\varepsilon,$$
$$||\boldsymbol{A}_\varepsilon||_{L^6(I\!R^3)} \le C||\nabla \boldsymbol{A}_\varepsilon||_{L^2(I\!R^3)} \le C.$$

*Proof.* The only inequality needs investigating is

$$||\boldsymbol{A}_\varepsilon||_{L^6(I\!R^3)} \le C||\nabla \boldsymbol{A}_\varepsilon||_{L^2(I\!R^3)}.$$

To achieve this, we start with smooth functions with compact support and use the fact that $|\nabla u| = |\nabla |u||$ a.e.:

$$\int_{I\!R^3} u^6 dx = \int dx_1 \int\int |u|^3|u|^3 dx_2 dx_3$$
$$\le \int dx_1 \Big(\max_{x_2} \int |u|^3 dx_3\Big)\Big(\max_{x_3} \int |u|^3 dx_2\Big)$$

$$\begin{aligned}
&\leq 9\int dx_1\Big(\int\int |u|^2|u_{x_2}|dx_2dx_3\Big)\Big(\int\int |u|^2|u_{x_3}|dx_2dx_3\Big)\\
&\leq 9\int dx_1\int\int |u|^4dx_2dx_3\sqrt{\int\int |u_{x_2}|^2dx_2dx_3\int\int |u_{x_3}|^2dx_2dx_3}\\
&\leq 9\max_{x_1}\int\int |u|^4dx_2dx_3\sqrt{\int\int\int |u_{x_2}|^2\int\int\int |u_{x_3}|^2}\\
&\leq 36\int\int\int |u^2u_{x_1}|dx||u_{x_2}||_{L^2}||u_{x_3}||_{L^2}\\
&\leq 36\sqrt{\int_{I\!R^3} u^6dx}||u_{x_1}||_{L^2}||u_{x_2}||_{L^2}||u_{x_3}||_{L^2}.
\end{aligned}$$

It is now easy to see the result.

In particular, if we restrict on the rescaled domain $\Omega$, we have

**Proposition 8.1.2** *There exists a constant $C > 0$ such that for any $\varepsilon > 0$,*

$$\begin{aligned}
&||\boldsymbol{A}_\varepsilon||_{H^1(\Omega)} \leq C,\\
&||\nabla'\psi_\varepsilon||_{L^2(\Omega)} \leq C,\\
&||\partial_z\psi_\varepsilon||_{L^2(\Omega)} \leq C\varepsilon,\\
&||\psi_\varepsilon||_{L^\infty(\Omega)} \leq 1,\\
&||\psi_\varepsilon||_{H^1(\Omega)} \leq C.
\end{aligned}\tag{1.9}$$

*Proof.* (1.9-1) follows from Corollary 1.2. (1.9-4) is the standard $L^\infty$ bound on the order parameter. (1.9-2) and (1.9-3) follow from the energy estimate, (1.9-1) and (1.9-4). (1.9-5) is a trivial consequence of (1.9-2)–(1.9-4).

### 8.1.2 Convergence properties of the rescaled variables

Firstly, we establish the convergence properties of $\boldsymbol{A}_\varepsilon$.

**Theorem 8.1.1** *There exists a subsequence $\{\varepsilon_k\}$ of $\{\varepsilon\}$ with $\varepsilon_k \to 0$ as $k \to \infty$ such that*

$$\boldsymbol{A}_{\varepsilon_k} \rightharpoonup \boldsymbol{0} \quad in \quad \boldsymbol{H}^1(\Omega).$$

*Consequently,*

$$\boldsymbol{A}_{\varepsilon_k} \to \boldsymbol{0} \quad in \quad \boldsymbol{L}^2(\Omega).$$

*Proof.* Step 1: We show that there is a subsequence such that $\nabla\boldsymbol{A}_{\varepsilon_k} \rightharpoonup 0$ in $L^2(I\!R^3)$. Note that now the domain is unbounded.

From $||\nabla'\boldsymbol{A}_\varepsilon||_{L^2(I\!R^3)} \leq C$ and $||\partial_z\boldsymbol{A}_\varepsilon||_{L^2(I\!R^3)} \leq C\varepsilon$, there exists a subsequence $\{\varepsilon_k\}$ such that

$$\nabla\boldsymbol{A}_{\varepsilon_k} \rightharpoonup Q = (\boldsymbol{Q}_1, \boldsymbol{Q}_2, \boldsymbol{Q}_3) \text{ in } L^2(I\!R^3).$$

From Corollary 1.2, we have $||\partial_z \boldsymbol{A}_\varepsilon||_{L^2(\mathbb{R}^3)} \le C\varepsilon$ which implies $\partial_z \boldsymbol{A}_\varepsilon \to \mathbf{0}$ as $\varepsilon \to 0$. The third component of each $\boldsymbol{Q}_j$ is zero. From the identity $\mathbf{curl}\nabla \boldsymbol{A}_{\varepsilon_k} = 0$, we obtain $\mathbf{curl}\boldsymbol{Q}_j = 0$ for $j = 1,2,3$. This implies that $\boldsymbol{Q}_j = \nabla g_j$ for some scalar function $g_j$. As the third component of $\boldsymbol{Q}_j$ (or equivalently, $\partial_z g_j$) is 0, $g_j = g_j(x,y)$. From the fact that

$$\int_{\mathbb{R}^3} |\boldsymbol{Q}_j| dxdydz = \int_{\mathbb{R}^3} |\nabla g_j(x,y)| dxdydz \le C,$$

we must have $\nabla g_j \equiv \mathbf{0}$ for $j = 1,2,3$.

Step 2: We now show that $\boldsymbol{A}_{\varepsilon_k} \rightharpoonup \mathbf{0}$ in $\boldsymbol{H}^1(\Omega)$. Note that now the domain is bounded.

Similar to Step 1, we know that $\boldsymbol{A}_{\varepsilon_k} \rightharpoonup \mathbf{C}$ in $\boldsymbol{H}^1(\tilde{\Omega})$ for some constant vector $\mathbf{C}$. Corollary 1.2 gives that

$$\begin{aligned} ||\mathbf{C}||_{L^6(\tilde{\Omega})} &\le \underline{\lim}_{k\to\infty} ||\boldsymbol{A}_{\varepsilon_k}||_{L^6(\tilde{\Omega})} \le \underline{\lim}_{k\to\infty} ||\boldsymbol{A}_{\varepsilon_k}||_{L^6(\mathbb{R}^3)} \\ &\le \underline{\lim}_{k\to\infty} C||\nabla \boldsymbol{A}_{\varepsilon_k}||_{L^2(\mathbb{R}^3)} \le C \end{aligned}$$

which implies that $\mathbf{C} = \mathbf{0}$.

Step 3: Finally, the strong convergence in $\boldsymbol{L}^2$ is a straightforward consequence of the weak convergence in $\boldsymbol{H}^1$. The theorem is proved.

Secondly, we establish the convergence properties of $\psi_\varepsilon$.

**Theorem 8.1.2** *There exists a subsequence $\{\varepsilon_k\}$ with $\varepsilon_k \to 0$ as $k \to \infty$ such that*

$$\begin{aligned} &\psi_{\varepsilon_k} \rightharpoonup (\to)\ \psi = \psi(x,y) \ \ in\ \mathcal{H}^1(\Omega)\ (\ in\ \mathcal{L}^2(\Omega))\ , \\ &||\partial_z \psi_{\varepsilon_k}||_{\mathcal{L}^2(\Omega)} \to 0. \end{aligned}$$

*Proof.* The conclusion that $||\partial_z \psi_{\varepsilon_k}||_{\mathcal{L}^2(\Omega)} \to 0$ comes from the fact that $||\partial_z \psi_\varepsilon||_{L^2(\Omega)} \le C\varepsilon$. The first convergence follows from the facts that $||\psi_\varepsilon||_{H^1(\Omega)} \le C$ and $\mathcal{H}^1(\Omega) \subset\subset \mathcal{L}^2(\Omega)$. The second convergence implies that the limit function must be a function of $x$ and $y$ only. The proof is therefore complete.

Finally, we give some consequences of our convergence results.

**Corollary 8.1.3** *As $k \to \infty$, we have*

$$\int_{-1}^{1} \boldsymbol{A}_{\varepsilon_k}(x,y,\frac{1}{2}\Big((\tau+1)a(x,y) + (\tau-1)b(x,y)\Big))d\tau \to \mathbf{0} \ \ in\ \boldsymbol{L}^2(\Omega_0),$$

$$\frac{1}{2}\int_{-1}^{1} \psi_{\varepsilon_k}(x,y,\frac{1}{2}\Big((\tau+1)a(x,y) + (\tau-1)b(x,y)\Big))d\tau \rightharpoonup \psi(x,y) \ \ in\ \mathcal{H}^1(\Omega_0),$$

$$\frac{1}{2}\int_{-1}^{1} \psi_{\varepsilon_k}(x,y,\frac{1}{2}\Big((\tau+1)a(x,y) + (\tau-1)b(x,y)\Big))d\tau \to \psi(x,y) \ \ in\ \mathcal{L}^2(\Omega_0).$$

*Proof.* We only prove the first convergence as an example.

$$\int_{\Omega_0} |\int_{-1}^{1} \boldsymbol{A}_{\varepsilon_k}(x,y,\frac{1}{2}\Big((\tau+1)a(x,y)+(\tau-1)b(x,y)\Big))d\tau|^2 dxdy$$

$$\leq C\int_{\Omega_0}\int_{-1}^{1} |\boldsymbol{A}_{\varepsilon_k}(x,y,\frac{1}{2}\Big((\tau+1)a(x,y)+(\tau-1)b(x,y)\Big))|^2 d\tau dxdy$$

$$\leq C\int_{\Omega_0}\int_{-b}^{a} \frac{2}{a+b}|\boldsymbol{A}_{\varepsilon_k}(x,y,z)|^2 dxdydz$$

$$\leq C\int_{\Omega_0}\int_{-b}^{a} |\boldsymbol{A}_{\varepsilon_k}(x,y,z)|^2 dxdydz$$

(use Theorem 1.1) $\to 0$.

### 8.1.3 Passing to the limit

In order to pass to the limit, we look at the variational formulation satisfied by the solution $(\psi_{\varepsilon_k}, \boldsymbol{A}_{\varepsilon_k})$. In order to simplify notation, we use $\varepsilon$ to denote $\varepsilon_k$. Before carrying on the computation, we point out that we have used the rescaling (1.7)–(1.8) without changing the notation of the functions. In the following, we stick to the same arrangement. That is if we have written

$$\frac{1}{\varepsilon}\int_{\Omega_\varepsilon} F(\psi_\varepsilon, \boldsymbol{A}_\varepsilon)dxdydz = \int_{\Omega} F(\psi_\varepsilon, \boldsymbol{A}_\varepsilon)dxdydz,$$

then in the first integration, we are referring to the unscaled original variables and functions whereas in the second integration, we are referring to the rescaled variables and functions. The rule of the rescaling is that defined by (1.7)–(1.8).

For $\boldsymbol{A}_\varepsilon$, we already know that $\mathbf{curl}\underline{\boldsymbol{A}}_\varepsilon \to \boldsymbol{H}$ (or equivalently, $\mathbf{curl}\boldsymbol{A}_\varepsilon \to \mathbf{0}$), so there is nothing to study. We only look at the equation satisfied by $\psi_\varepsilon$. The variational formulation for $\psi_\varepsilon$ is

$$\frac{1}{2\varepsilon}\int_{\Omega_\varepsilon} \mathrm{Re}\Big\{(i\nabla+\boldsymbol{A}_\varepsilon+\bar{\boldsymbol{A}})\psi_\varepsilon\cdot(-i\nabla+\boldsymbol{A}_\varepsilon+\bar{\boldsymbol{A}})\phi^*\Big\}dxdydz +$$

$$+\frac{1}{2\varepsilon}\int_{\Omega_\varepsilon} \mathrm{Re}\Big\{\kappa^2(|\psi_\varepsilon|^2-1)\psi_\varepsilon\phi^*\Big\}dxdydz = 0 \ \ \forall\phi\in\mathcal{H}^1(\Omega). \qquad (1.10)$$

Now we are going to take the limit $\varepsilon \to 0$ in (1.10) term by term. First, choose $\phi = \phi(x,y) \in \mathcal{H}^1(\Omega_0)$ in the above formulation.

By Theorem 1.2 and Corollary 1.3,

$$\frac{1}{2\varepsilon}\int_{\Omega_\varepsilon}(|\psi_\varepsilon|^2-1)\psi_\varepsilon\phi^* dxdydz = \frac{1}{2}\int_{\Omega_0}\int_{-1}^{1}\Big((|\psi_\varepsilon|^2-1)\psi\Big)(x,y,\frac{1}{2}\Big((\tau+1)a(x,y)$$

$$+(\tau-1)b(x,y)\Big))d\tau d(x,y)\phi^*(x,y)dxdy$$

$$\to \int_{\Omega_0}(|\psi_\varepsilon|^2(x,y)-1)\psi(x,y)d(x,y)\phi^*(x,y)dxdy. \qquad (1.11)$$

For the terms involving $\bar{\boldsymbol{A}}$, write $\bar{\boldsymbol{A}}(x,y,z) = \bar{\boldsymbol{A}}(x,y,0) + \bar{\boldsymbol{B}}(x,y,z)z$ with $\bar{\boldsymbol{B}}(x,y,z)$ uniformly bounded. Denoting $\bar{\boldsymbol{A}}(x,y,0)$ by $\bar{\boldsymbol{A}}_0(x,y)$, we have, by Corollary 1.3,

$$\begin{aligned}\frac{1}{2\varepsilon}\int_{\Omega_\varepsilon} \bar{\boldsymbol{A}}\psi_\varepsilon \cdot \bar{\boldsymbol{A}}\phi^* dxdydz &= \frac{1}{2\varepsilon}\int_{\Omega_\varepsilon} \bar{\boldsymbol{A}}_0\psi_\varepsilon \cdot \bar{\boldsymbol{A}}_0\phi^* dxdydz + o(1)\\ &= \frac{1}{2}\int_{\Omega_0}\int_{-1}^{1} |\bar{\boldsymbol{A}}_0|^2(x,y)\psi_\varepsilon(x,y,\frac{1}{2}\Big((\tau+1)a(x,y)\\ &\quad + (\tau-1)b(x,y)\Big))d\tau p(x,y)\phi^*(x,y)dxdy + o(1)\\ &\to \int_{\Omega_0} |\bar{\boldsymbol{A}}_0|^2\psi(x,y)\phi^*(x,y)p(x,y)dxdy. \end{aligned} \tag{1.12}$$

By Theorem 1.1 and Corollary 1.3,

$$\frac{1}{2\varepsilon}\int_{\Omega_\varepsilon} \boldsymbol{A}_\varepsilon\psi_\varepsilon \cdot \bar{\boldsymbol{A}}\phi^* dxdydz = \frac{1}{2}\int_{\Omega} \boldsymbol{A}_\varepsilon\psi_\varepsilon \cdot \bar{\boldsymbol{A}}\phi^* dxdydz \to 0. \tag{1.13}$$

$$\frac{1}{2\varepsilon}\int_{\Omega_\varepsilon} \bar{\boldsymbol{A}}\psi_\varepsilon \cdot \boldsymbol{A}_\varepsilon\phi^* dxdydz = \frac{1}{2}\int_{\Omega} \bar{\boldsymbol{A}}\psi_\varepsilon \cdot \boldsymbol{A}_\varepsilon\phi^* dxdydz \to 0. \tag{1.14}$$

$$\frac{1}{2\varepsilon}\int_{\Omega_\varepsilon} \boldsymbol{A}_\varepsilon\psi_\varepsilon \cdot \boldsymbol{A}_\varepsilon\phi^* dxdydz \to 0. \tag{1.15}$$

$$\frac{1}{2\varepsilon}\int_{\Omega_\varepsilon} \boldsymbol{A}_\varepsilon\psi_\varepsilon \cdot (-i)\nabla\phi^* dxdydz \to 0. \tag{1.16}$$

$$\frac{1}{2\varepsilon}\int_{\Omega_\varepsilon} i\nabla\psi_\varepsilon \cdot \boldsymbol{A}_\varepsilon\phi^* dxdydz \to 0. \tag{1.17}$$

By Theorem 1.2,

$$\begin{aligned}\frac{1}{2\varepsilon}\int_{\Omega_\varepsilon} i\nabla\psi_\varepsilon\bar{\boldsymbol{A}}\phi^* dxdydz &= \frac{i}{2}\int_{\Omega} \nabla\psi_\varepsilon(x,y,z)\cdot\bar{\boldsymbol{A}}_0\phi^* d(x,y)dxdydz + o(1)\\ &\to \int_{\Omega_0} i\nabla\psi\cdot\bar{\boldsymbol{A}}_0\phi^* d(x,y)dxdy,\end{aligned} \tag{1.18}$$

$$\frac{1}{2\varepsilon}\int_{\Omega_\varepsilon} \bar{\boldsymbol{A}}\psi_\varepsilon\cdot(-i)\nabla\phi^* dxdydz \to \int_{\Omega_0} \bar{\boldsymbol{A}}_0\psi\cdot(-i)\nabla\phi^* d(x,y)dxdy, \tag{1.19}$$

$$\frac{1}{2\varepsilon}\int_{\Omega_\varepsilon} i\nabla\psi_\varepsilon\cdot(-i)\nabla\phi^* dxdydz \to \int_{\Omega_0} i\nabla\psi\cdot(-i)\nabla\phi^* d(x,y)dxdy. \tag{1.20}$$

Combining (1.11)–(1.20), the limit of (1.10) becomes

$$\begin{aligned}&\int_{\Omega_0} \mathrm{Re}\Big\{(i\nabla + \bar{\boldsymbol{A}}_0)\psi(-i\nabla + \bar{\boldsymbol{A}}_0)\phi^*\Big\}p(x,y)dxdy +\\ &+\int_{\Omega_0} \mathrm{Re}\Big\{\kappa^2(|\psi|^2-1)\psi\phi^*\Big\}p(x,y)dxdy = 0 \ \ \forall\phi\in\mathcal{H}^1(\Omega).\end{aligned} \tag{1.21}$$

This leads to the reduced G-L equation as we predicted in (1.1)

## 8.2 The outside field is given – evolutionary case

To have a change, we work with the evolutionary model. We use the following G-L system:

$$
\begin{aligned}
&\frac{\partial \psi}{\partial t} - i\mathrm{div}\boldsymbol{A}\psi + \Big(i\nabla + \boldsymbol{A}\Big)^2\psi + \kappa^2\psi(|\psi|^2-1) = 0, \quad \text{in} \;\; Q_\varepsilon \\
&\frac{\partial \boldsymbol{A}}{\partial t} - \Delta\boldsymbol{A} + Re\Big[\Big(i\nabla\psi + \psi\boldsymbol{A}\Big)\bar{\psi}\Big] = \mathbf{curl}\boldsymbol{H}, \quad \text{in} \;\; Q_\varepsilon \\
&\mathrm{div}\boldsymbol{A} + \phi = 0, \quad \text{in} \;\; Q_\varepsilon \\
&\nabla\psi\cdot\boldsymbol{n} = 0, \quad \boldsymbol{A}\cdot\boldsymbol{n} = 0, \quad \mathbf{curl}\boldsymbol{A}\times\boldsymbol{n} = \boldsymbol{H}\times\boldsymbol{n}, \quad \text{on} \;\; \Gamma_\varepsilon \\
&\psi(x,0) = \psi_0(x), \quad \boldsymbol{A}(x,0) = \boldsymbol{A}_0(x), \quad \text{in} \;\; \Omega_\varepsilon
\end{aligned}
\tag{2.1}
$$

where (2.1-3) is the choice of gauge.

The assumptions on the data are:

(A1) $\psi_0 \in \mathcal{H}^1(\Omega_\varepsilon)$, $|\psi_0| \leq 1$ a.e. in $\Omega_\varepsilon$;

(A2) $\boldsymbol{A}_0 \in \boldsymbol{H}^1_n(\Omega_\varepsilon)$;

(A3) $\boldsymbol{H} \in \boldsymbol{H}^1(0,T;\boldsymbol{L}^2(\Omega_\varepsilon))$.

It is interesting to see that the convergence proof is slightly more complicated than the G-L + Maxwell case dealt with in Section 1.

### 8.2.1 Existence and uniqueness of solutions

First, the domain $\Omega_\varepsilon$ is not a $C^2$ domain, second, we have the given magnetic field on the boundary, so we give a quick description on how to prove the existence of solution in this case. As the thickness parameter $\varepsilon$ is considered to be a constant in this section, to simplify the notations, we use $\Omega$ for $\Omega_\varepsilon$ and drop the dependence of solutions on $\varepsilon$. To begin with, we define

$$
\begin{aligned}
\mathcal{W} &= \mathcal{L}^2(0,T;\mathcal{H}^1(\Omega)) \cap \mathcal{H}^1(0,T;(\mathcal{H}^1(\Omega))'), \\
\boldsymbol{W}_n &= \boldsymbol{L}^2(0,T;\boldsymbol{H}^1_n(\Omega)) \cap \boldsymbol{H}^1(0,T;(\boldsymbol{H}^1_n(\Omega))').
\end{aligned}
\tag{2.2}
$$

The weak formulation for the system (2.1) is then to find $(\psi,\boldsymbol{A}) \in \mathcal{W}\times\boldsymbol{W}_n$ such that

$$
\psi(x,0) = \psi_0(x), \;\; \boldsymbol{A}(x,0) = \boldsymbol{A}_0(x) \tag{2.3}
$$

and

$$
\begin{aligned}
&\int_0^T\int_\Omega \frac{\partial\psi}{\partial t}\omega dxdt - i\int_0^T\int_\Omega \mathrm{div}\boldsymbol{A}\psi\omega dxdt \\
+\;& \int_0^T\int_\Omega \Big(i\nabla\psi + \boldsymbol{A}\psi\Big)\Big(-i\nabla\omega + \boldsymbol{A}\omega\Big)dxdt \\
+\;& \int_0^T\int_\Omega \kappa^2(|\psi|^2-1)\psi\omega dxdt = 0, \quad \text{for any } \omega \in \mathcal{L}^2(0,T;\mathcal{H}^1(\Omega))
\end{aligned}
\tag{2.4}
$$

$$\begin{aligned}
&\int_0^T\int_\Omega \frac{\partial \boldsymbol{A}}{\partial t}\boldsymbol{B}dxdt + \int_0^T\int_\Omega \big[\mathbf{curl}\boldsymbol{A}\mathbf{curl}\boldsymbol{B} + \mathrm{div}\boldsymbol{A}\mathrm{div}\boldsymbol{B}\big]dxdt \\
+ &\int_0^T\int_\Omega Re\Big[\Big(i\nabla\psi + \boldsymbol{A}\psi\Big)\bar{\psi}\Big]\boldsymbol{B}dxdt \\
= &\int_0^T\int_\Omega \boldsymbol{H}\mathbf{curl}\boldsymbol{B}dxdt, \quad \text{for any } \boldsymbol{B}\in \boldsymbol{L}^2(0,T;\boldsymbol{H}^1_n(\Omega)). \qquad (2.5)
\end{aligned}$$

**Theorem 8.2.1** *Let the assumptions (A1)–(A3) be satisfied. Then (2.3)–(2.5) has a unique pair of solutions* $(\psi, \boldsymbol{A})$ *satisfying*

$$\psi \in \mathcal{L}^\infty(0,T;\mathcal{H}^1(\Omega)) \cap \mathcal{H}^1(0,T;\mathcal{L}^2(\Omega)); \qquad (2.6)$$
$$\boldsymbol{A} \in \boldsymbol{L}^\infty(0,T;\boldsymbol{H}^1_n(\Omega)) \cap \boldsymbol{H}^1(0,T;\boldsymbol{L}^2(\Omega)); \qquad (2.7)$$
$$|\psi| \le 1 \text{ almost everywhere on } \Omega\times(0,T). \qquad (2.8)$$

The theorem is the consequence of the lemmas in the following:

**Lemma 8.2.1** (UNIQUENESS) *The solution* $(\psi, \boldsymbol{A})$ *of (2.3)–(2.5) satisfying (2.6)–(2.8) is unique.*

*Proof.* Let $(\psi_1, \boldsymbol{A}_1)$ and $(\psi_2, \boldsymbol{A}_2)$ be two solutions of (2.3)–(2.5) satisfying (2.6)–(2.8) and set $\widetilde{\psi} = \psi_1 - \psi_2$, $\widetilde{\boldsymbol{A}} = \boldsymbol{A}_1 - \boldsymbol{A}_2$. Then, subtracting the corresponding equations, we have

$$\begin{aligned}
&\int_0^T\int_\Omega \frac{\partial\widetilde{\psi}}{\partial t}\omega dxdt + \int_0^T\int_\Omega \nabla\widetilde{\psi}\nabla\omega dxdt = i\int_0^T\int_\Omega (\mathrm{div}\widetilde{\boldsymbol{A}}\psi_1 + \mathrm{div}\boldsymbol{A}_2\widetilde{\psi})\omega dxdt \\
&-i\int_0^T\int_\Omega \Big(\widetilde{\boldsymbol{A}}\nabla\psi_1 + \boldsymbol{A}_2\nabla\widetilde{\psi}\Big)\omega dxdt + i\int_0^T\int_\Omega (\widetilde{\boldsymbol{A}}\psi_1 + \boldsymbol{A}_2\widetilde{\psi})\nabla\omega dxdt \\
&-\int_0^T\int_\Omega [\widetilde{\boldsymbol{A}}(\boldsymbol{A}_1 + \boldsymbol{A}_2)\psi_1 + |\boldsymbol{A}_2|^2\widetilde{\psi}]\omega dxdt \qquad (2.9) \\
&+\kappa^2\int_0^T\int_\Omega \Big[(|\psi_1|^2 + |\psi_2|^2 - 1)\widetilde{\psi} + \psi_1\psi_2\overline{\widetilde{\psi}}\Big]\omega dxdt \\
&= (\mathrm{I})_1 + (\mathrm{I})_2 + (\mathrm{I})_3 + (\mathrm{I})_4 + (\mathrm{I})_5. \qquad (2.10)
\end{aligned}$$

$$\begin{aligned}
&\int_0^T\int_\Omega \frac{\partial\widetilde{\boldsymbol{A}}}{\partial t}\boldsymbol{B}dxdt + \int_0^T\int_\Omega \big[\mathbf{curl}\widetilde{\boldsymbol{A}}\mathbf{curl}\boldsymbol{B} + \mathrm{div}\widetilde{\boldsymbol{A}}\mathrm{div}\boldsymbol{B}\big]dxdt \\
&= -\int_0^T\int_\Omega Re\Big[\Big(i\nabla\widetilde{\psi} + \widetilde{\boldsymbol{A}}\psi_1 + \boldsymbol{A}_2\widetilde{\psi}\Big)\bar{\psi}_1\Big]\boldsymbol{B}dxdt \\
&-\int_0^T\int_\Omega Re\Big[\Big(i\nabla\psi_2 + \boldsymbol{A}_2\psi_2\Big)\bar{\widetilde{\psi}}\Big]\boldsymbol{B}dxdt \\
&= (\mathrm{II})_1 + (\mathrm{II})_2. \qquad (2.11)
\end{aligned}$$

We note that since $(\psi_1, \boldsymbol{A}_1)$, $(\psi_2, \boldsymbol{A}_2)$ satisfy (2.6)–(2.8), we have

$$\|\psi_j\|_{L^\infty(\Omega)} \le 1, \quad \|\psi_j\|_{H^1(\Omega)} \le c, \quad \|\boldsymbol{A}_j\|_{L^4(\Omega)} \le c, \quad \|\boldsymbol{A}_j\|_{H^1(\Omega)} \le c$$

for $j = 1$, 2 and for almost every $t \in (0, T)$. Let $\omega = \bar{\widetilde{\psi}}\chi_{(0,t)}$ in (2.9) and take the real part of the equation, then

$$\begin{aligned}
|Re(\mathrm{I})_1| &\le \frac{1}{4}\int_0^t\int_\Omega |\mathrm{div}\widetilde{\boldsymbol{A}}|^2 dxdt + c\int_0^t\int_\Omega |\widetilde{\psi}|^2 dxdt;\\
|Re(\mathrm{I})_2| &\le \int_0^t \Big[\|\widetilde{\boldsymbol{A}}\|_{L^4(\Omega)}\|\nabla\psi_1\|_{L^2(\Omega)}\|\widetilde{\psi}\|_{L^4(\Omega)}\\
&\quad +\|\boldsymbol{A}_2\|_{L^4(\Omega)}\|\nabla\widetilde{\psi}\|_{L^2(\Omega)}\|\widetilde{\psi}\|_{L^4(\Omega)}\Big]dt\\
&\le \frac{1}{4}\int_0^t\int_\Omega |\nabla\widetilde{\psi}|^2 dxdt + c\int_0^t \Big[\|\widetilde{\boldsymbol{A}}\|^2_{L^4(\Omega)} + \|\widetilde{\psi}\|^2_{L^4(\Omega)}\Big]dt;\\
|Re(\mathrm{I})_3| &\le \int_0^t \Big[\|\widetilde{\boldsymbol{A}}\|_{L^2(\Omega)}\|\psi_1\|_{L^\infty(\Omega)}\|\nabla\widetilde{\psi}\|_{L^2(\Omega)}\\
&\quad +\|\boldsymbol{A}_2\|_{L^4(\Omega)}\|\widetilde{\psi}\|_{L^4(\Omega)}\|\nabla\widetilde{\psi}\|_{L^2(\Omega)}\Big]dt\\
&\le \frac{1}{4}\int_0^t\int_\Omega |\nabla\widetilde{\psi}|^2 dxdt + c\int_0^t\int_\Omega |\widetilde{\boldsymbol{A}}|^2 dxdt + c\int_0^t \|\widetilde{\psi}\|^2_{L^4(\Omega)}dt;\\
|Re(\mathrm{I})_4| &\le \int_0^t \Big[\|\widetilde{\boldsymbol{A}}\|_{L^4(\Omega)}\|\boldsymbol{A}_1 + \boldsymbol{A}_2\|_{L^4(\Omega)}\|\psi_1\|_{L^\infty(\Omega)}\|\widetilde{\psi}\|_{L^2(\Omega)}\\
&\quad +\|\boldsymbol{A}_2\|^2_{L^4(\Omega)}\|\widetilde{\psi}\|^2_{L^4(\Omega)}\Big]dt\\
&\le c\int_0^t \Big[\|\widetilde{\boldsymbol{A}}\|^2_{L^4(\Omega)} + \|\widetilde{\psi}\|^2_{L^4(\Omega)} + \|\widetilde{\psi}\|^2_{L^2(\Omega)}\Big]dt;\\
|Re(\mathrm{I})_5| &\le c\int_0^t\int_\Omega |\widetilde{\psi}|^2 dxdt.
\end{aligned}$$

By Nirenberg's inequality, we have

$$\|\widetilde{\psi}\|_{L^4(\Omega)} \le c\|\widetilde{\psi}\|^{3/4}_{H^1(\Omega)}\|\widetilde{\psi}\|^{1/4}_{L^2(\Omega)} \le \delta\|\nabla\widetilde{\psi}\|_{L^2(\Omega)} + c_\delta\|\widetilde{\psi}\|_{L^2(\Omega)},$$

and, by Lemma 2.11 in the following,

$$\begin{aligned}
\|\widetilde{\boldsymbol{A}}\|_{L^4(\Omega)} &\le c\|\widetilde{\boldsymbol{A}}\|^{3/4}_{H^1(\Omega)}\|\widetilde{\boldsymbol{A}}\|^{1/4}_{L^4(\Omega)}\\
&\le \delta\Big[\|\mathrm{div}\widetilde{\boldsymbol{A}}\|_{L^2(\Omega)} + \|\mathbf{curl}\widetilde{\boldsymbol{A}}\|_{L^2(\Omega)}\Big] + c_\delta\|\widetilde{\boldsymbol{A}}\|_{L^2(\Omega)}.
\end{aligned}$$

Consequently, by choosing $\delta$ appropriately, we have

$$\begin{aligned}
&\frac{1}{4}\int_\Omega |\widetilde{\psi}|^2(x,t)dx + \frac{1}{\kappa^2}\int_0^t\int_\Omega |\nabla\widetilde{\psi}|^2 dxdt\\
\le &\int_0^t\int_\Omega \Big[\frac{3}{4}|\nabla\widetilde{\psi}|^2 + \frac{1}{2}|\mathbf{curl}\widetilde{\boldsymbol{A}}|^2 + \frac{1}{2}|\mathrm{div}\widetilde{\boldsymbol{A}}|^2 + c(|\widetilde{\boldsymbol{A}}|^2 + |\widetilde{\psi}|^2)\Big]dxdt. \qquad (2.12)
\end{aligned}$$

Similarly, taking $\boldsymbol{B} = \widetilde{\boldsymbol{A}}$ in (2.10), we obtain

$$\frac{1}{2}\int_\Omega |\widetilde{\boldsymbol{A}}|^2(x,t)dx + \int_0^t\int_\Omega \Big[|\mathbf{curl}\widetilde{\boldsymbol{A}}|^2 + |\mathrm{div}\widetilde{\boldsymbol{A}}|^2\Big]dxdt$$
$$\leq \int_0^t\int_\Omega \Big[\frac{1}{8}|\nabla\widetilde{\psi}|^2 + \frac{1}{4}|\mathbf{curl}\widetilde{\boldsymbol{A}}|^2 + \frac{1}{4}|\mathrm{div}\widetilde{\boldsymbol{A}}|^2 + c(|\widetilde{\boldsymbol{A}}|^2 + |\widetilde{\psi}|^2)\Big]dxdt. \tag{2.13}$$

Now uniqueness follows from (2.11)–(2.12) by using Gronwall's inequality.

In order to show the existence of the solutions of (2.3)–(2.5), we introduce the following semi-discretized approximation problem: let $N \geq 1$ be an integer, $\Delta t = T/N$ be the step size, $\boldsymbol{H}_j = \boldsymbol{H}(x, j\Delta t)$ for $j = 0, 1, 2, \cdots, N$. The approximation problem is then to find $(\psi_j, \boldsymbol{A}_j) \in \mathcal{H}^1(\Omega) \times \boldsymbol{H}^1_n(\Omega), j = 1, 2, \ldots, N$ such that

$$\int_\Omega \frac{\psi_j - \psi_{j-1}}{\Delta t}\omega dx - i\int_\Omega \mathrm{div}\boldsymbol{A}_j\psi_j\omega dx$$
$$+\int_\Omega \Big(i\nabla\psi_j + \boldsymbol{A}_j\psi_j\Big)\Big(-i\nabla\omega + \boldsymbol{A}_j\omega\Big)dx$$
$$+\int_\Omega \kappa^2(|\psi_j|^2 - 1)\psi_j\omega dx = 0, \quad \text{for any } \omega \in \mathcal{H}^1(\Omega). \tag{2.14}$$

$$\int_\Omega \frac{\boldsymbol{A}_j - \boldsymbol{A}_{j-1}}{\Delta t}\boldsymbol{B}dx + \int_\Omega \Big[\mathbf{curl}\boldsymbol{A}_j\mathbf{curl}\boldsymbol{B} + \mathrm{div}\boldsymbol{A}_j\mathrm{div}\boldsymbol{B}\Big]dx$$
$$+\int_\Omega Re\Big[\Big(i\nabla\psi_{j-1} + \boldsymbol{A}_{j-1}\psi_{j-1}\Big)\bar{\psi}_{j-1}\Big]\boldsymbol{B}dx$$
$$= \int_\Omega \boldsymbol{H}_j\mathbf{curl}\boldsymbol{B}dx, \quad \text{for any } \boldsymbol{B} \in \boldsymbol{H}^1_n(\Omega), \tag{2.15}$$

where $(\psi_0, \boldsymbol{A}_0)$ is given by (2.3).

**Lemma 8.2.2** (EXISTENCE OF SOLUTIONS OF (2.13)–(2.14)) *Let $\Delta t > 0$ be sufficiently small. Then the problem (2.13)–(2.14) has a unique solution $(\psi_j, \boldsymbol{A}_j) \in \mathcal{H}^1(\Omega) \times \boldsymbol{H}^1_n(\Omega)$ for each $j = 1, \ldots, N$.*

*Proof.* We first notice that (2.13)–(2.14) are independent of each other. (2.14) is a standard linear elliptic problem for $\boldsymbol{A}_j$ with $\psi_{j-1}$ given by the previous step. Thus the existence and uniqueness of $\boldsymbol{A}_j \in \boldsymbol{H}^1_n(\Omega)$ follows from a standard argument. When $\boldsymbol{A}_j$ is determined, (2.13) is a semilinear elliptic problem with respect to $\psi_j$. The existence and uniqueness of $\psi_j$ again follows by a standard argument.

**Lemma 8.2.3** *For any $j \geq 1$, $|\psi_j| \leq 1$ for almost every $x \in \Omega$.*

*Proof.* This lemma can be proved by taking $\omega = (|\psi_j|^2 - 1)_+\bar{\psi}_j$ in (2.13) and using the method in Chapter 4.

In the following, $c$ is used to denote various constants independent of $N$, $\Delta t$ *and* $\varepsilon$. We show a number of lemmas which will enable us to take the limit in (2.13)–(2.14) when $N \to \infty$ and consequently, prove that (2.3)–(2.5) admits solutions.

**Lemma 8.2.4** *We have*

$$\max_{1\le j\le N}\int_\Omega |\psi_j|^2dx+\sum_{j=1}^{N}\Delta t\int_\Omega\left|\frac{i}{\kappa}\nabla\psi_j+\boldsymbol{A}_j\psi_j\right|^2dx\le c\int_\Omega|\psi_0|^2dx.$$

*Proof.* Take $\omega=\bar{\psi}_j\Delta t$ in (2.13).

**Lemma 8.2.5** *We have*

$$\max_{1\le j\le N}\int_\Omega|\boldsymbol{A}_j|^2dx+\sum_{j=1}^{N}\Delta t\int_\Omega\left[|\mathrm{div}\boldsymbol{A}_j|^2+|\mathbf{curl}\boldsymbol{A}_j|^2\right]dx$$
$$\le c\int_\Omega\left[|\boldsymbol{A}_0|^2+|\psi_0|^2\right]dx+c\left[\|\boldsymbol{H}\|^2_{H^1(0,T;L^2(\Omega))}+\|\boldsymbol{H}_0\|^2_{L^2(\Omega)}\right].$$

*Proof.* Taking $\boldsymbol{B}=\boldsymbol{A}_j\Delta t$ in (2.14) and applying Lemma 2.4, we obtain

$$\sum_{j=1}^{N}\Delta t\int_\Omega|\boldsymbol{H}_j|^2dx\le c\left[\|\boldsymbol{H}_0\|^2_{L^2(\Omega)}+\|\boldsymbol{H}\|^2_{H^1(0,T;L^2(\Omega))}\right].$$

This completes the proof.

**Lemma 8.2.6** *We have*

$$\sum_{j=1}^{N}\Delta t\int_\Omega|\nabla\psi_j|^2dx\le c\int_\Omega\left[|\boldsymbol{A}_0|^2+|\psi_0|^2\right]dx+c\left[\|\boldsymbol{H}\|^2_{H^1(0,T;L^2(\Omega))}+\|\boldsymbol{H}_0\|^2_{L^2(\Omega)}\right].$$

*Proof.* Use the results of Lemmas 2.4 and 2.5.

**Lemma 8.2.7** *Let $\partial\boldsymbol{A}_j=(\boldsymbol{A}_j-\boldsymbol{A}_{j-1})/\Delta t$. We have*

$$\sum_{j=1}^{N}\Delta t\int_\Omega|\partial\boldsymbol{A}_j|^2dx+\max_{1\le j\le N}\int_\Omega\left[|\mathrm{div}\boldsymbol{A}_j|^2+|\mathbf{curl}\boldsymbol{A}_j|^2\right]dx$$
$$\le\ c\left[\|\boldsymbol{A}_0\|^2_{H^1(\Omega)}+\|\psi_0\|^2_{L^2(\Omega)}+\|\boldsymbol{H}\|^2_{H^1(0,T;L^2(\Omega))}+\|\boldsymbol{H}_0\|^2_{L^2(\Omega)}\right].$$

*Proof.* Let $\boldsymbol{B}=\Delta t\partial\boldsymbol{A}_j$ in (2.14), then for any $m\le N$,

$$\sum_{j=1}^{m}\Delta t\int_\Omega\left[|\partial\boldsymbol{A}_j|^2+\frac{1}{2}|\mathrm{div}\boldsymbol{A}_m|^2+\frac{1}{2}|\mathbf{curl}\boldsymbol{A}_m|^2\right]dx$$
$$\le\frac{1}{2}\int_\Omega\left[|\mathrm{div}\boldsymbol{A}_0|^2+|\mathbf{curl}\boldsymbol{A}_0|^2\right]dx$$
$$-\sum_{j=1}^{m}\Delta t\int_\Omega Re\left[\left(i\nabla\psi_{j-1}+\boldsymbol{A}_{j-1}\psi_{j-1}\right)\bar{\psi}_{j-1}\right]\partial\boldsymbol{A}_jdx$$
$$+\sum_{j=1}^{m}\int_\Omega\boldsymbol{H}_j\mathbf{curl}(\boldsymbol{A}_j-\boldsymbol{A}_{j-1})dx$$
$$=\ (\mathrm{III})_1+(\mathrm{III})_2+(\mathrm{III})_3.$$

Use the results of Lemma 2.4, we have

$$
\begin{aligned}
(\mathrm{III})_2 &\le \frac{1}{2}\sum_{j=1}^{m}\Delta t\Big[\Big\|i\nabla\psi_{j-1}+\boldsymbol{A}_{j-1}\psi_{j-1}\Big\|^2_{L^2(\Omega)}+\|\partial\boldsymbol{A}_j\|^2_{L^2(\Omega)}\Big]\\
&\le \frac{1}{2}\sum_{j=1}^{m}\Delta t\|\partial\boldsymbol{A}_j\|^2_{L^2(\Omega)}+c\|\psi_0\|^2_{L^2(\Omega)}.
\end{aligned}
$$

$$
\begin{aligned}
(\mathrm{III})_3 &= \int_\Omega \boldsymbol{H}_m\mathbf{curl}\boldsymbol{A}_m dx-\int_\Omega \boldsymbol{H}_0\mathbf{curl}\boldsymbol{A}_0 dx-\sum_{n=1}^{m}\int_\Omega(\boldsymbol{H}_j-\boldsymbol{H}_{j-1})\mathbf{curl}\boldsymbol{A}_{j-1}dx\\
&\le \frac{1}{4}\int_\Omega|\mathbf{curl}\boldsymbol{A}_m|^2dx+\|\boldsymbol{H}_m\|^2_{L^2(\Omega)}+\|\boldsymbol{H}_0\|^2_{L^2(\Omega)}+\|\mathbf{curl}\boldsymbol{A}_0\|^2_{L^2(\Omega)}\\
&\quad+\sum_{n=1}^{m}\Delta t\Big[\|\partial\boldsymbol{H}_j\|^2_{L^2(\Omega)}+\frac{1}{4}\|\mathbf{curl}\boldsymbol{A}_{j-1}\|^2_{L^2(\Omega)}\Big]\\
&\le \frac{1}{4}\int_\Omega|\mathbf{curl}\boldsymbol{A}_m|^2dx+\frac{1}{4}\sum_{n=1}^{m}\Delta t\|\mathbf{curl}\boldsymbol{A}_{j-1}\|^2_{L^2(\Omega)}\\
&\quad+c\Big[\|\boldsymbol{H}_0\|^2_{L^2(\Omega)}+\|\boldsymbol{A}_0\|^2_{H^1(\Omega)}+\|\boldsymbol{H}\|^2_{H^1(0,T;L^2(\Omega))}\Big].
\end{aligned}
$$

The Lemma then follows from the Gronwall's inequality.

**Lemma 8.2.8** *Let* $\partial\psi_j=(\psi_j-\psi_{j-1})/\Delta t$. *We have*

$$
\begin{aligned}
&\sum_{j=1}^{N}\Delta t\int_\Omega|\partial\psi_j|^2dx+\max_{1\le j\le N}\int_\Omega|\nabla\psi_j|^2dx\\
\le\ & c\Big[\|\boldsymbol{A}_0\|^2_{H^1(\Omega)}+\|\psi_0\|^2_{H^1(\Omega)}+\|\boldsymbol{H}\|^2_{H^1(0,T;L^2(\Omega))}+\|\boldsymbol{H}_0\|^2_{L^2(\Omega)}\Big].
\end{aligned}
$$

*Proof.* Take $w=\partial\bar{\psi}_j\Delta t$ in (2.13) and apply Lemmas 2.3–2.7.

*Proof of Theorem 2.1.* The uniqueness has been proved in Lemma 2.1. The existence of the solutions $(\psi,\boldsymbol{A})$ satisfying (2.3)–(2.5) can be proved from the estimates given in Lemmas 2.3–2.8 by employing standard convergence argument.

As a direct consequence of Theorem 2.1 and Lemmas 2.3–2.8 we obtain the following result.

**Corollary 8.2.1** *The solution* $(\psi,\boldsymbol{A})$ *of (2.3)–(2.5) satisfies the following estimate*

$$
\begin{aligned}
&\mathrm{esssup}_{0\le t\le T}\int_\Omega\Big[|\psi|^2+|\boldsymbol{A}|^2+|\nabla\psi|^2+|\mathrm{div}\boldsymbol{A}|^2+|\mathbf{curl}\boldsymbol{A}|^2\Big]dx\\
&\quad+\int_0^T\int_\Omega\Big[\Big|\frac{\partial\psi}{\partial t}\Big|^2+\Big|\frac{\partial\boldsymbol{A}}{\partial t}\Big|^2\Big]dxdt\\
&\le c\Big[\|\psi_0\|^2_{H^1(\Omega)}+\|\boldsymbol{A}_0\|^2_{H^1(\Omega)}+\|\boldsymbol{H}_0\|^2_{L^2(\Omega)}\Big]+c\int_0^T\int_\Omega\Big|\frac{\partial\boldsymbol{H}}{\partial t}\Big|^2dxdt,
\end{aligned}
$$

*where the constant $c$ is independent of $\varepsilon$.*

### 8.2.2 The limit when $\varepsilon \to 0$

From now on, $\Omega = \Omega_\varepsilon$ and we denote by $(\psi_\varepsilon, \boldsymbol{A}_\varepsilon)$ the solutions of the system (2.3)–(2.5). Let $\rho(x') = a(x') + b(x')$. For any (real, complex and/or vector valued) function $f(x,t) = f(x', x_3, t)$ defined on $\Omega_\varepsilon \times (0,T)$, we define the average operator

$$[\![ f ]\!](x',t) = \frac{1}{\varepsilon\rho(x')} \int_{-\varepsilon b(x')}^{\varepsilon a(x')} f(x', x_3, t) dx_3.$$

The purpose of this section is to prove that as $\varepsilon \to 0$, the average $([\![ \psi_\varepsilon ]\!], [\![ \boldsymbol{A}_\varepsilon ]\!])$ of the solutions $(\psi_\varepsilon, \boldsymbol{A}_\varepsilon)$ of the three dimensional problem (2.3)–(2.5) converge to solutions of a two dimensional problem.

We first describe precisely the assumptions on the boundary $\partial\Omega_\varepsilon$. Let

$$\begin{aligned}
\Gamma_1 &= \{ (x', x_3) \in \partial\Omega_\varepsilon : \ x_3 = \varepsilon a(x'), \ x' \in \Omega_0 \}, \\
\Gamma_2 &= \{ (x', x_3) \in \partial\Omega_\varepsilon : \ x_3 = -\varepsilon b(x'), \ x' \in \Omega_0 \}, \\
\Gamma_3 &= \{ (x', x_3) \in \partial\Omega_\varepsilon : \ x_3 \in (-\varepsilon b(x'), \varepsilon a(x')), \ x' \in \partial\Omega_0 \},
\end{aligned}$$

then it is clear that

$$\partial\Omega_\varepsilon = \Gamma_1 \cup \Gamma_2 \cup \Gamma_3 \cup (\bar\Gamma_1 \cap \bar\Gamma_3) \cup (\bar\Gamma_2 \cap \bar\Gamma_3).$$

We impose the following corner conditions for the boundary parts $\bar\Gamma_1 \cap \bar\Gamma_3, \bar\Gamma_2 \cap \bar\Gamma_3$. We point out that each $\Gamma_j$ depends on $\varepsilon$, just for notational convenience, we drop the subscript $\varepsilon$.

**Hypothesis (H)**: *For every $x_0 \in \bar\Gamma_1 \cap \bar\Gamma_3$ (similar for $x_0 \in \bar\Gamma_2 \cap \bar\Gamma_3$) there exist a neighbourhood $V$ of $x_0 \in I\!R^3$ and a mapping $\boldsymbol{\eta} = (\eta_1, \eta_2, \eta_3)$ from $V$ into $I\!R^3$ such that $\boldsymbol{\eta}$ is injective, $\boldsymbol{\eta}$ and $\boldsymbol{\eta}^{-1}$ (defined on $\boldsymbol{\eta}(V)$) are continuously differentiable with their $C^1$ norms bounded independent of $\varepsilon$, and*

$$\begin{aligned}
\Omega_\varepsilon \cap V &= \{ x \in \Omega_\varepsilon \ : \ \eta_1(x) > 0, \eta_3(x) < 0 \} \\
\Gamma_1 \cap V &= \{ x \in \partial\Omega_\varepsilon \ : \ \eta_1(x) > 0, \eta_3(x) = 0 \} \\
\Gamma_3 \cap V &= \{ x \in \partial\Omega_\varepsilon \ : \ \eta_1(x) = 0, \eta_3(x) < 0 \} \\
\bar\Gamma_1 \cap \bar\Gamma_3 \cap V &= \{ x \in \partial\Omega_\varepsilon \ : \ \eta_1(x) = 0, \eta_3(x) = 0 \}.
\end{aligned}$$

*Furthermore, we assume that the matrices $W = (\nabla_x \boldsymbol{\eta})^{-1}$ and $W^{-1} \in C^1(\bar V)$, det $W \neq 0$ such that*

$$\begin{aligned}
&W|_{\Gamma_1 \cap V} = (-\varepsilon a_x, -\varepsilon a_y, 1)^T / \sqrt{\varepsilon^2 |\nabla a|^2 + 1} = (0,0,1)^T, \\
&W|_{\Gamma_3 \cap V} \mathbf{n} = (-1, 0, 0)^T
\end{aligned}$$

*where $\mathbf{n}$ denotes the unit outward normal vector of $\partial\Omega_\varepsilon$ along $\Gamma_3 \cap V$.*

The geometric meaning of this hypothesis is fairly obvious. We just point out here that if $a(x') \equiv$ constant, $b(x') \equiv$ constant (the case of a plate), it is straightforward to verify that Hypothesis (H) is satisfied provided that $\partial\Omega_0$ is sufficiently smooth.

The second point we want to make is that the assumption $\det W > 0$ in (H) prevents that the edge of the plate forms a re-entrant angle. This is automatically satisfied provided that $a$ and $b$ are $C^1(\bar{\Omega}_0)$ functions.

We note that $(V, \boldsymbol{\eta})$ and $W$ are all functions depending on $\varepsilon$. We dropped this dependence for notational convenience. In the following proofs, we have made sure that the proofs will not be affected by this point.

In order to derive the two-dimensional model, we make the following assumptions in this section in addition to (A1)–(A3):

(A1)′ $\psi_0 = \psi_0(x')$;
(A2)′ $\boldsymbol{A}_0 = (\boldsymbol{A}_0^1(x'), \boldsymbol{A}_0^2(x'), 0)^T$;
(A3)′ $\boldsymbol{H} = (0, 0, H(x', t))^T$.

The following theorem is the main result of this section.

**Theorem 8.2.2** *Let the assumptions (A1)–(A3) and (A1)′–(A3)′ be satisfied. For any $\varepsilon > 0$, let $(\psi_\varepsilon, \boldsymbol{A}_\varepsilon)$ denote the solution of (2.2)–(2.5). Then, there exist two functions*

$$
\begin{aligned}
&\psi \in \mathcal{H}^1(0,T;\mathcal{L}^2(\Omega_0)) \cap \mathcal{L}^\infty(0,T;\mathcal{H}^1(\Omega_0))\\
&\textit{and}\\
&\boldsymbol{A}' \in \boldsymbol{H}^1(0,T;\boldsymbol{L}^2(\Omega_0)) \cap \boldsymbol{L}^\infty(0,T;\boldsymbol{H}_n^1(\Omega_0))
\end{aligned}
$$

*such that as $\varepsilon \to 0$*

$$
\begin{aligned}
[\![\, \psi_\varepsilon \,]\!] \to \psi \qquad & \textit{weakly in } \mathcal{H}^1(0,T;\mathcal{L}^2(\Omega_0))\\
& \textit{and weakly* in } \mathcal{L}^\infty(0,T;\mathcal{H}^1(\Omega_0)), \qquad (2.16)\\
[\![\, \boldsymbol{A}_\varepsilon \,]\!] \to \boldsymbol{A} \qquad & \textit{weakly in } \boldsymbol{H}^1(0,T;\boldsymbol{L}^2(\Omega_0))\\
& \textit{and weakly* in } \boldsymbol{L}^\infty(0,T;\boldsymbol{H}_n^1(\Omega_0)), \qquad (2.17)
\end{aligned}
$$

*where $\boldsymbol{A} = (\boldsymbol{A}', 0)$, and $(\psi, \boldsymbol{A}')$ satisfies*

$$
\psi(x',0) = \psi_0(x'), \quad \boldsymbol{A}'(x',0) = \boldsymbol{A}_0'(x') \tag{2.18}
$$

*and*

$$
\begin{aligned}
&\int_0^T \int_{\Omega_0} \rho(x') \frac{\partial \psi}{\partial t} \omega dx' dt - i \int_0^T \int_{\Omega_0} \operatorname{div}' \big(\rho(x') \boldsymbol{A}'\big) \psi \omega dx' dt\\
+ &\int_0^T \int_{\Omega_0} \rho(x') \Big( i \nabla' \psi + \boldsymbol{A}' \psi \Big) \Big( - i \nabla' \omega + \boldsymbol{A}' \omega \Big) dx' dt\\
+ &\int_0^T \int_{\Omega_0} \kappa^2 \rho(x') (|\psi|^2 - 1) \psi \omega dx' dt = 0\\
&\textit{for any } \omega \in \mathcal{L}^2(0,T;\mathcal{H}^1(\Omega_0)). \qquad (2.19)
\end{aligned}
$$

$$\begin{aligned}
&\int_0^T\int_{\Omega_0}\rho(x')\frac{\partial \boldsymbol{A}'}{\partial t}\boldsymbol{B}dx'dt+\int_0^T\int_{\Omega_0}\rho(x')\mathrm{curl}\boldsymbol{A}'\mathrm{curl}\boldsymbol{B}dx'dt\\
+&\int_0^T\int_{\Omega_0}\mathrm{div}'\big(\rho(x')\boldsymbol{A}'\big)\mathrm{div}'\boldsymbol{B}dx'dt\\
+&\int_0^T\int_{\Omega_0}\rho(x')Re\Big[\Big(i\nabla'\psi+\boldsymbol{A}'\psi\Big)\bar{\psi}\Big]\boldsymbol{B}dx'dt\\
=&\int_0^T\int_{\Omega_0}\rho(x')H\mathrm{curl}\boldsymbol{B}dx'dt,\quad \textit{for any } \boldsymbol{B}\in \boldsymbol{L}^2(0,T;\boldsymbol{H}^1_n(\Omega_0)). \qquad (2.20)
\end{aligned}$$

*Moreover, the electric potential $\phi_\varepsilon$ given in (2.1-3) satisfies, as $\varepsilon\to 0$,*

$$[\![\,\phi_\varepsilon\,]\!]\to-\frac{1}{\rho(x')}\mathrm{div}'\big(\rho(x')\boldsymbol{A}'\big)\quad \textit{weakly in } L^2(0,T;L^2(\Omega_0)). \qquad (2.21)$$

**Remark 8.2.1** *(2.18)–(2.20) is the weak formulation of the following problem:*

$$\begin{aligned}
&\rho(x')\frac{\partial\psi}{\partial t}-i\mathrm{div}'\big(\rho(x')\boldsymbol{A}'\big)\psi+\Big(i\nabla'+\boldsymbol{A}'\Big)\Big[\rho(x')\Big(i\nabla'+\boldsymbol{A}'\Big)\Big]\psi\\
&\qquad+\kappa^2\rho(x')(|\psi|^2-1)\psi=0,\quad in\ Q_0\\
&\rho(x')\frac{\partial \boldsymbol{A}'}{\partial t}+\mathbf{curl}'\Big[\rho(x')\mathrm{curl}\boldsymbol{A}'\Big]-\nabla'\Big[\mathrm{div}'\big(\rho(x')\boldsymbol{A}'\big)\Big]\\
&\qquad+\rho(x')Re\Big[\Big(i\nabla'\psi+\boldsymbol{A}'\psi\Big)\bar{\psi}\Big]=\mathbf{curl}'\big(\rho(x')H\big),\quad in\ Q_0\\
&\nabla'\psi\cdot\boldsymbol{n}'=0,\quad \boldsymbol{A}'\cdot\boldsymbol{n}'=0,\quad \mathrm{curl}\boldsymbol{A}'=H,\quad on\ \Gamma_0\\
&\psi(x',0)=\psi_0(x'),\quad \boldsymbol{A}'(x',0)=\boldsymbol{A}'_0(x'),\quad in\ \Omega_0
\end{aligned}$$

*where $Q_0=\Omega_0\times(0,T)$, $\Gamma_0=\partial\Omega_0\times(0,T)$, and $\boldsymbol{n}'$ is the unit outer normal of $\partial\Omega_0$.*

**Remark 8.2.2** *It is clear from (2.18)–(2.20) that when $\rho(x')$ is a constant, we are able to derive the standard 2-dimensional G-L equations.*

The proof of Theorem 2.2 will be given at the end of this section. We close this subsection by stating the following theorem concerning the existence and uniqueness of solutions for (2.18)–(2.20).

**Theorem 8.2.3** *Let $\psi_0\in\mathcal{H}^1(\Omega_0)$, $\boldsymbol{A}'_0\in\boldsymbol{H}^1_n(\Omega_0)$ satisfying $|\psi_0|\le 1$ a.e. in $\Omega_0$. Assume that $H\in H^1(0,T;L^2(\Omega_0))$. Then (2.18)–(2.20) has a unique pair of solutions $(\psi,\boldsymbol{A}')$ satisfying*

$$\begin{aligned}
&\psi\in\mathcal{L}^\infty(0,T;\mathcal{H}^1(\Omega_0))\cap\mathcal{H}^1(0,T;\mathcal{L}^2(\Omega_0));\\
&\boldsymbol{A}'\in\boldsymbol{L}^\infty(0,T;\boldsymbol{H}^1_n(\Omega_0))\cap\boldsymbol{H}^1(0,T;\boldsymbol{L}^2(\Omega_0));
\end{aligned}$$

*and $|\psi|\le 1$ a.e. in $\Omega_0\times(0,T)$. Moreover, if $H\in H^1(0,T;H^1(\Omega_0))$, then we also have the regularity*

$$\psi\in\mathcal{L}^2(0,T;\mathcal{H}^2(\Omega_0)),\quad \boldsymbol{A}'\in\boldsymbol{L}^2(0,T;\boldsymbol{H}^2(\Omega_0)).$$

### 8.2.3 Some estimates

In the following, $c$ is used to denote the various constants independent of $\varepsilon$. We begin this subsection with some elementary results on the average operator $[\![ \cdot ]\!]$.

**Lemma 8.2.9** *Let $f \in L^2(\Omega_\varepsilon)$ be a given function, we have*

$$\|[\![ f ]\!]\|_{L^2(\Omega_0)} \leq \frac{c}{\sqrt{\varepsilon}} \|f\|_{L^2(\Omega_\varepsilon)}. \tag{2.22}$$

*Let $f \in H^1(\Omega_\varepsilon)$ be a given function. For almost every $x' \in \Omega_0$, we have*

$$|f(x', x_3) - [\![ f ]\!](x')| \leq c\sqrt{\varepsilon} \Big\| \frac{\partial f}{\partial x_3} \Big\|_{L^2(-\varepsilon b(x'), \varepsilon a(x'))}(x') \tag{2.23}$$

*for all $x_3 \in [-\varepsilon b(x'), \varepsilon a(x')]$ where $c$ is a constant independent of both $\varepsilon$ and $x'$. Consequently,*

$$\|f - [\![ f ]\!]\|_{L^2(\Omega_\varepsilon)} \leq c\varepsilon \Big\| \frac{\partial f}{\partial x_3} \Big\|_{L^2(\Omega_\varepsilon)}. \tag{2.24}$$

*Proof.* First, we note that by a regularity theorem due to Morrey and Necas (cf. [Mo 66], Chapter 3.1 or [Ne 67], Theorem 2.2.2, see also [MTY 93]): any $W^{1,1}(\Omega_\varepsilon)$ function $f(x', x_3)$ is absolutely continuous in $x_3$ for almost every $x' \in \Omega_0$ and the derivative in $x_3$ (which exists in an almost everywhere sense with respect to the Lebesgue measure of $\mathbb{R}^1$) coincides with the generalized derivative almost everywhere. Hence we have

$$f(x', x_3) - [\![ f ]\!](x') = \int_\xi^{x_3} \frac{\partial f}{\partial x_3} dx_3$$

for some $\xi \in (-\varepsilon b(x'), \varepsilon a(x'))$. Consequently, using the Cauchy-Schwartz inequality, we obtain

$$|f(x', x_3) - [\![ f ]\!](x')| \leq c\sqrt{\varepsilon} \Big\| \frac{\partial f}{\partial x_3} \Big\|_{L^2(-\varepsilon b(x'), \varepsilon a(x'))}(x').$$

It is straightforward now to obtain (2.24).

It is easy to see from (2.23) that for any $f \in H^1(\Omega_\varepsilon)$ and almost every $x' \in \Omega_0$ and $x_3 \in [-\varepsilon b(x'), \varepsilon a(x')]$, we have

$$|f(x', x_3)| \leq |[\![ f ]\!]|(x') + c\sqrt{\varepsilon} \Big\| \frac{\partial f}{\partial x_3} \Big\|_{L^2(-\varepsilon b(x'), \varepsilon a(x'))}(x'). \tag{2.25}$$

**Lemma 8.2.10** *Under Hypothesis (H), $\boldsymbol{H}^1_n(\Omega_\varepsilon) \cap \boldsymbol{H}^2(\Omega_\varepsilon)$ is dense in $\boldsymbol{H}^1_n(\Omega_\varepsilon)$.*

*Proof.* For any $\boldsymbol{Q} \in \boldsymbol{H}^1_n(\Omega_\varepsilon)$, with the help of a family of local charts and of the partition of unity, we need to consider only $\phi \boldsymbol{Q}$ with $\phi$ being a nonnegative

$C_0^\infty(\mathbb{R}^3)$ function such that $\mathrm{supp}\phi \cap \bar{\Gamma}_1 \cap \bar{\Gamma}_3 \neq \emptyset$ (same for $\mathrm{supp}\phi \cap \bar{\Gamma}_2 \cap \bar{\Gamma}_3 \neq \emptyset$ and simpler for $\mathrm{supp}\phi \cap \Gamma_j \neq \emptyset, j = 1, 2, 3$). In this case, let $(V, \boldsymbol{\eta})$ be the corresponding local chart in Hypothesis (H) and $\boldsymbol{P}(\cdot) = (W^{-T}\phi \boldsymbol{Q})(\boldsymbol{\eta}^{-1}(\cdot))$, the condition that $\boldsymbol{Q} \cdot \boldsymbol{n} = \boldsymbol{Q}^T \boldsymbol{n} = 0$ on $\partial\Omega_\varepsilon$ is transformed to (note that $\boldsymbol{n}$ is transformed to $W\boldsymbol{n}$) $P_1 = 0$ on $y_1 = 0$, $P_3 = 0$ on $y_3 = 0$. Our problem is then to approximate such $H^1$ functions while keeping these boundary conditions.

We just sketch how to approximate $P_1$ and the rest of the proof can be carried out in a similar way. We define a new function $P_1'$ as follows:

$$P_1' = \begin{cases} P_1(y_1, y_2, y_3) & \text{when } y_1 > 0,\ y_3 < 0, \\ P_1(y_1, y_2, -y_3) & \text{when } y_1 > 0,\ y_3 > 0, \\ -P_1(-y_1, y_2, -y_3) & \text{when } y_1 < 0,\ y_3 > 0, \\ -P_1(-y_1, y_2, y_3) & \text{when } y_1 < 0,\ y_3 < 0. \end{cases}$$

$P_1'$ can be regarded as a function defined in $\mathbb{R}^3$ with compact support. Use a radially symmetric mollifier $\sigma_\mu(\cdot)$ where $\mu$ is the standard mollifier parameter which goes to zero, we have obviously $P_1' * \sigma_\mu|_{y_1=0} = 0$ ($*$ denotes the standard convolution) and the sequence $\{P_1' * \sigma_\mu\}_\mu$ is the desired approximation of $P_1$ when restricted to the region $\{y \in \mathbb{R}^3 : y_1 > 0, y_3 < 0\}$. This completes the proof.

**Lemma 8.2.11** *Let $\varepsilon$ be sufficiently small. Then for any $\boldsymbol{Q} \in \boldsymbol{H}_n^1(\Omega_\varepsilon)$, we have*

$$\|\boldsymbol{Q}\|_{H^1(\Omega_\varepsilon)} \leq \widehat{C}(\|\boldsymbol{Q}\|_{L^2(\Omega_\varepsilon)} + \|\mathrm{div}\boldsymbol{Q}\|_{L^2(\Omega_\varepsilon)} + \|\mathbf{curl}\boldsymbol{Q}\|_{L^2(\Omega_\varepsilon)}),\ \forall \boldsymbol{Q} \in \boldsymbol{H}_n^1(\Omega_\varepsilon)$$

*where the constant $\widehat{C}$ is independent of $\varepsilon$.*

*Proof.* By Lemma 2.10, it is obvious that we only need to prove the lemma for $\boldsymbol{Q} \in \boldsymbol{H}^2(\Omega_\varepsilon) \cap \boldsymbol{H}_n^1(\Omega_\varepsilon)$. For any $\boldsymbol{Q} \in \boldsymbol{H}^2(\Omega_\varepsilon) \cap \boldsymbol{H}_n^1(\Omega_\varepsilon)$, we have

$$\int_{\Omega_\varepsilon} |\mathrm{div}\boldsymbol{Q}|^2 dx + \int_{\Omega_\varepsilon} |\mathbf{curl}\boldsymbol{Q}|^2 dx - \sum_{i,j=1}^{3} \int_{\Omega_\varepsilon} \left|\frac{\partial Q_i}{\partial x_j}\right|^2 dx$$
$$= \int_{\Gamma_1 \cup \Gamma_2 \cup \Gamma_3} \mathcal{B}(\boldsymbol{Q}; \boldsymbol{Q}) d\sigma, \tag{2.26}$$

where $\mathcal{B}$ is the second fundamental quadratic form of $\Gamma_j, j = 1, 2, 3$, and $d\sigma$ is the surface element of $\partial\Omega_\varepsilon$. An elementary definition of $\mathcal{B}$ is recalled in [G 85], p. 133. If $x_0$ is a point of $\Gamma_j$, we consider a related new orthogonal co-ordinate system $\{y_1, y_2, y_3\}$ with origin at $x_0$ defined as follows: there exist a cube $V = \{(y_1, y_2, y_3) : -a_j < y_j < a_j, j = 1, 2, 3\}$ and a function $\varphi$ of class $C^2$ in $V' = \{(y_1, y_2) : -a_j < y_j < a_j, j = 1, 2\}$ such that $|\varphi(y')| \leq a_3/2$ for every $y' = (y_1, y_2) \in V'$, $\Omega_\varepsilon \cap V = \{y = (y', y_3) \in V : y_3 < \varphi(y')\}, \Gamma_j \cap V = \{y = (y', y_3) \in V : y_3 = \varphi(y')\}$. We can even choose the new co-ordinates so that the plane $y_3 = 0$ is tangent to $\Gamma_j$ at $x_0$, which implies $\nabla\varphi(0) = 0$. Then, if $\xi, \eta$ are tangent vectors to $\Gamma_j$ at $x_0$ with components $(\xi_1, \xi_2)$ and $(\eta_1, \eta_2)$ in the direction of $\{y_1, y_2\}$, we have

$$\mathcal{B}_{x_0}(\xi, \eta) = \sum_{j,k=1}^{2} \frac{\partial^2 \varphi}{\partial y_j \partial y_k}(0) \xi_k \xi_j.$$

It is obvious that

$$|\mathcal{B}_{x_0}(\xi,\eta)| \le c|\xi||\eta|, \quad \text{for all } x_0 \in \Gamma_3,$$

for any tangent vectors $\xi, \eta$ to $\Gamma$ at $x_0$, and the constant $c$ is independent of $\varepsilon$. Thus

$$\Big|\int_{\Gamma_3} \mathcal{B}(\boldsymbol{Q};\boldsymbol{Q})d\sigma\Big| \le c\int_{\partial\Omega_0}\int_{-\varepsilon b(x')}^{\varepsilon a(x')} |\boldsymbol{Q}|^2 dx_3 d\sigma',$$

where $d\sigma'$ is the integral element on $\partial\Omega_0$. By applying the trace theorem

$$\int_{\partial\Omega_0} |u|d\sigma' \le C(\Omega_0)\Big[\int_{\Omega_0} |\nabla' u|dx' + \int_{\Omega_0} |u|dx'\Big]$$

for all $u \in W^{1,1}(\Omega_0)$, we get

$$\Big|\int_{\Gamma_3} \mathcal{B}(\boldsymbol{Q};\boldsymbol{Q})d\sigma\Big| \le c\int_{\Omega_0}\Big|\nabla'\Big[\int_{-\varepsilon b(x')}^{\varepsilon a(x')} |\boldsymbol{Q}|^2 dx_3\Big]\Big|dx' + c\int_{\Omega_0}\int_{-\varepsilon b(x')}^{\varepsilon a(x')} |\boldsymbol{Q}|^2 dx_3 dx'.$$

It is easy to see that

$$\begin{aligned}\nabla'\Big[\int_{-\varepsilon b(x')}^{\varepsilon a(x')} |\boldsymbol{Q}|^2 dx_3\Big] &= \int_{-\varepsilon b(x')}^{\varepsilon a(x')} \nabla'|\boldsymbol{Q}|^2 dx_3 \\ &+ \varepsilon|\boldsymbol{Q}(x',\varepsilon a(x'))|^2\nabla' a(x') - \varepsilon|\boldsymbol{Q}(x',-\varepsilon b(x'))|^2\nabla' b(x').\end{aligned}$$

As in the proof of Lemma 2.9, we get

$$\begin{aligned}&\int_{\Omega_0}\Big|\varepsilon|\boldsymbol{Q}(x',\varepsilon a(x'))|^2\nabla' a(x')\Big|dx' \le c\varepsilon\int_{\Omega_0} |\boldsymbol{Q}(x',\varepsilon a(x'))|^2 dx' \\ &\le c\varepsilon\int_{\Omega_0} [\![\, |\boldsymbol{Q}|^2 \,]\!] dx' + c\varepsilon\int_{\Omega_0}\Big||\boldsymbol{Q}(x',\varepsilon a(x'))|^2 - [\![\, |\boldsymbol{Q}|^2 \,]\!](x')\Big|dx' \\ &\le c\int_{\Omega_\varepsilon} |\boldsymbol{Q}|^2 dx' + c\varepsilon\int_{\Omega_0}\int_{\xi}^{\varepsilon a(x')} \partial_3(|\boldsymbol{Q}|^2)dx_3 dx'\end{aligned}$$

for some $\xi \in (-\varepsilon b(x'), \varepsilon a(x'))$. Thus we get

$$\begin{aligned}\int_{\Omega_0} \varepsilon|\boldsymbol{Q}(x',\varepsilon a(x')|^2|\nabla a|^2 dx' &\le c\int_{\Omega_\varepsilon} |\boldsymbol{Q}|^2 dx' + c\varepsilon\int_{\Omega_\varepsilon} |\boldsymbol{Q}\partial_3\boldsymbol{Q}|dx \\ &\le \varepsilon^2\delta\sum_{j,k=1}^{3}\int_{\Omega_\varepsilon} |\partial_k Q_j|^2 dx + c(1+\frac{1}{\delta})||\boldsymbol{Q}||^2_{L^2(\Omega_\varepsilon)}\end{aligned}$$

for any $\delta > 0$, where the constant $c$ is independent of $\varepsilon$ and $\delta$.

Similarly, we get

$$\int_{\Omega_0} |\boldsymbol{Q}(x',-\varepsilon b(x'))|^2|\nabla b|^2 dx' \le \varepsilon^2\delta\sum_{j,k=1}^{3}\int_{\Omega_\varepsilon} |\partial_k Q_j|^2 dx + c(1+\frac{1}{\delta})||\boldsymbol{Q}||^2_{L^2(\Omega_\varepsilon)}$$

for any $\delta > 0$, where the constant $c$ is independent of $\varepsilon$ and $\delta$. By applying Cauchy's inequality and Young's inequality, we have

$$\begin{aligned}\int_{\Omega_0}\int_{-\varepsilon b(x')}^{\varepsilon a(x')} \nabla' |\boldsymbol{Q}|^2 dx_3 dx' &\le c||\boldsymbol{Q}||_{L^2(\Omega_\varepsilon)}\Big[\sum_{j,k=1}^{3}\int_{\Omega_\varepsilon} |\partial_k Q_j|^2 dx\Big]^{1/2} \\ &\le \frac{1}{2}\sum_{j,k=1}^{3}\int_{\Omega_\varepsilon} |\partial_k Q_j|^2 dx + C||\boldsymbol{Q}||^2_{L^2(\Omega_\varepsilon)},\end{aligned}$$

so that

$$\Big|\int_{\Gamma_3} \mathcal{B}(\boldsymbol{Q};\boldsymbol{Q})d\sigma\Big| \le (\frac{1}{2} + 2\varepsilon^2\delta)\sum_{j,k=1}^{3}\int_{\Omega_\varepsilon} |\partial_k Q_j|^2 dx + C(1+\frac{1}{\delta})||\boldsymbol{Q}||^2_{L^2(\Omega_\varepsilon)}, \quad (2.27)$$

for any $\delta > 0$, where the constant $c$ is independent of $\varepsilon$ and $\delta$. On $\Gamma_1$, it is easy to obtain that

$$|\mathcal{B}_{x_0}(\xi,\eta)| \le c\varepsilon|\xi|\,|\eta|, \quad \text{for all } \; x_0 \in \Gamma_3$$

for any tangent vectors $\xi, \eta$ to $\Gamma$ at $x_0$, and the constant $c$ is independent of $\varepsilon$. Therefore

$$\begin{aligned}\Big|\int_{\Gamma_1} \mathcal{B}(\boldsymbol{Q};\boldsymbol{Q})d\sigma\Big| &\le c\varepsilon\int_{\Omega_0} |\boldsymbol{Q}(x',\varepsilon a(x'))|^2 dx' \\ &\le \varepsilon^2\delta\sum_{j,k=1}^{3}\int_{\Omega_\varepsilon}\Big|\frac{\partial Q_j}{\partial x_k}\Big|^2 dx + c(1+\frac{1}{\delta})||\boldsymbol{Q}||^2_{L^2(\Omega_\varepsilon)}. \quad (2.28)\end{aligned}$$

for any $\delta > 0$ with the constant $c$ independent of $\varepsilon$ and $\delta$. In deriving (2.28), we have used the argument leading to (2.27). Similarly, we have

$$\Big|\int_{\Gamma_2} \mathcal{B}(\boldsymbol{Q};\boldsymbol{Q})d\sigma\Big| \le \varepsilon^2\delta\sum_{j,k=1}^{3}\int_{\Omega_\varepsilon}\Big|\frac{\partial Q_j}{\partial x_k}\Big|^2 dx + c(1+\frac{1}{\delta})||\boldsymbol{Q}||^2_{L^2(\Omega_\varepsilon)} \quad (2.29)$$

for any $\delta > 0$ with the constant $c$ independent of $\varepsilon$ and $\delta$. Now the lemma follows from (2.26)–(2.29) by choosing $\delta = 1/16$ for $\varepsilon < 1$ and $\delta = 1/(16\varepsilon^2)$ for $\varepsilon \ge 1$.

In the following, we will always assume that $\varepsilon$ is sufficiently small. From Lemma 2.11 and Corollary 2.1 we obtain the following lemma.

**Lemma 8.2.12** *We have*

$$\operatorname{esssup}_{0\le t\le T}\Big[||\psi_\varepsilon||^2_{H^1(\Omega_\varepsilon)} + ||\boldsymbol{A}_\varepsilon||^2_{H^1(\Omega_\varepsilon)}\Big] + \int_0^T\int_{\Omega_\varepsilon}\Big[\Big|\frac{\partial\psi_\varepsilon}{\partial t}\Big|^2 + \Big|\frac{\partial\boldsymbol{A}_\varepsilon}{\partial t}\Big|^2\Big]dxdt \le c\varepsilon.$$

To proceed further, note that

$$
\begin{aligned}
\nabla'[\![ f ]\!] &= \nabla'\Big[\frac{1}{\varepsilon\rho(x')}\Big]\int_{-\varepsilon b(x')}^{\varepsilon a(x')} f(x',x_3)dx_3 + \frac{1}{\varepsilon\rho(x')}\int_{-\varepsilon b(x')}^{\varepsilon a(x')}\nabla' f(x',x_3)dx_3 \\
&+ \frac{1}{\varepsilon\rho(x')}\Big[\varepsilon f(x',\varepsilon a(x'))\nabla' a(x') + \varepsilon f(x',-\varepsilon b(x'))\nabla' b(x')\Big] \\
&= \frac{1}{\rho(x')}\nabla' a(x')\Big[f(x',\varepsilon a(x')) - [\![ f ]\!](x')\Big] \\
&+ \frac{1}{\rho(x')}\nabla' b(x')\Big[f(x',-\varepsilon b(x')) - [\![ f ]\!](x')\Big] + [\![ \nabla' f ]\!]. \qquad (2.30)
\end{aligned}
$$

Therefore, by (2.22) and (2.23), we get

$$
\|\nabla'[\![ f ]\!]\|_{L^2(\Omega_0)} \leq \frac{c}{\sqrt{\varepsilon}}\|\nabla' f\|_{L^2(\Omega_\varepsilon)} + c\sqrt{\varepsilon}\Big\|\frac{\partial f}{\partial x_3}\Big\|_{L^2(\Omega_\varepsilon)}. \qquad (2.31)
$$

**Lemma 8.2.13** *We have*

$$
\begin{aligned}
\mathrm{esssup}_{0\leq t\leq T}\|[\![ \psi_\varepsilon ]\!]\|^2_{H^1(\Omega_0)} + \int_0^T \Big\|\frac{\partial}{\partial t}[\![ \psi_\varepsilon ]\!]\Big\|^2_{L^2(\Omega_0)} dt \leq c, \\
\mathrm{esssup}_{0\leq t\leq T}\|[\![ \boldsymbol{A}_\varepsilon ]\!]\|^2_{H^1(\Omega_0)} + \int_0^T \Big\|\frac{\partial}{\partial t}[\![ \boldsymbol{A}_\varepsilon ]\!]\Big\|^2_{L^2(\Omega_0)} dt \leq c.
\end{aligned}
$$

*Proof.* It is obvious from Lemma 2.12 and (2.31).

**Lemma 8.2.14** *Let* $\boldsymbol{A}_\varepsilon = (A^1_\varepsilon, A^2_\varepsilon, A^3_\varepsilon)$. *We have*

$$
\mathrm{esssup}_{0\leq t\leq T}\Big[\|A^3_\varepsilon(\cdot,\varepsilon a(\cdot),t)\|_{L^2(\Omega_0)} + \|A^3_\varepsilon(\cdot,-\varepsilon b(\cdot),t)\|_{L^2(\Omega_0)} + \|[\![ A^3_\varepsilon ]\!](\cdot,t)\|_{L^2(\Omega_0)}\Big] \leq c\varepsilon.
$$

*Proof.* From the boundary condition $\boldsymbol{A}_\varepsilon\cdot\boldsymbol{n} = 0$, we obtain

$$
A^3_\varepsilon(x',\varepsilon a(x'),t) = \varepsilon\nabla' a(x')\cdot\boldsymbol{A}'_\varepsilon(x',\varepsilon a(x'),t), \qquad (2.32)
$$

$$
A^3_\varepsilon(x',-\varepsilon b(x'),t) = -\varepsilon\nabla' b(x')\cdot\boldsymbol{A}'_\varepsilon(x',-\varepsilon b(x'),t). \qquad (2.33)
$$

By (2.25) and Lemma 2.12, we get

$$
\|A^3_\varepsilon(\cdot,\varepsilon a(\cdot),t)\|_{L^2(\Omega_0)} \leq c\varepsilon\|\boldsymbol{A}'_\varepsilon(\cdot,\varepsilon a(\cdot),t)\|_{L^2(\Omega_0)} \leq c\varepsilon. \qquad (2.34)
$$

Similarly, we have

$$
\|A^3_\varepsilon(\cdot,-\varepsilon b(\cdot),t)\|_{L^2(\Omega_0)} \leq c\varepsilon\|\boldsymbol{A}'_\varepsilon(\cdot,-\varepsilon b(\cdot),t)\|_{L^2(\Omega_0)} \leq c\varepsilon. \qquad (2.35)
$$

Finally, by (2.23), Lemma 2.12 and (2.34), we have

$$\begin{aligned}\|[\![ A_\varepsilon^3 ]\!](\cdot,t)\|_{L^2(\Omega_0)} &\le \|[\![ A_\varepsilon^3 ]\!](\cdot,t) - A_\varepsilon^3(\cdot,\varepsilon a(\cdot),t)\|_{L^2(\Omega_0)} \\ &\quad + \|A_\varepsilon^3(\cdot,\varepsilon a(\cdot),t)\|_{L^2(\Omega_0)} \\ &\le c\sqrt{\varepsilon}\Big\|\frac{\partial A_\varepsilon^3}{\partial x_3}\Big\|_{L^2(\Omega_\varepsilon)} + c\varepsilon \le c\varepsilon,\end{aligned}$$

which completes the proof.

### 8.2.4 The convergence

In the weak formulation (2.17)–(2.19), choosing the test functions $\omega, \boldsymbol{B}$ as

$$\omega = \omega(x',t), \quad \boldsymbol{B} = (\boldsymbol{B}^1(x',t), \boldsymbol{B}^2(x',t), 0)$$

and noting that

$$\mathrm{div}\boldsymbol{B} = \mathrm{div}'\boldsymbol{B}', \quad \mathbf{curl}\boldsymbol{B} = (0,0,\mathrm{curl}\boldsymbol{B}'),$$

we have

$$\begin{aligned}&\int_0^T\int_{\Omega_0}\rho(x')\Big[\!\!\Big[ \frac{\partial\psi_\varepsilon}{\partial t}\Big]\!\!\Big]\omega dx'dt - i\int_0^T\int_{\Omega_0}\rho(x')[\![ \mathrm{div}\boldsymbol{A}_\varepsilon\psi_\varepsilon ]\!]\omega dx'dt \\ +\ &\int_0^T\int_{\Omega_0}\rho(x')\Big[\!\!\Big[ \Big(i\nabla'\psi_\varepsilon + \boldsymbol{A}_\varepsilon'\psi_\varepsilon\Big)\Big(-i\nabla'\omega + \boldsymbol{A}_\varepsilon'\omega\Big)\Big]\!\!\Big] dx'dt \\ +\ &\int_0^T\int_{\Omega_0}\rho(x')\Big[\!\!\Big[ \Big(i\frac{\partial\psi_\varepsilon}{\partial x_3} + A_\varepsilon^3\psi_\varepsilon\Big)A_\varepsilon^3\Big]\!\!\Big]\omega dx'dt \\ +\ &\int_0^T\int_{\Omega_0}\kappa^2\rho(x')[\![ (|\psi_\varepsilon|^2-1)\psi_\varepsilon ]\!]\omega dx'dt = 0,\end{aligned}$$

for any $\omega \in \mathcal{L}^2(0,T;\mathcal{H}^1(\Omega_0))$ (2.36)

$$\begin{aligned}&\int_0^T\int_{\Omega_0}\rho(x')\Big[\!\!\Big[ \frac{\partial\boldsymbol{A}_\varepsilon'}{\partial t}\Big]\!\!\Big]\boldsymbol{B}'dx'dt + \int_0^T\int_{\Omega_0}\rho(x')[\![ \mathrm{curl}\boldsymbol{A}_\varepsilon' ]\!]\mathrm{curl}\boldsymbol{B}'dx'dt \\ +\ &\int_0^T\int_{\Omega_0}\rho(x')[\![ \mathrm{div}\boldsymbol{A}_\varepsilon ]\!]\mathrm{div}'\boldsymbol{B}'dx'dt \\ +\ &\int_0^T\int_{\Omega_0}\rho(x')\Big[\!\!\Big[ Re\Big[\Big(i\nabla'\psi_\varepsilon + \boldsymbol{A}_\varepsilon'\psi_\varepsilon\Big)\bar{\psi}_\varepsilon\Big]\Big]\!\!\Big]\boldsymbol{B}'dx'dt \\ =\ &\int_0^T\int_{\Omega_0}\rho(x')H\mathrm{curl}\boldsymbol{B}'dx'dt, \quad \text{for any } \boldsymbol{B}' \in \boldsymbol{L}^2(0,T;\boldsymbol{H}_n^1(\Omega_0)). \qquad (2.37)\end{aligned}$$

In order to let $\varepsilon \to 0$ in (2.36)–(2.37), we prove several lemmas.

**Lemma 8.2.15** *We have*

$$[\![ \mathrm{div}\boldsymbol{A}_\varepsilon ]\!] = \frac{1}{\rho(x')}\mathrm{div}'\big(\rho(x')[\![ \boldsymbol{A}_\varepsilon' ]\!]\big)$$

*Proof.* By (2.30) and (2.32)–(2.33), we have

$$\begin{aligned}
[\![\operatorname{div}\boldsymbol{A}_\varepsilon]\!] &= [\![\operatorname{div}'\boldsymbol{A}'_\varepsilon]\!] + \frac{1}{\varepsilon\rho(x')}\int_{-\varepsilon b(x')}^{\varepsilon a(x')} \frac{\partial A^3_\varepsilon}{\partial x_3}dx_3 \\
&= [\![\operatorname{div}'\boldsymbol{A}'_\varepsilon]\!] + \frac{1}{\varepsilon\rho(x')}\Big[A^3_\varepsilon(x',\varepsilon a(x'),t) - A^3_\varepsilon(x',-\varepsilon b(x'),t)\Big] \\
&= [\![\operatorname{div}'\boldsymbol{A}'_\varepsilon]\!] + \frac{1}{\rho(x')}\Big[\nabla' a(x')\cdot\boldsymbol{A}'_\varepsilon(x',\varepsilon a(x'),t) \\
&\quad +\nabla' b(x')\cdot\boldsymbol{A}'_\varepsilon(x',-\varepsilon b(x'),t)\Big] \\
&= \operatorname{div}'[\![\boldsymbol{A}'_\varepsilon]\!] + \frac{1}{\rho(x')}\Big[\nabla' a(x') + \nabla' b(x')\Big][\![\boldsymbol{A}'_\varepsilon]\!] \\
&= \frac{1}{\rho(x')}\operatorname{div}'\big(\rho(x')[\![\boldsymbol{A}'_\varepsilon]\!]\big).
\end{aligned}$$

This completes the proof.

**Lemma 8.2.16** *We have*

$$\operatorname{esssup}_{0\le t\le T}\|[\![\operatorname{curl}\boldsymbol{A}'_\varepsilon]\!] - \operatorname{curl}[\![\boldsymbol{A}'_\varepsilon]\!]\|_{L^2(\Omega_0)} \le c\varepsilon; \tag{2.38}$$
$$\operatorname{esssup}_{0\le t\le T}\|[\![\nabla'\psi_\varepsilon]\!] - \nabla'[\![\psi_\varepsilon]\!]\|_{L^2(\Omega_0)} \le c\varepsilon. \tag{2.39}$$

*Proof.* We only prove (2.38). The estimate (2.39) can be proved similarly. By (2.30), we have

$$\begin{aligned}
[\![\operatorname{curl}\boldsymbol{A}'_\varepsilon]\!] - \operatorname{curl}[\![\boldsymbol{A}'_\varepsilon]\!] &= \frac{1}{\rho(x')}\mathbf{curl}' a(x')\Big[\boldsymbol{A}'_\varepsilon(x',\varepsilon a(x'),t) - [\![\boldsymbol{A}'_\varepsilon]\!](x',t)\Big] \\
&+ \frac{1}{\rho(x')}\mathbf{curl}' b(x')\Big[\boldsymbol{A}'_\varepsilon(x',-\varepsilon b(x'),t) - [\![\boldsymbol{A}'_\varepsilon]\!](x',t)\Big].
\end{aligned}$$

Now (2.38) follows from (2.23) and Lemma 2.12.

**Lemma 8.2.17** *We have*

$$\operatorname{esssup}_{0\le t\le T}\|\rho[\![\operatorname{div}\boldsymbol{A}_\varepsilon\psi_\varepsilon]\!] - \operatorname{div}'\big(\rho[\![\boldsymbol{A}'_\varepsilon]\!]\big)[\![\psi_\varepsilon]\!]\|_{L^1(\Omega_0)} \le c\varepsilon; \tag{2.40}$$
$$\operatorname{esssup}_{0\le t\le T}\|[\![\nabla'\psi_\varepsilon\cdot\boldsymbol{A}'_\varepsilon]\!] - \nabla'[\![\psi_\varepsilon]\!]\cdot[\![\boldsymbol{A}'_\varepsilon]\!]\|_{L^1(\Omega_0)} \le c\varepsilon. \tag{2.41}$$

*Proof.* We only prove (2.40). The estimate (2.41) can be proved similarly. By Lemma 2.15 we have

$$\rho[\![\operatorname{div}\boldsymbol{A}_\varepsilon\psi_\varepsilon]\!] - \operatorname{div}'\big(\rho[\![\boldsymbol{A}'_\varepsilon]\!]\big)[\![\psi_\varepsilon]\!] = \rho[\![\operatorname{div}\boldsymbol{A}_\varepsilon(\psi_\varepsilon - [\![\psi_\varepsilon]\!])]\!].$$

Note that by (2.24) and Lemma 2.12, we have

$$\begin{aligned}
\|\rho[\![\operatorname{div}\boldsymbol{A}_\varepsilon(\psi_\varepsilon - [\![\psi_\varepsilon]\!])]\!]\|_{L^1(\Omega_0)} &\le \frac{c}{\varepsilon}\|\operatorname{div}\boldsymbol{A}_\varepsilon\|_{L^2(\Omega_\varepsilon)}\|\psi_\varepsilon - [\![\psi_\varepsilon]\!]\|_{L^2(\Omega_\varepsilon)} \\
&\le \frac{c}{\varepsilon}\cdot c\sqrt{\varepsilon}\cdot c\varepsilon\Big\|\frac{\partial\psi_\varepsilon}{\partial x_3}\Big\|_{L^2(\Omega_\varepsilon)} \le c\varepsilon.
\end{aligned}$$

This completes the proof.

**Lemma 8.2.18** *Let $Q_0 = \Omega_0 \times (0,T)$. We have*

$$\|[\![ \boldsymbol{A}'_\varepsilon \psi_\varepsilon ]\!] - [\![ \boldsymbol{A}'_\varepsilon ]\!][\![ \psi_\varepsilon ]\!]\|_{L^2(Q_0)} \leq c\varepsilon; \tag{2.42}$$

$$\|[\![ |\boldsymbol{A}'_\varepsilon|^2 \psi_\varepsilon ]\!] - |[\![ \boldsymbol{A}'_\varepsilon ]\!]|^2[\![ \psi_\varepsilon ]\!]\|_{L^1(Q_0)} \leq c\varepsilon; \tag{2.43}$$

$$\|[\![ (|\psi_\varepsilon|^2 - 1)\psi_\varepsilon ]\!] - (|[\![ \psi ]\!]|^2 - 1)[\![ \psi_\varepsilon ]\!]\|_{L^1(Q_0)} \leq c\varepsilon. \tag{2.44}$$

*Proof.* We only prove (2.42) and (2.43). The estimate (2.44) can be proved similarly. Since $|\psi_\varepsilon| \leq 1$ a.e. in $Q_\varepsilon$, we have

$$\begin{aligned}
&\int_0^T \int_{\Omega_0} \Big|[\![ \boldsymbol{A}'_\varepsilon \psi_\varepsilon ]\!] - [\![ \boldsymbol{A}'_\varepsilon ]\!][\![ \psi_\varepsilon ]\!]\Big|^2 dx'dt \\
= &\int_0^T \int_{\Omega_0} \Big|[\![ (\boldsymbol{A}'_\varepsilon - [\![ \boldsymbol{A}'_\varepsilon ]\!])\psi_\varepsilon ]\!]\Big|^2 dx'dt \\
\leq &\int_0^T \int_{\Omega_0} \frac{1}{\varepsilon^2\rho(x')^2}\Big[\int_{-\varepsilon b(x')}^{\varepsilon a(x')} |\boldsymbol{A}'_\varepsilon - [\![ \boldsymbol{A}'_\varepsilon ]\!]| \cdot |\psi_\varepsilon| dx_3\Big]^2 dx'dt \\
\leq &\int_0^T \int_{\Omega_0} \frac{1}{\varepsilon^2\rho(x')^2}\Big[\int_{-\varepsilon b(x')}^{\varepsilon a(x')} \Big|\frac{\partial \boldsymbol{A}'_\varepsilon}{\partial x_3}\Big| dx_3\Big]^2 \Big[\int_{-\varepsilon b(x')}^{\varepsilon a(x')} |\psi_\varepsilon| dx_3\Big]^2 dx'dt \\
\leq &\int_0^T \int_{\Omega_0} \Big[\int_{-\varepsilon b(x')}^{\varepsilon a(x')} \Big|\frac{\partial \boldsymbol{A}'_\varepsilon}{\partial x_3}\Big| dx_3\Big]^2 dx'dt \\
\leq &\; c\varepsilon \int_0^T \int_{\Omega_\varepsilon} \Big|\frac{\partial \boldsymbol{A}'_\varepsilon}{\partial x_3}\Big|^2 dxdt \leq c\varepsilon^2,
\end{aligned}$$

where in the last inequality we have used Lemma 2.12. This proves (2.42). In order to show (2.43) we observe first that

$$[\![ |\boldsymbol{A}'_\varepsilon|^2 \psi_\varepsilon ]\!] - |[\![ \boldsymbol{A}'_\varepsilon ]\!]|^2[\![ \psi_\varepsilon ]\!] = [\![ |\boldsymbol{A}'_\varepsilon|^2(\psi_\varepsilon - [\![ \psi_\varepsilon ]\!]) ]\!] + [\![ \boldsymbol{A}'_\varepsilon(\boldsymbol{A}'_\varepsilon - [\![ \boldsymbol{A}'_\varepsilon ]\!]) ]\!][\![ \psi_\varepsilon ]\!] \tag{2.45}$$

But

$$\begin{aligned}
&\int_{\Omega_0} \Big|[\![ |\boldsymbol{A}'_\varepsilon|^2(\psi_\varepsilon - [\![ \psi_\varepsilon ]\!]) ]\!]\Big| dx' \\
\leq &\int_{\Omega_0} \frac{1}{\varepsilon\rho(x')} \int_{-\varepsilon b(x')}^{\varepsilon a(x')} |\boldsymbol{A}'_\varepsilon|^2 |\psi_\varepsilon - [\![ \psi_\varepsilon ]\!]| dx_3 dx' \\
\leq &\int_{\Omega_0} \Big[\frac{1}{\varepsilon\rho(x')} \int_{-\varepsilon b(x')}^{\varepsilon a(x')} |\boldsymbol{A}'_\varepsilon|^2 dx_3\Big] \cdot \Big[\int_{-\varepsilon b(x')}^{\varepsilon a(x')} \Big|\frac{\partial \psi_\varepsilon}{\partial x_3}\Big| dx_3\Big] dx' \\
\leq &\; \|[\![ |\boldsymbol{A}'_\varepsilon|^2 ]\!]\|_{L^2(\Omega_0)} \Big[\int_{\Omega_0} \Big|\int_{-\varepsilon b(x')}^{\varepsilon a(x')} \Big|\frac{\partial \psi_\varepsilon}{\partial x_3}\Big| dx_3\Big|^2 dx'\Big]^{1/2} \\
\leq &\; c\varepsilon \|[\![ |\boldsymbol{A}'_\varepsilon|^2 ]\!]\|_{L^2(\Omega_0)},
\end{aligned} \tag{2.46}$$

where we have used Lemma 2.12. By the embedding theorem $W^{1,1}(\Omega_0) \hookrightarrow L^2(\Omega_0)$, we get

$$\|[\![ |\boldsymbol{A}'_\varepsilon|^2 ]\!]\|_{L^2(\Omega_0)} \leq C(\Omega_0)\Big[\|\nabla'[\![ |\boldsymbol{A}'_\varepsilon|^2 ]\!]\|_{L^1(\Omega_0)} + \|[\![ |\boldsymbol{A}'_\varepsilon|^2 ]\!]\|_{L^1(\Omega_0)}\Big].$$

It is easy to see that

$$\|[\![ |\boldsymbol{A}'_\varepsilon|^2 ]\!]\|_{L^1(\Omega_0)} \leq \frac{c}{\varepsilon}\|\boldsymbol{A}'_\varepsilon\|^2_{L^2(\Omega_\varepsilon)} \leq c, \tag{2.47}$$

where we have used Lemma 2.12. By (2.30), we have

$$\begin{aligned}
&\|\nabla'[\![ |\boldsymbol{A}'_\varepsilon|^2 ]\!]\|_{L^1(\Omega_0)} \\
\leq\ & \|[\![ \nabla'|\boldsymbol{A}'_\varepsilon|^2 ]\!]\|_{L^1(\Omega_0)} + c\||\boldsymbol{A}'_\varepsilon|^2(\cdot,\varepsilon a(\cdot),t) - [\![ |\boldsymbol{A}'_\varepsilon|^2 ]\!](\cdot,t)\|_{L^1(\Omega_0)} \\
& + c\||\boldsymbol{A}'_\varepsilon|^2(\cdot,-\varepsilon b(\cdot),t) - [\![ |\boldsymbol{A}'_\varepsilon|^2 ]\!](\cdot,t)\|_{L^1(\Omega_0)} \\
\leq\ & \frac{c}{\varepsilon}\|\boldsymbol{A}'_\varepsilon\|_{L^2(\Omega_\varepsilon)}\|\boldsymbol{A}'_\varepsilon\|_{H^1(\Omega_\varepsilon)} + c\|\boldsymbol{A}'_\varepsilon\|_{L^2(\Omega_\varepsilon)}\left\|\frac{\partial \boldsymbol{A}'_\varepsilon}{\partial x_3}\right\|_{L^2(\Omega_\varepsilon)} \\
\leq\ & \frac{c}{\varepsilon}\cdot c\sqrt{\varepsilon}\cdot c\sqrt{\varepsilon} + c\sqrt{\varepsilon}\cdot c\sqrt{\varepsilon} \leq c,
\end{aligned} \tag{2.48}$$

where we have used Lemma 2.12. Thus, by (2.46)–(2.48), we have

$$\|[\![ |\boldsymbol{A}'_\varepsilon|^2(\psi_\varepsilon - [\![ \psi_\varepsilon ]\!]) ]\!]\|_{L^1(Q_0)} \leq c\varepsilon. \tag{2.49}$$

On the other hand, since $|\psi_\varepsilon| \leq 1$ a.e. in $\Omega_\varepsilon \times (0,T)$, we have $\|[\![ \psi_\varepsilon ]\!]\|_{L^\infty(Q_0)} \leq 1$. Thus

$$\begin{aligned}
&\|[\![ \boldsymbol{A}'_\varepsilon(\boldsymbol{A}'_\varepsilon - [\![ \boldsymbol{A}'_\varepsilon ]\!]) ]\!][\![ \psi_\varepsilon ]\!]\|_{L^1(Q_0)} \\
\leq\ & \int_0^T \|[\![ \boldsymbol{A}'_\varepsilon(\boldsymbol{A}'_\varepsilon - [\![ \boldsymbol{A}'_\varepsilon ]\!]) ]\!]\|_{L^1(\Omega_0)} dt \\
\leq\ & \frac{c}{\varepsilon}\int_0^T \|\boldsymbol{A}'_\varepsilon\|_{L^2(\Omega_\varepsilon)}\|\boldsymbol{A}'_\varepsilon - [\![ \boldsymbol{A}'_\varepsilon ]\!]\|_{L^2(\Omega_\varepsilon)} dt \\
\leq\ & \frac{c}{\varepsilon}\int_0^T c\sqrt{\varepsilon}\cdot c\varepsilon\left\|\frac{\partial \boldsymbol{A}'_\varepsilon}{\partial x_3}\right\|_{L^2(\Omega_\varepsilon)} dt \leq c\varepsilon,
\end{aligned} \tag{2.50}$$

where we have used Lemma 2.12 and (2.24). Now (2.43) follows from (2.45) and (2.49)–(2.50). This completes the proof.

**Lemma 8.2.19** *We have*

$$\operatorname{esssup}_{0\leq t\leq T}\left\|\left[\!\!\left[ \left(\frac{i}{\kappa}\frac{\partial \psi_\varepsilon}{\partial x_3} + A^3_\varepsilon\psi_\varepsilon\right)A^3_\varepsilon \right]\!\!\right]\right\|_{L^1(\Omega_0)} \leq c\sqrt{\varepsilon}.$$

*Proof.* At first, by using (2.24), Lemma 2.12 and Lemma 2.14, we have

$$\begin{aligned}
\|A^3_\varepsilon\|_{L^2(\Omega_\varepsilon)} &\leq \|A^3_\varepsilon - [\![ A^3_\varepsilon ]\!]\|_{L^2(\Omega_\varepsilon)} + \|[\![ A^3_\varepsilon ]\!]\|_{L^2(\Omega_\varepsilon)} \\
&\leq c\varepsilon\left\|\frac{\partial A^3_\varepsilon}{\partial x_3}\right\|_{L^2(\Omega_\varepsilon)} + c\varepsilon \leq c\varepsilon\cdot c\sqrt{\varepsilon} + c\varepsilon \leq c\varepsilon.
\end{aligned}$$

Thus

$$\left\|\left[\!\!\left[ \left(\frac{i}{\kappa}\frac{\partial \psi_\varepsilon}{\partial x_3} + A^3_\varepsilon\psi_\varepsilon\right)A^3_\varepsilon \right]\!\!\right]\right\|_{L^1(\Omega_0)} \leq \frac{c}{\varepsilon}\left\|\frac{i}{\kappa}\frac{\partial \psi_\varepsilon}{\partial x_3} + A^3_\varepsilon\psi_\varepsilon\right\|_{L^2(\Omega_\varepsilon)}\|A^3_\varepsilon\|_{L^2(\Omega_\varepsilon)} \leq c\sqrt{\varepsilon}.$$

**Lemma 8.2.20** *We have*

$$\big\|[\![ \nabla'\psi_\varepsilon \cdot \bar{\psi}_\varepsilon ]\!] - \nabla'[\![ \psi_\varepsilon ]\!] \cdot [\![ \bar{\psi}_\varepsilon ]\!]\big\|_{L^2(Q_0)} \le c\varepsilon; \tag{2.51}$$

$$\|[\![ \boldsymbol{A}'_\varepsilon|\psi_\varepsilon|^2 ]\!] - [\![ \boldsymbol{A}'_\varepsilon ]\!]|[\![ \psi_\varepsilon ]\!]|^2\|_{L^2(0,T;L^{\frac{4}{3}}(\Omega_0))} \le c\varepsilon. \tag{2.52}$$

*Proof.* The estimate (2.51) can be proved by using (2.30), (2.23), Lemma 2.12 and the fact that $|\psi_\varepsilon| \le 1$ a.e. in $Q_\varepsilon$. To prove (2.52) we observe first that

$$\begin{aligned} & [\![ \boldsymbol{A}'_\varepsilon|\psi_\varepsilon|^2 ]\!] - [\![ \boldsymbol{A}'_\varepsilon ]\!]|[\![ \psi_\varepsilon ]\!]|^2 \\ = \; & [\![ (\boldsymbol{A}'_\varepsilon - [\![ \boldsymbol{A}'_\varepsilon ]\!])|\psi_\varepsilon|^2 ]\!] + [\![ \boldsymbol{A}'_\varepsilon ]\!][\![ (\psi_\varepsilon - [\![ \psi_\varepsilon ]\!])\bar{\psi}_\varepsilon ]\!]. \end{aligned} \tag{2.53}$$

On the one hand, similar to the proof of (2.42), we can obtain

$$\|[\![ (\boldsymbol{A}'_\varepsilon - [\![ \boldsymbol{A}'_\varepsilon ]\!])|\psi_\varepsilon|^2 ]\!]\|_{L^2(Q_0)} \le c\varepsilon. \tag{2.54}$$

On the other hand, by Hölder's inequality, the embedding theorem $H^1(\Omega_0) \hookrightarrow L^4(\Omega_0)$ and Lemma 2.13, we have

$$\begin{aligned} \|[\![ \boldsymbol{A}'_\varepsilon ]\!][\![ (\psi_\varepsilon - [\![ \psi_\varepsilon ]\!])\bar{\psi}_\varepsilon ]\!]\|_{L^{\frac{4}{3}}(\Omega_0)} & \le \|[\![ \boldsymbol{A}'_\varepsilon ]\!]\|_{L^4(\Omega_0)}\|[\![ (\psi_\varepsilon - [\![ \psi_\varepsilon ]\!])\bar{\psi}_\varepsilon ]\!]\|_{L^2(\Omega_0)} \\ & \le c\|[\![ (\psi_\varepsilon - [\![ \psi_\varepsilon ]\!])\bar{\psi}_\varepsilon ]\!]\|_{L^2(\Omega_0)}. \end{aligned}$$

Again similar to the proof of (2.42), we can prove

$$\|[\![ (\psi_\varepsilon - [\![ \psi_\varepsilon ]\!])\bar{\psi}_\varepsilon ]\!]\|_{L^2(Q_0)} \le c\varepsilon. \tag{2.55}$$

Now the estimate (2.52) follows from (2.53)–(2.55).

The following is the main result of this section:

**Theorem 8.2.4** *Let the assumptions (A1)–(A3) and (A1)′–(A3)′ be satisfied. For any $\varepsilon > 0$, let $(\psi_\varepsilon, \boldsymbol{A}_\varepsilon)$ denote the solution of (3.2)–(3.4). Then, there exist two functions*

$$\psi \in \mathcal{H}^1(0,T;\mathcal{L}^2(\Omega_0)) \cap \mathcal{L}^\infty(0,T;\mathcal{H}^1(\Omega_0))$$

*and*

$$\boldsymbol{A}' \in \boldsymbol{H}^1(0,T;\boldsymbol{L}^2(\Omega_0)) \cap \boldsymbol{L}^\infty(0,T;\boldsymbol{H}^1_n(\Omega_0))$$

*such that as $\varepsilon \to 0$*

$$\begin{aligned} [\![ \psi_\varepsilon ]\!] \to \psi \quad & \text{weakly in } \mathcal{H}^1(0,T;\mathcal{L}^2(\Omega_0)) \\ & \text{and weakly* in } \mathcal{L}^\infty(0,T;\mathcal{H}^1(\Omega_0)), \end{aligned} \tag{2.56}$$

$$\begin{aligned} [\![ \boldsymbol{A}_\varepsilon ]\!] \to \boldsymbol{A} \quad & \text{weakly in } \boldsymbol{H}^1(0,T;\boldsymbol{L}^2(\Omega_0)) \\ & \text{and weakly* in } \boldsymbol{L}^\infty(0,T;\boldsymbol{H}^1_n(\Omega_0)), \end{aligned} \tag{2.57}$$

*where $\boldsymbol{A} = (\boldsymbol{A}', 0)$, and $(\psi, \boldsymbol{A}')$ satisfies*

$$\psi(x',0) = \psi_0(x'), \quad \boldsymbol{A}'(x',0) = \boldsymbol{A}'_0(x') \tag{2.58}$$

*and*

$$\eta\int_0^T\int_{\Omega_0}\rho(x')\frac{\partial\psi}{\partial t}\omega dx'dt - i\eta\kappa\int_0^T\int_{\Omega_0}\operatorname{div}'\big(\rho(x')\boldsymbol{A}'\big)\psi\omega dx'dt$$
$$+\int_0^T\int_{\Omega_0}\rho(x')\Big(\frac{i}{\kappa}\nabla'\psi+\boldsymbol{A}'\psi\Big)\Big(-\frac{i}{\kappa}\nabla'\omega+\boldsymbol{A}'\omega\Big)dx'dt$$
$$+\int_0^T\int_{\Omega_0}\rho(x')(|\psi|^2-1)\psi\omega dx'dt = 0,$$
$$\text{for any } \omega\in\mathcal{L}^2(0,T;\mathcal{H}^1(\Omega_0)) \tag{2.59}$$
$$\int_0^T\int_{\Omega_0}\rho(x')\frac{\partial\boldsymbol{A}'}{\partial t}\boldsymbol{B}dx'dt+\int_0^T\int_{\Omega_0}\rho(x')\mathrm{curl}\boldsymbol{A}'\mathrm{curl}\boldsymbol{B}dx'dt$$
$$+\int_0^T\int_{\Omega_0}\operatorname{div}'\big(\rho(x')\boldsymbol{A}'\big)\operatorname{div}'\boldsymbol{B}dx'dt$$
$$+\int_0^T\int_{\Omega_0}\rho(x')Re\Big[\Big(\frac{i}{\kappa}\nabla'\psi+\boldsymbol{A}'\psi\Big)\bar\psi\Big]\boldsymbol{B}dx'dt$$
$$=\int_0^T\int_{\Omega_0}\rho(x')H\mathrm{curl}\boldsymbol{B}dx'dt,\quad \text{for any } \boldsymbol{B}\in\boldsymbol{L}^2(0,T;\boldsymbol{H}_n^1(\Omega_0)). \tag{2.60}$$

*Moreover, the electric potential $\phi_\varepsilon$ given in (2.7) satisfies, as $\varepsilon\to 0$,*

$$[\![\,\phi_\varepsilon\,]\!]\to-\frac{1}{\rho(x')}\operatorname{div}'\big(\rho(x')\boldsymbol{A}'\big)\quad \text{weakly in } L^2(0,T;L^2(\Omega_0)). \tag{2.61}$$

*Proof.* From Lemma 2.13 we know that there exist two functions $\psi,\boldsymbol{A}$ such that after possibly extracting a subsequence, (2.16)–(2.17) are satisfied. From Lemma 2.14 we have that $A^3=0$, that is, $\boldsymbol{A}=(\boldsymbol{A}',0)$. By letting $\varepsilon\to 0$ in (2.36) and using Lemmas 2.13-2.16, we know that $(\psi,\boldsymbol{A}')$ satisfies (2.19) for any $\omega\in\mathcal{L}^2(0,T;\mathcal{H}^1(\Omega_0))\cap\mathcal{L}^\infty(Q_0)$ thus also for any $\omega\in\mathcal{L}^2(0,T;\mathcal{H}^1(\Omega_0))$ by the standard density argument. That $(\psi,\boldsymbol{A}')$ satisfies (2.20) can be obtained by letting $\varepsilon\to 0$ in (2.37) by using Lemmas 2.13–2.20, the embedding theorem $\boldsymbol{H}_n^1(\Omega_0)\hookrightarrow\boldsymbol{L}^4(\Omega_0)$ and employing standard convergence argument.

## 8.3 The outside field is given – formal analysis

As in Section 2, we look at the following G-L system

$$\begin{cases}\psi_t-i\mathrm{div}\boldsymbol{A}\psi+\Big(i\nabla+\boldsymbol{A}\Big)^2\psi+\kappa^2\psi(|\psi|^2-1)=0, & \text{in } Q_\varepsilon\\ \boldsymbol{A}_t-\Delta\boldsymbol{A}+\mathrm{Re}\Big[\Big(i\nabla\psi+\psi\boldsymbol{A}\Big)\bar\psi\Big]=\mathbf{curl}\boldsymbol{H}, & \text{in } Q_\varepsilon\\ \nabla\psi\cdot\boldsymbol{n}=0,\quad \boldsymbol{A}\cdot\boldsymbol{n}=0,\quad \mathbf{curl}\boldsymbol{A}\times\boldsymbol{n}=\boldsymbol{H}\times\boldsymbol{n}, & \text{on } \Gamma_\varepsilon\\ \psi(x,0)=\psi_0(x),\quad \boldsymbol{A}(x,0)=\boldsymbol{A}_0(x), & \text{in } \Omega_\varepsilon\end{cases} \tag{3.1}$$

with the assumption that $\boldsymbol{H}$: $\Omega_\varepsilon\to I\!R^3$ is a general given function. We assume that the central surface of the plate lies in the plane $x_3=0$.

### 8.3.1 Variational formulation

Similar to Section 2, we define

$$\begin{aligned}\mathcal{W}^\varepsilon &= \mathcal{L}^2(0,T;\mathcal{H}^1(\Omega_\varepsilon)) \cap \mathcal{H}^1(0,T;(\mathcal{H}^1(\Omega_\varepsilon))'),\\ \boldsymbol{W}_n^\varepsilon &= \boldsymbol{L}^2(0,T;\boldsymbol{H}_n^1(\Omega_\varepsilon)) \cap \boldsymbol{H}^1(0,T;(\boldsymbol{H}_n^1(\Omega_\varepsilon))').\end{aligned} \tag{3.2}$$

The weak formulation for the system (3.1) is then to find $(\psi_\varepsilon, \boldsymbol{A}_\varepsilon) \in \mathcal{W}^\varepsilon \times \boldsymbol{W}_n^\varepsilon$ such that

$$\psi_\varepsilon(x,0) = \psi_{\varepsilon 0}(x), \;\; \boldsymbol{A}_\varepsilon(x,0) = \boldsymbol{A}_{\varepsilon 0}(x) \tag{3.3}$$

and

$$\begin{aligned}&\int_0^T\int_{\Omega_\varepsilon} \frac{\partial \psi_\varepsilon}{\partial t}\omega dx_\varepsilon dt - i\int_0^T\int_{\Omega_\varepsilon} \mathrm{div}_\varepsilon \boldsymbol{A}_\varepsilon \psi_\varepsilon \omega dx_\varepsilon dt\\ &+ \int_0^T\int_{\Omega_\varepsilon} \Big(i\nabla_\varepsilon\psi_\varepsilon + \boldsymbol{A}_\varepsilon\psi_\varepsilon\Big)\Big(-i\nabla_\varepsilon\omega + \boldsymbol{A}_\varepsilon\omega\Big)dx_\varepsilon dt\\ &+ \int_0^T\int_{\Omega_\varepsilon} \kappa^2(|\psi_\varepsilon|^2-1)\psi_\varepsilon\omega dx_\varepsilon dt = 0, \;\; \text{for any } \omega \in \mathcal{L}^2(0,T;\mathcal{H}^1(\Omega_\varepsilon)),\end{aligned} \tag{3.4}$$

$$\begin{aligned}&\int_0^T\int_{\Omega_\varepsilon} \boldsymbol{A}_t \boldsymbol{B} dx_\varepsilon dt + \int_0^T\int_{\Omega_\varepsilon}\Big[\mathbf{curl}_\varepsilon\, \boldsymbol{A}_\varepsilon \mathbf{curl}_\varepsilon\, \boldsymbol{B} + \mathrm{div}_\varepsilon\boldsymbol{A}_\varepsilon \mathrm{div}_\varepsilon \boldsymbol{B}\Big]dx_\varepsilon dt\\ &+ \int_0^T\int_{\Omega_\varepsilon} Re\Big[\Big(i\nabla_\varepsilon\psi_\varepsilon + \boldsymbol{A}_\varepsilon\psi_\varepsilon\Big)\bar{\psi_\varepsilon}\Big]\boldsymbol{B} dx_\varepsilon dt\\ &= \int_0^T\int_{\Omega_\varepsilon} \boldsymbol{H}_\varepsilon \mathbf{curl}_\varepsilon\, \boldsymbol{B} dx_\varepsilon dt, \;\; \text{for any } \boldsymbol{B} \in \boldsymbol{L}^2(0,T;\boldsymbol{H}_n^1(\Omega_\varepsilon)).\end{aligned} \tag{3.5}$$

Here $\boldsymbol{H}_\varepsilon$ is a given vector valued function which depends on $x_\varepsilon$, $\varepsilon$ and $t$. We remark that $\omega$ and $\boldsymbol{B}$ are functions depending on $\varepsilon$, but we will only specify their dependence when it is necessary.

Here $\nabla_\varepsilon$, $\mathbf{curl}_\varepsilon$ and $\mathrm{div}_\varepsilon$ denote the gradient, curl and divergence operators respectively in $\Omega_\varepsilon$. $x_\varepsilon$ denotes the points in $\Omega_\varepsilon$.

The proofs are the same as in the previous section and we omit the details.

### 8.3.2 Formal asymptotic analysis when $\varepsilon \to 0$

The purpose of this section is to give a motivation for the study of the thin plate model subject to large parallel field. We adopt a formal analysis approach for the rigorous proofs are similar to the computations in Section 2 (or Section 4 in the following).

Since the solutions $\psi_\varepsilon$ and $\boldsymbol{A}_\varepsilon$ are smooth functions, from the boundary condition $\boldsymbol{A}_\varepsilon \cdot \boldsymbol{n}_\varepsilon = 0$ or equivalently,

$$\boldsymbol{A}_{\varepsilon 1}\varepsilon a_{x_1} + \boldsymbol{A}_{\varepsilon 2}\varepsilon a_{x_2} - \boldsymbol{A}_{\varepsilon 3} = 0 \tag{3.6}$$

where $x_3 = \varepsilon a(x)$ and

$$\boldsymbol{A}_{\varepsilon 1}\varepsilon b_{x_1} + \boldsymbol{A}_{\varepsilon 2}\varepsilon b_{x_2} + \boldsymbol{A}_{\varepsilon 3} = 0 \tag{3.7}$$

where $x_3 = -\varepsilon b(x)$, using the average function introduced in Section 2, we give the following formal expansion:

$$\begin{cases} \boldsymbol{A}_{\varepsilon 1}(x) = [\![\, \boldsymbol{A}_\varepsilon \,]\!]_1(x') + (x_3 - \xi_{\varepsilon 1}(x'))P_{\varepsilon 1}(x') + O((x_3 - \xi_{\varepsilon 1})^2), \\ \boldsymbol{A}_{\varepsilon 2}(x) = [\![\, \boldsymbol{A}_\varepsilon \,]\!]_2(x') + (x_3 - \xi_{\varepsilon 2}(x'))P_{\varepsilon 2}(x') + O((x_3 - \xi_{\varepsilon 2})^2), \\ \boldsymbol{A}_{\varepsilon 3}(x) = [\![\, \boldsymbol{A}_\varepsilon \,]\!]_3(x') + o((x_3 - \xi_{\varepsilon 3}(x'))). \end{cases} \tag{3.8}$$

Here $-\varepsilon b(x') < \xi_{\varepsilon j}(x') < \varepsilon a(x')$ is the number such that

$$\boldsymbol{A}_{\varepsilon j}(x', \xi_{\varepsilon j}(x')) = [\![\, \boldsymbol{A}_\varepsilon \,]\!]_j(x').$$

From (3.6) and (3.7), it is understood that $[\![\, \boldsymbol{A}_\varepsilon \,]\!]_3$ tends to zero (at least formally) as $\varepsilon$ tends to zero. In the same spirit, we set

$$\psi_\varepsilon = [\![\, \psi_\varepsilon \,]\!] + (x_3 - \xi_{\varepsilon 4}(x'))\eta_\varepsilon + O((x_3 - \xi_4)^2) \tag{3.9}$$

Where $\eta_\varepsilon$ is some bounded function.

Taking $\omega$ as a complex valued test function depending only on $x'$, substituting the expansions of $\psi_\varepsilon$ and $\boldsymbol{A}_\varepsilon$ into the variational formulation (3.4) and (3.5). Assume that as $\varepsilon \to 0$,

$$[\![\, \boldsymbol{A}_\varepsilon \,]\!] \to \boldsymbol{A}, \ \ [\![\, \psi_\varepsilon \,]\!] \to \psi, \ \ \eta_\varepsilon \to \eta, \ \ P_{\varepsilon 1} \to P_1, \ \ P_{\varepsilon 2} \to P_2,$$

dropping the lower order terms and taking the limit $\varepsilon \to 0$ formally in the remaining parts, we obtain the following variational problems satisfied by the limits of the leading order terms:

$$\begin{aligned} &\int_0^T \int_{\Omega_0} \rho(x') \tfrac{\partial \psi}{\partial t} \omega dx' dt - i\kappa \int_0^T \int_{\Omega_0} \operatorname{div}'(\rho(x')\boldsymbol{A}')\psi\omega dx' dt \\ &+ \int_0^T \int_{\Omega_0} \rho(x') \Big( \tfrac{i}{\kappa} \mathbf{grad}'\psi + \boldsymbol{A}'\psi \Big) \Big( -\tfrac{i}{\kappa} \nabla'\omega + \boldsymbol{A}'\omega \Big) dx' dt \\ &+ \int_0^T \int_{\Omega_0} \rho(x')(|\psi|^2 - 1)\psi\omega dx' dt = 0, \end{aligned} \tag{3.10}$$

for any $\omega \in \mathcal{L}^2(0, T; \mathcal{H}^1(\Omega_0))$ and

$$\begin{aligned} &\int_0^T \int_{\Omega_0} \rho(x') \tfrac{\partial \boldsymbol{A}'}{\partial t} \boldsymbol{B} dx' dt + \int_0^T \int_{\Omega_0} \rho(x') \boldsymbol{W} \mathbf{curl} \boldsymbol{B} dx' dt \\ &+ \int_0^T \int_{\Omega_0} \tfrac{1}{\rho(x')} \operatorname{div}'(\rho(x')\boldsymbol{A}') \operatorname{div}'(\rho(x')\boldsymbol{B}) dx' dt \\ &+ \int_0^T \int_{\Omega_0} \rho(x') Re\Big[ \Big( i\boldsymbol{V} + \boldsymbol{A}'\psi \Big) \bar{\psi} \Big] \boldsymbol{B} dx' dt \\ &= \int_0^T \int_{\Omega_0} \rho(x') \boldsymbol{H} \mathbf{curl} \boldsymbol{B} dx' dt, \end{aligned} \tag{3.11}$$

for any $\boldsymbol{B} \in \boldsymbol{L}^2(0, T; \boldsymbol{H}^1_n(\Omega_0))$ with $\boldsymbol{B}_3 = 0$. Here $\boldsymbol{W} = (-P_2, P_1, \mathbf{curl}'\boldsymbol{A}')^T$, $\boldsymbol{V} = (\nabla'\psi^T, \eta)^T$.

Note that $\boldsymbol{B} = \boldsymbol{B}(x')$,

$$\mathbf{curl}\boldsymbol{B} = (\partial_2 \boldsymbol{B}_3,\ -\partial_1 \boldsymbol{B}_3,\ \partial_1 \boldsymbol{B}_2 - \partial_2 \boldsymbol{B}_1). \tag{3.12}$$

Because $\boldsymbol{B}_3 = 0$, $\boldsymbol{B}_1$, $\boldsymbol{B}_2$ and $\omega$ arbitrary, we derive the differential equations:

$$\begin{aligned}&\rho(x')\tfrac{\partial\psi}{\partial t} - i\mathrm{div}'(\rho(x')\boldsymbol{A}')\psi + \left(i\nabla' + \boldsymbol{A}'\right)\left[\rho(x')\left(i\nabla' + \boldsymbol{A}'\right)\right]\psi \\ &+\kappa^2\rho(x')(|\psi|^2 - 1)\psi = 0, \quad \text{in } Q_0,\end{aligned} \tag{3.13}$$

$$\begin{aligned}&\rho(x')\tfrac{\partial \boldsymbol{A}'}{\partial t} + \mathbf{curl}'\left[\rho(x')\mathbf{curl}\boldsymbol{A}'\right] - \rho(x')\nabla'(\tfrac{1}{\rho(x')}\mathrm{div}'(\rho(x')\boldsymbol{A}')) \\ &+\rho(x')Re\left[\left(i\nabla'\psi + \boldsymbol{A}'\psi\right)\bar{\psi}\right] = \mathbf{curl}'(\rho(x')H_3), \quad \text{in } Q_0,\end{aligned} \tag{3.14}$$

$$\nabla'\psi \cdot \boldsymbol{n}' = 0, \quad \boldsymbol{A}' \cdot \boldsymbol{n}' = 0, \quad \mathbf{curl}'\boldsymbol{A}' = H_3, \quad \text{on } \Gamma_0, \tag{3.15}$$

$$\psi(x', 0) = \psi_0(x'), \quad \boldsymbol{A}'(x', 0) = \boldsymbol{A}_0'(x'), \quad \text{in } \Omega_0. \tag{3.16}$$

The result is interesting because we see that for the leading order terms, the limit functions $\psi$ and $\boldsymbol{A}'$ are independent of $H_1$ and $H_2$. The $\boldsymbol{H}'$ relevant components of $\boldsymbol{V}$ and $\boldsymbol{W}$ do not come into the final equation. These results suggest that finite horizontal applied field plays a minor role in the thin plate model as it only gets involved in the lower order terms of the formal asymptotic expansion. In order that the effects of $H_1$ and $H_2$ are felt, we have to increase their sizes while taking the limit $\varepsilon \to 0$. This is reasonable to expect because the physical dimension of the sample confronting the components $H_1$ and $H_2$ of the applied field is the area of the side of the cylindrical domain $\Omega_0 \times (-\varepsilon b(x'), a(x'))$ which is only proportional to $\varepsilon$, it is therefore natural that the strength of the horizontal applied field is proportional to $1/\varepsilon$ so that the combined effect of applied field and sample area would be a finite quantity. More precisely, of the order $|\boldsymbol{H}|\times$ area$\propto \frac{1}{\varepsilon} \cdot \varepsilon = 1$.

## 8.4 Bibliographical remarks

The justification of a 2-d G-L thin film superconductivity model from the 3-d G-L superconductivity model was established in [CDG 96]. The results implied total magnetic penetration without destroying superconductivity. To try to understand this unusual behaviour, [CET 98] tried a different approach based on the assumption that when the film is really thin, the outside magnetic field is almost identical to the applied field. The derived limit model demonstrated some very interesting behaviour of the PDE system involved. We include it here and believe that the limiting behaviour is interesting and could have other application perspectives.

# Chapter 9
# Pinning Theory

In Chapter 6, we have discussed the location of vortices when the G-L parameter $\kappa$ tends to $\infty$. We have obtained a renormalized energy and we claimed that any configuration of the vortices must minimize the corresponding renormalized energy.

In this Chapter, we discuss another factor that affects the location of vortices – the pinning effect. A physical explanation is as follows: provided we put some impurities into a superconductor sample, or because of the non-homogeneous (or non-isotropic) properties of the superconducting crystals, the governing G-L energy involves a function which depends on $x$. More precisely, the G-L energy becomes

$$\int_\Omega a(x)\Big[|(i\nabla + \boldsymbol{A})\psi|^2 + |\mathbf{curl}\boldsymbol{A}|^2 + \frac{\kappa^2}{2}(|\psi|^2 - 1)^2\Big]dx \tag{0.1}$$

where $a(x)$ is a smooth and strictly positive function of $x$. The expected mathematical results is that as $\kappa \to \infty$, the vortices will be forced into (pinned to) the locations where $a(x)$ achieves minimum value. In particular, we minimize the energy in the following admissible function class (see also Chapter 6):

$$\begin{aligned} V = \quad & \{(\psi, \boldsymbol{A}) \in \mathcal{H}^1(\Omega) \times \boldsymbol{H}^1(\Omega) : \\ & |\psi| = 1 \text{ on } \partial\Omega,\ \deg(\psi, \partial\Omega) = d > 0,\ J \cdot \boldsymbol{\tau} = g \text{ on } \partial\Omega\} \end{aligned}$$

where $d$ is a given integer, $g$ is a given smooth function, $J = (\psi_\varepsilon, (i\nabla - \boldsymbol{A}_\varepsilon)\psi_\varepsilon)$ and $(a, b) = \frac{1}{2}(ab^* + a^*b)$ for complex numbers $a$ and $b$. The choice of the set $V$ makes it impossible for the vortices to appear on the boundary (similar argument has been established in Chapter 6 for the non-pinning case). In addition, many proofs in this Chapter are similar or parallel to that in Chapter 6. When such case occurs, we will omit the details and refer to Chapter 6. We mainly follow the work [DLi 97], see also [DLi1], [DLi2], [DLi3], [DLY, 97] and [BHa, 95] for relevent works.

The complex G-L vortex structured phase transition as we know it has only been proved rigorously in the 2-d case, the pinning results are therefore also restricted to the 2-d situation. As a consequence of the discussions in Chapter 1 on the 2-d model, the outside magnetic field is assumed to be a given function.

The mathematical expectation is that vortices will be attracted to the locations where $a(x)$ achieves its minimum value because the energy spent to maintain a vortex at these locations is smaller than elsewhere. It is interesting to notice that the magnetic field is not playing an important role in the location of vortices in this situation at all (in this rescaling regime).

We notice that in (0.1), there is no applied magnetic field. In fact, the presence of the applied field plays little role in the pinning of vortices. This becomes clearer as we go through the proofs. However, the proofs presented in this chapter can be easily extended to the case where there is an applied field involved. We prove that the vortices are effectively pinned to the places where $a(x)$ achieves its minimum. This is independent of the presence of $\boldsymbol{A}$. That is, in this rescaling scheme, magnetic field is becoming secondary in affecting the location of vortices.

After studying (0.1), we will briefly discuss some other related problems. We discuss the pinning effect of the following energy while disregarding the presence of the magnetic potential:

$$\int_\Omega a_{jk}(x)\partial_j u \cdot \partial_k u + \frac{1}{2\varepsilon^2}(|u|^2 - \beta(x))^2 dx \tag{0.2}$$

where $u$ is regarded as a mapping from $\Omega \subset \mathbb{R}^2$ to $\mathbb{R}^2$, $\beta(x)$, $a_{jk}(x) \in C^2(\overline{\Omega})$, $j,k = 1,2$. $\beta(x)$ is strictly positive: $0 < m_0 \leq \beta(x) \leq M_0 < \infty$ and $(a_{jk})$ is a symmetric, positive definite matrix in the standard sense. We claim that the vortices will be pinned to the locations where a function of $\beta$ and the eigenvalues of the matrix $(a_{jk})$ achieves minimum value. The physical explanation of $\beta(x)$ is either that the temperature field is not uniform, or that a thin plate is subject to large horizontal applied field.

We also briefly present the pinning result related to the thin film superconductivity model (but with non-physical boundary conditions due to technical difficulties) derived in Section 1 of Chapter 8.

In terms of notation, we use the operator $\nabla - i\boldsymbol{A}$ to replace $i\nabla + \boldsymbol{A}$ and $-(\nabla - i\boldsymbol{A})^2$ to replace $(i\nabla + \boldsymbol{A})^2$ in the G-L system and energy as in Chapter 6.

## 9.1 Local Pohozaev-type identity

Before proving the local Pohozaev-type identity, we give some definitions and remarks. We have already noticed, in Chapter 6, that the Pohozaev-type identity is of vital importance in studying the location of vortices.

The first result concerns the choice of gauge. Due to the nature of the pinning problem, we need to specify a gauge choice that is different from the standard choices of gauge dealt with so far in this book.

**Lemma 9.1.1** *Let $(\psi, \boldsymbol{A}) \in \mathcal{H}^1(\Omega) \times \boldsymbol{H}^1(\Omega)$. There is $(\tilde{\psi}, \tilde{\boldsymbol{A}}) \in \mathcal{H}^1(\Omega) \times \boldsymbol{H}^1(\Omega)$, gauge equivalent to $(\psi, \boldsymbol{A})$, such that*

$$\begin{cases} \operatorname{div}(a(x)\tilde{\boldsymbol{A}}) = 0 & \text{in } \Omega, \\ \tilde{\boldsymbol{A}} \cdot \boldsymbol{n} = 0 & \text{on } \partial\Omega. \end{cases} \tag{1.1}$$

*Proof.* Consider the problem

$$\begin{cases} -\operatorname{div}(a(x)\nabla\phi) = \operatorname{div}(a(x)\boldsymbol{A}) \text{ in } \Omega, \\ \partial_{\boldsymbol{n}}\phi = -\boldsymbol{A} \cdot \boldsymbol{n} \text{ on } \partial\Omega. \end{cases}$$

It is easy to see that $(\tilde{\psi}, \tilde{\boldsymbol{A}}) = (\psi e^{i\phi}, \boldsymbol{A} + \nabla\phi)$ satisfies (1.1).

In the following we often need a local version of Lemma 1.1.

**Lemma 9.1.2** *Let $G$ be a smooth sub-domain of $\Omega$ and let $(\psi, \boldsymbol{A}) \in \mathcal{H}^1(G) \times \boldsymbol{H}^1(G)$. There is $(\tilde{\psi}, \tilde{\boldsymbol{A}}) \in \mathcal{H}^1(\Omega) \times \boldsymbol{H}^1(\Omega)$, gauge equivalent to $(\psi, \boldsymbol{A})$, such that*

$$\begin{cases} \operatorname{div}(a(x)\tilde{\boldsymbol{A}}) = 0 & \text{in } G, \\ \tilde{\boldsymbol{A}} \cdot \boldsymbol{n} = 0 & \text{on } \partial G. \end{cases} \tag{1.2}$$

Using the notations of Chapter 6, we have the following results (with similar proofs, hence we do not give details).

**Lemma 9.1.3** *Any minimizer $(\psi_\varepsilon, \boldsymbol{A}_\varepsilon)$ satisfying (1.1) solves the following G-L equations*

$$\begin{cases} -(\nabla - i\boldsymbol{A}_\varepsilon) \cdot a(x)(\nabla - i\boldsymbol{A}_\varepsilon)\psi_\varepsilon + \frac{1}{\varepsilon^2}a(x)(|\psi_\varepsilon|^2 - 1)\psi_\varepsilon = 0 \text{ in } \Omega, \\ \mathbf{curl}(a(x)\mathbf{curl}\boldsymbol{A}) + a(x)(i\psi_\varepsilon, (\nabla - i\boldsymbol{A}_\varepsilon)\psi_\varepsilon) = 0 \text{ in } \Omega, \\ \operatorname{div}(a(x)\boldsymbol{A}_\varepsilon) = 0 \text{ in } \Omega, \\ \boldsymbol{A}_\varepsilon \cdot \boldsymbol{n} = 0,\ |\psi_\varepsilon| = 1,\ J_\varepsilon \cdot \boldsymbol{\tau} = g \text{ on } \partial\Omega. \end{cases} \tag{1.3}$$

**Lemma 9.1.4** *Any minimizer $(\psi_\varepsilon, \boldsymbol{A}_\varepsilon)$ satisfying (1.3) satisfies*

$$|\psi_\varepsilon| \le 1 \text{ in } \Omega. \tag{1.4}$$

Now we give the main result of this section:

**Theorem 9.1.1 (Local Pohozaev-type Identity))** *Let $G$ be a sub-domain of $\Omega$. We have, for any solution $(\psi, \boldsymbol{A})$ of the G-L system (1.3),*

$$\begin{aligned} &\int_G a(x)[\frac{(|\psi|^2-1)^2}{\varepsilon^2} - 2|\mathbf{curl}\boldsymbol{A}|^2] + \int_G [(x - x_0) \cdot \nabla a[|(\nabla - i\boldsymbol{A})\psi|^2 \\ &+2|\mathbf{curl}\boldsymbol{A}|^2 + \frac{(|\psi|^2-1)^2}{2\varepsilon^2}] \\ = \ &\int_{\partial G} ((x - x_0) \cdot \boldsymbol{n}) a(x)[|\boldsymbol{\tau} \cdot (\nabla - i\boldsymbol{A})\psi|^2 - |\boldsymbol{n} \cdot (\nabla - i\boldsymbol{A})\psi|^2] \\ &-2\int_{\partial G} ((x - x_0) \cdot \boldsymbol{\tau}) a(x)(\boldsymbol{n} \cdot (\nabla - i\boldsymbol{A})\psi, \boldsymbol{\tau} \cdot (\nabla - i\boldsymbol{A})\psi) \\ &+\frac{1}{2}\int_{\partial G} ((x - x_0) \cdot \boldsymbol{n}) a(x)[\frac{(|\psi|^2-1)^2}{\varepsilon^2} - 2|\mathbf{curl}\boldsymbol{A}|^2], \quad \text{for } x_0 \in G. \end{aligned} \tag{1.5}$$

*Proof.* In this proof, and the rest of this chapter, we will use the notation $h$ for $\mathbf{curl}\boldsymbol{A}$ (or $h_\varepsilon$ for $\mathbf{curl}\boldsymbol{A}_\varepsilon$, or $h_*$ for $\mathbf{curl}\boldsymbol{A}_*$ etc.). Assume $x_0 = 0$ for simplicity. Multiplying (1.3-1) by $x \cdot \nabla\psi^*$ and the conjugate of (1.3-1) by $x \cdot \nabla\psi$, we have

$$\begin{aligned} &-\int_G [\Big((\nabla - i\boldsymbol{A})\big(a(x)(\nabla - i\boldsymbol{A})\psi\big)\Big)(x \cdot \nabla\psi^*) \\ &+\Big((\nabla + i\boldsymbol{A})\big(a(x)(\nabla + i\boldsymbol{A})\psi^*\big)\Big)(x \cdot \nabla\psi)] \\ =\ & \frac{1}{\varepsilon^2}\int_G a(x)[\psi(1-|\psi|^2)(x\cdot\nabla\psi^*) + \psi^*(1-|\psi|^2)(x\cdot\nabla\psi)]. \end{aligned} \tag{1.6}$$

The right-hand side of (1.6) yields

$$-\frac{1}{2\varepsilon^2}\int_{\partial G} a(x)(x\cdot\boldsymbol{n})(1-|\psi|^2)^2 + \frac{1}{2\varepsilon^2}\int_G \operatorname{div}(xa(x))(1-|\psi|^2)^2$$

and the left-hand side is equal to

$$\begin{aligned} &-\int_{\partial G} a(x)[(\boldsymbol{n}\cdot(\nabla + i\boldsymbol{A})\psi^*)(x\cdot\nabla\psi) + (\boldsymbol{n}\cdot(\nabla - i\boldsymbol{A})\psi)(x\cdot\nabla\psi^*)] \\ &-\int_G (x\cdot\nabla a(x))|(\nabla - i\boldsymbol{A})\psi|^2 + \int_{\partial G} a(x)(x\cdot\boldsymbol{n})|(\nabla - i\boldsymbol{A})\psi|^2 \\ &+2\int_G a(x)x_k\partial_k\boldsymbol{A}_j(i\psi, \partial_j\psi - i\boldsymbol{A}_j\psi) + 2\int_G a(x)\boldsymbol{A}_j(i\psi, \partial_j\psi - i\boldsymbol{A}_j\psi). \end{aligned}$$

From $\mathbf{curl}(a\mathbf{curl}\boldsymbol{A}) = -a(x)(i\psi, (\nabla - i\boldsymbol{A})\psi)$, we have

$$\big(\partial_2(a\mathbf{curl}\boldsymbol{A}), -\partial_1(a\mathbf{curl}\boldsymbol{A})\big) = a(x)\big((i\psi, \partial_1\psi - i\boldsymbol{A}_1\psi), (i\psi, \partial_2\psi - i\boldsymbol{A}_2\psi)\big).$$

Hence

$$\begin{aligned} & 2\int_G a(x)x_k\partial_k\boldsymbol{A}_j(i\psi, \partial_j\psi - i\boldsymbol{A}_j\psi) \\ =\ & 2\int_G x_k\partial_k\boldsymbol{A}_1\partial_2(a\mathbf{curl}\boldsymbol{A}) - 2\int_G x_k\partial_k\boldsymbol{A}_2\partial_1(a\mathbf{curl}\boldsymbol{A}) \\ =\ & 2\int_{\partial G} ah[x_k\partial_k\boldsymbol{A}_1\boldsymbol{n}_2 - x_k\partial_k\boldsymbol{A}_2\boldsymbol{n}_1] - 2\int_G (x\cdot\nabla a)h^2 + \int_{\partial G}(x\cdot\boldsymbol{n})ah^2. \end{aligned}$$

Similarly, we have

$$2\int_G a(x)\boldsymbol{A}_j(i\psi, \partial_j\psi - i\boldsymbol{A}_j\psi) = 2\int_{\partial G} ah[\boldsymbol{A}_1\boldsymbol{n}_2 - \boldsymbol{A}_2\boldsymbol{n}_1] + 2\int_G ah^2.$$

Combining above relations with (1.6), we deduce that

$$\begin{aligned} &-\int_{\partial G} a(x)[(\boldsymbol{n}\cdot(\nabla + i\boldsymbol{A})\psi^*)(x\cdot\nabla\psi) + (\boldsymbol{n}\cdot(\nabla - i\boldsymbol{A})\psi)(x\cdot\nabla\psi^*)] \\ &-\int_G (x\cdot\nabla a(x))|(\nabla - i\boldsymbol{A})\psi|^2 + \int_{\partial G} a(x)(x\cdot\boldsymbol{n})|(\nabla - i\boldsymbol{A})\psi|^2 \end{aligned}$$

$$
\begin{aligned}
&+2\int_{\partial G} ah[x_k\partial_k \boldsymbol{A}_1\boldsymbol{n}_2 - x_k\partial_k\boldsymbol{A}_2\boldsymbol{n}_1] - 2\int_G (x\cdot\nabla a)h^2\\
&+\int_{\partial G}(x\cdot\boldsymbol{n})ah^2 + 2\int_{\partial\Omega} ah[\boldsymbol{A}_1\boldsymbol{n}_2 - \boldsymbol{A}_2\boldsymbol{n}_1] + 2\int_G ah^2\\
= &-\frac{1}{2\varepsilon^2}\int_{\partial G} a(x)(x\cdot\boldsymbol{n})(1-|\psi|^2)^2 + \frac{1}{2\varepsilon^2}\int_G \operatorname{div}(xa)(1-|\psi|^2)^2. \qquad (1.7)
\end{aligned}
$$

Again,

$$
\begin{aligned}
&(\boldsymbol{n}\cdot(\nabla - i\boldsymbol{A})\psi)(x\cdot\nabla\psi^*) + (\boldsymbol{n}\cdot(\nabla + i\boldsymbol{A})\psi^*)(x\cdot\nabla\psi)\\
= &(\boldsymbol{n}\cdot(\nabla - i\boldsymbol{A})\psi)(x\cdot(\nabla + i\boldsymbol{A})\psi^*)\\
&+(\boldsymbol{n}\cdot(\nabla + i\boldsymbol{A})\psi^*)(x\cdot(\nabla - i\boldsymbol{A})\psi)\\
&+[(\boldsymbol{n}\cdot(\nabla + i\boldsymbol{A})\psi^*)ix\boldsymbol{A}\psi - (\boldsymbol{n}\cdot(\nabla - i\boldsymbol{A})\psi)ix\boldsymbol{A}\psi^*]\\
= &2(x\cdot\boldsymbol{n})|(\boldsymbol{n}\cdot(\nabla - i\boldsymbol{A})\psi)|^2\\
&+2(x\cdot\boldsymbol{\tau})(\boldsymbol{n}\cdot(\nabla - i\boldsymbol{A})\psi, \boldsymbol{\tau}\cdot(\nabla - i\boldsymbol{A})\psi)\\
&+2(x\cdot\boldsymbol{A})(i\psi, \boldsymbol{n}\cdot(\nabla - i\boldsymbol{A})\psi). \qquad (1.8)
\end{aligned}
$$

From (1.7) and (1.8) we obtain, using the fact that $\operatorname{div}\mathbf{curl}\boldsymbol{B} = 0$ for any $\boldsymbol{B}$,

$$
\begin{aligned}
&-2\int_{\partial G} a(x)(x\cdot\boldsymbol{A})(i\psi, \boldsymbol{n}\cdot(\nabla - i\boldsymbol{A})\psi)\\
= &-2\int_G \operatorname{div}\big(a(x)(x\cdot\boldsymbol{A})(i\psi, (\nabla - i\boldsymbol{A})\psi)\big)\\
= &-2\int_G \operatorname{div}\big((x\cdot\boldsymbol{A})\mathbf{curl}(ah)\big) = -2\int_G \nabla(x\cdot\boldsymbol{A})\cdot\mathbf{curl}(ah)\\
= &-2\int_G \{(\boldsymbol{A}_1 + x_1\partial_1\boldsymbol{A}_1 + x_2\partial_1\boldsymbol{A}_2)\partial_2(ah)\\
&-(\boldsymbol{A}_2 + x_1\partial_2\boldsymbol{A}_1 + x_2\partial_2\boldsymbol{A}_2)\partial_1(ah)\}\\
= &\int_{\partial G} 2ah(\boldsymbol{A}_2\boldsymbol{n}_1 - \boldsymbol{A}_1\boldsymbol{n}_2) + 2ah[(x_1\partial_2\boldsymbol{A}_1 + x_2\partial_2\boldsymbol{A}_2)\boldsymbol{n}_1 \qquad (1.9)\\
&-(x_1\partial_1\boldsymbol{A}_1 + x_2\partial_1\boldsymbol{A}_2)\boldsymbol{n}_2]\\
= &\,2\int_{\partial G} ah(\boldsymbol{A}_2\boldsymbol{n}_1 - \boldsymbol{A}_1\boldsymbol{n}_2) + 2\int_{\partial G} ah[(x_1\partial_1\boldsymbol{A}_2 + x_2\partial_2\boldsymbol{A}_2)\boldsymbol{n}_1\\
&-(x_1\partial_1\boldsymbol{A}_1 + x_2\partial_2\boldsymbol{A}_1)\boldsymbol{n}_2]\\
&+2\int_{\partial G} ah[(x_1\partial_2\boldsymbol{A}_1 - x_1\partial_1\boldsymbol{A}_2)\boldsymbol{n}_1 - (x_2\partial_1\boldsymbol{A}_2 - x_2\partial_2\boldsymbol{A}_1)\boldsymbol{n}_2]\\
= &\,2\int_{\partial G} ah(\boldsymbol{A}_2\boldsymbol{n}_1 - \boldsymbol{A}_1\boldsymbol{n}_2) + 2\int_{\partial G} ah[(x_k\partial_k\boldsymbol{A}_2)\boldsymbol{n}_1\\
&-(x_k\partial_k\boldsymbol{A}_1)\boldsymbol{n}_2] - 2\int_{\partial G} ah^2(x\cdot\boldsymbol{n}). \qquad (1.10)
\end{aligned}
$$

Finally, substituting the first term in the left-hand side of (1.7) into (1.8) and (1.9), we have

$$\begin{aligned}
&\int_{\partial G} a(x)(x\cdot\boldsymbol{n})[|\boldsymbol{\tau}\cdot(\nabla-i\boldsymbol{A})\psi|^2-|\boldsymbol{n}\cdot(\nabla-i\boldsymbol{A})\psi|^2] \\
&-2\int_{\partial\Omega} a(x)(x\cdot\boldsymbol{\tau})(\boldsymbol{n}\cdot(\nabla-i\boldsymbol{A})\psi,\boldsymbol{\tau}\cdot(\nabla-i\boldsymbol{A})\psi) \\
&-\int_{\partial G} a(x\cdot\boldsymbol{n})h^2-\int_G(x\cdot\nabla a)|(\nabla-i\boldsymbol{A})\psi|^2-2\int_G(x\cdot\nabla a)h^2+2\int_G ah^2 \\
= &\ \frac{1}{\varepsilon^2}\int_G a(x)(|\psi|^2-1)^2+\frac{1}{2\varepsilon^2}\int_G(x\cdot\nabla a)(|\psi|^2-1)^2 \\
&-\frac{1}{2\varepsilon^2}\int_{\partial G} a(x)(x\cdot\boldsymbol{n})(|\psi|^2-1)^2.
\end{aligned}$$

This gives (1.5).

**Remark 9.1.1** *Choose $a(x)\equiv 1$ in (1.5), we obtain Proposition 1.2 in Chapter 6.*

## 9.2 Estimate the energy of minimizers

**Lemma 9.2.1** *For any given $\sigma>0$, $0\le\varepsilon\le 1$, we have*

$$G_\varepsilon(\psi_\varepsilon,\boldsymbol{A}_\varepsilon)\le\pi d(m+\sigma)|\ln\varepsilon|+C, \tag{2.1}$$

*where $m=\min a(x)$, $C$ depends only on $\Omega$, $g$, $a$ and $\sigma$.*

*Moreover, if $N=Card(a^{-1}(m)\cap\Omega)\ge d$, then*

$$G_\varepsilon(\psi_\varepsilon,\boldsymbol{A}_\varepsilon)\le\pi dm|\ln\varepsilon|+\tilde{C}(\Omega,g,a). \tag{2.2}$$

*Proof.* Since $a(x)$ is smooth, there exists a ball $B_\rho(x_0)$ with radius $\rho>0$ such that $\max_{\overline{\Omega\cap B_\rho(x_0)}} a(x)\le m+\sigma$.

Fix $d$ distinct points $b_1,\dots,b_d$ in $B_\rho(x_0)\cap\Omega$ and let $R>0$ so small that

$$B_R(b_j)\subset\subset\Omega\cap B_\rho(x_0),\ \overline{B_R(b_j)}\cap\overline{B_R(b_k)}=\emptyset\ \text{ if }\ j\neq k.$$

Set $\tilde\Omega=\Omega\setminus\cup_{j=1}^d B_R(b_j)$ and let $w_0$ be a smooth map from $\tilde\Omega$ to $S^1$ verifying for all $j$,

$$w_0=e^{i\theta}=\frac{z-b_j}{|z-b_j|}\quad\text{on }\ \partial B_R(b_j).$$

It is clear that

$$\partial_1 w_0\times\partial_2 w_0=0.$$

Let $\boldsymbol{B}_0 = (\boldsymbol{B}_1^0, \boldsymbol{B}_2^0)$, where

$$\boldsymbol{B}_1^0 = w_0 \times \partial_1 w_0 = (iw_0, \partial_1 w_0),$$
$$\boldsymbol{B}_2^0 = w_0 \times \partial_2 w_0 = (iw_0, \partial_2 w_0).$$

It follows from a simple calculation that $(\nabla - i\boldsymbol{B}_0)w_0 = 0$ in $\tilde{\Omega}$ and that $\mathbf{curl}\boldsymbol{B}_0 = 0$ in $\tilde{\Omega}$. Let $\boldsymbol{B}_1$ be a solution of

$$\begin{cases} \Delta \boldsymbol{B}_1 = 0 & \text{in} \quad \Omega, \\ \boldsymbol{B}_1 \cdot \boldsymbol{\tau} = -g, \ \boldsymbol{B}_1 \cdot \boldsymbol{n} = 0 & \text{on} \quad \partial\Omega. \end{cases}$$

We deduce from elliptic estimates that

$$\int_\Omega (|\boldsymbol{B}_1|^2 + |\nabla \boldsymbol{B}_1|^2) \leq C.$$

Moreover, let

$$\boldsymbol{B}_\varepsilon = \boldsymbol{B}_0 + \boldsymbol{B}_1 \ \text{ and } \ v_\varepsilon = w_0 \ \text{ on } \tilde{\Omega}.$$

One easily gets

$$|v_\varepsilon| = 1, \ (\nabla - i\boldsymbol{B}_\varepsilon)v_\varepsilon = -i\boldsymbol{B}_1 v_\varepsilon \ \text{ on } \tilde{\Omega}$$

and

$$\mathbf{curl}\boldsymbol{B}_\varepsilon = \mathbf{curl}\boldsymbol{B}_1 \ \text{ on } \ \tilde{\Omega}.$$

Hence, we have

$$\frac{1}{2}\int_{\tilde{\Omega}} a(x)[|(\nabla - i\boldsymbol{B}_\varepsilon)v_\varepsilon|^2 + |\mathbf{curl}\boldsymbol{B}_\varepsilon|^2 + \frac{1}{2\varepsilon^2}(|v_\varepsilon|^2 - 1)^2$$
$$= \frac{1}{2}\int_{\tilde{\Omega}} a(x)[|\boldsymbol{B}_1|^2 + |\nabla \boldsymbol{B}_1|^2] \leq C.$$

It remains to define $(v_\varepsilon, \boldsymbol{B}_\varepsilon)$ on $\cup_{j=1}^d B_R(b_j)$. Let $\xi$ be a smooth function from $I\!R$ to $I\!R_+$ such that

$$\xi(t) = 1 \ \text{ if } \ t \geq 1; \ \xi(t) = 0 \ \text{ if } t \leq 1/2.$$

On $B_R(b_j)$ and for $\varepsilon < R$, set

$$v_\varepsilon = \frac{x - b_j}{|x - b_j|}\xi(\frac{|x - b_j|}{\varepsilon}), \ \boldsymbol{B}_\varepsilon(x) = \boldsymbol{B}_\varepsilon(\frac{x - b_j}{|x - b_j|}R)\xi(\frac{|x - b_j|}{R})$$

to yield

$$\frac{1}{2}\int_{B_R(b_j)} a(x)[|(\nabla - i\boldsymbol{B}_\varepsilon)v_\varepsilon|^2 + |\mathbf{curl}\boldsymbol{B}_\varepsilon|^2$$
$$+ \frac{1}{2\varepsilon^2}(|v_\varepsilon|^2 - 1)^2] \leq (m + \sigma)\pi|\ln \varepsilon| + C.$$

It is clear that $(v_\varepsilon, \boldsymbol{B}_\varepsilon) \in \mathcal{H}^1 \times \boldsymbol{H}^1$ and

$$G_\varepsilon(v_\varepsilon, \boldsymbol{B}_\varepsilon) \le \pi d(m+\sigma)|\ln\varepsilon| + C \tag{2.3}$$

On the other hand, we have

$$\deg(v_\varepsilon, \partial\Omega) = \deg(w_0, \partial\Omega) = d, \quad |v_\varepsilon| = |w_0| = 1 \ \text{ on } \ \partial\Omega$$

and

$$J_\varepsilon \cdot \boldsymbol{\tau} = -\boldsymbol{B}_1 \cdot \boldsymbol{\tau} = g \ \text{ on } \ \partial\Omega.$$

Hence $(v_\varepsilon, \boldsymbol{B}_\varepsilon) \in V$ and (2.1) follows from (2.3).

Note that if $N \ge d$, then one may choose $b_j \in a^{-1}(m) \cap \Omega$ in the proof of (2.1). Since $a(x) \le m + a_0|x-b_j|^2$ in $B_R(b_j)$, (2.2) follows from a simple computation.

**Remark 9.2.1** *This lemma states that the energy is controlled by $m|\ln\varepsilon|$. It is this fact that implies the pinning of vortices as we shall see in Section 6.*

**Lemma 9.2.2** *We have, for some constant $C$ depending only on $g$, $\Omega$ and $a(x)$,*

$$|(\nabla - i\boldsymbol{A}_\varepsilon)\psi_\varepsilon| \le C\varepsilon^{-1}, \quad |\nabla(a\,\mathbf{curl}\boldsymbol{A}_\varepsilon)| \le C\varepsilon^{-1} \ \text{ in } \ \Omega. \tag{2.4}$$

The proof is the same as the proof of Proposition 1.4 in Chapter 6. We omit the details.

## 9.3 Local estimates

In this section, to simplify notation, we drop the dependence of $\varepsilon$ from $\psi_\varepsilon$, $\boldsymbol{A}_\varepsilon$ and use $\psi$ and $\boldsymbol{A}$ instead.

**Lemma 9.3.1** *We have, for any $0 < \alpha < 1$ and $x_0 \in \Omega$,*

$$\int_{B(x_0,\varepsilon^\alpha)\cap\Omega} |\mathbf{curl}\boldsymbol{A}|^2 \le C\varepsilon^\alpha|\ln\varepsilon| \tag{3.1}$$

*with $C$ independent of $\varepsilon$.*

*Proof.* The G-L equation in magnetic potential implies that

$$|\nabla(a\,\mathbf{curl}\boldsymbol{A})| = |\mathbf{curl}(a\,\mathbf{curl}\boldsymbol{A})| \le C|(\nabla - i\boldsymbol{A})\psi|,$$

we have

$$\int_\Omega |\nabla(a\,\mathbf{curl}\boldsymbol{A})|^2 \le C|\ln\varepsilon|.$$

On the other hand, we also have, by standard energy estimate,

$$\int_\Omega |\mathbf{curl}\boldsymbol{A}|^2 \le C|\ln\varepsilon|.$$

Hence

$$\int_\Omega |\nabla \mathbf{curl} \boldsymbol{A}|^2 \le C\Big(\int_\Omega |\mathbf{curl}\boldsymbol{A}|^2 + \int_\Omega |\nabla(a\mathbf{curl}\boldsymbol{A})|^2\Big) \le C|\ln\varepsilon|.$$

By Sobolev embedding, we deduce that for any $p > 2$,

$$\Big(\int_\Omega |\mathbf{curl}\boldsymbol{A}|^p\Big)^{1/p} \le C_p|\ln\varepsilon|^{1/2}.$$

It follows that

$$\begin{aligned}\Big(\int_{\Omega\cap B(x_0,\varepsilon^\alpha)} |\mathbf{curl}\boldsymbol{A}|^2\Big)^{1/2} &\le \Big(\int_\Omega |\mathbf{curl}\boldsymbol{A}|^p\Big)^{1/p}(\mathrm{meas}B(x_0,\varepsilon^\alpha))^{1/2-1/p}\\ &\le C|\ln\varepsilon|^{1/2}\varepsilon^{\alpha(1-2/p)}.\end{aligned}$$

This gives (3.1).

For $\rho > 0$, set

$$\begin{aligned} f(\rho) &= f(\rho, x_0, \varepsilon, \psi, \boldsymbol{A})\\ &= \frac{\rho}{2}\int_{\partial B_\rho(x_0)\cap\Omega} a(x)\big[|(\nabla - i\boldsymbol{A})\psi|^2 + |\mathbf{curl}\boldsymbol{A}|^2 + \frac{1}{2\varepsilon^2}(|\psi|^2-1)^2\big]ds.\end{aligned}$$

**Lemma 9.3.2** *:*

*(i) For* $0 < \varepsilon < e^{-1}$, *we have*

$$\inf_{\varepsilon^{1/2}\le\rho\le\varepsilon^{1/4}} f(\rho) \le 4\frac{G(\psi,\boldsymbol{A},\Omega\cap B_{\varepsilon^{1/4}}(x_0))}{|\ln\varepsilon|} \le C_1,$$

*and*

$$\inf_{5\varepsilon^{1/4}\le\rho\le5\varepsilon^{1/8}} f(\rho) \le 2C_1,$$

*where* $C_1 = C_1(\Omega, g, a)$ *and*

$$G(\psi,\boldsymbol{A},\Omega\cap B_\rho(x_0)) = \\ \frac{1}{2}\int_{\Omega\cap B_\rho(x_0)} a(x)\big[|(\nabla - i\boldsymbol{A})\psi|^2 + |\mathbf{curl}\boldsymbol{A}|^2 + \frac{1}{2\varepsilon^2}(|\psi|^2-1)^2\big]dx.$$

*(ii) There are constants* $\gamma = \gamma(\Omega, g, a, \delta)$ *and* $\varepsilon_0 = \varepsilon_0(\Omega, g, a) > 0$ *such that for* $0 < \varepsilon < \varepsilon_0$, *there holds*

$$\inf_{B_\rho(x_0)} |\psi(x)| \ge \frac{1}{2}, \quad \textit{and} \quad G_\varepsilon(\psi,\boldsymbol{A},\Omega\cap B_\rho(x_0)) \le \delta$$

*whenever* $\varepsilon^{1/2} \le \rho \le \varepsilon^{1/4}$ *and* $f(\rho) \le \gamma$. *($\delta > 0$ will be determined in Lemma 5.2).*

*Proof.* (i) It follows from Fubini's theorem that

$$\begin{aligned} G_\varepsilon(\psi, \boldsymbol{A}) &\geq G_\varepsilon(\psi, \boldsymbol{A}, \Omega \cap B_{\varepsilon^{1/4}}(x_0)) \geq \int_{\varepsilon^{1/2}}^{\varepsilon^{1/4}} \frac{f(\rho)}{\rho} d\rho \\ &\geq \ln\left(\frac{\varepsilon^{1/4}}{\varepsilon^{1/2}}\right) \inf_{\varepsilon^{1/2} \leq \rho \leq \varepsilon^{1/4}} f(\rho) = \frac{1}{4} |\ln \varepsilon| \inf_{\varepsilon^{1/2} \leq \rho \leq \varepsilon^{1/4}} f(\rho) \end{aligned}$$

and similarly for the 2nd inequality, here we have used Lemma 3.1.

(ii) Choose $\rho_1 = \rho_1(\Omega) > 0$ such that for $0 < \rho < \rho_1$ the domain $D = \Omega \cap B_\rho(x_0)$ is strongly star-shaped with respect to some point $x_1 \in D$ in the following sense:

$$\boldsymbol{n} \cdot (x - x_1) \geq \frac{\rho}{4} \quad \text{for all} \quad x \in \partial D.$$

Denote $\partial D = (\partial B_\rho(x_0) \cap \Omega) \cup \Gamma_\rho)$ where $\Gamma_\rho = \partial\Omega \cap B_\rho(x_0)$. Applying Theorem 1.1 on $D$ we have

$$\begin{aligned} \frac{1}{\varepsilon^2} \int_D (|\psi|^2 - 1)^2 dx &\leq C\rho G_\varepsilon(\psi, \boldsymbol{A}) + C f(\rho) \\ &\quad + C\rho \int_{\Gamma_\rho} |\boldsymbol{\tau} \cdot (\nabla - i\boldsymbol{A})\psi|^2 + C \int_D |h|^2 dx \\ &\leq C\varepsilon^{1/4} |\ln \varepsilon| + C\varepsilon^{1/4} \int_{\Gamma_\rho} g^2 + C\gamma \leq C(\varepsilon^{1/8} + \gamma), \end{aligned} \tag{3.2}$$

since $|(x - x_1) \cdot \nabla a| \leq \rho a_1$ and $\rho \leq \rho_1$ small enough. Therefore, if $|u(x_2)| \leq \frac{1}{2}$ for some $x_2 \in D$, by (2.3) we have

$$|\nabla \psi| \leq C|(\nabla - i\boldsymbol{A})\psi| \leq C_2 \varepsilon^{-1}.$$

Hence

$$|u(y)| \leq \frac{3}{4} \quad \text{for} \quad |y - x_2| \leq \frac{\varepsilon}{4C_2}$$

and

$$\frac{1}{\varepsilon^2} \int_D (|\psi|^2 - 1)^2 \geq C_3 > 0$$

which contradicts (3.2) when $\varepsilon$, $\gamma$ are choosen sufficiently small.

Now, if $|\psi| \geq 1/2$ in $D$, we are able to prove the 2nd assertion of (ii). In fact, note that $f(\rho) \leq \gamma$, $D = \Omega \cap B_\rho(x_0)$, $\varepsilon^{1/2} \leq \rho \leq \varepsilon^{1/4}$. We know, by constructing a comparison function as in (1.87)–(1.88) in Chapter 6, that

$$G_\varepsilon(\psi, \boldsymbol{A}, D) \leq C f(\rho) < \delta$$

whenever $0 < \varepsilon \leq \varepsilon_0$, $C\gamma < \delta$.

For $0 < \varepsilon \leq \varepsilon_0$ and minimizers $(\psi, \boldsymbol{A})$ of $G_\varepsilon$, consider the set

$$S_\varepsilon = \{x \in \Omega : \ |\psi_\varepsilon| \leq \frac{1}{2} \ \text{or} \ G_\varepsilon(\psi, \boldsymbol{A}, \Omega \cap B_\rho) \geq \delta\}$$

and its cover $\{B_{\varepsilon^{1/4}}\}_{x\in S_\varepsilon}$. For $x \in S_\varepsilon$, let $\varepsilon^{1/4} \le \rho(x) \le \varepsilon^{1/2}$ be determined as in Lemma 3.2 such that

$$\frac{4G_\varepsilon(\psi, \boldsymbol{A}, \Omega \cap B_{\varepsilon^{1/4}}(x))}{|\ln \varepsilon|} \ge f(\rho(x), x, \varepsilon, \psi, \boldsymbol{A}) \ge \gamma. \tag{3.3}$$

By Vitali's covering lemma, there is a collection of disjoint balls $B_j = B_{\varepsilon^{1/4}}(x_j)$, $x_j \in S_\varepsilon$, $1 \le j \le I = I(\psi_\varepsilon, \boldsymbol{A}_\varepsilon)$ such that

$$\Omega \cap \bigcup_{x\in S_\varepsilon} B_{\varepsilon^{1/4}}(x) \subset \bigcup_{j=1}^{I} B_{5\varepsilon^{1/4}}(x_j).$$

Moreover, by (3.3) we obtain a uniform bound

$$I \le \sum_j 4\frac{G_\varepsilon(\psi, \boldsymbol{A}, \Omega \cap B_j)}{\gamma|\ln \varepsilon|} \le C\gamma^{-1} = I_0 \tag{3.4}$$

on the number of such balls $B_j$.

For $1 \le j \le I$, let $\rho_j \in [5\varepsilon^{1/4}, 5\varepsilon^{1/8}]$ such that

$$f(\rho_j, x_j, \varepsilon, \psi, \boldsymbol{A}) \le 2C_1$$

and denote $D_j = \Omega \cap B_{\rho_j}(x_j)$. Using Theorem 1.1 as in the proof of Lemma 3.2 (ii), we have

**Lemma 9.3.3** *There exists a constant $C_4$ such that*

$$\frac{1}{\varepsilon^2}\int_{D_j} (|\psi|^2 - 1)^2 \le C_4$$

*uniformly in $0 < \varepsilon < \varepsilon_0$, $1 \le j \le I$.*

Combining Lemmas 2.2 and 3.3 we obtain

**Lemma 9.3.4** *There is a number $J_0 = J_0(\Omega, g, a) \in \mathbb{N}$ such that for any collection of disjoint balls $\{B_{\varepsilon/5}(x)\}_{x\in S_\varepsilon}$, $x_j \in \Omega$, $1 \le j \le J$ with $|\psi(x_j)| < 1/2$, there holds $J \le J_0$.*

Now consider the cover $\{B_{\varepsilon/5}(x)\}_{x\in S_\varepsilon}$ of $S_\varepsilon$. Again by Vitali's covering lemma, we may find a collection of disjoint balls $B_{\varepsilon/5}(x_j)$, $x_j \in S_\varepsilon$, $1 \le j \le J$ such that $S_\varepsilon \subset \cup_j B_\varepsilon(x_j)$. By Lemma 3.4, we have $J \le J_0$ independent of $\varepsilon$.

**Corollary 9.3.1** *There exists a constant $J_0$ independent of $\varepsilon$ and points $x_1^\varepsilon$, ..., $x_{N_\varepsilon}^\varepsilon$ in $\Omega$ such that*

$$N_\varepsilon \le J_0,\ |\psi(x)| \ge \frac{1}{2} \text{ on } \Omega \setminus \bigcup_{j=1}^{N_\varepsilon} B(x_j^\varepsilon, \varepsilon).$$

However, we have not excluded the possibility that some of the discs $B(x_j^\varepsilon,\varepsilon)$ may intersect. To avoid this situation, we argue as in Section 1.6 of Chapter 6 and replace $B(x_j^\varepsilon,\varepsilon)$ by slightly larger discs $B(x_j^\varepsilon,\lambda\varepsilon)$ for $j\in\{1,\dots,l_\varepsilon\}\subset\{1,\dots,N_\varepsilon\}$ where $\lambda\ge 1$ is independent of $\varepsilon$ such that $\cup_1^{N_\varepsilon}B(x_j^\varepsilon,\varepsilon)\subset\cup_{j=1}^{l_\varepsilon}B(x_j^\varepsilon,\lambda\varepsilon)$ and

$$B(x_j^\varepsilon,4\lambda\varepsilon)\cap B(x_k^\varepsilon,4\lambda\varepsilon)=\emptyset,\ j\neq k, \tag{3.5}$$

$$|\psi_\varepsilon(x)\ge\frac{1}{2},\quad \text{on } \Omega\setminus\cup_{j=1}^{l_\varepsilon}B(x_j^\varepsilon,\lambda\varepsilon). \tag{3.6}$$

Set $J_\varepsilon=\{1,\dots,l_\varepsilon\}$ and drop $\varepsilon$ if no confusion occurs. Denote

$$\omega_j=B(x_j,\lambda\varepsilon),\ j\in J_\varepsilon,$$
$$\Gamma_j=\partial B(x_j,\lambda\varepsilon),\ j\in J_\varepsilon.$$

Define $J_1=\{j\in J_\varepsilon:\ B(x_j,\lambda\varepsilon)\cap\partial\Omega=\emptyset\}$, $J_2=J_\varepsilon\setminus J_1$ and

$$\begin{aligned}\tilde\Omega&=\Omega\setminus\cup_{j\in J_\varepsilon}B(x_j^\varepsilon,\lambda\varepsilon),\\ \hat\Omega&=\Omega\setminus\cup_{j\in J_\varepsilon}B(x_j^\varepsilon,2\lambda\varepsilon),\\ \Omega_1&=\Omega\setminus\cup_{j\in J_2}B(x_j^\varepsilon,\lambda\varepsilon)\end{aligned} \tag{3.7}$$

and $\Gamma_0=\partial\Omega_1$. We call such balls $B(x_j^\varepsilon,\lambda\varepsilon)$ $(j\in J_\varepsilon)$ *bad* discs of $\psi$.

## 9.4 Global Estimates

**Theorem 9.4.1** *There are constants $C_p$ and $C_q$ depending only on $d$, $a$, $g$, $\Omega$, $p$ and $q$ such that*

$$\int_\Omega|\nabla\mathbf{curl}\boldsymbol{A}_\varepsilon|^p\le C_p,\quad\forall p<2, \tag{4.1}$$

$$\int_\Omega|\mathbf{curl}\boldsymbol{A}_\varepsilon|^q\le C_q,\quad\forall q<+\infty, \tag{4.2}$$

$$\int_\Omega|(\nabla-i\boldsymbol{A}_\varepsilon)\psi_\varepsilon|^p\le C_p,\quad\forall p<2. \tag{4.3}$$

As the proof is almost the same as that for Theorem 1.3 in Section 1.6, Chapter 6, we omit the details.

## 9.5 Convergence of solutions and the term $\frac{1}{\varepsilon^2}\int_\Omega\left(|\psi_\varepsilon|^2-1\right)^2$

In this section, we define

$$m=\min_{x\in\overline\Omega}a(x), \tag{5.1}$$

$$a^{-1}(m)=\{x\in\overline\Omega|\ a(x)=m\}. \tag{5.2}$$

Recall that

$$d=\deg(\psi,\partial\Omega) \tag{5.3}$$

is a positive integer. In the following we will distinguish two cases in our proofs:

**Case I:** $a^{-1}(m)\subset\Omega$:

$$N=\text{Card } a^{-1}(m)\geq d, \tag{5.4}$$

or

$$N=\text{Card } a^{-1}(m)< d. \tag{5.5}$$

**Case II:** $a^{-1}(m)\cap\partial\Omega\neq\emptyset$:

$$N=\text{Card } \left(a^{-1}(m)\cap\Omega\right)\geq d, \tag{5.6}$$

or

$$N=\text{Card } \left(a^{-1}(m)\cap\Omega\right)< d. \tag{5.7}$$

After giving the notation, we now announce the main results:

**Theorem 9.5.1** *Suppose that (5.4) (or (5.5), or (5.6), or (5.7)) holds. Let* $(\psi_\varepsilon,\boldsymbol{A}_\varepsilon)$ *be any sequence of minimizers of (0.1) satisfying (1.1). There are a subsequence* $(\psi_{\varepsilon_n},\boldsymbol{A}_{\varepsilon_n})$*, points* $a_1,\dots,a_{N_0}\in\overline{\Omega}$ $(N_0<\infty)$ *and* $(\psi_*,\boldsymbol{A}_*)$ *smooth except at* $a_1,\dots,a_{N_0}$ *such that*

$$\begin{cases}\operatorname{div}(a(x)\boldsymbol{A}_*)=0 & \textit{in } \Omega,\\ \boldsymbol{A}_*\cdot\boldsymbol{n}=0 & \textit{on } \partial\Omega\end{cases} \tag{5.8}$$

*and*

$$\psi_{\varepsilon_n}\to\psi_* \textit{ strongly in } \mathcal{H}^1_{loc}(\Omega\setminus\cup_{j=1}^{N_0}\{a_j\}) \textit{ and in } \mathcal{W}^{1,p}(\Omega),\ \forall p<2. \tag{5.9}$$

*Set* $h_{\varepsilon_n}=\mathbf{curl}\boldsymbol{A}_{\varepsilon_n}$, $h_*=\mathbf{curl}\boldsymbol{A}_*$. *We have*

$$h_{\varepsilon_n}\to h_* \textit{ strongly in } \boldsymbol{H}^1_{loc}(\Omega\setminus\cup_{j=1}^{N_0}\{a_j\}) \textit{ and in } \boldsymbol{W}^{1,p}(\Omega),\ \forall p<2. \tag{5.10}$$

The following of this section is dedicated to the proof of this result:

Extract a subsequence $\varepsilon_n\to 0$ such that $|J_{\varepsilon_n}|$ =constant= $N_1$, $x_j^{\varepsilon_n}\to l_j\in\overline{\Omega}$, $\forall j\in J=J_{\varepsilon_n}$. Denote by $a_1,\dots,a_{N_0}$ $(N_0\leq N_1)$ the collection of distinct points in $\{l_1,\dots,l_{N_1}\}$. For simplicity we will use the notation $h_n=h_{\varepsilon_n}$, $\boldsymbol{A}_n=\boldsymbol{A}_{\varepsilon_n}$, $\psi_n=\psi_{\varepsilon_n}$. Recall that the gauge is chosen so that

$$\begin{aligned}&\operatorname{div}(a(x)\boldsymbol{A}_n)=0 \text{ in } \Omega,\\ &\boldsymbol{A}_n\cdot\boldsymbol{n}=0 \text{ on } \partial\Omega,\end{aligned} \tag{5.11}$$

then $a\boldsymbol{A}_n=\mathbf{curl}\xi_n$ for some scalar valued function $\xi_n$ which satisfies

$$\begin{aligned}&\Delta\xi_n-\tfrac{1}{a}\nabla a\cdot\mathbf{curl}\xi_n=ah_n, \text{ in } \Omega,\\ &\xi_n=0 \text{ on } \partial\Omega.\end{aligned} \tag{5.12}$$

From Theorem 4.1, by extracting a further subsequence if necessary, there is some $h_*\in W^{1,p}(\Omega,\mathbb{R})$, $\boldsymbol{A}_*\in\boldsymbol{W}^{2,p}(\Omega)$ and $\xi_*\in W^{3,p}(\Omega,\mathbb{R})$ such that

$$h_n\rightharpoonup h_* \text{ weakly in } W^{1,p}(\Omega),\ \forall p<2, \tag{5.13}$$

$$\boldsymbol{A}_n\rightharpoonup\boldsymbol{A}_* \text{ weakly in } W^{2,p}(\Omega),\ \forall p<2, \tag{5.14}$$

$$\xi_n\rightharpoonup\xi_* \text{ weakly in } W^{3,p}(\Omega),\ \forall p<2. \tag{5.15}$$

Finally, we have by (4.3) and the fact (follows from (4.2))

$$\int_\Omega |\nabla \boldsymbol{A}_n|^q \le C_q,\ \forall q < +\infty, \tag{5.16}$$

that

$$\int_\Omega |\nabla \psi_n|^p \le C_p,\ \forall p < 2. \tag{5.17}$$

Hence, passing to a further subsequence if necessary, we may assume that there is $\psi_* \in W^{1,p}(\Omega, \mathbb{R}^2)$ such that

$$\psi_n \rightharpoonup \psi_* \text{ weakly in } W^{1,p}(\Omega),\ \forall p < 2. \tag{5.18}$$

Since

$$\int_\Omega (|\psi_n|^2 - 1)^2 \le C\varepsilon_n^2 |\ln \varepsilon_n| \to 0,$$

we obtain

$$|\psi_*| = 1, \quad \text{a.e. in } \Omega. \tag{5.19}$$

**Proposition 9.5.1** *Let $K$ be a compact subset of $\Omega \setminus \cup_{j=1}^{N_0}\{a_j\}$. For some $C_K > 0$ independent of $n$, we have*

$$\int_K (|\nabla h_n|^2 + |\nabla \psi_n|^2) \le C_K. \tag{5.20}$$

*Moreover, we have*

$$h_n \to h_* \text{ strongly in } H^1(K), \tag{5.21}$$

$$\psi_n \to \psi_* \text{ strongly in } \mathcal{H}^1(K), \tag{5.22}$$

$$\frac{1}{\varepsilon_n^2}\int_K (|\psi_n|^2 - 1)^2 \to 0. \tag{5.23}$$

*Proof.* Let $\eta$ be a smooth function compactly supported in $\Omega \setminus \cup_{j=1}^{N_0}\{a_j\}$ such that $\eta \equiv 1$ in $K$. Recall that on $\tilde{\Omega}_{\varepsilon_n}$, $h_n$ solves

$$-\mathrm{div}\left(\frac{1}{a\rho_n^2}\nabla(ah_n)\right) + h_n = 0, \ \text{ in } \tilde{\Omega}_{\varepsilon_n}. \tag{5.24}$$

For $n$ large enough, the support of $\eta$ is contained in $\tilde{\Omega}_{\varepsilon_n}$. Hence we may multiply (5.24) by $\eta a h_n$ and integrate to obtain

$$\int_K |\nabla(ah_n)|^2 + |h_n|^2 \le C\int_K |\nabla\eta||\nabla h_n||h_n| \le C$$

and then

$$\int_K |\nabla h_n|^2 = \int_K |\nabla(\frac{1}{a}ah_n)|^2 \le C\int_K |\nabla(ah_n)|^2 + |h_n|^2 \le C.$$

Since $h_n \rightharpoonup h_*$, we deduce that

$$\int_K |\nabla h_*|^2 + |h_*|^2 \le C. \tag{5.25}$$

Next, multiplying the both sides of (5.24) by $\eta(ah_n - ah_*)$, we have

$$\begin{aligned}\int_\Omega \frac{\eta}{a\rho_n^2} & |\nabla(ah_n)|^2 + a\eta h_n(h_n-h_*) \\ &= \int_\Omega a(h_n-h_*)\frac{1}{a\rho_n^2}\nabla(ah_n)\cdot\nabla\eta + \int_\Omega \frac{1}{a\rho_n^2}\eta\nabla(ah_n)\cdot\nabla(ah_*).\end{aligned}$$

Since $h_n \to h_*$ weakly in $H^1(K)$, we deduce that

$$||h_n - h_*||_{L^q(K)} \to 0,\ \forall q < +\infty.$$

Hence

$$\int_\Omega \frac{1}{a\rho_n^2}\eta|\nabla(ah_n)|^2 \to \int_\Omega \frac{1}{a}\eta|\nabla(ah_*)|^2.$$

Since $\rho_n \le 1$, this implies, by lower semicontinuity that

$$\nabla(ah_n) \to \nabla(ah_*) \text{ strongly in } L^2(K).$$

Therefore,

$$\nabla h_n \to \nabla h_* \text{ strongly in } L^2(K),$$

this establishes (6.21).

Since $\rho_n$ verifies

$$-\text{div}(a\nabla\rho_n) + \frac{1}{a\rho_n^3}|\nabla(ah_n)|^2 = \frac{a}{\varepsilon_n^2}\rho_n(1-|\rho_n|^2) \text{ in } \tilde{\Omega}_{\varepsilon_n},$$

multiplying it by $\eta(1-\rho_n)$ yields

$$\begin{aligned}\int_\Omega \eta(|\nabla\rho_n|^2 &+ \frac{\rho_n}{\varepsilon_n^2}(1-\rho_n^2)^2) \\ &\le C\int_\Omega \eta(1-\rho_n)\frac{1}{a\rho_n^3}|\nabla h_n|^2 + C\int_\Omega (1-\rho_n)|\nabla\eta||\nabla\rho_n|.\end{aligned} \tag{5.26}$$

Since $\rho_n \to 1$ and $\rho_n$ is bounded in $W^{1,p}$, we deduce from (5.21) and the dominated convergence theorem that the right-hand side of (5.26) goes to zero as $n \to \infty$. Hence,

$$\int_K \eta(|\nabla\rho_n|^2 + \frac{1}{\varepsilon_n^2}(1-\rho_n^2)^2) \to 0, \text{ as } n \to \infty. \tag{5.27}$$

This proves (5.23). Moreover, we have

$$|(\nabla - i\boldsymbol{A}_n)\psi_n|^2 = \frac{1}{\rho_n^2}(i\psi_n, (\nabla - i\boldsymbol{A}_n)\psi_n)^2 + |\nabla\rho_n|^2 = \frac{1}{a\rho_n^2}|\nabla(ah_n)|^2 + |\nabla\rho_n|^2$$

and then

$$(\nabla - i\boldsymbol{A}_n)\psi_n \to (\nabla - i\boldsymbol{A}_*)\psi_* \text{ strongly in } L^2(K). \tag{5.28}$$

Hence, (5.22) follows from (5.28) and the fact that $\boldsymbol{A}_n \to \boldsymbol{A}_*$ strongly in $L^\infty$.

Finally, we try to bound the quantity $\frac{1}{\varepsilon^2}\int_\Omega(1-|\psi_\varepsilon|^2)^2$. We first need the following preliminary results:

Consider for $k = 1,2,3,4$, the sets $G_\varepsilon^k = \Omega \setminus \cup_{j=1}^I B_{5\varepsilon^{1/4}-k\varepsilon^{1/2}}(x_j)$. We may assume that $4\varepsilon^{1/2} < \varepsilon^{1/4}$ for $\varepsilon \le \varepsilon_1$. Hence, by Lemma 3.2, there holds $|\psi_\varepsilon| \ge \frac{1}{2}$ on $G_\varepsilon^{(k)}$, $1 \le k \le 4$. Moreover, we have

$$\sup_{x_0 \in G_\varepsilon^{(4)}} \int_{B_{\sqrt{\varepsilon}}(x_0)} |\nabla\psi_\varepsilon|^2 \le 2\delta.$$

**Lemma 9.5.1** *There hold for any $\phi \in C^\infty(\overline{\Omega})$, $0 \le \phi \le 1$,*
*(i)* $\int_{\partial\Omega} |\nabla\psi_\varepsilon|^2 \le C|\ln\varepsilon|$,
*(ii)* $|\int_{\partial\Omega} \partial_{\boldsymbol{n}}(|\nabla\psi_\varepsilon|^2)\phi^2| \le C|\ln\varepsilon| + C\int_{\partial\Omega}|\nabla\phi|^2$
*with $C$ independent of $\varepsilon$.*

*Proof.* (i) Let $\boldsymbol{n}$ be a smooth extension of the outer unit normal vector field to $\Omega$. Multiplying (1.3-1) by $\boldsymbol{n}\cdot\nabla\psi_\varepsilon$ and noting (1.3-3), we have, dropping subscript $\varepsilon$, that

$$\begin{aligned}
&\operatorname{div}\{a(x)\nabla\psi(\boldsymbol{n}\cdot\nabla\psi) - \boldsymbol{n}a(x)[\frac{|\nabla\psi|^2}{2} + \frac{(|\psi|^2-1)^2}{4\varepsilon^2}]\}\\
&-\sum_{j,k,l} a(x)\partial_j\psi_k\partial_j\boldsymbol{n}_l\partial_l\psi_k + a(x)(\operatorname{div}\boldsymbol{n})\{\frac{|\nabla\psi|^2}{2} + \frac{(|\psi|^2-1)^2}{4\varepsilon^2}\}\\
&+\{\frac{1}{2}|\nabla\psi|^2 + \frac{(|\psi|^2-1)^2}{4\varepsilon^2}\}(\boldsymbol{n}\cdot\nabla a) + a|\boldsymbol{A}|^2\psi(\boldsymbol{n}\cdot\nabla\psi)\\
&+\sum_{j,k}(a\boldsymbol{A}_j\partial_j\psi_1\boldsymbol{n}_k\partial_k\psi_2 - a\boldsymbol{A}_j\partial_j\psi_2\boldsymbol{n}_k\partial_k\psi_1) = 0.
\end{aligned}$$

Noting also that $||\boldsymbol{A}_\varepsilon||_{L^\infty(\Omega)} \le C$ (from $||\boldsymbol{A}_\varepsilon||_{W^{2,p}(\Omega)} \le C$, $\forall p < 2$), one finds, by integrating over $\Omega$ and temporarily dropping the subscript $\varepsilon$, that

$$\int_{\partial\Omega} a(x)[|\partial_{\boldsymbol{n}}\psi|^2 - \frac{1}{2}|\nabla\psi|^2] \le CG_\varepsilon(\psi,\boldsymbol{A}) + C\int_\Omega|\nabla\psi|^2 + C,$$

and the claim follows, since $|\nabla\psi|^2 = |\partial_{\boldsymbol{n}}\psi|^2 + |\partial_{\boldsymbol{\tau}}\psi|^2 = |\partial_{\boldsymbol{n}}\psi|^2 + |\boldsymbol{A}\cdot\boldsymbol{\tau} + g|^2$ on $\partial\Omega$.

(ii) Let $(\boldsymbol{\tau}, \boldsymbol{n})$ be a smooth orthonormal frame in a neighbourhood of $\partial\Omega$ such that $\partial_{\boldsymbol{n}}\boldsymbol{n} = \partial_{\boldsymbol{\tau}}\boldsymbol{\tau} = 0$ on $\partial\Omega$. Then near $\partial\Omega$ we have

$$|\nabla\psi|^2 = |\partial_{\boldsymbol{n}}\psi|^2 + |\partial_{\boldsymbol{\tau}}\psi|^2,$$
$$\partial_{\boldsymbol{n}}(\frac{1}{2}|\nabla\psi|^2) = \partial^2_{\boldsymbol{n}}\psi\partial_{\boldsymbol{n}}\psi + \partial_{\boldsymbol{n}}\partial_{\boldsymbol{\tau}}\psi\partial_{\boldsymbol{\tau}}\psi$$

while on $\partial\Omega$, we have

$$\begin{aligned} 0 &= \Delta\psi + \frac{1}{a}\nabla a\cdot\nabla\psi - |\boldsymbol{A}|^2\psi - 2i\boldsymbol{A}\cdot\nabla\psi \\ &= \partial^2_{\boldsymbol{n}}\psi + \partial^2_{\boldsymbol{\tau}}\psi - \partial_{\boldsymbol{\tau}}\boldsymbol{\tau}\cdot\nabla\psi + \frac{1}{a}\nabla a\nabla\psi - |\boldsymbol{A}|^2\psi - 2i\boldsymbol{A}\cdot\nabla\psi. \end{aligned}$$

Since $\partial^2_{\boldsymbol{\tau}}\psi = \partial_{\boldsymbol{\tau}}(i\psi(\boldsymbol{A}\cdot\boldsymbol{\tau}+g)) + \partial_{\boldsymbol{\tau}}\boldsymbol{\tau}\cdot\nabla\psi$ on $\partial\Omega$, along $\partial\Omega$, we have

$$\begin{aligned} \partial_{\boldsymbol{n}}(\frac{1}{2}|\nabla\psi|^2) &= \Delta\psi\partial_{\boldsymbol{n}}\psi + \partial_{\boldsymbol{n}}\partial_{\boldsymbol{\tau}}\psi\partial_{\boldsymbol{\tau}}\psi - \partial_{\boldsymbol{n}}\psi\partial^2_{\boldsymbol{\tau}}\psi + \partial_{\boldsymbol{n}}\psi(\partial_{\boldsymbol{\tau}}\boldsymbol{\tau}\cdot\nabla\psi) \\ &= (-\frac{1}{a}\nabla a\cdot\nabla\psi + |\boldsymbol{A}|^2\psi + 2i\boldsymbol{A}\cdot\nabla\psi)\partial_{\boldsymbol{n}}\psi \\ &\quad + \partial_{\boldsymbol{\tau}}(\partial_{\boldsymbol{n}}\psi\cdot i\psi(\boldsymbol{A}\cdot\boldsymbol{\tau}+g)) - (\partial_{\boldsymbol{\tau}}\boldsymbol{n}\cdot\nabla\psi)i\psi(\boldsymbol{A}\cdot\boldsymbol{\tau}+g) \\ &\quad - 2\partial_{\boldsymbol{n}}\psi\partial_{\boldsymbol{\tau}}(i\psi(\boldsymbol{A}\cdot\boldsymbol{\tau}+g)) - \partial_{\boldsymbol{n}}\psi(\partial_{\boldsymbol{\tau}}\boldsymbol{\tau}\cdot\nabla\psi). \end{aligned}$$

Hence, by integrating by parts, we have

$$|\int_{\partial\Omega}\partial_{\boldsymbol{n}}(\frac{1}{2}|\nabla\psi|^2)\phi^2| \le C\int_{\partial\Omega}\{|\nabla\psi||\phi||\nabla\phi| + \phi^2(1+|\nabla\psi|^2) + |\partial_{\boldsymbol{\tau}}\boldsymbol{A}|^2\phi^2\}$$

here we have used $||\boldsymbol{A}||_{L^\infty(\Omega)} \le C$. Notice that

$$\left(\int_{\partial\Omega}|\partial_{\boldsymbol{\tau}}\boldsymbol{A}|^2\phi^2\right)^{1/2} \le C||\boldsymbol{A}||_{W^{3/2,2}(\Omega)} \le C||\boldsymbol{A}||_{W^{2,4/3}(\Omega)} \le C.$$

The assertion (ii) then follows from (i).

Using (1.3-1), $|\psi(x)| \ge \frac{1}{2}$ on $G^{(4)}_\varepsilon$ and ($H\psi$ is the Hessian matrix of $\psi$)

$$\Delta(\frac{1}{2}|\nabla\psi|^2) = |H\psi|^2 + \sum_{j=1}^{2}\partial_j\psi\Delta\partial_j\psi,$$

we obtain that

$$-\Delta(\frac{1}{2}|\nabla\psi|^2) + \frac{1}{4}|H\psi|^2 \le C_1|\nabla\psi|^4 + C_2|\nabla\psi|^2 + C_3|h|^2 \text{ on } G^{(4)}_\varepsilon. \tag{5.29}$$

**Lemma 9.5.2** *There hold*

$$\int_{G^{(3)}_\varepsilon}(|H\psi_\varepsilon|^2 + |\nabla\psi_\varepsilon|^4)dx \le \frac{C}{\varepsilon}|\ln\varepsilon|, \tag{5.30}$$

*and*

$$1 - |\psi_\varepsilon|^2 \le C\varepsilon^{1/8}|\ln\varepsilon| \text{ on } G^{(2)}_\varepsilon, \tag{5.31}$$

*where* $C = C(g, \Omega, a)$.

*Proof.* Take $\phi \in C^\infty(\mathbb{R}^2)$ satisfying $0 \le \phi \le 1$, $\phi \equiv 1$ in $G_\varepsilon^{(3)}$, $\phi \equiv 0$ in $\Omega \setminus G_\varepsilon^{(4)}$, $|\nabla^k \phi| \le 2\varepsilon^{-k/2}$. By a Sobolev inequality, we have, dropping subscript $\varepsilon$, that for some absolute constant $C_0 > 0$,

$$\int_\Omega |\nabla\psi|^4\phi^2 \le C_0 \sup_{x_0 \in G_\varepsilon^{(4)}} \Big(\int_{B_{\sqrt{\varepsilon}}(x_0)} |\nabla\psi|^2\Big)\Big\{\int_\Omega |H\psi|^2\phi^2 + \frac{1}{\varepsilon}\int_\Omega |\nabla\psi|^2\Big\}. \quad (5.32)$$

Choosing $16C_1\delta C_0 \le 1$ and multiplying (5.29) by $\phi^2$, using Lemmas 2.1 and 5.1, we obtain

$$\begin{aligned}\int_\Omega |H\psi|^2\phi^2 &\le 2\int_\Omega |\nabla\psi|^2\Delta\phi^2 + 2\int_{\partial\Omega}(\partial_{\boldsymbol{n}}|\nabla\psi|^2\phi^2 - |\nabla\psi|^2\partial_{\boldsymbol{n}}\phi^2)\\ &\quad +4C_1\int_\Omega |\nabla\psi|^4\phi^2 + 4C_2\int_\Omega |\nabla\psi|^2\phi^2 + 4C_3\int_\Omega h^2\phi^2\\ &\le \frac{1}{2}\int_\Omega |H\psi|^2\phi^2 + \frac{C}{\varepsilon}\int_\Omega |\nabla\psi|^2 + \frac{C}{\varepsilon}|\ln\varepsilon| + C|\ln\varepsilon| + C.\end{aligned}$$

Hence

$$\int_\Omega |H\psi|^2\phi^2 \le \frac{C}{\varepsilon}|\ln\varepsilon|.$$

Together with (5.32), this yields (5.30). Using (5.30) and Sobolev embedding, we deduce (5.31).

Now we give the 2nd main result of this section:

**Proposition 9.5.2** *We have for any minimizers* $(\psi_\varepsilon, \boldsymbol{A}_\varepsilon)$,

$$\frac{1}{\varepsilon^2}\int_\Omega (1-|\psi_\varepsilon|^2)^2 \le C, \ \text{as } \varepsilon \le \varepsilon_1 \text{ for some } \varepsilon_1 > 0, \quad (5.33)$$

*where $C$ depends only on $g$, $d$, $a(x)$ and $\Omega$.*

*Proof.* It follows from (1.3-1) that on $G_\varepsilon^{(2)}$

$$\begin{aligned}&\Delta\Big(\frac{1-|\psi|^2}{2}\Big) + |\nabla\psi|^2 - \frac{1}{4a}\nabla a\cdot\nabla\Big(\frac{1-|\psi|^2}{2}\Big) + a|(\nabla - i\boldsymbol{A})\psi|^2\\ &= \frac{1}{\varepsilon^2}(1-|\psi|^2)|\psi|^2 \ge \frac{(1-|\psi|^2)}{4\varepsilon^2}.\end{aligned}$$

Multiplying both sides of the equation by $\phi^2(1-|\psi|^2)$ with $\phi$ satisfying $\phi \in C^\infty(\overline{\Omega})$, $0 \le \phi \le 1$, $\phi \equiv 1$ on $G_\varepsilon^{(1)}$, $\phi \equiv 0$ on $\Omega \setminus G_\varepsilon^{(2)}$, $|\nabla^k\phi| \le 2\varepsilon^{-k/2}$, we obtain

$$\begin{aligned}&\frac{1}{4\varepsilon^2}\int_\Omega (1-|\psi|^2)^2\phi^2 + \frac{1}{2}\int_\Omega |\nabla(1-|\psi|^2)|^2\phi^2\\ &+\frac{1}{16}\int_\Omega \frac{1}{a}\nabla a\cdot\nabla(1-|\psi|^2)(1-|\psi|^2)\phi^2\end{aligned}$$

$$\begin{aligned}
&\leq M\int_\Omega(1-|\psi|^2)\phi^2|(\nabla-i\boldsymbol{A})\psi|^2 \\
&\quad+\int_\Omega|\nabla\psi|^2(1-|\psi|^2)\phi^2+\frac{1}{4}\int_\Omega(1-|\psi|^2)^2\Delta\phi^2 \\
&\leq \sup_{G_\varepsilon^{(2)}}(1-|\psi|^2)[\int_\Omega|\nabla\psi|^2+M\int_\Omega|(\nabla-i\boldsymbol{A})\psi|^2]+\frac{4}{\varepsilon}\int_\Omega(1-|\psi|^2)^2.
\end{aligned}$$

Then the conclusion follows from Lemmas 2.1, 3.3, 3.4, 5.2 and the fact that

$$|\frac{1}{16}\int_\Omega\frac{1}{a}\nabla a\cdot\nabla(1-|\psi|^2)^2\phi^2|=|\frac{-1}{16}(1-|\psi|^2)^2\mathrm{div}(\phi^2\frac{\nabla a}{a})|\leq C.$$

## 9.6 Properties of $(\psi_*, \boldsymbol{A}_*)$

The following theorem determines the singularities of $\psi_*$ which is a kind of harmonic map.

**Theorem 9.6.1** *Let $\psi_*$, $h_*$, $a_1,\ldots,a_{N_0}$ be as in Theorem 5.1. We have*

*(i)*

$$\begin{array}{l} a_1,\ldots,a_{N_0}\in a^{-1}(m)\cap\Omega \text{ if (5.4) (or (5.5), or (5.6)) holds;} \\ a_1,\ldots,a_{N_0}\in a^{-1}(m) \text{ if (5.7) holds.}\end{array} \tag{6.1}$$

*(ii) Let $d_j=\deg(\psi_*,a_j)$, then*

$$d_j=1,\ N_0=d,\quad \text{if } N\geq d, \tag{6.2}$$

$$d_j\geq 1,\sum_{j=1}^{N_0}d_j=d,\quad \text{if } N<d. \tag{6.3}$$

*(iii) When one of (5.4) or (5.5) or (5.6) holds, $\psi_*$ and $h_*$ solves, respectively, the following equations:*

$$\begin{cases} -\mathrm{div}(a(x)\nabla\psi_*)=a(x)\psi_*|\nabla\psi_*|^2, & \text{in } \Omega\setminus\cup_{j=1}^{N_0}\{a_j\}, \\ |\psi_\varepsilon|=1, & a.e.,\end{cases} \tag{6.4}$$

$$\begin{cases} -\mathrm{div}(\frac{1}{a}\nabla(ah_*))+h_*=2\pi\sum_{j=1}^{N_0}d_j\delta_{a_j}, & \text{in } \Omega, \\ \partial_{\boldsymbol{n}}(ah_*)=-ag, & \text{on } \partial\Omega.\end{cases} \tag{6.5}$$

*Proof.* First, we prove (6.4). In fact, it is clear from (1.3-2) that

$$\partial_j\Big(\psi_{\varepsilon_n}\times\big(a(x)(\partial_j-i\boldsymbol{A}_{\varepsilon_n j})\psi_{\varepsilon_n}\big)\Big)=0 \text{ in } \mathcal{D}'(\Omega). \tag{6.6}$$

Passing to the limit in (6.6), we are led to

$$\partial_j\Big(\psi_*\times\big(a(x)(\partial_j-i\boldsymbol{A}_{*j})\psi_*\big)\Big)=0 \text{ in } \mathcal{D}'(\Omega). \tag{6.7}$$

Since $\operatorname{div}(a\boldsymbol{A}_*) = 0$ and $|\psi_*| = 1$, we deduce that

$$\partial_j\Big(\psi_* \times (a(x)\partial_j\psi_*)\Big) = 0 \ \text{ in } \ \mathcal{D}'(\Omega). \tag{6.8}$$

Hence we obtain (6.4).

Second, we prove (6.1)–(6.3). The first step is to show that $a_1, \dots, a_{N_0} \in a^{-1}(m)$. Since $\partial_{\boldsymbol{T}}\psi_n = i\psi_n(\boldsymbol{A}_n \cdot \boldsymbol{\tau} + g)$ on $\partial\Omega$, it follows from (5.14) that

$$\int_{\partial\Omega} |\partial_{\boldsymbol{T}}\psi_n|^2 \leq C, \ \text{ uniformly in } n.$$

This implies that $\psi_n$ can be extended from $\Omega$ to $\Omega'$ which contains $\overline{\Omega}$ (still denoted by $\psi_n$) such that $\psi_n \in \mathcal{H}^1(\Omega' \setminus \Omega, S^1)$ and

$$\int_{\Omega'\setminus\Omega} |\nabla\psi_n|^2 \leq C, \ \text{ uniformly in } n. \tag{6.9}$$

Set $\delta \leq \frac{1}{4}\min_{j\neq k}|a_j - a_k|$, $j,k = 1,2,\dots,N_0$ and $\Lambda_j = \{j \in J : \ x_j^{\varepsilon_n} \to a_j\}$, $1 \leq j \leq N_0$. Then for $\varepsilon_n$ small enough, one has

$$\begin{aligned}
&B(x_j^{\varepsilon_n}, \lambda\varepsilon_n) \subset\subset B(a_j, \delta), \quad \text{for } j \in \Lambda_j,\\
&|\psi_{\varepsilon_n}| \geq \frac{1}{2} \ \text{ on } \Omega' \setminus \cup_{j=1}^{N_0} B(a_j, \delta),\\
&d_j = \sum_{k\in\Lambda_j} \deg(\psi_{\varepsilon_n}, \partial B(x_k, \lambda\varepsilon_n)) = \deg(\psi_*, \partial B(a_j, \delta)),\\
&d = \sum_{j=1}^{N_0} d_j.
\end{aligned}$$

We have $d_j \neq 0$ for $j = 1, \dots, N_0$. Otherwise, as in Chapter 6, one deduces that $|\psi_{\varepsilon_n}| \geq \frac{1}{2}$ on $B(a_j, \delta)$, $\forall 0 < \varepsilon_n \leq \varepsilon_2$ for some $\varepsilon_2 > 0$. This contradicts the definition of $x_j^{\varepsilon_n}$.

If on the contrary, there is $a_j \in a^{-1}(m)$, say $a_1 \notin a^{-1}(m)$, then there exists $\sigma_0 > 0$, $\delta_0 \in (0, \delta)$ such that $a(x) \geq m + \sigma_0\frac{d}{|d_1|}$ on $B(a_1, \delta_0)$. Note that $\overline{B(a_j, \delta_0)} \cap \overline{B(a_k, \delta_0)} = \emptyset$ when $j \neq k$ and $a(x) \geq m$ on $B(a_j, \delta)$ for $j = 1, \dots, N_0$. Similar to Theorem 1.4 in Section 1.8, Chapter 6, one concludes that

$$\begin{aligned}
&\frac{1}{2}\sum_{j=1}^{N_0}\int_{B(a_j,\delta_0)\setminus\cup_{j\in\Lambda}B(x_j^{\varepsilon_n},\lambda\varepsilon_n)} a(x)|\nabla\psi_{\varepsilon_n}|^2\\
&\geq (m + \frac{d\sigma_0}{|d_1|})\pi|d_1||\ln\varepsilon| + \sum_{j=2}^{N_0} m\pi|d_j||\ln\varepsilon_n| - C\\
&\geq \pi d m|\ln\varepsilon_n| + \pi\sigma_0 d|\ln\varepsilon_n| - C.
\end{aligned} \tag{6.10}$$

On the other hand, we have from $||\boldsymbol{A}_n||_{L^\infty(\Omega)} \leq C$ and (6.9) that

$$\begin{aligned}
&\frac{1}{2}\sum_{j=1}^{N_0}\int_{B(a_j,\delta_0)} a(x)|\nabla\psi_n|^2 \leq \frac{1}{2}\int_{\Omega'\setminus\Omega} a(x)|\nabla\psi_n|^2 + \frac{1}{2}\int_\Omega |\nabla\psi_n|^2 \\
\leq\ & \frac{1}{2}\int_\Omega a(x)|(\nabla - i\boldsymbol{A}_{\varepsilon_n})\psi_n|^2 + C \leq C + G_{\varepsilon_n}(\psi_n, \boldsymbol{A}_n) \\
\leq\ & C + \pi d(m+\sigma_1)|\ln\varepsilon_n|. \qquad (6.11)
\end{aligned}$$

Combining (6.10) with (6.11), taking $\sigma_1 = \sigma_0/2$, we are led to

$$\frac{\pi}{2}\sigma_0 d|\ln\varepsilon_n| \leq C \text{ uniformly in } n.$$

This is impossible. Hence, $a_1, \ldots, a_{N_0} \in a^{-1}(m)$.

Secondly, from $a_1, \ldots, a_{N_0} \in a^{-1}(m)$, (6.10) transforms into

$$\frac{1}{2}\sum_{j=1}^{N}\int_{\Omega_j} a(x)|\nabla\psi_n|^2 \geq \pi m\sum_{j=1}^{N_0}|d_j||\ln\varepsilon_n| - C - \pi m d|\ln\delta_0| \qquad (6.12)$$

where $\Omega_j = B(a_j, \delta_0) \setminus \cup_{k\in\Lambda_j} B(x_k^{\varepsilon_n}, \lambda\varepsilon_n)$. This together with (6.11) yields

$$\sum_{j=1}^{N_0}|d_j| \leq d + \frac{\sigma_1}{m} + \frac{C}{|\ln\varepsilon_n|}. \qquad (6.13)$$

Send $\varepsilon_n \to 0$ and then $\sigma_1 \to 0$ in (6.13) to give

$$\sum_{j=1}^{N_0}|d_j| \leq d.$$

But $\sum_{j=1}^{N_0} d_j = d$, this implies $d_j \geq 1$ in view of the fact $d_j \neq 0$.

Thirdly, if $\mathrm{Card}(a^{-1}(m)\cap\Omega) \geq d$, one may use (2.2) to deduce by (6.12) that

$$\frac{1}{2}\int_{\Omega'\setminus\cup_{j=1}^{N_0}B(a_j,\delta_0)} a(x)|\nabla\psi_n|^2 \leq \pi m d|\ln\delta_0| + C \quad \forall\delta_0 \in (0,\delta) \qquad (6.14)$$

with $C = O(\delta_0)$. Send $n \to \infty$ in (6.14), we have

$$\frac{1}{2}\int_{\Omega'\setminus\cup_{j=1}^{N_0}B(a_j,\delta_0)} a(x)|\nabla\psi_*|^2 \leq \pi m d|\ln\delta_0| + C. \qquad (6.15)$$

On the other hand, from Section 1.8 Chapter 6, it is easy to see that

$$\frac{1}{2}\int_{\Omega'\setminus\cup_{j=1}^{N_0}B(a_j,\delta_0)} a(x)|\nabla\psi_*|^2 \geq \frac{1}{2}\int_{\Omega'\setminus\cup_{j=1}^{N_0}B(a_j,\delta_0)} m|\nabla\psi_*|^2 \geq \pi m|\ln\delta_0|\sum_{j=1}^{N_0} d_j^2 - C,$$

therefore,

$$\sum_{j=1}^{N_0}(d_j^2 - d_j)|\ln\delta_0| \le C \text{ for any } \delta_0 \in (0,\delta).$$

This implies that $d_j = 1$ and $N_0 = d$.

Finally, under (5.4) and (5.5), it is clear that $a_j \notin \partial\Omega$ for $j = 1,\dots,N_0$. On the other hand, under condition (6.15), one also deduce that $a_j \notin \partial\Omega$ by adopting the argument of Section 1.5, Chapter 6.

The proof of (6.5) is just the same as in Section 1.8, Chapter 6.

The following theorem gives a precise statement for the pinning mechanism: as $\varepsilon$ small enough, the vortices (zeros of $|\psi_\varepsilon|$) are accumulated near the places where $a$ attains its minimum.

**Theorem 9.6.2** *Let $Z_\varepsilon = \{x \in \Omega|\ |\psi_\varepsilon(x)| = 0\}$. Then for any $\delta > 0$, there exists $\delta_0 > 0$ such that for $0 < \varepsilon \le \varepsilon_0$,*

$$Z_\varepsilon \subset O_\delta(a^{-1}(m)) = \{p \in \Omega|\ dist(p, a^{-1}(m)) < \delta\}. \tag{6.16}$$

*Proof.* We again argue by contradiction. If the theorem is false, there exists $\varepsilon_n \to 0$ and $x^{\varepsilon_n} \in Z_{\varepsilon_n}$ such that $x^{\varepsilon_n} \to x_0 \notin a^{-1}(m)$, that is, for some $\beta > 0$,

$$\text{dist}(x_0, a^{-1}(m)) \ge \beta > 0. \tag{6.17}$$

For $\psi_{\varepsilon_n}$, applying the proofs of Theorem 5.1 and 6.1, we know that there is a further subsequence of $\psi_{\varepsilon_n}$, denoted by $\psi_{n_k}$, such that

$$Z_{n_k} = \{x \in \Omega|\ |\psi_{n_k}(x) = 0\} \subset \cup_{j=1}^{J_k} B(b_j^{n_k}, \lambda\varepsilon_{n_k}),$$

where $J_k \le J_0$ and $B(b_j^{n_k}, \lambda\varepsilon_{n_k})$ satisfies (3.5) and (3.6).

Since $J_k \le J_0$ is independent of $k$, we may assume that

$$x^{n_k} \in B(b_{s_0}^{n_k}, \lambda\varepsilon_{n_k}) \text{ for some fixed } s_0 \in \{1,\dots,J_k\}.$$

Again, from the proofs of Theorems 5.1 and 6.1, we have

$$b_{s_0}^{n_k} \to \overline{a}_{j_0} \in a^{-1}(m) \text{ as } k \to +\infty.$$

Hence

$$|x^{n_k} - \overline{a}_{j_0}| \le |x^{n_k} - b_{s_0}^{n_k}| + |b_{s_0}^{n_k} - \overline{a}_{j_0}| \le \lambda\varepsilon_{n_k} + |b_{s_0}^{n_k} - \overline{a}_{j_0}| \to 0$$

as $k \to \infty$. This contradicts (6.16) since

$$|x_0 - \overline{a}_{j_0}| \le |x_0 - x^{n_k}| + |x^{n_k} - \overline{a}_{j_0}| \to 0.$$

## 9.7 Renormalized energy

The purpose of this section is to further locate $\{a_j\}$, the set of singular points of the limit order parameter function $\psi_*$, when $\text{Card}(a^{-1}(m)\cap\Omega) \geq d$. As in Chapter 6, we introduce the so-called renormalized energy with weight and show that $(a_j)$ must be an energy minimizer of the renormalized energy in $(a^{-1}(m)\cap\Omega)^d$. The renormalized energy for the G-L functional without $\boldsymbol{A}$ has been discussed in [DLi3] and [BHa 95]. In the case with magnetic field, it is not so different.

In order to give the renormalized energy, we need some preliminary lemmas as in Chapter 6 to describe the energy expressions for the energy-minimizing harmonic maps with weight. The proofs of these lemmas are exactly parallel to those in Chapter 6.

We first change the notion of the so-called canonical harmonic maps in Chapter 6.

**Definition 9.7.1** *Let $\Omega$ be a smooth, bounded and simply connected open domain in $\mathbb{R}^2$. Fix $n$ points $a_1,\ldots,a_n$ in $\Omega\cap a^{-1}(m)$ and $n$ integers $d_1,\ldots,d_n$. Set $d = \sum_{j=1}^n d_j$ and let $g$ be a map from $\partial\Omega$ into $S^1$, such that $\deg(g,\partial\Omega) = d$.*

*Let $\Phi_0$ be the solution of*

$$\begin{cases} \operatorname{div}(\frac{1}{a(x)}\nabla\Phi_0) = \sum_{j=1}^n 2\pi d_j\delta_{a_j} & \text{in} \quad \Omega, \\ \frac{1}{a(x)}\partial_{\boldsymbol{n}}\Phi_0 = g\times\partial_{\boldsymbol{\tau}}g & \text{on} \quad \partial\Omega, \\ \int_{\partial\Omega}\Phi_0 = 0. \end{cases} \tag{7.1}$$

*Then there exists a unique harmonic map $u_0 : \Omega\setminus\{a_1,\ldots,a_n\}\to S^1$ such that*
*1) $\deg(u_0,a_j) = d_j$ ($\deg(u_0,a_j)$ is the degree of $u_0$ restricted to any small circle centred at $a_j$ not containing other $a_k'$s),*
*2) $u_0$ is continuous up to $\partial\Omega$ and $u_0|_{\partial\Omega} = g$ which is associated to $\Phi_0$, namely,*

$$\begin{cases} u_0\times\partial_1 u_0 = -\frac{1}{a(x)}\partial_2\Phi_0 & \text{in} \quad \Omega\setminus\{a_1,\ldots,a_n\}, \\ u_0\times\partial_2 u_0 = \frac{1}{a(x)}\partial_1\Phi_0 & \text{in} \quad \Omega\setminus\{a_1,\ldots,a_n\}. \end{cases} \tag{7.2}$$

*The function $u_0$ is called the* **canonical harmonic map with weight** *associated to the configuration $(g,a_j,d)$ where $d = (d_1,\ldots,d_n)$.*

Then we establish the following

**Theorem 9.7.1** *Let $u_0$ be a canonical harmonic map associated to $(g,a_j,d)$ with $a_j\in\Omega\cap a^{-1}(m)$, then as $\rho\to 0$*

$$\frac{1}{2}\int_{\Omega_\rho} a(x)|\nabla u_0|^2 = m\pi\sum_{j=1}^n d_j^2\log(1/\rho) + W + O(\rho^2) \tag{7.3}$$

*where* $\Omega_\rho = \Omega \setminus \overline{\cup_{j=1}^n B(a_j, \rho)}$ *and* $W = W(g, a_j, a, d)$ *is the weighted renormalized energy defined as follows*

$$\begin{aligned} W &= W(g, a_j, a, d) \\ &= -\pi \sum_{i \neq j} a(a_j) d_i d_j \log |a_i - a_j| + \frac{1}{2} \int_{\partial\Omega} \Phi_0 (g \times g_\tau) - \pi \sum_{j=1}^n d_j R_0(a_j) \end{aligned} \quad (7.4)$$

*where*

$$R_0(x) = \Phi_0(x) - \sum_{j=1}^n a(a_j) d_j \log |x - a_j|.$$

*Proof.* The proof is similar to that of Theorem 8.1 of Chapter 6.

Here the canonical harmonic map can be expressed as

$$u_0(z) = \prod_{j=1}^n \frac{z - a_j}{|z - a_j|} e^{i\phi}$$

and

$$-\text{div}(a \nabla \phi) = \nabla a(x) \cdot (\nabla \theta_1 + \cdots + \nabla \theta_n), \text{ in } \Omega,$$

with $\phi = \phi_0$ on $\partial\Omega$ where $\phi_0 : \partial\Omega \to R$ is such that

$$g(z) = \prod_{j=1}^n \frac{z - a_j}{|z - a_j|} e^{i\phi_0}$$

and

$$\nabla \theta_k = \left( \frac{-(y - \beta_k)}{|z - a_k|}, \frac{x - \alpha_k}{|z - a_k|} \right)$$

where $z = x + iy$, $a_k = \alpha_k + i\beta_k$.

Modifying the discussions in Chapter 6 we have

**Lemma 9.7.1** *Let $G$ be a bounded simply connected smooth domain in $R^2$ and $\omega_i$ $(i = 1, \ldots, n)$ be $n$ sub-domains in $G$ such that $\overline{\omega}_i \cap \overline{\omega}_j = \emptyset$, $i \neq j$ and $\overline{\omega}_i \subset G$, $\Omega = G \setminus \bigcup_{i=1}^n \omega_i$. Define*

$$\mathcal{E}_1 = \{\psi \in H^1(\Omega, S^1) : \psi \times \partial_{\boldsymbol{\tau}} \psi = g, \text{ on } \partial G, \deg(\psi, \partial\omega_i) = d_i\},$$

*where $g$ is a given fnction. If $v \in \mathcal{E}_1$ such that*

$$E(v) = \int_\Omega a(x) |\nabla v|^2 = E_1 = \inf_{\psi \in \mathcal{E}_1} \int_\Omega a(x) |\nabla \psi|^2,$$

*then*

$$E_1 = E(v) = \int_\Omega \frac{1}{a(x)} |\nabla \Phi_1|^2,$$

*where $\Phi_1$ solves the following problem*

$$\begin{cases} \operatorname{div}(\frac{1}{a(x)}\nabla\Phi_1) = 0, & \text{in } \Omega, \\ \Phi_1 = Const. = C_i, & \text{on } \partial\omega_i, i = 1, \dots, n, \\ \int_{\partial\omega_i} \frac{1}{a(x)}\frac{\partial\Phi_1}{\partial \boldsymbol{n}} = 2\pi d_i, i = 1, \dots, n, \\ \frac{1}{a(x)}\frac{\partial\Phi_1}{\partial \boldsymbol{n}} = g, \text{ on } \partial\Omega. \end{cases}$$

*Proof.* Since $v$ takes its values in $S^1$ we have

$$v_{x_1} \times v_{x_2} = 0, \quad |\nabla v|^2 = |v \times v_{x_1}|^2 + |v \times v_{x_2}|^2,$$

and hence

$$\frac{\partial}{\partial x_1}(v \times v_{x_2}) + \frac{\partial}{\partial x_2}(-v \times v_{x_1}) = 0.$$

Setting $D = (-v \times v_{x_2} + \frac{1}{a(x)}\Phi_{1x_1}, v \times v_{x_1} + \frac{1}{a(x)}\Phi_{1x_2})$, we have

$$\operatorname{div} D = 0, \quad \int_{\omega_i} D \cdot \boldsymbol{n} = 0, i = 1, \dots, n.$$

Then there exists a function $f$ on $\Omega$ such that $D = (\frac{\partial f}{\partial x_2}, -\frac{\partial f}{\partial x_1})$, that is

$$\begin{cases} v \times v_{x_1} = -\frac{\partial f}{\partial x_1} - \frac{1}{a(x)}\frac{\partial \Phi_1}{\partial x_2}, \\ v \times v_{x_2} = -\frac{\partial f}{\partial x_2} + \frac{1}{a(x)}\frac{\partial \Phi_1}{\partial x_1}, \end{cases}$$

and

$$a(x)|\nabla v|^2 = \frac{1}{a(x)}|\nabla\Phi_1|^2 + a(x)|\nabla f|^2 + 2(\frac{\partial f}{\partial x_1}\frac{\partial \Phi_1}{\partial x_2} - \frac{\partial f}{\partial x_2}\frac{\partial \Phi_1}{\partial x_1}).$$

Noting that

$$\begin{aligned} &\int_\Omega (\frac{\partial f}{\partial x_1}\frac{\partial \Phi_1}{\partial x_2} - \frac{\partial f}{\partial x_2}\frac{\partial \Phi_1}{\partial x_1}) \\ &= \int_\Omega \operatorname{div}(f\frac{\partial \Phi_1}{\partial x_2}, -f\frac{\partial \Phi_1}{\partial x_1}) \\ &= \int_{\partial G} f\frac{\partial \Phi_1}{\partial \tau} - \sum_{i=1}^n \int_{\partial\omega_i} f\frac{\partial \Phi_1}{\partial \tau} \\ &= \int_{\partial G} f\frac{\partial \Phi_1}{\partial \tau} = -\int_{\partial G} \Phi_1\frac{\partial f}{\partial \tau} \\ &= -\int_{\partial G} \Phi_1(-(v \times v_\tau) + \frac{1}{a(x)}\frac{\partial \Phi_1}{\partial \boldsymbol{n}}) = 0, \end{aligned}$$

we have

$$\int_\Omega a(x)|\nabla v|^2 = \int_\Omega \frac{1}{a(x)}|\nabla\Phi_1|^2 + a(x)|\nabla f|^2 \geq \int_\Omega \frac{1}{a(x)}|\nabla\Phi_1|^2, \text{ for any } v \in \mathcal{E}_1.$$

On the other hand, the problem

$$\begin{cases} v \times v_{x_1} = F_1, \\ v \times v_{x_2} = F_2 \end{cases}$$

has a solution $v : \Omega \to S^1$ if and only if

$$\frac{\partial F_1}{\partial x_2} = \frac{\partial F_2}{\partial x_1}$$

and

$$\int_{\Gamma_i} \mathbf{F} \cdot \tau \in 2\pi Z$$

for each connected component $\Gamma_i$ of $\partial\Omega$ where $Z$ denotes the set of integers. Hence it follows from the definition of $\Phi_1$ that the problem

$$\begin{cases} v \times v_{x_1} = -\frac{1}{a(x)} \frac{\partial \Phi_1}{\partial x_2}, \\ v \times v_{x_2} = \frac{1}{a(x)} \frac{\partial \Phi_1}{\partial x_1} \end{cases}$$

admits a solution $v : \Omega \to S^1$ and we have

$$\int_\Omega a(x)|\nabla v|^2 = \int_\Omega \frac{1}{a(x)}|\nabla \Phi_1|^2.$$

Moreover, we may see that $v \times v_\tau = g$ on $\partial G$ and

$$2\pi \deg(v, \partial\omega_i) = \int_{\partial\omega_i} v \times v_\tau = \int_{\partial\omega_i} \frac{1}{a(x)} \frac{\partial \Phi_1}{\partial \boldsymbol{n}} = 2\pi d_i$$

which means that $v \in \mathcal{E}_1$ and $E_1$ is attained by $v \in \mathcal{E}_1$ and $E_1 = \int_\Omega \frac{1}{a(x)}|\nabla\Phi_1|^2$. The lemma is proved. Now we consider a particular case that $\omega_i = B(a_i, \delta)$ and further consider the limit as $\delta \to 0$. Denote $\Omega_\delta = \Omega \setminus \cup_{i=1}^n B(a_i, \delta)$ where $\Omega$ is now a simply connected domain. Define

$$\mathcal{E}_\delta = \{\psi \in H^1(\Omega_\delta, S^1) : \psi \times \partial_{\boldsymbol{\tau}}\psi = g, \quad \text{on } \partial\Omega, \deg(\psi, \partial B(a_i, \delta)) = d_i\}.$$

We deduce from the above lemma that the minimum

$$\nu_\delta = \inf_{\psi \in \mathcal{E}_\delta} \int_{\Omega_\delta} a(x)|\nabla\psi|^2$$

is attained by a function in $\mathcal{E}_\delta$ and

$$\nu_\delta = \int_{\Omega_\delta} \frac{1}{a(x)}|\nabla\Phi_\delta|^2$$

where $\Phi_\delta$ is determined by the above lemma with $\omega_i$ replaced by $B(a_i, \delta)$.

For the minimizer $v_\delta$ obtained above and the canonical harmonic map $v_0$, we can prove the following two lemmas in the same way as that of Lemma 9.2 and Lemma 9.3 in Chapter 6.

**Lemma 9.7.2** *As $\delta \to 0$, $v_\delta$ converges to $v_0$ in $C^k(K)$ on every compact subset $K$ of $(\Omega \setminus \{a_1, \dots, a_n\}) \cup \partial\Omega$.*

**Lemma 9.7.3** *As $\delta \to 0$,*

$$\frac{1}{2}\int_{\Omega_\delta} a(x)|\nabla v_\delta|^2 = \pi m \sum_{i=1}^{n} d_i^2 \log\frac{1}{\delta} + W + O(\delta)$$

*where $W$ is defined as above.*

Parallel to Lemma 9.4 of Chapter 6, we have

**Lemma 9.7.4** *Let $v_\delta$ and $v$ be the minimizers of the minimization problems*

$$\inf_{\psi \in \mathcal{E}_\delta^1} \int_{\Omega_\delta} a(x)|\nabla\psi|^2$$

*and*

$$\inf_{\psi \in \mathcal{E}_\delta^2} \int_{\Omega_\delta} a(x)|\nabla\psi|^2$$

*respectively, where*

$$\mathcal{E}_\delta{}^1 = \{\psi \in H^1(\Omega_\delta) : \psi \times \partial_{\boldsymbol{\tau}}\psi|_{\partial\Omega} = g + \frac{1}{a(x)}\partial_{\boldsymbol{n}}\xi_\delta,\ \deg(\psi, \partial B(a_i,\delta)) = d_i\},$$

$$\mathcal{E}_\delta{}^2 = \{\psi \in H^1(\Omega_\delta) : \psi \times \partial_{\boldsymbol{\tau}}\psi|_{\partial\Omega} = g + \frac{1}{a(x)}\partial_{\boldsymbol{n}}\xi_*,\ \deg(\psi, \partial B(a_i,\delta)) = d_i\}.$$

*If $\xi_\delta$ converges to $\xi_*$ in $W^{2,2}(\Omega)$, then as $\delta \to 0$ we have*

$$\int_{\Omega_\delta} a(x)|\nabla v_\delta|^2 = \int_{\Omega_\delta} a(x)|\nabla v|^2 + o(1). \tag{7.5}$$

Now we are in a position to derive the renormalized energy for our problem. Consider $d$ distinct points $b_1, \dots, b_d$ in $\Omega \cap a^{-1}(m)$ and $\delta > 0$. We choose $\delta$ sufficiently small so that

$$\begin{array}{l} B(b_j,\delta) \subset \Omega \ \ \forall j \in \{1,\dots,d\}, \\ B(b_j,\delta) \cap B(b_k,\delta) = \emptyset \ \text{ for } \ j \neq k. \end{array} \tag{7.6}$$

As before, we define $\Omega_\delta = \Omega \setminus \cup_{j=1}^d B(a_j,\delta)$ and consider the space $W_\delta$ defined by

$$W_\delta = \left\{ \begin{array}{c} (\psi, \boldsymbol{A}) \in H^1(\Omega_\delta, S^1) \times H^1(\Omega, \mathbb{R}^2) \ \text{such that} \\ (i\psi, \boldsymbol{\tau}\cdot((\nabla - i\boldsymbol{A})\psi)) = g \ \text{ on } \ \partial\Omega \\ \deg(\psi, \partial B(b_j,\delta)) = 1. \end{array} \right\}$$

and the functional (note that now, $(|\psi|^2-1)^2=0$)

$$E_\delta(\psi,\boldsymbol{A})=\frac{1}{2}\int_{\Omega_\delta}a(x)|(\nabla-i\boldsymbol{A})\psi|^2+\frac{1}{2}\int_\Omega a(x)|\mathbf{curl}\boldsymbol{A}|^2.$$

We define

$$\mu_\delta(b)=\mathrm{Inf}_{(\psi,\boldsymbol{A})\in W_\delta}E_\delta(\psi,\boldsymbol{A}). \tag{7.7}$$

Note that the condition

$$(i\psi,\boldsymbol{\tau}\cdot((\nabla-i\boldsymbol{A})\psi))=g \text{ on } \partial\Omega$$

is equivalent to

$$\psi\times\psi_{\boldsymbol{\tau}}=g+\boldsymbol{A}\cdot\boldsymbol{\tau} \text{ on } \partial\Omega.$$

**Proposition 9.7.1** *$\mu_\delta(b)$ is achieved for $\delta$ sufficiently small so that (7.6) holds and moreover,*

$$\mu_\delta(b)\le m\pi d\log\frac{1}{\delta}+C \tag{7.8}$$

*where $C$ depends only on the configuration $(b_j)$ and $\delta_0$.*

*Proof.* The proof is similar to that of (2.2).

**Proposition 9.7.2** *When $\delta$ is sufficiently small so that (7.6) holds, for the minimizers $(v_\delta,\boldsymbol{B}_\delta)$ of (7.7), we have*

$$\int_\Omega a(x)|\mathbf{curl}\boldsymbol{B}_\delta|^2\le C, \tag{7.9}$$

$$\int_{\Omega_\delta}a(x)|(\nabla-i\boldsymbol{B}_\delta)v_\delta|^2\ge m\pi d\log\frac{1}{\delta}-C \tag{7.10}$$

*where $C$ depends only on $(b_j)$ and $\delta_0$.*

*Proof.* In this proof we omit the subscript $\delta$ for simplicity. Let $(v,\boldsymbol{B})$ be a pair of minimizers of (7.7) satisfying the gauge condition

$$\begin{cases}\mathrm{div}(a(x)\boldsymbol{B})=0 & \text{in} \quad \Omega,\\ \boldsymbol{B}\cdot\boldsymbol{n}=0 & \text{on} \quad \partial\Omega.\end{cases} \tag{7.11}$$

We know that

$$a(x)\boldsymbol{B}=\mathbf{curl}\xi=(-\xi_{x_2},\xi_{x_1}) \tag{7.12}$$

for the scalar function $\xi$ solving

$$\begin{cases}\Delta\xi-\frac{1}{a(x)}\nabla a(x)\cdot\nabla\xi=a(x)h=a(x)\mathbf{curl}\boldsymbol{B} & \text{in} \quad \Omega,\\ \xi=0 & \text{on} \quad \partial\Omega.\end{cases} \tag{7.13}$$

We have then

$$\int_{\Omega_\delta} a(x)|(\nabla - i\boldsymbol{B})v|^2 = \int_{\Omega_\delta} a(x)|\nabla v|^2 + \frac{1}{a(x)}|\nabla\xi|^2 + 2(i\nabla v \cdot \mathbf{curl}\xi, v), \quad (7.14)$$

$$\int_{\Omega_\delta} a(x)|\mathbf{curl}\boldsymbol{B}|^2 = \int_{\Omega_\delta} \frac{1}{a(x)}|\Delta\xi - \frac{1}{a(x)}\nabla a(x) \cdot \nabla\xi|^2 \quad (7.15)$$

here we have used the relation $|v(x)| = 1$ on $\Omega_\delta$, $a(x)|\boldsymbol{B}|^2 = \frac{1}{a(x)}|\nabla\xi|^2$ and

$$a(x)[i\boldsymbol{B}v^*\nabla v - i\boldsymbol{B}v\nabla v^*] = (i\nabla v \cdot \mathbf{curl}\xi, v).$$

We have

$$\begin{aligned} &\int_{\Omega_\delta} (i\nabla v \cdot \mathbf{curl}\xi, v) \\ = &-\int_{\Omega_\delta} \xi\mathbf{curl}(iv, \nabla v) + \sum_{j=1}^{d} \int_{\partial B(b_j,\delta)} \xi(iv, \partial_{\boldsymbol{\tau}} v) \\ = &\sum_{j=1}^{d} \int_{\partial B(b_j,\delta)} \xi(iv, \partial_{\boldsymbol{\tau}} v) \end{aligned} \quad (7.16)$$

where $\boldsymbol{\tau}$ is the unit tangential vector to $\partial B(b_j, \delta)$, $\boldsymbol{n}$ is the outward unit normal vector to $\partial B(b_j, \delta)$. Here we have used the fact that $v = e^{i\phi}$ for some local function $\phi$ ($|v| = 1$ by assumption), so $\mathbf{curl}(iv, \nabla v) = \mathbf{curl}\nabla\phi = 0$ locally.

To estimate the right-hand side of (7.16), we proceed as follows.

From $\deg(v, \partial B(b_j, \delta)) = 1$ and $|v(x)| = 1$ on $\Omega_\delta$, we may write (let $\theta_j$ be the standard polar angle variable with respect to $b_j$),

$$v = e^{i(\theta_j + \eta)} \quad \text{on} \quad U_j \setminus B(b_j, \delta)$$

in some fixed neighbourhood $U_j$ of $b_j$. Here $\eta$ is some real single valued function on $U_j \setminus B(b_j, \delta)$. We now have, for $x \in U_j \setminus B(b_j, \delta)$,

$$\nabla v = i(\nabla\theta + \nabla\eta)e^{i(\theta+\eta)}$$

which implies

$$\nabla\eta = -\nabla\theta - ie^{-i(\theta+\eta)}\nabla v$$

and finally

$$|\nabla\eta|^2 \le 2(|\nabla v|^2 + \frac{1}{|x - b_j|^2}). \quad (7.17)$$

On the boundary of $B(b_j, \delta)$, we have

$$(iv, \partial_{\boldsymbol{T}} v) = \frac{1}{\delta} + \partial_{\boldsymbol{T}} \eta.$$

We deduce then, by integrating by parts along $\partial B(b_j, \delta)$,

$$\int_{\partial B(b_j,\delta)} \xi(iv, \partial_{\boldsymbol{T}} v) = 2\pi \overline{\xi}_j + \int_{\partial B(b_j,\delta)} \partial_{\boldsymbol{T}} \xi (\eta - \overline{\eta}_j) \tag{7.18}$$

where

$$\overline{\xi}_j = \frac{1}{|\partial B(b_j, \delta)|} \int_{\partial B(b_j,\delta)} \xi, \quad \overline{\eta}_j = \frac{1}{|\partial B(b_j, \delta)|} \int_{\partial B(b_j,\delta)} \eta.$$

For any $\xi$ solving (7.13), by Sobolev embedding and elliptic estimates, we have for any $\alpha < 1$,

$$|\xi|_{C^{0,\alpha}} \leq C_\alpha |\Delta \xi|_{L^2}. \tag{7.19}$$

From this we have

$$\begin{aligned} &\textstyle\int_{B(b_j,\delta)} \partial_j \xi = \int_{\partial B(b_j,\delta)} \xi \boldsymbol{e}_j \cdot \boldsymbol{n} = \int_{\partial B(b_j,\delta)} (\xi - \xi(b_j)) \boldsymbol{e}_j \cdot \boldsymbol{n}, \\ &\tfrac{1}{|B(b_j,\delta)|} \Big| \textstyle\int_{B(b_j,\delta)} \partial_j \xi \Big| \leq 2\pi C_\alpha \delta^{\alpha - 1} |\Delta \xi|_{L^2}. \end{aligned} \tag{7.20}$$

By the standard trace theorem, we obtain

$$\int_{\partial B(b_j,\delta)} |\partial_j \xi - \frac{1}{|B(b_j, \delta)|} \int_{B(b_j,\delta)} \partial_j \xi|^2 \leq C\delta \int_\Omega |H\xi|^2 \leq C\delta \int_\Omega |\Delta \xi|^2. \tag{7.21}$$

Combining (7.20)–(7.21), we have

$$\int_{\partial B(b_j,\delta)} |\nabla \xi|^2 \leq C(\delta + \delta^{2\alpha - 1}) \int_\Omega |\Delta \xi|^2. \tag{7.22}$$

Similar to $\xi$, we have also

$$\int_{\partial B(b_j,\delta)} |\eta - \overline{\eta}_j|^2 \leq C\delta \int_U |\nabla \eta|^2 \leq C\delta \Big( \int_{\Omega_\delta} |\nabla v|^2 + 2\pi \log(1/\delta) \Big). \tag{7.23}$$

Combining (7.18), (7.22) and (7.23), we conclude

$$\Big| \int_{\partial B(b_j,\delta)} \partial_{\boldsymbol{T}} \xi (\eta - \overline{\eta}_j) \Big| \leq C\delta^\alpha |\Delta \xi|_{L^2(\Omega)} \Big( |\nabla v|_{L^2(\Omega_\delta)} + \sqrt{\log(1/\delta)} \Big). \tag{7.24}$$

This finishes the estimate for (7.16).

Turning to the first term on the right-hand side of (7.14), we have, by Theorem 7.1,

$$\int_{\Omega_\delta} a(x)|\nabla v|^2 \geq 2m\pi d\log(1/\delta) - C. \tag{7.25}$$

Combining (7.17), (7.18), (7.20), (7.25) and Proposition 7.1, we obtain

$$|\int_{\Omega_\delta}(\frac{1}{a(x)}|\Delta\xi - \frac{1}{a(x)}\nabla a(x)\cdot\nabla\xi|^2 + \frac{1}{a(x)}|\nabla\xi|^2) + 4\pi\sum_{j=1}^{d}\xi(b_j) + R(\delta)| \leq C \tag{7.26}$$

where

$$R(\delta) = 2\sum_{j=1}^{d}\Big[(\bar{\xi}_j - \xi(b_j)) + \int_{\partial B(b_j,\delta)} \partial_{\boldsymbol{T}}\xi(\eta - \bar{\eta}_j)\Big]. \tag{7.27}$$

Combining (7.24) and (7.19), we verify that

$$|R(\delta)| \leq C|\Delta\xi|_{L^2(\Omega)}\Big(\delta^\alpha|\nabla v|_{L^2(\Omega_\delta)} + \delta^{\alpha/2}\Big). \tag{7.28}$$

Using the equality $\nabla v = (\nabla - i\boldsymbol{B})v + iv\boldsymbol{B}$, we obtain

$$\begin{aligned}\int_{\Omega_\delta} a(x)|\nabla v|^2 &\leq 2\int_{\Omega_\delta} a(x)\Big[|(\nabla - i\boldsymbol{B})v|^2 + |\boldsymbol{B}|^2\Big] \\ &\leq 2\int_{\Omega_\delta} a(x)\Big[|(\nabla - i\boldsymbol{B})v|^2 + \frac{1}{a^2(x)}|\nabla\xi|^2\Big] \\ \text{(by Proposition 7.1)} &\leq C\log(1/\delta)\end{aligned} \tag{7.29}$$

Going back to (7.28), we deduce that

$$R(\delta) \to 0 \ \text{ as } \delta \to 0. \tag{7.30}$$

By standard elliptic estimates, we have

$$\Big|\sum_{j=1}^{d}\xi(b_j)\Big| \leq C|\Delta\xi|_{L^2(\Omega)}. \tag{7.31}$$

(7.30), (7.31) and (7.26) yield

$$|\Delta\xi|_{L^2(\Omega)} \leq C.$$

(7.9) and (7.10) then follow.

**Proposition 9.7.3** *Let $\xi_*$ be the solution of*

$$\begin{cases} -\nabla\cdot(\frac{1}{a(x)}\nabla(\Delta\xi_* - \frac{1}{a(x)}\nabla a(x)\cdot\nabla\xi_*)) \\ +\frac{1}{a(x)}(\Delta\xi_* - \frac{1}{a(x)}\nabla a(x)\cdot\nabla\xi_*) = 2\pi\sum_{j=1}^d \delta_{b_j} & \text{in} \quad \Omega, \\ \xi_* = 0 & \text{on} \quad \partial\Omega, \\ \frac{1}{a(x)}\partial_{\boldsymbol{n}}(\Delta\xi_* - \frac{1}{a(x)}\nabla a(x)\cdot\nabla\xi_*) = -g & \text{on} \quad \partial\Omega, \end{cases}$$

*this is equivalent to*

$$\begin{cases} -\nabla\cdot(\frac{1}{a(x)}\nabla(a(x)h_*)) + h_* = 2\pi\sum_{j=1}^d \delta_{b_j} & \text{in} \quad \Omega, \\ \xi_* = 0 & \text{on} \quad \partial\Omega, \\ \partial_{\boldsymbol{n}}(a(x)h_*) = -g & \text{on} \quad \partial\Omega, \end{cases}$$

*where* $h_* = \Delta\xi_* - \frac{1}{a(x)}\nabla a(x)\cdot\nabla\xi_* = \text{curl}\boldsymbol{A}_*$. *We have* $\xi_\delta \to \xi_*$ *in* $W^{2,2}(\Omega)$ *as* $\delta \to 0$.

*Proof.* First we remark that the equations in this proposition come from (6.5) and (7.12).

On $\Omega_\delta$, we have, by the fact that $(v_\delta, \boldsymbol{B}_\delta)$ are energy minimizers,

$$\begin{cases} -\nabla\cdot(\frac{1}{a(x)}\nabla(\Delta\xi_\delta - \frac{1}{a(x)}\nabla a(x)\cdot\nabla\xi_\delta)) \\ +\frac{1}{a(x)}(\Delta\xi_\delta - \frac{1}{a(x)}\nabla a(x)\cdot\nabla\xi_\delta) = 0 & \text{in} \quad \Omega, \\ \frac{1}{a(x)}\partial_{\boldsymbol{n}}(\Delta\xi_\delta - \frac{1}{a(x)}\nabla a(x)\cdot\nabla\xi_\delta) = -g & \text{on} \quad \partial\Omega, \end{cases} \tag{7.32}$$

On the other hand,

$$\begin{aligned} &\int_{\partial B(a_j,\delta)} \frac{1}{a(x)}\partial_{\boldsymbol{n}}(\Delta\xi_\delta - \frac{1}{a(x)}\nabla a(x)\cdot\nabla\xi_\delta) \\ &= -\int_{\partial B(a_j,\delta)} (iv_\delta, \boldsymbol{\tau}\cdot(\nabla - i\boldsymbol{B})v_\delta) = -2\pi + o(1). \end{aligned} \tag{7.33}$$

From Proposition 7.2, we know that

$$|\Delta\xi_\delta|_{L^2(\Omega)} + |\nabla\xi_\delta|_{L^2(\Omega)} \le |\Delta\xi_\delta|_{L^2(\Omega)} \le C$$

and hence (passing to a subsequence if necessary) we may assert that

$$\xi_\delta \to \bar{\xi} \ \text{ in } \ W^{2,2}(\Omega) \ \text{ weakly.}$$

Passing to the limit in (7.32) and using (7.33), we see that $\bar{\xi}$ verifies the desired equation and hence $\bar{\xi} = \xi_*$. The full sequence therefore converges by uniqueness of the limit. By the fact that $(v_\delta, \boldsymbol{B}_\delta)$ is an energy minimizer, we have

$$\begin{aligned} &\int_{\Omega_\delta} a(x)|(\nabla - i\boldsymbol{B}_\delta)v_\delta|^2 + \int_\Omega a(x)|\mathbf{curl}\boldsymbol{B}_\delta|^2 \\ \le &\int_{\Omega_\delta} a(x)|(\nabla - i\boldsymbol{A}_*)v_\delta|^2 + \int_\Omega a(x)|\mathbf{curl}\boldsymbol{A}_*|^2. \end{aligned} \tag{7.34}$$

Expanding the first term on both sides of above inequality as in (7.14), we obtain

$$\begin{aligned}&\int_{\Omega_\delta} a(x)|\mathbf{curl}\boldsymbol{B}_\delta|^2+\frac{1}{a(x)}|\nabla\xi_\delta|^2+2[\xi_\delta,v_\delta]\\ \leq\ &\int_{\Omega} a(x)|\mathbf{curl}\boldsymbol{A}_*|^2+\frac{1}{a(x)}|\nabla\xi_*|^2+2[\xi_*,v_*].\end{aligned}$$

Since $\xi_\delta\to\xi_*$ weakly in $W^{2,2}(\Omega)$, we obtain

$$\int_\Omega a(x)|\mathbf{curl}\boldsymbol{B}_\delta|^2\leq\int_\Omega a(x)|\mathbf{curl}\boldsymbol{A}_*|^2+o(1),$$

that is (by (7.15))

$$\int_\Omega a(x)|\Delta\xi_\delta|^2\leq\int_\Omega a(x)|\Delta\xi_*|^2+o(1).$$

This implies the strong convergence of $\Delta\xi_\delta$ to $\Delta\xi_*$ in $L^2$ in view of the weak lower semi-continuity (with $dx$ replaced by $d\nu=a(x)dx$) and then Proposition 7.3 follows.

**Proposition 9.7.4** *Let*

$$\begin{aligned}\overline{\mu}_\delta&=Inf_{\psi\in G_\delta}\frac{1}{2}\int_{\Omega_\delta} a(x)|\nabla\psi|^2,\\ G_\delta\ &=\ \{\psi\in H^1(\Omega_\delta,S^1),\ \deg(\psi,\partial B(b_j,\delta))=1,\\ &\quad\ \partial_{\boldsymbol{T}}\psi=i(g+\frac{1}{a(x)}\partial_{\boldsymbol{n}}\xi_*)\psi\ \ on\ \ \partial\Omega\}.\end{aligned}$$

*We have then*

$$\left|\frac{1}{2}\int_{\Omega_\delta} a(x)|\nabla v_\delta|^2-\overline{\mu_\delta}\right|\to 0\ \ as\ \delta\to 0. \tag{7.35}$$

*Proof.* Since

$$\partial_{\boldsymbol{T}}v_\delta=i(g+\frac{1}{a(x)}\partial_{\boldsymbol{n}}\xi_\delta)v_\delta\ \ \text{on}\ \ \partial\Omega,$$

we see that $v_\delta\in\mathcal{E}^1_\delta$ (see Lemma 7.4) and then

$$\frac{1}{2}\int_{\Omega_\delta} a(x)|\nabla v_\delta|^2\geq\inf_{\psi\in\mathcal{E}^1_\delta}\frac{1}{2}\int_{\Omega_\delta} a(x)|\nabla\psi|^2:=\overline{\mu}^\delta_\delta.$$

On the other hand we have from Lemma 7.4 that

$$\overline{\mu}^\delta_\delta=\overline{\mu}_\delta+o(1).$$

This implies

$$\frac{1}{2}\int_{\Omega_\delta} a(x)|\nabla v_\delta|^2 \geq \overline{\mu}_\delta + o(1). \tag{7.36}$$

Recall that

$$R(\delta) \to 0 \text{ as } \delta \to 0,$$

we deduce

$$\begin{aligned} E_\delta(v_\delta, \boldsymbol{B}_\delta) &= \frac{1}{2}\int_{\Omega_\delta} a(x)|\nabla v_\delta|^2 + \frac{1}{2}\int_\Omega (\frac{1}{a(x)}|\nabla \xi_*|^2 \\ &+ \frac{1}{a(x)}|\Delta\xi_* - \frac{1}{a(x)}\nabla a \cdot \nabla \xi_*|^2) + 2\pi\sum_{j=1}^d \xi_*(b_j) + o(1).(7.37) \end{aligned}$$

Hence, in view of (7.37), we have uncoupled $v$ and $\boldsymbol{B}$. Consequently, in order to minimize $E_\delta(v, \boldsymbol{B})$, it suffices to minimize $\overline{\mu}_\delta$, and this yields

$$\frac{1}{2}\int_{\Omega_\delta} a(x)|\nabla v_\delta|^2 \leq \overline{\mu}_\delta + o(1).$$

Proposition 7.4 follows.

**Proposition 9.7.5** *Let $1 < p < 2$, the map $v_\delta$ remains bounded in $W^{1,p}$ and $v_\delta$ converges to $v_*$ strongly in $W^{1,p}$, and in $C^k_{loc}(\Omega \setminus \cup_{j=1}^d \{a_j\})$, where $v_*$ is defined on $\Omega$ by*

$$v_* = \prod_{j=1}^d \frac{z - a_j}{|z - a_j|} e^{i\phi_*} \tag{7.38}$$

*and where $\phi_*$ solves*

$$-\text{div}(a(x)\nabla\phi_*) = \nabla a(x) \cdot (\nabla\theta_1 + \cdots + \nabla\theta_d), x \in \Omega, \tag{7.39}$$

*with boundary condition*

$$\partial_{\boldsymbol{\tau}}\phi_* = -(i\prod_{j=1}^d \frac{z - a_j}{|z - a_j|}, \partial_{\boldsymbol{\tau}}(\frac{z - a_j}{|z - a_j|})) + \frac{1}{a(x)}\partial_{\boldsymbol{n}}\xi_* + g, \tag{7.40}$$

*where $\nabla\theta_k = (\frac{-(y-\beta_k)}{|z-a_k|}, \frac{x-\alpha_k}{|z-a_k|})$ with $z = x + iy$, $a_k = \alpha_k + i\beta_k$, $k = 1, \ldots, d$.*

*Proof.* Since $v_\delta$ takes its value into $S^1$, we have

$$\text{div}(iv_\delta, \textbf{curl} v_\delta) = 0 \text{ in } \Omega_\delta.$$

Let $\Phi$ be the solution of

$$\begin{cases} \operatorname{div}(\frac{1}{a(x)}\nabla\Phi) = 2\pi\sum_{j=1}^{d}\delta_{b_j}, \\ \frac{1}{a(x)}\partial_{\boldsymbol{n}}\Phi = g + \frac{1}{a(x)}\partial_{\boldsymbol{n}}\xi_\delta. \end{cases} \tag{7.41}$$

We have

$$\operatorname{div}((iv_\delta, \mathbf{curl}v_\delta) + \frac{1}{a(x)}\nabla\Phi) = 0,$$

and

$$\begin{aligned} &\int_{\partial B(b_j,\delta)} ((iv_\delta, \mathbf{curl}v_\delta) + \frac{1}{a(x)}\nabla\Phi)\cdot\boldsymbol{n} \\ = & -\int_{\partial B(b_j,\delta)} (iv_\delta, \partial_{\boldsymbol{\tau}}v_\delta) + \int_{\partial B(b_j,\delta)} \frac{1}{a(x)}\partial_{\boldsymbol{n}}\Phi = 0. \end{aligned}$$

Hence there is some $f_\delta \in H^1(\Omega_\delta, \mathbb{R})$ such that

$$(iv_\delta, \mathbf{curl}v_\delta) = -\frac{1}{a(x)}\nabla\Phi + \mathbf{curl}f_\delta,$$

that is

$$\begin{cases} v_\delta \times v_{\delta x_1} = -\frac{\partial f_\delta}{\partial x_1} - \frac{1}{a(x)}\frac{\partial\Phi}{\partial x_2}, \\ v_\delta \times v_{\delta x_2} = -\frac{\partial f_\delta}{\partial x_2} + \frac{1}{a(x)}\frac{\partial\Phi}{\partial x_1}. \end{cases}$$

We have from this equality that

$$\begin{aligned} \int_{\Omega_\delta} a(x)|\nabla v_\delta|^2 &= \int_{\Omega_\delta} a(x)(|v_\delta \times v_{\delta x_1}|^2 + |v_\delta \times v_{\delta x_2}|^2) \\ &= \int_{\Omega_\delta} \frac{1}{a(x)}|\nabla\Phi|^2 + \int_{\Omega_\delta} a(x)|\nabla f_\delta|^2 + 2\int_{\Omega_\delta}(\frac{\partial f_\delta}{\partial x_1}\frac{\partial\Phi}{\partial x_2} - \frac{\partial f_\delta}{\partial x_2}\frac{\partial\Phi}{\partial x_1}). \end{aligned}$$

As in the proof of Lemma 7.1, we have

$$\begin{aligned} &\int_{\Omega_\delta}(\frac{\partial f_\delta}{\partial x_1}\frac{\partial\Phi}{\partial x_2} - \frac{\partial f_\delta}{\partial x_2}\frac{\partial\Phi}{\partial x_1}) \\ &= -\int_{\partial\Omega}\Phi\frac{\partial f_\delta}{\partial\tau} + \sum_{i=1}^{d}\int_{\partial B(a_i,\delta)}\Phi\frac{\partial f_\delta}{\partial\tau} \\ &= -\int_{\partial\Omega}\Phi(-v_\delta \times v_{\delta\tau} + \frac{1}{a(x)}\frac{\partial\Phi}{\partial\boldsymbol{n}}) = o(1), \end{aligned}$$

here we have used $\xi_\delta \to \xi$ in $W^{2,2}(\Omega)$ and

$$-\int_{\partial B(b_j,\delta)}(iv_\delta, \partial_{\boldsymbol{\tau}}v_\delta) + \int_{\partial B(b_j,\delta)}\frac{1}{a(x)}\partial_{\boldsymbol{n}}\Phi = 0.$$

Therefore

$$\int_{\Omega_\delta} a(x)|\nabla v_\delta|^2 = \int_{\Omega_\delta} \frac{1}{a(x)}|\nabla\Phi|^2 + \int_{\Omega_\delta} a(x)|\nabla f_\delta|^2 + o(1).$$

It follows from Lemma 7.1 that $\int_{\Omega_\delta} \frac{1}{a(x)}|\nabla\Phi|^2$ is the minimum defined in Lemma 7.1 with $g$ replaced by $g + \frac{1}{a(x)}\partial_{\boldsymbol{n}}\xi_\delta$. Therefore it follows from Lemma 7.1 that

$$|\frac{1}{2}\int_{\Omega_\delta} \frac{1}{a(x)}|\nabla\Phi|^2 - \overline{\mu}_\delta| \to 0 \text{ as } \delta \to 0. \tag{7.42}$$

From Proposition 7.4, we deduce that

$$\int_{\Omega_\delta} |\nabla f_\delta|^2 \to 0.$$

Hence

$$||(iv_\delta, \mathbf{curl} v_\delta) + \frac{1}{a(x)}\nabla\Phi||_{L^2(\Omega_\delta)} \to 0 \text{ as } \delta \to 0$$

and the convergence in $C^k_{\mathrm{loc}}(\Omega \setminus \cup_{j=1}^d\{b_j\})$ of Proposition 7.5 follows from Lemma 7.2 and the convergence in $W^{1,p}$ follows from the fact that

$$|\nabla\Phi| \sim \frac{d_j}{|x-b_j|}, \quad \text{near } b_j.$$

**Proposition 9.7.6** *We have for any configuration $(b_j) \in (\Omega \cap a^{-1}(m))^d$,*

$$\mu_\delta(b) = W(b) + m\pi d \log\frac{1}{\delta} + o(1) \quad as \quad \delta \to 0 \tag{7.43}$$

*where $W$ is defined by the renormalized energy*

$$\begin{aligned} W(b) \quad &= \quad -\pi\sum_{j\neq k} a(b_j)\log|b_j - b_k| - \pi\sum_{j=1}^d R(b_j) \\ &\quad +\pi\sum_{j=1}^d \xi_*(b_j) + \frac{1}{2}\int_{\partial\Omega}\Phi(g + \frac{1}{a(x)}\partial_{\boldsymbol{n}}\xi_*) \\ &\quad +\frac{1}{2}\int_{\partial\Omega}\frac{1}{a(x)}(\Delta\xi_* - \frac{1}{a(x)}\nabla a(x)\cdot\nabla\xi_*)\partial_{\boldsymbol{n}}\xi_* \end{aligned} \tag{7.44}$$

*where $\xi_*$ is the solution of the following problem*

$$\begin{cases} -\nabla\cdot(\frac{1}{a(x)}\nabla(\Delta\xi_* - \frac{1}{a(x)}\nabla a(x)\cdot\nabla\xi_*)) & \\ +\frac{1}{a(x)}(\Delta\xi_* - \frac{1}{a(x)}\nabla a(x)\cdot\nabla\xi_*) = 2\pi\sum_{j=1}^d \delta_{b_j} & in \quad \Omega, \\ \xi_* = 0 & on \quad \partial\Omega, \\ \frac{1}{a(x)}\partial_{\boldsymbol{n}}(\Delta\xi_* - \frac{1}{a(x)}\nabla a(x)\cdot\nabla\xi_*) = -g & on \quad \partial\Omega, \end{cases} \tag{7.45}$$

*and $R$ is given by*

$$R(x) = \Phi(x) - m \sum_{j=1}^{d} \log |x - b_j|. \tag{7.46}$$

*In addition, $\Phi$ is the solution of*

$$\begin{cases} \operatorname{div}(\frac{1}{a(x)} \nabla \Phi) = 2\pi \sum_{j=1}^{d} \delta_{b_j} & in \quad \Omega, \\ \frac{1}{a(x)} \partial_{\boldsymbol{n}} \Phi = g + \frac{1}{a(x)} \partial_{\boldsymbol{n}} \xi & on \quad \partial\Omega. \end{cases} \tag{7.47}$$

*Proof.* We deduce from Propositions 7.3–7.5 and (7.37) that,

$$\begin{aligned} \mu_\delta(b) &= \frac{1}{2} \int_{\Omega_\delta} a(x) |\nabla v_\delta|^2 + \frac{1}{2} \int_\Omega (\frac{1}{a(x)} |\nabla \xi_*|^2 \\ &\quad + \frac{1}{a(x)} |\Delta \xi_* - \frac{1}{a(x)} \nabla a \cdot \nabla \xi_*|^2) + 2\pi \sum_{j=1}^{d} \xi_*(b_j) + o(1). \end{aligned} \tag{7.48}$$

Moreover, multipying (7.45) by $\xi_*$, and then integrating over $\Omega$, we have,

$$\begin{aligned} 2\pi \sum_{j=1}^{d} \xi_*(b_j) &= \int_{\partial\Omega} \frac{1}{a(x)} (\Delta \xi_* - \frac{1}{a(x)} \nabla a(x) \cdot \nabla \xi_*) \partial_{\boldsymbol{n}} \xi_* \\ &\quad - \int_\Omega \frac{1}{a(x)} |\Delta \xi_* - \frac{1}{a(x)} \nabla a(x) \cdot \nabla \xi_*|^2 - \int_\Omega \frac{1}{a(x)} |\nabla \xi_*|^2 \end{aligned} \tag{7.49}$$

Lemma 7.4 and Lemma 7.3 imply

$$\frac{1}{2} \int_{\Omega_\delta} a(x) |\nabla v_\delta|^2 = m\pi d \log \frac{1}{\delta} + \omega(b, d, g) + O(\delta) \tag{7.50}$$

where

$$\omega(b, d, g) = -m\pi \sum_{j \neq k} \log |b_j - b_k| - \pi \sum_{j=1}^{d} R(b_j) + \frac{1}{2} \int_{\partial\Omega} \frac{1}{a(x)} \Phi \partial_{\boldsymbol{n}} \Phi.$$

Here $\Phi$ is a function verifying (7.41) and $R$ denotes

$$R(x) = \Phi(x) - m \sum_{j=1}^{d} \log |x - b_j|.$$

Combining (7.48)–(7.50), we obtain the desired result.

Finally, we conclude with

**Theorem 9.7.2** *Let $a_j$ be as in Theorem 6.1 of this chapter, and we assume that $d \leq Card(\Omega \cap a^{-1}(m))$. Then the configuration $a = (a_j)$ minimizes $W(b)$ in $(\Omega \cap a^{-1}(m))^d$ where $W(b)$ is defined by (7.44).*

**Remark 9.7.1** *Theorem 7.2 is the main result of this section.*

*Proof of Theorem 7.2.* For $b_0 \in \Omega \cap a^{-1}(m)$, we define an auxiliary minimization problem

$$I(\varepsilon,\delta) = \mathrm{Inf}_{v\in Q}\frac{1}{2}\int_{B(b_0,\delta)} a(x)|\nabla v|^2 + \frac{1}{4\varepsilon^2}a(x)(1-|v|^2)^2,$$
$$Q = \{v \in H^1(B(b_0,\delta), I\!R^2),\ v|_{\partial B(b_0,\delta)} = e^{i\theta}\}.$$

Let $b = (b_j)$ be a configuration of $d$ points in $\Omega \cap a^{-1}(m)$. We construct a comparison function $v_{\varepsilon,\delta}$ in $H^1(\Omega)$ as follows

(1) $v_{\varepsilon,\delta} = v_\delta$ on $\Omega_\delta$.

(2) On $B(b_i,\delta)\setminus B(b_i,\delta/2)$, $v_{\varepsilon,\delta}$ is a harmonic map from $B(b_i,\delta)\setminus B(b_i,\delta/2)$ to $S^1$ with boundary condition $v_{\varepsilon,\delta}|_{\partial B(b_i,\delta)} = v_\delta$ and $v_{\varepsilon,\delta}|_{\partial B(b_i,\delta/2)} = \frac{z-b_i}{|z-b_i|}$.

(3) On $B(b_i,\delta/2)$, $v_{\varepsilon,\delta}$ is the minimizer of $I(\varepsilon,\delta/2)$.

Then, we have

$$\frac{1}{2}\int_{B(b_j,\delta)} a(x)|\nabla v_{\varepsilon,\delta}|^2 + \frac{a(x)}{2\varepsilon^2}(1-|v_{\varepsilon,\delta}|^2)^2 \le I(\varepsilon,\delta/2) + C(\delta) \qquad (7.51)$$

where $|C(\delta)| \to 0$ as $\delta \to 0$. It follows that

$$\begin{aligned} G_\varepsilon(\psi_\varepsilon, \boldsymbol{A}_\varepsilon) &\le G_\varepsilon(v_{\varepsilon,\delta}, \boldsymbol{A}_\varepsilon)\\ &\le E_\varepsilon(v_\delta, \boldsymbol{A}_*) + dI(\varepsilon,\delta/2) + C(\delta)\\ &\le \mu_\delta(b) + dI(\varepsilon,\delta/2) + C(\delta)\\ &\le W(b) + m\pi d\log(1/\delta) + dI(\varepsilon,\delta/2) + C(\delta). \end{aligned} \qquad (7.52)$$

Here we have used Proposition and $|C(\delta)| \to 0$ as $\delta \to 0$.

On the other hand, it follows from the definition of the functional $G_\varepsilon(u, \boldsymbol{A})$ that

$$\begin{aligned} &G_\varepsilon(\psi_\varepsilon, \boldsymbol{A}_\varepsilon)\\ = &\ \frac{1}{2}\int_\Omega a(x)|(\nabla - i\boldsymbol{A}_\varepsilon)\psi_\varepsilon|^2 + \frac{1}{2}\int_\Omega a(x)|\mathbf{curl}\boldsymbol{A}_\varepsilon|^2 + \frac{1}{4\varepsilon^2}\int_\Omega a(x)(|\psi_\varepsilon|^2-1)^2\\ = &\ \frac{1}{2}\int_{\Omega'_\delta} a(x)|(\nabla - i\boldsymbol{A}_\varepsilon)\psi_\varepsilon|^2 + \frac{1}{2}\int_\Omega a(x)|\mathbf{curl}\boldsymbol{A}_\varepsilon|^2 + \frac{1}{4\varepsilon^2}\int_{\Omega'_\delta} a(x)(|\psi_\varepsilon|^2-1)^2\\ &+\sum_{j=1}^d\left[\frac{1}{2}\int_{B(a_j,\delta)} a(x)|(\nabla - i\boldsymbol{A}_\varepsilon)\psi_\varepsilon|^2 + \frac{1}{4\varepsilon^2}\int_{B(a_j,\delta)} a(x)(|\psi_\varepsilon|^2-1)^2\right]\\ \ge &\ \frac{1}{2}\int_{\Omega'_\delta} a(x)|(\nabla - i\boldsymbol{A}_\varepsilon)\psi_\varepsilon|^2 + \frac{1}{2}\int_\Omega a(x)|\mathbf{curl}\boldsymbol{A}_\varepsilon|^2\\ &+\sum_{j=1}^d\left[\frac{1}{2}\int_{B(a_j,\delta)} a(x)|\nabla\psi_\varepsilon|^2 + \frac{1}{4\varepsilon^2}\int_{B(a_j,\delta)} a(x)(|\psi_\varepsilon|^2-1)^2\right] + O(\delta), \end{aligned} \qquad (7.53)$$

here we have used the fact $|\psi_\varepsilon| \le 1$ and $||\boldsymbol{A}_\varepsilon||_{L^\infty} \le C$, $\Omega'_\delta = \Omega \setminus \cup_{j=1}^d B(a_j,\delta)$.

Arguing as in the proof of Lemma VIII.2 of [BBH 94] and using Theorem 7.1, we know that

$$G_\varepsilon(\psi_\varepsilon, \boldsymbol{A}_\varepsilon) \geq W(a) + m\pi d \log\frac{1}{\delta} + \sum_{j=1}^{d}[\frac{1}{2}\int_{B(a_j,\delta/2)} a(x)|\nabla\psi_\varepsilon|^2 + \frac{1}{4\varepsilon^2}\int_{B(a_j,\delta/2)} a(x)(|\psi_\varepsilon|^2-1)^2] + O(\delta), \tag{7.54}$$

where $a = (a_j)$ is determined in Theorem 6.1. The summation term on the right-hand side of (7.54) can be dealt with by using the same arguments as in the proof of Theorem 9.1 in Chapter 6 (see also [BHa 95]), that is

$$\sum_{j=1}^{d}[\frac{1}{2}\int_{B(a_j,\delta/2)} a(x)|\nabla\psi_\varepsilon|^2 + \frac{1}{4\varepsilon^2}\int_{B(a_j,\delta/2)} a(x)(|\psi_\varepsilon|^2-1)^2] \geq dI(\varepsilon,\delta/2).$$

We finally have

$$G_\varepsilon(\psi_\varepsilon, \boldsymbol{A}_\varepsilon) \geq dI(\varepsilon,\delta/2) + W(a) + m\pi d\log\frac{1}{\delta} + C'(\delta) \tag{7.55}$$

where $C'(\delta) \to 0$ as $\delta \to 0$. Combining (7.52) and (7.55), we obtain, letting $\delta \to 0$,

$$W(a) \leq W(b) \qquad \forall\, b = (b_j).$$

The theorem is proved.

## 9.8 Pinning of vortices in other circumstances

In the following, we give some pinning results in other circumstances. As the proofs are more or less the same, we only announce the results by omitting the details of proofs.

### 9.8.1 G-L model subject to thermo-perturbation or large horizontal field

As we have seen in Chapter 1, provided that the temperature field is not a constant across the sample, we have the following steady-state G-L energy

$$E_\varepsilon(\psi) = \frac{1}{2}\int_\Omega [|\nabla\psi|^2 + \frac{1}{2\varepsilon^2}(|\psi|^2 - \beta(x))^2].$$

Here the distinct feature is that $\beta$ is a function of $x$ rather than a constant. $\Omega$ is a smooth simply connected domain in $\mathbb{R}^2$. Let's look at the case when we minimize the energy subject to the Dirichlet type boundary condition $\psi|_{\partial\Omega} = g\beta(x)$ with smooth $g$ and $|g|(x) = 1$ for almost all $x \in \partial\Omega$, $\deg(g, \partial\Omega) = d > 0$. We also assume that $\beta$ is a smooth function such that

$$0 < m = \min_{x\in\overline{\Omega}} \beta(x) < \max_{x\in\overline{\Omega}} \beta(x) = M < 1.$$

Let $\beta^{-1}(m)=\{x\in\overline{\Omega}:\ \beta(x)=m\}$, and suppose that $N=$Card$(\beta^{-1}(m)\cap\Omega)$ is a finite number. We distinguish the following 4 situations

$$\beta^{-1}(m)\subset\Omega,\ N\geq d, \tag{8.1}$$
$$\beta^{-1}(m)\subset\Omega,\ N<d, \tag{8.2}$$
$$\beta^{-1}(m)\cap\partial\Omega\neq\emptyset,\ N\geq d, \tag{8.3}$$
$$\beta^{-1}(m)\cap\partial\Omega\neq\emptyset,\ N<d. \tag{8.4}$$

Our main results are as follows:

**Theorem 9.8.1** *Under one of the conditions (8.1)–(8.3), for any minimizing sequence $\{\psi_\varepsilon\}$, there is a subsequence $\{\psi_{\varepsilon_n}\}$ (denoted by $\{\psi_n\}$ in the following) such that there exists a function $\psi_*$ such that*

$$\psi_n\to\psi_*\ \ in\ \ C^{1,\alpha}_{loc}(\overline{\Omega}\setminus\cup_{j=1}^{N_0}\{a_j\})\cup C^k_{loc}(\Omega\setminus\cup_{j=1}^{N_0}\{a_j\})$$

*for some $k\geq 1$. Moreover, $\psi_*$ is smooth (depending on $\beta$), $|\psi_*|=\beta(x)$ almost everywhere in $\Omega$ and*

$$-\Delta\psi_*=\beta^{-2}(x)\big(|\nabla\psi_*|^2-\frac{1}{2}\Delta\beta^2(x)\big)\psi_*\ \ in\ \ \Omega\setminus\cup_{j=1}^{N_0}\{a_j\}.$$

*In addition, let $d_j=\deg(\psi_*,a_j)$, then,*

$$d=\sum_{j=1}^{N_0}d_j$$

*and $a_1,\dots,a_{N_0}\in\beta^{-1}(m)\cap\Omega$. Furthermore, we have*

$$d_j=1,\ \ N_0=d\ \ \text{under the conditions (8.1) or (8.3),}$$
$$d_j\geq 1,\ \ N_0<d\ \ \text{under the condition (8.2).}$$

**Theorem 9.8.2** *Under the conditions (8.1) or (8.3), denote*

$$\hat{a}=(a_1,\dots,a_d),\ \ \hat{d}=(1,\dots,1),$$

*where the notations are the same as in Theorem 8.1. Then $(\hat{a},\hat{d})$ is the minimizer of the following renormalized energy:*

$$\begin{aligned}W_g(\hat{a},\hat{d},\beta)\ &=\ -\sum_{j\neq k}\beta^2(a_j)d_jd_k\ln|a_j-a_k|\\ &\quad+\frac{1}{2}\int_{\partial\Omega}\Phi_0(g\times\partial_{\mathcal{T}}g)-\sum_{j=1}^{d}\pi d_jR_0(a_j)\end{aligned}$$

*where* $R_0(x) = \Phi_0(x) - \sum_{j=1}^d \beta^2(a_j) d_j \ln|x - a_j|$, $\Phi_0$ *is the solution of the following problem:*

$$\begin{aligned}
&\operatorname{div}\left(\frac{1}{\beta^2}\nabla\Phi_0\right) = 2\pi\sum_{j=1}^{d} d_j\delta_{a_j} \quad \text{in } \Omega,\\
&\frac{1}{\beta^2}\partial_{\boldsymbol{n}}\Phi_0 = g\times\partial_{\boldsymbol{T}} g \quad \text{on } \partial\Omega,\\
&\int_{\partial\Omega}\Phi_0 = 0.
\end{aligned}$$

The first theorem is about the convergence of the solutions and the second theorem is about the location of the vortices. Notice that only (8.1)–(8.3) have been covered by Theorem 8.1 and (8.1) and (8.3) have been covered by Theorem 8.2.

### 9.8.2 An anisotropic G-L model

Let $\Omega$ be a bounded, smooth and simply connected domain in $I\!R^2$, let

$$\begin{aligned}
&\beta(x), a_{jk}(x) \in C^2(\overline{\Omega}),\ j,k = 1,2,\\
&0 < m_0 = \min_{x\in\overline{\Omega}}\beta(x) < \max_{x\in\overline{\Omega}}\beta(x) = 1 < \infty,\\
&\lambda_0|\xi|^2 \le a_{jk}\xi_j\xi_k^* \le \Lambda_0|\xi|^2,\ \forall x\in\Omega, \forall\xi\in \mathbb{C}^2
\end{aligned}$$

where $\lambda_0$ and $\Lambda_0$ are positive constants. Let $g(x)$: $\partial\Omega \to S^1$ (the unit circle in the complex plane) be a smooth function satisfying $\deg(g,\partial\Omega) = d > 0$. Let $W = \{v \in \mathcal{H}^1(\Omega, I\!R^2) : \psi|_{\partial\Omega} = g\beta(x)\}$.

For $\varepsilon > 0$, consider the anisotropic G-L functional

$$E_\varepsilon(\psi) = \frac{1}{2}\int_\Omega \left[a_{jk}(x)\partial_j\psi\partial_k\psi^* + \frac{1}{2\varepsilon^2}(|\psi|^2 - \beta(x))^2\right].$$

It is clear that for each $\varepsilon > 0$, the infimum

$$\inf_{\psi\in W} E_\varepsilon(\psi) \tag{8.5}$$

is attained by a minimizer $\psi_\varepsilon \in W$.

To state the results, let $\lambda_1(x)$, $\lambda_2(x)$ be the two eigenvalues of the $2\times 2$ matrix $(a_{jk}(x))$. Denote

$$a(x) = \sqrt{\lambda_1(x)\lambda_2(x)}\beta^2(x), \tag{8.6}$$

$$a^{-1}(m) = \{x\in\Omega :\ a(x) = \inf_{x\in\Omega} a(x) = m\}. \tag{8.7}$$

We again look at 4 possibilities:

$$a^{-1}(m)\subset\Omega,\ N = \text{ Card } a^{-1}(m) \ge d, \tag{8.8}$$

$$a^{-1}(m)\subset\Omega,\ N = \text{ Card } a^{-1}(m) < d, \tag{8.9}$$

$$a^{-1}(m)\cap\partial\Omega\neq\emptyset,\ N = \text{ Card } \left(a^{-1}(m)\cap\Omega\right) \ge d, \tag{8.10}$$

$$a^{-1}(m)\cap\partial\Omega\neq\emptyset,\ N = \text{ Card } \left(a^{-1}(m)\cap\Omega\right) < d. \tag{8.11}$$

Under all these assumptions, we have

**Theorem 9.8.3** *For any energy minimizer sequence $\{\psi_\varepsilon\}$ of the problem (8.5), there is a subsequence denoted by $\{\psi_n\}$ such that under one of the conditions (8.8)–(8.10), we have*

$$\psi_n \to \psi_* \ \ in \ \ C^{1,\alpha}_{loc}(\overline{\Omega} \setminus \cup_{j=1}^{N_0}\{a_j\}), \ \forall \alpha \in (0,1).$$

*Moreover, we have*

$$d = \sum_{j=1}^{N_0} d_j \quad where \ \ d_j = \deg(\psi_*, a_j)$$

*and*

$$\begin{aligned} &a_1, \ldots, a_{N_0} \in a^{-1}(m) \cap \Omega, \\ &d_j = 1, \ N_0 = d \ \ under \ (8.8) \ or \ (8.10), \\ &d_j \geq 1, \ N_0 < d \ \ under \ (8.9), \end{aligned}$$

*where $\psi_*$ is smooth on $\Omega \setminus \cup_{j=1}^{N_0}\{a_j\}$, $|\psi_*| = \beta(x)$ a.e. in $\Omega$ and solves*

$$-\partial_j(a_{jk}\partial_k\psi_*) = \beta^{-2}\big[a_{jk}\partial_j\psi_*\partial_k\psi_* - \frac{1}{2}\partial_j(a_{jk}\partial_k\beta^2)\big]\psi_* \ \ in \ \ \Omega \setminus \cup_{j=1}^{N_0}\{a_j\}.$$

**Theorem 9.8.4** *Under all our assumptions and (8.8) (or (8.9), or (8.10)), let $Z_\varepsilon = \{x \in \Omega, \ \psi_\varepsilon(x) = 0\}$, then for any given $\delta > 0$, there is an $\varepsilon_0 > 0$ such that for $0 < \varepsilon < \varepsilon_0$, $Z_\varepsilon$ is contained in a $\delta$-neighbourhood of $a^{-1}(m) \cap \Omega$.*

This last theorem gives a better picture of what is happening to the location of the vortices when thermo-effect is taken into account.

### 9.8.3 A thin film G-L model

In this section, we discuss briefly the pinning results for the thin film model derived in Section 1 Chapter 8. Let $\Omega$, $a(x)$ be as in the beginning of this chapter, $\boldsymbol{A}_0(x)$ be the vector potential of the applied magnetic field which is a given smooth vector. Let $g(x)$: $\partial\Omega \to S^1$ be a smooth function satisfying $\deg(g, \partial\Omega) = d > 0$. Let $W = \{v \in H^1(\Omega, \mathbb{R}^2) : \psi|_{\partial\Omega} = g\}$.

For $\varepsilon > 0$, consider the following G-L functional

$$E_\varepsilon(\psi) = \frac{1}{2}\int_\Omega a(x)\big[|(\nabla - i\boldsymbol{A}_0)\psi|^2 + \frac{1}{2\varepsilon^2}(|\psi|^2 - 1)^2\big].$$

It is clear that for each $\varepsilon > 0$, the infimum

$$\inf_{\psi \in W} E_\varepsilon(\psi) \tag{8.12}$$

is attained by a minimizer $\psi_\varepsilon \in W$.

To state the results, we look at four possibilities:

$$a^{-1}(m) \subset \Omega, \quad N = \text{ Card } a^{-1}(m) \geq d, \tag{8.13}$$
$$a^{-1}(m) \subset \Omega, \quad N = \text{ Card } a^{-1}(m) < d, \tag{8.14}$$
$$a^{-1}(m) \cap \partial\Omega \neq \emptyset, \quad N = \text{ Card } \left(a^{-1}(m) \cap \Omega\right) \geq d, \tag{8.15}$$
$$a^{-1}(m) \cap \partial\Omega \neq \emptyset, \quad N = \text{ Card } \left(a^{-1}(m) \cap \Omega\right) < d. \tag{8.16}$$

Under all these assumptions, we have

**Theorem 9.8.5** *For any energy minimizer sequence $\{\psi_\varepsilon\}$ of the problem (8.12), there is a subsequence denoted by $\{\psi_n\}$ such that under one of the conditions (8.13)–(8.15), we have*

$$\psi_n \to \psi_* \quad in \quad C^{1,\alpha}_{loc}(\overline{\Omega} \setminus \cup_{j=1}^{N_0}\{a_j\}), \ \forall \alpha \in (0,1).$$

*Moreover, we have*

$$d = \sum_{j=1}^{N_0} d_j \quad where \quad d_j = \deg(\psi_*, a_j)$$

*and*

$$\begin{aligned} &a_1, \ldots, a_{N_0} \in a^{-1}(m) \cap \Omega, \\ &d_j = 1, \ N_0 = d \quad under \ (8.13) \ or \ (8.15), \\ &d_j \geq 1, \ N_0 < d \quad under \ (8.14), \end{aligned}$$

*where $\psi_*$ is smooth on $\Omega \setminus \cup_{j=1}^{N_0}\{a_j\}$, $|\psi_*| = 1$ a.e. in $\Omega$ and solves*

$$-\nabla \cdot \left(a(x)\nabla\psi_*\right) = a(x)\left[|\nabla\psi_*|^2 + i\nabla \cdot \boldsymbol{A}_0\right]\psi_*$$

**Theorem 9.8.6** *Under all our assumptions and (8.13) (or (8.14), or (8.15)), let $Z_\varepsilon = \{x \in \Omega, \ \psi_\varepsilon(x) = 0\}$, then for any given $\delta > 0$, there is an $\varepsilon_0 > 0$ such that for $0 < \varepsilon < \varepsilon_0$, $Z_\varepsilon$ is contained in a $\delta$-neighbourhood of $a^{-1}(m) \cap \Omega$.*

## 9.9 Bibliographical remarks

The pinning of vortices is mathematically realized by having non-constant coefficients as weights in the G-L energy. Many papers have been devoted to this topic in recent years (see [ASh 95], [Bha 95], [Dli 97], [DLi1], [DLi2], [DLi3], [DLY] and many others).

# Chapter 10
# Numerical Analysis

In the existence proofs, we have already used the discretization method and proved convergence. Although we did not give specific a priori error estimates, they are not very difficult to obtain. In this chapter, however, we discuss the a posteriori error analysis in the numerical analysis of the two-d G-L system that is more interesting from the numerical analysis point of view. We discuss both the Galerkin method and the finite element method.

## 10.1 Introduction

As usual, we deal with the following system

$$\eta\partial_t\psi + i\eta\kappa\phi\psi + \Big(\frac{i}{\kappa}\nabla + \boldsymbol{A}\Big)^2\psi + (|\psi|^2 - 1)\psi = 0 \quad \text{in } Q_T, \tag{1.1}$$

$$\partial_t\boldsymbol{A} + \nabla\phi + \mathbf{curl}^2\boldsymbol{A} + \mathrm{Re}\Big[\Big(\frac{i}{\kappa}\nabla\psi + \boldsymbol{A}\psi\Big)\bar{\psi}\Big] = 0 \quad \text{in } Q_T, \tag{1.2}$$

$$\Big(\frac{i}{\kappa}\nabla\psi + \boldsymbol{A}\psi\Big)\cdot\mathbf{n} = 0, \quad \mathbf{curl}\boldsymbol{A} = H \quad \text{on } \Gamma_T, \tag{1.3}$$

$$\psi(\cdot,0) = \psi_0(\cdot), \quad \boldsymbol{A}(\cdot,0) = \boldsymbol{A}_0(\cdot) \quad \text{on } \Omega, \tag{1.4}$$

under the Lorentz gauge

$$\phi = -\div\boldsymbol{A} \quad \text{in } Q_T \quad \text{and} \quad \boldsymbol{A}\cdot\mathbf{n} = 0 \quad \text{on } \Gamma_T. \tag{1.5}$$

With this gauge choice, a semi-implicit finite element scheme is proposed and analysed in [CHo 95]. A mixed finite element method is studied in [Che 97] which was used to approximate the potential $\boldsymbol{A}$, the magnetic field $\mathbf{curl}\boldsymbol{A}$ and the electric potential $\varphi = \div\boldsymbol{A}$. Other works on the numerical analysis include, for example, [Du 94], [DGP 92] and [Mu 99]. More comments on a posteriori estimates are included in the bibliographical remarks at the end of this chapter.

In this chapter, we mainly follow the works of [CHo 95] and [CDa 99] to introduce and analyse a numerical discretization of the G-L system using semi-implicit finite difference in time and a new method to derive a posteriori estimates for the numerical scheme.

For the discretization method, we show some stability estimates and error estimates for the scheme.

For the a posteriori estimates, we analyse their role in providing necessary information for modifying the mesh and time-step according to varying external magnetic field and corresponding motion of vortices. The estimates, which exhibit the same characteristics as that of linear parabolic equations [EJo 91], are based on the analysis of a dual problem which is the dual of the linearization of the original G-L system with gauge choice (1.5). We show that the additional terms in the error representation formula due to the change of dual problem are of higher order, and thus can be absorbed under a suitable non-degeneracy assumption (see Section 4 for details). We remark that the non-degeneracy assumption, which is not very restrictive in practices, is used in [BMi 87], [Noc 93], [Noc 95] to obtain both upper and lower bound of a posteriori error estimates for linear elliptic problems. The numerical experiments in Section 9 show clearly the reliability and flexibility of the adaptive algorithm based on our a posteriori error estimators.

This chapter is organized as follows. In Section 2, we propose a fully discrete finite element scheme to solve the problem. We will use backward Euler method in time with fixed time step length and simple linear finite element method in space. In Section 3, we deduce some stability estimates for the discrete solutions. These estimates will be used in Section 4 to deduce an error bound for our method. In Section 5 we report a numerical example to show the effectiveness of the non-adaptive method. In Section 6, we discuss the semi-implicit finite element scheme, discretization using non-uniform time step length and adaptive space mesh. In Section 7, we introduce the parabolic dual problem and prove the strong stability estimates. In Section 8, we prove the a posteriori error estimates. In Section 9, we present the numerical simulation while comparing the adaptive and non-adaptive method.

The space dimension is assumed to be 2 throughout this chapter. $\Omega$ is always assumed to be a smooth, bounded and simply connected domain in $\mathbb{R}^2$.

We now introduce some of the notation convention to be used in this Chapter. If $X$ denotes some Banach space of real scalar functions, the corresponding space of complex scalar functions will be denoted by its calligraphic form $\mathcal{X}$ and the corresponding space of real vector-valued functions, each of its components belonging to $X$, will be denoted by its boldfaced form $\mathbf{X}$. However, we shall use $\|\cdot\|_X$ to denote the norms of the Banach spaces $X$, $\mathcal{X}$ or $\mathbf{X}$.

We also recall the following inequality (see Chapter 8)

$$\|\mathbf{B}\|_{H^1(\Omega)} \leq C\Big[\|\mathbf{B}\|_{L^2(\Omega)} + \|\div \mathbf{B}\|_{L^2(\Omega)} + \|\mathbf{curl}\mathbf{B}\|_{L^2(\Omega)}\Big] \quad \forall \mathbf{B} \in \mathbf{H}^1_n(\Omega), \tag{1.6}$$

where the constant $C$ depends on the domain $\omega$. Hence, in this chapter, the norm of $\mathbf{H}^1_n(\Omega)$ is defined by

$$\|\boldsymbol{Q}\|_{\boldsymbol{H}^1_n(\Omega)} = \left[\|\div \boldsymbol{Q}\|^2_{L^2(\Omega)} + \|\mathbf{curl}\boldsymbol{Q}\|^2_{L^2(\Omega)} + \|\boldsymbol{Q}\|^2_{L^2(\Omega)}\right]^{1/2}$$

for any $\boldsymbol{Q} \in \mathbf{H}_n^1(\Omega)$. We will also use the semi-norm on $\boldsymbol{H}_n^1(\Omega)$

$$|\boldsymbol{Q}|_{\boldsymbol{H}_n^1(\Omega)} = \left[\| \div \boldsymbol{Q}\|_{L^2(\Omega)}^2 + \|\mathbf{curl}\boldsymbol{Q}\|_{L^2(\Omega)}^2\right]^{1/2}$$

which is equivalent to the norm $\|\boldsymbol{Q}\|_{H^1(\Omega)}$ when the domain $\Omega$ is in addition simply connected (see [GRa 86]).

Now, we give some preliminary regularity results for the G-L problem (1.1)–(1.5) which are useful in the subsequent numerical analysis.

**Theorem 10.1.1** *Let $(\psi_0, \boldsymbol{A}_0) \in \mathcal{H}^2(\Omega)$ such that $|\psi_0| \leq 1$ a.e. on $\Omega$ and $\boldsymbol{A}_0 \cdot \boldsymbol{n} = 0$ on $\partial\Omega$. Assume further that $H \in H^1(0,T; H(\,\mathbf{curl};\,\Omega))$. Then the unique solution $(\psi, \boldsymbol{A})$ of the Problem (1.1)–(1.5) satisfies*

$$\begin{aligned}
&\psi \in C(0,T;\mathcal{H}^2(\Omega)) \cap H^1(0,T;\mathcal{H}^1(\Omega)) \cap C^1(0,T;\mathcal{L}^2(\Omega)),\\
&\psi_t \in \mathcal{L}^4(Q_T), && (1.7)\\
&\boldsymbol{A} \in C(0,T;\boldsymbol{H}^2(\Omega)) \cap H^1(0,T;\boldsymbol{H}_n^1(\Omega)) \cap C^1(0,T;\boldsymbol{L}^2(\Omega)),\\
&\boldsymbol{A}_t \in \boldsymbol{L}^4(Q_T) && (1.8)
\end{aligned}$$

*where $Q_T = \Omega \times (0,T)$.*

The proof of this theorem depends on the following lemmas.

**Lemma 10.1.1** *For any $u \in H^1(\Omega)$, the inequality*

$$\|u\|_{L^4(\Omega)} \leq C\|u\|_{H^1(\Omega)}^{1/2}\|u\|_{L^2(\Omega)}^{1/2}$$

*holds with the constant $C$ depending only on $\Omega$.*

This lemma is a special case of the so-called multiplicative inequality in [LSU 68], p. 62–63.

**Lemma 10.1.2** *Consider the parabolic equation*

$$\frac{\partial \boldsymbol{A}}{\partial t} - \Delta \boldsymbol{A} = \mathbf{curl}H + f(x,t) \quad in \;\; Q_T \tag{1.9}$$

*with the boundary condition $\boldsymbol{A}\cdot\boldsymbol{n} = 0$ and $\mathbf{curl}\boldsymbol{A} = H$ on $\Gamma$ and the initial condition $\boldsymbol{A}(x,0) = \boldsymbol{A}_0(x)$. Let $f \in \boldsymbol{L}^2(Q_T) \cap H^1(0,T;(\boldsymbol{H}_n^1(\Omega))')$, $H \in H^1(0,T;H(\mathbf{curl};\Omega))$ and $\boldsymbol{A}_0 \in \boldsymbol{H}^2(\Omega) \cap \boldsymbol{H}_n^1(\Omega))$. Then the unique solution of (1.1)–(1.5) satisfies*

$$\boldsymbol{A} \in H^1(0,T;\boldsymbol{H}_n^1(\Omega)) \cap C^1(0,T;\boldsymbol{L}^2(\Omega)) \;\; and \;\; \boldsymbol{A}_t \in \boldsymbol{L}^4(Q_T). \tag{1.10}$$

*Moreover, if in addition $f \in C(0,T;\boldsymbol{L}^2(\Omega))$, we have $\boldsymbol{A} \in C(0,T;\boldsymbol{H}^2(\Omega))$.*

*Proof.* We know that there exists a unique solution $\boldsymbol{A} \in H^1(0,T;\boldsymbol{H}^1(\Omega)) \cap L^2(0,T;\boldsymbol{H}^2(\Omega) \cap \boldsymbol{H}^1_n(\Omega))$ to the problem (1.1)–(1.5). In order to prove (1.10) we consider the following parabolic equation

$$\frac{\partial \boldsymbol{B}}{\partial t} - \Delta \boldsymbol{B} = \mathbf{curl} H_t + f_t(x,t) \text{ in } Q_T \tag{1.11}$$

with the boundary conditions $\boldsymbol{B} \cdot \boldsymbol{n} = 0$ and $\mathbf{curl}\boldsymbol{B} = H_t$ on $\Gamma$ and the initial condition $\boldsymbol{B}(x,0) = f(x,0) + \mathbf{curl}H(x,0) + \Delta \boldsymbol{A}_0(x)$. It is easy to show that (1.11) has a unique solution $\boldsymbol{B} \in L^2(0,T;\boldsymbol{H}^1_n(\Omega)) \cap H^1(0,T;(\boldsymbol{H}^1_n(\Omega))')$. Thus, by embedding theorem, $\boldsymbol{B} \in C(0,T;\boldsymbol{L}^2(\Omega))$. As a matter of fact, $\boldsymbol{B} = \boldsymbol{A}_t$. Therefore, $\boldsymbol{A} \in H^1(0,T;\boldsymbol{H}^1_n(\Omega)) \cap C^1(0,T;\boldsymbol{L}^2(\Omega))$. Now by virtue of Lemma 1.1 we have $\boldsymbol{A}_t \in \boldsymbol{L}^4(Q_T)$. The fact that $\boldsymbol{A} \in C(0,T;\boldsymbol{H}^2(\Omega))$ follows from the equation (1.9) and Theorem 3.1 in [CHL 93].

Analogously, we can show the following results concerning $\psi$.

**Lemma 10.1.3** *Consider the parabolic equation*

$$\eta \frac{\partial \psi}{\partial t} - \frac{1}{\kappa^2} \Delta \psi = g(x,t) \quad \text{in } Q_T \tag{1.12}$$

*with the boundary conditions* $\nabla \psi \cdot \boldsymbol{n} = 0$ *on* $\Gamma$ *and the initial condition* $\psi(x,0) = \psi_0(x)$. *Let* $g \in \boldsymbol{L}^2(Q_T) \cap H^1(0,T;(\mathcal{H}^1(\Omega))')$ *and* $\psi_0 \in \mathcal{H}^2(\Omega)$. *Then the unique solution of (1.12) fulfills*

$$\psi \in H^1(0,T;\mathcal{H}^1(\Omega)) \cap C^1(0,T;\mathcal{L}^2(\Omega)) \quad \text{and} \quad \psi_t \in \mathcal{L}^4(Q_T). \tag{1.13}$$

*Moreover, if in addition* $g \in C(0,T;\mathcal{L}^2(\Omega))$, *we have* $\psi \in C(0,T;\mathcal{H}^2(\Omega))$.

*Proof of Theorem 2.1.* From the standard existence results, we know that $(\psi, \boldsymbol{A}) \in \mathcal{W}^{2,1}_2(Q_T) \times \boldsymbol{W}^{2,1}_2(Q_T)$. Thus, by using embedding theorem, we have $\psi \in C(0,T;\mathcal{H}^1(\Omega))$ and $\boldsymbol{A} \in C(0,T;\boldsymbol{H}^1_n(\Omega))$. In order to use Lemmas 1.2 and 1.3 to prove the theorem, we denote first

$$\begin{aligned}
\boldsymbol{f} &= -\frac{i}{2\kappa}(\overline{\psi}\nabla\psi - \psi\nabla\overline{\psi}) - |\psi|^2 \boldsymbol{A}, \\
g &= i(\eta\kappa + \frac{1}{\kappa} \div \boldsymbol{A}\psi - \frac{2i}{\kappa}\nabla(\boldsymbol{A}\psi) - |\boldsymbol{A}|^2\psi - (|\psi|^2 - 1)\psi.
\end{aligned}$$

It is easy to check that $\boldsymbol{f} \in \boldsymbol{L}^2(Q_T) \cap H^1(0,T;(\boldsymbol{H}^1_n(\Omega))')$, thus by applying Lemma 1.2 to equation (1.2), we obtain $\boldsymbol{A} \in H^1(0,T;\boldsymbol{H}^1_n(\Omega)) \cap C^1(0,T;\boldsymbol{L}^2(\Omega))$ and $\boldsymbol{A}_t \in \boldsymbol{L}^4(Q_t)$. Now we can show $g \in \mathcal{L}^2(Q_T) \cap H^1(0,T;(\mathcal{H}^1(\Omega))')$, hence by applying Lemma 1.3 to equation (1.1), we get $\psi \in H^1(0,T;\mathcal{H}^1(\Omega)) \cap C^1(0,T;\mathcal{L}^2(\Omega))$ and $\psi_t \in \mathcal{L}^4(Q_T)$. Since $\psi \in H^1(0,T;\mathcal{H}^1(\Omega))$, we have $\psi \in C(0,T;\mathcal{L}^p(\Omega))$ for any $p \geq 1$. Similarly, $\boldsymbol{A} \in C(0,T;\boldsymbol{L}^p(\Omega))$ for any $p \geq 1$. Therefore, $|\psi|^2\boldsymbol{A} \in C(0,T;\boldsymbol{L}^2(\Omega))$. By virtue of $\psi \in C(0,T;\mathcal{L}^2(\Omega)) \cap L^2(0,T;\mathcal{H}^2(\Omega))$ and by using Lemma 1.1, we know that $\nabla\psi \in \mathcal{L}^4(Q_T))$ which, together with $|\psi| \leq 1$ a.e. in

$Q_T$ and $\psi \in L^2(0,T;\mathcal{H}^2(\Omega))$, yields $\overline{\psi}\nabla\psi - \psi\nabla\overline{\psi} \in L^2(0,T;\boldsymbol{H}^1(\Omega))$. Furthermore, from $\psi \in H^1(0,T;\mathcal{H}^1(\Omega)) \cap \mathcal{L}^\infty(\Omega)$, we have $\overline{\psi}\nabla\psi_t - \psi\nabla\overline{\psi}_t \in \boldsymbol{L}^2(Q_T)$. Moreover, it is clear that $\overline{\psi}_t\nabla\psi - \psi_t\nabla\overline{\psi} \in L^2(0,T;(\boldsymbol{H}^0(\Omega))')$. Thus $\overline{\psi}\nabla\psi - \psi\nabla\overline{\psi} \in H^1(0,T;(\boldsymbol{H}^1(\Omega))')$ which, along with $\overline{\psi}\nabla\psi - \psi\nabla\overline{\psi} \in L^2(0,T;\boldsymbol{H}^1(\Omega))$, implies $\overline{\psi}\nabla\psi - \psi\nabla\overline{\psi} \in C(0,T;\boldsymbol{L}^2(\Omega))$. Hence $\boldsymbol{f} \in C(0,T;\boldsymbol{L}^2(\Omega))$ which implies $\boldsymbol{A} \in C(0,T;\boldsymbol{H}^2(\Omega))$ by applying Lemma 1.2 to equation (1.2). Similarly, we get $\psi \in C(0,T;\mathcal{H}^2(\Omega))$. This completes the proof.

## 10.2 Discretization

### 10.2.1 Weak formulation

We first state the hypotheses concerning the data.

(H1) $\psi_0 \in \mathcal{H}^2(\Omega), \boldsymbol{A}_0 \in \mathbf{H}^2(\Omega) \cap \mathbf{H}^1_n(\Omega)$.satisfying $|\psi_0| \le 1$ on $\Omega$;

(H2) $H \in H^1(0,T;H^{1/2}(\partial\Omega))$.

In view of (H2), we extended $H$ to $\Omega$ in such a way that $H \in H^1(0,T;H^1(\Omega))$. In the sequel we assume for convenience that

(H3) Given $\mathbf{f} \in \mathbf{L}^2(\Omega)$ and $H_{\text{ext}} \in H^1(\Omega)$, the linear elliptic problem

$$-\Delta\mathbf{Q} = \mathbf{f} \quad \text{in } \Omega, \qquad \mathbf{Q}\cdot\mathbf{n} = 0, \quad \mathbf{curl}\mathbf{Q} = H_{\text{ext}} \quad \text{on } \partial\Omega$$

has a unique solution $\mathbf{Q} \in \mathbf{H}^2(\Omega) \cap \mathbf{H}^1_n(\Omega)$ which satisfies the a priori estimate

$$\|\mathbf{Q}\|_{H^2(\Omega)} \le C(\|\mathbf{f}\|_{L^2(\Omega)} + \|H_{\text{ext}}\|_{H^1(\Omega)}).$$

This property holds, for instance, when the domain $\Omega$ has a smooth boundary $C^{2,1}$ (see [CHL 93]) or when $\Omega$ is a rectangular domain and in fact, when $\Omega$ is a convex polygonal domain (see [Gri 85]).

We set $\mathcal{W}(0,T) = L^2(0,T;\mathcal{H}^1(\Omega)) \cap H^1(0,T;(\mathcal{H}^1(\Omega))')$ and $\mathbf{W}_n(0,T) = L^2(0,T;\mathbf{H}^1_n(\Omega)) \cap H^1(0,T;(\mathbf{H}^1_n(\Omega))')$, where $X'$ stands for the dual space of $X$. The weak formulation of the problem (1.1)–(1.4) under the gauge choice (1.5) becomes

**Continuous Problem (P):** Find a pair $(\psi, \boldsymbol{A}) \in \mathcal{W}(0,T) \times \mathbf{W}_n(0,T)$ such that

$$\psi(\cdot,0) = \psi_0(\cdot), \quad \boldsymbol{A}(\cdot,0) = \boldsymbol{A}_0(\cdot) \tag{2.1}$$

and

$$\begin{aligned} &\eta\int_0^T\int_\Omega \partial_t\psi\omega dxdt - i\eta\kappa\int_0^T\int_\Omega \div\boldsymbol{A}\psi\omega dxdt \\ &+ \int_0^T\int_\Omega \Big(\frac{i}{\kappa}\nabla\psi + \boldsymbol{A}\psi\Big)\Big(-\frac{i}{\kappa}\nabla\omega + \boldsymbol{A}\omega\Big)dxdt \\ &+ \int_0^T\int_\Omega (|\psi|^2 - 1)\psi\omega dxdt = 0 \quad \forall\omega \in L^2(0,T;\mathcal{H}^1(\Omega)), \end{aligned} \tag{2.2}$$

$$\begin{aligned}
&\int_0^T\int_\Omega \partial_t \boldsymbol{A}\mathbf{B}dxdt + \int_0^T\int_\Omega (\div \boldsymbol{A} \div \mathbf{B} + \mathbf{curl}\boldsymbol{A}\mathbf{curl}\mathbf{B})dxdt \\
+ &\int_0^T\int_\Omega \mathrm{Re}\Big[\Big(\frac{i}{\kappa}\nabla\psi + \boldsymbol{A}\psi\Big)\bar{\psi}\Big]\mathbf{B}dxdt \\
= &\int_0^T\int_\Omega H\mathbf{curl}\mathbf{B}dxdt \quad \forall \mathbf{B} \in L^2(0,T;\mathbf{H}^1_n(\Omega)). \qquad (2.3)
\end{aligned}$$

Assume from now on that $\Omega$ is a convex polygon satisfying (H3). The problem has a unique solution $(\psi, \boldsymbol{A}) \in \mathcal{W}_2^{2,1}(Q_T) \times \mathbf{W}_2^{2,1}(Q_T)$ such that $|\psi| \le 1$ a.e. in $Q_T$. This can be established in the same way as in Chapter 8.

For convenience, we set

$$\begin{aligned}
f(\psi, \boldsymbol{A}) &= (\frac{i}{\kappa} - i\eta\kappa)\div \boldsymbol{A}\psi + \frac{2i}{\kappa}\boldsymbol{A}\cdot\nabla\psi + (|\boldsymbol{A}|^2 + |\psi|^2 - 1)\psi, \\
\mathbf{g}(\psi, \boldsymbol{A}) &= \mathrm{Re}\Big[\Big(\frac{i}{\kappa}\nabla\psi + \boldsymbol{A}\psi\Big)\bar{\psi}\Big].
\end{aligned}$$

Then (1.1)–(1.4) with gauge choice (1.5) can be written as

$$\eta\partial_t\psi - \frac{1}{\kappa^2}\Delta\psi + f(\psi, \boldsymbol{A}) = 0 \ \text{ in } Q_T, \qquad (2.4)$$

$$\partial_t\boldsymbol{A} - \Delta\boldsymbol{A} + \mathbf{g}(\psi, \boldsymbol{A}) = 0 \ \text{ in } Q_T. \qquad (2.5)$$

### 10.2.2 Discretization

We shall make use of backward Euler scheme to discretize the Problem (P) in time. Let $M$ be a positive integer and $\Delta t = T/M$ be the time step. For any $n = 1, 2, \ldots, M$, we define $t^n = n\Delta t$ and $I^n = (t^{n-1}, t^n]$. Furthermore, we denote $\delta\eta^n = (\eta^n - \eta^{n-1})/\Delta t$ for any given sequence $\{\eta^n\}_{n=0}^M$ and $\eta^n := \eta(\cdot, t^n)$ for any given function $\eta \in C(0,T;X)$ with some Banach space $X$.

In space we shall utilize linear finite element approximations. Let $\{\Delta_h\}_{h>0}$ be a family of triangulations of $\Omega$. For $K \in \Delta_h$ we define $\rho(K)$ (respectively $\sigma(K)$) to be the radius of the smallest ball containing $K$ (respectively largest ball contained in $K$). For convenience we shall assume $\Omega$ is a convex polygonal domain in $\mathrm{R}^2$ and thus we can assume that

$$\Omega = \Omega_h = \cup_{K\in\Delta_h} K.$$

We shall also assume that the underlying triangulations are regular and quasi-uniform, i.e. for all $h > 0$,

$$\sup_{K\in\Delta_h}\frac{\rho(K)}{\sigma(K)} \le \gamma_0, \qquad \inf_{K\in\Delta_h}\frac{\sigma(K)}{h} > \gamma_1 > 0,$$

where $\gamma_0$, $\gamma_1$ are constants independent of $h$.

Define the finite element spaces

$$\mathcal{S}_h = \{\zeta \in \mathcal{C}(\overline{\Omega}) \ : \ \zeta|_K \ \text{ is linear for all } \ K \in \Delta_h\}$$

and

$$\mathcal{V}_h = \{\boldsymbol{Q} \in \boldsymbol{C}(\overline{\Omega}) \ : \ \boldsymbol{Q}|_K \ \text{is linear for all} \ K \in \Delta_h \ \text{and} \ \boldsymbol{Q} \cdot \boldsymbol{n} = 0\},$$

where $h = \sup_{K\in\Delta_h} \rho(K)$. We denote by $\zeta_1 \in \mathcal{S}_h$ (respectively $\boldsymbol{Q}_I \in \boldsymbol{V}_h$) the interpolate of $\zeta \in \mathcal{C}(\overline{\Omega})$ in $\mathcal{S}_h$ respectively $\boldsymbol{Q} \in \boldsymbol{C}(\overline{\Omega})$ in $\boldsymbol{V}_h$). We shall assume that $\{\mathcal{S}_h\}_{h>0}$, $\{\boldsymbol{V}_h\}_{h>0}$ have the following approximation properties (see [SFi 73], pp. 162–163))

$$\|\zeta - \zeta_1\|_{H^j(\Omega)} \leq C h^{2-j} \|\zeta\|_{H^2(\Omega)}, \quad \forall \zeta \in \mathcal{H}^2(\Omega),\ j = 0, 1, \tag{2.6}$$

and

$$\|\boldsymbol{Q} - \boldsymbol{Q}_1\|_{H^j(\Omega)} \leq C h^{2-j} \|\boldsymbol{Q}\|_{H^2(\Omega)}, \quad \forall \zeta \in \boldsymbol{H}^2(\Omega),\ j = 0, 1. \tag{2.7}$$

In practical computations, we shall use the techniques of numerical integration to calculate the integration in the weak formulations. However, in this paper we shall not consider the effects of numerical integration because we only want to concentrate on the new features resulting from the nonlinearities of the equations (2.2)–(2.3). One can certainly extend the results in the rest of this paper to these practical situations by the method in [Cia 78]. In the remainder of this paper we shall always assume that the solution $(\psi, \boldsymbol{A})$ of the Problem (P) has the regularities (1.7)–(1.8). We also remark that one can use the same method in this paper to treat the case when $\partial\Omega$ is smooth by employing the penalty method (see Remark 4.3).

Now we are in the position to introduce the following discrete problem.

**Problem (DP).** For $n = 1, 2, \ldots, M$, find $(\psi_h^n, \boldsymbol{A}_h^n) \in \mathcal{S}_h \times \boldsymbol{V}$ such that

$$\psi_h^0 = (\psi_0)_I, \quad \boldsymbol{A}_h^0 - (\boldsymbol{A}_0)_I, \tag{2.8}$$

and

$$\begin{aligned}
&\eta \int_\Omega \partial \psi_h^n \omega dx - i\eta\kappa \int_\Omega \div \boldsymbol{A}_n^h \psi_h^n \omega dx \\
&\int_\Omega \Big(-\frac{i}{\kappa}\nabla\psi_h^n - \boldsymbol{A}_h^n\psi_h^n\Big)\Big(\frac{i}{\kappa}\nabla\omega - \boldsymbol{A}_h^n\omega\Big) dx \\
&+ \int_\Omega (|\psi_h^n|^2 - 1)\psi_h^n \omega dx = 0, \quad \forall \omega \in \mathcal{S}_h
\end{aligned} \tag{2.9}$$

$$\begin{aligned}
&\int_\Omega \partial \boldsymbol{A}_h^n \boldsymbol{B}\, dx + \int_\Omega (\div \boldsymbol{A}_h^n \div \boldsymbol{B} + \mathbf{curl}\boldsymbol{A}_h^n \mathbf{curl}\boldsymbol{B}) dx \\
&+ \int_\Omega \frac{i}{2\kappa}(\overline{\psi_h}^{n-1} \nabla\psi_h^{n-1} - \psi_h^{n-1}\nabla\psi_h^{n-1})\boldsymbol{B} dx + \int_\Omega |\psi_h^{n-1}|^2 \boldsymbol{A}_h^n \cdot \boldsymbol{B}\, dx \\
= \ &\int_\Omega H^n \mathbf{curl}\boldsymbol{B}\, dx dt, \ \forall \boldsymbol{B} \in \boldsymbol{V}_h.
\end{aligned} \tag{2.10}$$

We note that at each time step $n$, (2.10) is a linear system of equations with positive definite coefficient matrix, which can be solved by standard methods. As soon as we know $\boldsymbol{A}_h^n$ from (2.10), we substitute it into (2.9) and solve the non-linear system of equations to obtain $\psi_h^n$. The existence and uniqueness of the solution $\psi_h^n$ to (2.9) are guaranteed by the following theorem.

**Theorem 10.2.1** *Let $\Delta t \leq \eta/2$. Then (2.9) has a unique solution.*

The uniqueness can be proved by standard methods. The existence can be shown by using Brouwer's fixed point theorem. We omit the details. A similar problem on phase field model has been discussed in [CHo 94].

Any of the many iterative methods for the solution of non-linear system of equations may be employed to solve the non-linear discrete system (2.9). For example, we can solve (2.9) by the following simple iterative method.

1°) Select some initial iterative function $\psi_h^{n,0}$ (e.g. $\psi_h^{n,0} = \psi_h^{n-1}$);

2°) For $k = 1, 2, \ldots$, solve the following linear system of equations

$$\begin{aligned} &\eta \int_\Omega \partial \psi_h^{n,k} \omega dx - i\eta\kappa \int_\Omega \div \boldsymbol{A}_h^n \psi_h^{n,k} \omega dx \\ + &\int_\Omega (-\frac{i}{\kappa} \nabla \psi_h^{n,k} - \boldsymbol{A}_h^n \psi_h^{n,k})(\frac{i}{k} \nabla \omega - \boldsymbol{A}_h^n \omega) dx \\ + &\int_\Omega (|\psi_h^{n,k-1}|^2 - 1)\psi_h^{n,k} \omega dx = 0, \quad \forall \omega \in \mathcal{S}_h \end{aligned}$$

to obtain $\psi_h^{n,k} \in \mathcal{S}_h$;

3°) If $\|\psi_h^{n,k} - \psi_h^{n,k-1}\|_{L^2(\Omega)} \leq \epsilon$, stop; otherwise, go to 2°.

One can easily use the methods in [CHo 94] to show that when $\Delta t/h^2$ is sufficiently small, the above iterative sequence $\{\psi_h^{n,k}\}_{k=0}^\infty$ converges to the solution $\psi_h^n$ of (3.4) as $k \to \infty$. This algorithm converges however rather slow in practical computations. To accelerate the convergence we use the following Newton's method to solve (2.9).

1°) Select some initial iterative function $\psi_h^{n,0}$;

2°) For $k = 1, 2, \ldots$, solve the following linear system of equations

$$\begin{aligned} &\eta \int_\Omega \psi_h^{n,k} \omega dx - i\eta\kappa\Delta t \int_\Omega \div \boldsymbol{A}_h^n \psi_h^{n,k} \omega dx \\ &+\Delta t \int_\Omega (-\frac{i}{\kappa} \nabla \psi_h^{n,k} - \boldsymbol{A}_h^n \psi_h^{n,k})(\frac{i}{\kappa} \nabla \omega - \boldsymbol{A}_h^n \omega) dx \\ &+\Delta t \int_\Omega [2|\psi_h^{n,k-1}|^2 \psi_h^{n,k} + \psi_h^{n,k-1} \psi_h^{n,k-1} \overline{\psi}_h^{n,k} - \psi_h^{n,k}] \omega dx \\ = \;&\Delta t \int_\Omega (2|\psi_h^{n,k-1}|^2 \psi_h^{n,k-1}) \omega dx + \eta \int_\Omega \psi_h^{n-1} \omega \, dx, \quad \forall \omega \in \mathcal{S}_h \end{aligned}$$

to obtain $\psi_h^{n,k} \in \mathcal{S}_h$;

3°) If $\|\psi_h^{n,k} - \psi_h^{n,k-1}\|_{L^2(\Omega)} \leq \epsilon$, stop; otherwise, go to 2°.

To use above Newton's iterative method, we need to choose the initial guess $\psi_h^{n,0}$ reasonably close to the real solution $\psi_0^n$ such that the iteration is convergent. A natural idea is to combine the Newton's method with the simple iterative method together, since the latter is globally convergent.

## 10.3 Stability estimates

In this section we deduce some stability estimates which will be useful in the error analysis. We begin with the following version of the discrete Gronwall's inequality.

**Lemma 10.3.1** *Let $\{a_n\}$, $\{b_n\}$, $\{c_n\}$, $\{\alpha_n\}$ and $\{\beta_n\}$ are nonnegative sequences such that $\alpha_n \le \alpha < 1$ for any $n = 1, 2, \ldots$. Furthermore, assume that*

$$(1-\alpha_n)a_n - (1+\beta_n)a_{n-1} + b_n \le c_n, \quad \text{for any } n = 1, 2, \ldots, \tag{3.1}$$

*then we have*

$$a_n + \sum_{k=1}^n b_k \le \exp\Big[\frac{1}{1-\alpha}\sum_{k=1}^n (\alpha_k + \beta_k)\Big]\Big(a_0 + \sum_{k=1}^n c_k\Big), \quad \text{for any } n = 1, 2, \ldots.$$

*Proof.* From (3.1) we have

$$\Big(\frac{1-\alpha_n}{1+\beta_n}\Big)a_n - a_{n-1} + \frac{1}{1+\beta_n}b_n \le \frac{1}{1+\beta_n}c_n \le c_n.$$

Multiplying the above inequality by $\prod_{k=1}^{n-1}(\frac{1-\alpha_k}{1+\beta_k})$ we get

$$\Big[\prod_{k=1}^{n}\Big(\frac{1-\alpha_k}{1+\beta_k}\Big)\Big]a_n - \Big[\prod_{k=1}^{n-1}\Big(\frac{1-\alpha_k}{1+\beta_k}\Big)\Big]a_{n-1} + \Big[\prod_{k=1}^{n-1}\Big(\frac{1-\alpha_k}{1+\beta_k}\Big)\Big]\frac{1}{1+\beta_n}b_n$$
$$\le \Big[\prod_{k=1}^{n-1}\Big(\frac{1-\alpha_k}{1+\beta_k}\Big)\Big]c_n$$

which yields, since $\alpha_k, \beta_k \ge 0$ for any $k = 1, 2, \ldots$,

$$\Big[\prod_{k=1}^{n}\Big(\frac{1-\alpha_k}{1+\beta_k}\Big)\Big]a_n - \Big[\prod_{k=1}^{n-1}\Big(\frac{1-\alpha_k}{1+\beta_k}\Big)\Big]a_{n-1} + \Big[\prod_{k=1}^{n}\Big(\frac{1-\alpha_k}{1+\beta_k}\Big)\Big]b_n \le c_n.$$

Summing the above inequality on $n$ from 1 to some integer $m \ge 1$ we obtain

$$\Big[\prod_{k=1}^{m}\Big(\frac{1-\alpha_k}{1+\beta_k}\Big)\Big]a_m - a_0 + \sum_{n=1}^m\Big[\prod_{k=1}^{n}\Big(\frac{1-\alpha_k}{1+\beta_k}\Big)\Big]b_n \le \sum_{n=1}^m c_n.$$

Therefore, by dividing both sides of the inequality by $\prod_{n+1}^m(\frac{1-\alpha_k}{1+\beta_k})$ and noticing that $\prod_{n+1}^m(\frac{1+\beta_k}{1-\alpha_k}) \ge 1$, we have

$$a_m + \sum_{n=1}^m b_n \quad \le \quad \Big[\prod_{k=1}^{m}\Big(\frac{1+\beta_k}{1-\alpha_k}\Big)\Big]\Big(a_0 + \sum_{n=1}^m c_n\Big)$$

$$\begin{aligned}
&\leq \left[\frac{1}{m}\sum_{k=1}^{m}\Big(\frac{1+\beta_k}{1-\alpha_k}\Big)\right]^m \Big(a_0+\sum_{n=1}^{m} c_n\Big)\\
&= \left[1+\frac{1}{m}\sum_{k=1}^{m}\Big(\frac{\alpha_k+\beta_k}{1-\alpha_k}\Big)\right]^m \Big(a_0+\sum_{n=1}^{m} c_n\Big)\\
&\leq \exp\left[\sum_{k=1}^{m}\Big(\frac{\alpha_k+\beta_k}{1-\alpha_k}\Big)\right]\Big(a_0+\sum_{n=1}^{m} c_n\Big)\\
&\leq \exp\left[\frac{1}{1-\alpha}\sum_{k=1}^{m}(\alpha_k+\beta_k)\right]\Big(a_0+\sum_{k=1}^{m} c_k\Big),
\end{aligned}$$

where we have used the famous arithmetic-geometric-mean inequality and the elementary inequality $(1+x)\leq e^x$ for $x\geq 0$. This completes the proof.

The following theorem is the main result of this section.

**Theorem 10.3.1** *Let $\Delta t \leq \eta/4$. Then the solution $\{(\psi_h^n, \boldsymbol{A}_h^n)\}_{n-1}^M$ of the Problem (DP) fulfils the following estimates*

$$\begin{aligned}
&\max_{1\leq n\leq M}\big[\|\psi_h^n\|^2_{L^2(\Omega)}+\|\boldsymbol{A}_h^n\|^2_{L^2(\Omega)}\big]\\
&+\sum_{n=1}^{M}\big[\|\psi_h^n-\psi_h^{n-1}\|^2_{L^2(\Omega)}+\|\boldsymbol{A}_h^n-\boldsymbol{A}_h^{n-1}\|^2_{L^2(\Omega)}\big]\\
&+\sum_{n=1}^{M}\Delta t\big[\|\nabla\psi_h^n\|^2_{L^2(\Omega)}+\|\div \boldsymbol{A}_h^n\|_{L^2(\Omega)}+\|\mathbf{curl}\boldsymbol{A}_h^n\|^2_{L^2(\Omega)}\big]\\
&+\sum_{n=1}^{M}\Delta t\big[\|\psi_h^n\|^4_{L^4(\Omega)}+\|\boldsymbol{A}_h^n\|^4_{L^4(\Omega)}\big]\leq C,
\end{aligned}$$

*where the constant $C$ is independent of $h$, $\Delta t$ and $M$.*

The proof of this theorem is the consequence of several lemmas that we discuss in the following.

**Lemma 10.3.2** *Let $\Delta t \leq \eta/4$. We have*

$$\begin{aligned}
&\max_{1\leq n\leq M}\|\psi_h^n\|^2_{L^2(\Omega)}+\sum_{n=1}^{M}\|\psi_h^n-\psi_h^{n-1}\|^2_{L^2(\Omega)}+\sum_{n=1}^{M}\Delta t\|\psi_h^n-\psi_h^{n-1}\|^4_{L^4(\Omega)}\\
&+\ \sum_{n=1}^{M}\Delta t\|-\frac{i}{\kappa}\nabla\psi_h^n-\boldsymbol{A}_h^n\psi_h^n\|^2_{L^2(\Omega)}\leq C, \qquad (3.2)
\end{aligned}$$

*where the constant $C$ is independent of $h$, $\Delta t$ and $M$.*

*Proof.* Choosing $\omega=\Delta t\overline{\psi}_h^n$ as the test function in (2.9) and taking the real part of the equation, then since

$$\mathrm{Re}\left[\eta\int_\Omega \partial\psi_h^n\omega dx\right]=\frac{\eta}{2}\left[\int_\Omega|\psi_h^{n-1}|^2dx-\int_\Omega|\psi_h^n-\psi_h^{n-1}|^2dx\right],$$

$$\begin{aligned}
&\mathrm{Re}\Big[-i\eta\kappa\int_\Omega \div \boldsymbol{A}_h^n\psi_h^n\omega dx\Big] = 0,\\
&\mathrm{Re}\Big[\int_\Omega (|\psi_h^n|^2-1)\psi_h^n\omega dx\Big] = \Delta t\Big[\int_\Omega |\psi_h^n|^4dx - \int_\Omega |\psi_h^n|^2dx\Big],\\
&\mathrm{Re}\Big[\int_\Omega (-\frac{i}{\kappa}\nabla\psi_h^n - \boldsymbol{A}_h^n\psi_h^n)(\frac{i}{\kappa}\nabla\omega - \boldsymbol{A}_h^n\omega)dx\Big]\\
&= \Delta t\| -\frac{i}{\kappa}\nabla\psi_h^n - \boldsymbol{A}_h^n\psi_h^n\|^2_{L^2(\Omega)},
\end{aligned}$$

we have

$$\begin{aligned}
&\frac{\eta}{2}\big[\|\psi_h^n\|^2_{L^2(\Omega)} - \|\psi_h^{n-1}\|^2_{L^2(\Omega)} + \|\psi_h^n-\psi_h^{n-1}\|^2_{L^2(\Omega)}\big] + \Delta t\|\psi_h^n\|^4_{L^4(\Omega)}\\
&+\Delta t\| -\frac{i}{\kappa}\nabla\psi_h^n - \boldsymbol{A}_h^n\psi_h^n\|^2_{L^2(\Omega)} \le \Delta t\|\psi_h^n\|^2_{L^2(\Omega)}.
\end{aligned}$$

This implies (3.2) by using Lemma 3.1. This completes the proof.

**Lemma 10.3.3** *Let $\Delta t \le \eta/4$. We have*

$$\begin{aligned}
&\max_{1\le n\le M}\|\boldsymbol{A}_h^n\|^2_{L^2(\Omega)} + \sum_{n=1}^M \|\boldsymbol{A}_h^n - \boldsymbol{A}_h^{n-1}\|^2_{L^2(\Omega)}\\
&+ \sum_{n=1}^M \Delta t\big[\|\div \boldsymbol{A}_h^n\|^2_{L^2(\Omega)} + \|\mathbf{curl}\boldsymbol{A}_h^n\|^2_{L^2(\Omega)}\big] \le C, \qquad (3.3)
\end{aligned}$$

*where the constant $C$ is independent of $h$, $\Delta t$ and $M$.*

*Proof.* At first we notice that $\frac{i}{2\kappa}(\overline{\psi}\nabla\psi - \psi\nabla\overline{\psi}) = \mathrm{Re}(\frac{i}{2\kappa}\overline{\psi}\nabla\psi)$. Taking $\boldsymbol{B} = \Delta t\boldsymbol{A}_h^n$ in (2.10) we get

$$\begin{aligned}
&\frac{1}{2}\Big[\|A_h^n\|^2_{L^2(\Omega)} - \|A_h^{n-1}\|^2_{L^2(\Omega)} + \|\boldsymbol{A}_h^n - \boldsymbol{A}_h^{n-1}\|^2_{L^2(\Omega)}\Big]\\
&+ \Delta t\Big[|\div \boldsymbol{A}_h^n|^2_{L^2(\Omega)} + \|\mathbf{curl}\boldsymbol{A}_h^n\|^2_{L^2(\Omega)}\Big]\\
&+ \Delta t\int_\Omega \mathrm{Re}\big[\frac{i}{\kappa}\overline{\psi}_h^{n-1}\nabla\psi_h^{n-1}\big]\boldsymbol{A}_h^n dx + \Delta t\int_\Omega |\psi_h^{n-1}|^2|\boldsymbol{A}_h^n|^2dx\\
&= \Delta t\int_\Omega H^n\mathbf{curl}\boldsymbol{A}_h^n dx. \qquad (3.4)
\end{aligned}$$

However, we have

$$\begin{aligned}
&\int_\Omega \mathrm{Re}\Big[\frac{i}{\kappa}\overline{\psi}_h^{n-1}\nabla\psi_h^{n-1}\Big]\boldsymbol{A}_h^n dx \int_\Omega |\psi_h^{n-1}|^2dx\\
=\ &\int_\Omega \mathrm{Re}\Big[\overline{\psi}_h^{n-1}(\frac{i}{\kappa}\nabla\psi_h^{n-1} + \boldsymbol{A}_h^{n-1}\psi_h^{n-1})\Big]\boldsymbol{A}_h^n dx\\
&+\int_\Omega |\psi_h^{n-1}|^2(\boldsymbol{A}_h^n - \boldsymbol{A}_h^{n-1})\boldsymbol{A}_h^n dx
\end{aligned}$$

$$
\begin{aligned}
&= \int_\Omega \mathrm{Re}\Big[\overline{\psi}_h^{n-1}(\frac{i}{\kappa}\nabla\psi_h^{n-1} + \boldsymbol{A}_h^{n-1}\psi_h^{n-1})\Big](\boldsymbol{A}_h^n - \boldsymbol{A}_h^{n-1})dx \\
&\quad + \int_\Omega \mathrm{Re}\Big[\overline{\psi}_h^{n-1}(\frac{i}{\kappa}\nabla\psi_h^{n-1} + \boldsymbol{A}_h^{n-1}\psi_h^{n-1})\Big]\boldsymbol{A}_h^{n-1}dx \\
&\quad + \int_\Omega |\psi_h^{n-1}|^2|\boldsymbol{A}_h^n - \boldsymbol{A}_h^{n-1}|^2 dx + \int_\Omega |\psi_h^{n-1}|^2(\boldsymbol{A}_h^n - \boldsymbol{A}_h^{n-1})\boldsymbol{A}_h^{n-1}dx \\
&\geq \int_\Omega |\psi_h^{n-1}|^2|\boldsymbol{A}_h^n - \boldsymbol{A}_h^{n-1}|^2 dx - \frac{3}{2}\int_\Omega |\psi_h^{n-1}|^2|\boldsymbol{A}_h^{n-1}|^2 dx \\
&\quad -\frac{3}{2}\int_\Omega |\frac{i}{\kappa}\nabla\psi_h^{n-1}| + |\boldsymbol{A}_h^{n-1}\psi_h^{n-1}|^2 dx, \qquad (3.5)
\end{aligned}
$$

where we have used Young's inequality in last estimate.

By employing Young's inequality again we also have

$$
\int_\Omega H^n \mathbf{curl}\boldsymbol{A}_h^n dx \leq 2\|H^n\|_{L^2(\Omega)}^2 + \frac{1}{8}\|\mathbf{curl}\boldsymbol{A}_h^n\|_{L^2(\Omega)}^2. \qquad (3.6)
$$

Moreover, by Lemma 1.1 and the equivalence of the norms $\|\cdot\|_{\boldsymbol{H}_n^1(\Omega)}$ and $\|\cdot\|_{H^1(\Omega)}$ on $\boldsymbol{H}_n^1(\Omega)$, we find

$$
\begin{aligned}
\int_\Omega |\psi_h^{n-1}|^2|\boldsymbol{A}_h^{n-1}|^2 dx &\leq \|\psi_h^{n-1}\|_{L^4(\Omega)}^2\|\boldsymbol{A}_h^{n-1}\|_{L^4(\Omega)}^2 \\
&\leq C\|\psi_h^{n-1}\|_{L^4(\Omega)}^2\|\boldsymbol{A}_h^{n-1}\|_{H^1(\Omega)}\|\boldsymbol{A}_h^{n-1}\|_{L^2(\Omega)} \\
&\leq C\|\psi_h^{n-1}\|_{L^4(\Omega)}^2\|\boldsymbol{A}_h^{n-1}\|_{\boldsymbol{H}^1(\Omega)}\|\boldsymbol{A}_h^{n-1}\|_{L^2(\Omega)} \\
&\leq \frac{1}{3}[\| \div \boldsymbol{A}_h^n\|_{L^2(\Omega)}^2 + \mathbf{curl}\boldsymbol{A}_h^n\|_{L^2(\Omega)}^2] \\
&+ C[\|\psi_h^{n-1}\|_{L^4(\Omega)}^4 + \|\psi_h^{n-1}\|_{L^2(\Omega)}^2]\|\boldsymbol{A}_h^{n-1}\|_{L^2(\Omega)}^2. \qquad (3.7)
\end{aligned}
$$

Combining (3.5)–(3.7) with (3.4) we obtain

$$
\begin{aligned}
&\frac{1}{2}\big[\|\boldsymbol{A}_h^n\|_{L^2(\Omega)}^2 - \|\boldsymbol{A}_h^{n-1}\|_{L^2(\Omega)}^2 + \|\boldsymbol{A}_h^n - \boldsymbol{A}_h^{n-1}\|_{L^2(\Omega)}\big] \\
&+\frac{\Delta t}{2}\big[\| \div \boldsymbol{A}_h^n\|_{L^2(\Omega)}^2 + \mathbf{curl}\boldsymbol{A}_h^n\|_{L^2(\Omega)}^2\big] \\
\leq\ & C\Delta t(\|\psi_h^{n-1}\|_{L^4(\Omega)}^4 + \|\psi_h^{n-1}\|_{L^2(\Omega)}^2)\|\boldsymbol{A}_h^{n-1}\|_{L^2(\Omega)}^2 \\
&+C\Delta t\big[\|H^n\|_{L^2(\Omega)}^2 + \|\frac{i}{\kappa}\nabla\psi_h^{n-1} + \boldsymbol{A}_h^{n-1}\psi_h^{n-1}\|_{L^2(\Omega)}^2\big].
\end{aligned}
$$

Now (3.3) follows from (3.2) by Lemma 3.1. This finishes the proof.

**Lemma 10.3.4** *Let $\Delta t \leq \eta/4$. We have*

$$
\sum_{n-1}^{M} \Delta t\|A_h^n\|_{L^4(\Omega)}^4 \leq C, \qquad (3.8)
$$

*where the constant $C$ is independent of $h$, $\Delta t$ and $M$.*

*Proof.* By Lemma 1.1 and Lemma 3.3, we have

$$\begin{aligned}\sum_{n-1}^{M}\Delta t\|A_h^n\|_{L^4(\Omega)}^4 &\le C\sum_{n-1}^{M}\Delta t\|A_h^n\|_{H^1(\Omega)}^2\|\boldsymbol{A}_h^n\|_{L^2(\Omega)}^2\\ &\le C\sum_{n-1}^{M}\Delta t\|A_h^n\|_{H^1(\Omega)}^2 \le C.\end{aligned}$$

This completes the proof.

**Lemma 10.3.5** *Let $\Delta t \le \eta/4$. We have*

$$\sum_{n-1}^{M}\Delta t\|\nabla\psi_h^n\|_{L^2(\Omega)}^2 \le C, \tag{3.9}$$

*where the constant $C$ is independent of $h$, $\Delta t$ and $M$.*

*Proof.* By (3.2) and (3.8), we get

$$\begin{aligned}\sum_{n-1}^{M}\Delta t\|A_h^n\psi_h^n\|_{L^2(\Omega)}^2 &\le \sum_{n-1}^{M}\Delta t\|A_h^n\|_{L^4(\Omega)}^2\|\psi_h^n\|_{L^4(\Omega)}^2\\ &\le \sum_{n-1}^{M}\Delta t\|A_h^n\|_{L^4(\Omega)}^4+\sum_{n=1}^{M}\Delta t\|\psi_h^n\|_{L^4(\Omega)}^4\\ &\le C\end{aligned}$$

which, together with (3.2) implies (3.9), completes the proof.

## 10.4 Error estimates

In this section we are going to prove the following theorem.

**Theorem 10.4.1** *Let $\Delta t$ be sufficiently small. Then under the regularity assumptions (1.7)–(1.8) we have the error estimate*

$$\begin{aligned}&\max_{1\le n\le M}\left[\|\psi^n-\psi_h^n\|_{L^2(\Omega)}^2+\|\boldsymbol{A}^n-\boldsymbol{A}_h^n\|_{L^2(\Omega)}^2\right]\\ &+\sum_{n=1}^{M}\Delta t\|\boldsymbol{A}^n-\boldsymbol{A}_h^n\|_{H^1(\Omega)}^2 \le C(h^2+\Delta t^2).\end{aligned} \tag{4.1}$$

*Moreover, if in addition $\Delta t \le Ch$ for some constant $C$ independent of $h$ and $\Delta t$, we have*

$$\sum_{n=1}^{M}\Delta t\|\psi^n-\psi_h^n\|_{H^1(\Omega)}^2 \le C(h^2+\Delta t^2). \tag{4.2}$$

*Here the constant $C$ is independent of $h$, $\Delta t$ and $M$.*

*Proof.* Let $\psi_1^n \in \mathcal{S}_h$ denote the interpolate of $\psi^n$ in $\mathcal{S}_h$ and set $\zeta_h^n = \psi_h^n - \psi_1^n$. Then from (2.2) and (2.9) we obtain

$$
\begin{aligned}
&\eta \int_\Omega \partial\zeta_h^n \omega dx + \int_\Omega \big(-\frac{i}{\kappa}\nabla\zeta_h^n - \boldsymbol{A}_h^n\zeta_h^n\big)\big(\frac{i}{\kappa}\nabla\omega - \boldsymbol{A}_h^n\omega\big)dx \\
= \;& \eta \int_\Omega \partial(\psi^n - \psi_I^n)\omega dx \\
&+ i\eta\kappa\Big[\int_\Omega \div \boldsymbol{A}_h^n\psi_h^n\omega dx - \frac{1}{\Delta t}\int_{I_n}\int_\Omega \div \boldsymbol{A}\psi\omega dxdt\Big] \\
&+ \Big[\int_\Omega (1 - |\psi_h^n|^2)\psi_h^n\omega dx - \frac{1}{\Delta t}\int_{I_n}\int_\Omega (1 - |\psi|^2)\psi\omega dxdt\Big] \\
&+ \Big[\int_\Omega \big(-\frac{i}{\kappa}\nabla\psi_I^n - \boldsymbol{A}_h^n\psi_h^n\big)\big(\frac{i}{\kappa}\nabla\omega - \boldsymbol{A}_h^n\omega\big)dxdt \\
&- \frac{1}{\Delta t}\int_{I_n}\int_\Omega \big(-\frac{i}{\kappa}\nabla\psi - \boldsymbol{A}\psi\big)\big(\frac{i}{\kappa}\nabla\omega - \boldsymbol{A}\omega\big)dxdt\Big] \\
= \;& (\mathrm{I}) + \cdots + (\mathrm{IV}), \quad \forall\omega \in \mathcal{S}_h.
\end{aligned}
$$

Choosing $\omega = \Delta t\overline{\zeta}_h^n$ as the test function and taking the real part of the equation, we have

$$
\begin{aligned}
&\frac{\eta}{2}\big[\|\zeta_h^n\|_{L^2(\Omega)}^2 - \|\zeta_h^{n-1}\|_{L^2(\Omega)}^2\big] + \Delta t\| - \frac{i}{\kappa}\nabla\zeta_h^n - \boldsymbol{A}_h^n\zeta_h^n\|_{L^2(\Omega)}^2 \\
&\le \mathrm{Re}\big[(\mathrm{I}) + \cdots + (\mathrm{IV})\big].
\end{aligned} \tag{4.3}
$$

Let $\boldsymbol{A}_I^n \in \boldsymbol{V}_h$ denote the interpolate of $\boldsymbol{A}^n$ in $\boldsymbol{V}_h$ and set $\boldsymbol{E}_h^n = \boldsymbol{A}_h^n - \boldsymbol{A}_I^n$. Then from (2.3) and (2.10) we can get

$$
\begin{aligned}
&\int_\Omega \partial\boldsymbol{E}_h^n\boldsymbol{B}dx + \int_\Omega \big[\div\boldsymbol{E}_h^n \div\boldsymbol{B} + \mathbf{curl}\boldsymbol{E}_h^n\mathbf{curl}\boldsymbol{B}\big]dx \\
= \;& \int_\Omega \partial(\boldsymbol{A}^n - \boldsymbol{A}_I^n)\boldsymbol{B}dx \\
&+ \Big[\int_\Omega (\div\boldsymbol{A}_I^n \div\boldsymbol{B} + \mathbf{curl}\boldsymbol{A}_I^n\mathbf{curl}\boldsymbol{B})dx \\
&- \frac{1}{\Delta t}\int_{I_n}\int_\Omega (\div\boldsymbol{A} \div\boldsymbol{B} + \mathbf{curl}\boldsymbol{A}\mathbf{curl}\boldsymbol{B})dxdt\Big] \\
&+ \mathrm{Re}\Big[\int_\Omega \overline{\psi}_h^{n-1}\big(-\frac{i}{\kappa}\nabla\psi_h^{n-1} - \boldsymbol{A}_h^{n-1}\psi_h^{n-1}\big)\boldsymbol{B}dx \\
&- \frac{1}{\Delta t}\int_{I_n}\int_\Omega \overline{\psi}\big(-\frac{i}{\kappa}\nabla\psi - \boldsymbol{A}\psi\big)\boldsymbol{B}dxdt\Big] \\
&- \int_\Omega |\psi_h^{n-1}|^2(\boldsymbol{A}_h^n - \boldsymbol{A}_h^{n-1})\boldsymbol{B}dx \\
&+ \Big[\int_\Omega H^n\mathbf{curl}\boldsymbol{B} - \frac{1}{\Delta t}\int_{I_n}\int_\Omega H\mathbf{curl}\boldsymbol{B}dxdt\Big] \\
= \;& (\mathrm{I}) + \cdots + (\mathrm{IV}), \quad \forall\boldsymbol{B} \subset \boldsymbol{V}_h.
\end{aligned} \tag{4.4}
$$

Taking $B = \Delta t \boldsymbol{E}_h^n$ in above equation, we find

$$\frac{1}{2}\big[\|\boldsymbol{E}_h^n\|_{L^2(\Omega)}^2 - \|\boldsymbol{E}_h^{n-1}\|_{L^2(\Omega)}^2\big] + \Delta t\big[\| \div \boldsymbol{E}_h^n\|_{L^2(\Omega)}^2 + \|\mathbf{curl}\boldsymbol{E}_h^n\|_{L^2(\Omega)}^2\big] \le (\mathrm{V}) + \cdots + (\mathrm{IX}). \tag{4.5}$$

Now we are going to estimate the terms (I), ..., (IV) in (5.3). At first we note that by a standard argument we can bound (I) by (cf. [CHo 94])

$$|\mathrm{Re}(I)| \le Ch\sqrt{\Delta t}\Big[\int_{I^n} \|\frac{\partial \psi}{\partial t}\|_{H^1(\Omega)}^2 dt\Big]^{1/2} \|\zeta_h^n\|_{L^2(\Omega)}. \tag{4.6}$$

The term (II) can be decomposed as follows

$$\begin{aligned}
(\mathrm{II}) &= i\eta\kappa \int_\Omega \div \boldsymbol{A}_h^n \zeta_h^n \omega dx + i\eta\kappa \int_\Omega \div \boldsymbol{A}_h^n (\psi_I^n - \psi^n)\omega dx \\
&+ i\eta\kappa \int_\Omega \div \boldsymbol{E}_h^n \psi_h^n \omega dx + i\eta\kappa \int_\Omega \div (\boldsymbol{A}_I^n - \boldsymbol{A}^n)\psi^n \omega dx \\
&+ i\eta\kappa\Big[\int_\Omega \div \boldsymbol{A}^n \psi^n \omega dx - \frac{1}{\Delta t}\int_{I^n}\int_\Omega \div \boldsymbol{A}\psi\omega dx dt\Big] \\
&= \sum_{j=1}^5 (\mathrm{II})_j.
\end{aligned}$$

It is obvious that $\mathrm{Re(II)}_1 = 0$ since $\omega = \Delta t \overline{\zeta}_h^n$. In view of (3.1) and by using inverse estimates, we get

$$\begin{aligned}
|\mathrm{Re(II)}_2| &\le C\Delta t\| \div \boldsymbol{A}_h^n\|_{L^2(O)} \|\psi_I^n - \psi^n\|_{L^2(\Omega)} \|\zeta_h^n\|_{L^\infty(\Omega)} \\
&\le Ch^2\Delta t \| \div \boldsymbol{A}_h^n\|_{L^2(O)} \|\psi^n\|_{H^2(\Omega)} \|\zeta_h^n\|_{L^\infty(\Omega)} \\
&\le Ch\Delta t \| \div \boldsymbol{A}_h^n\|_{L^2(O)} \|\zeta_h^n\|_{L^2(\Omega)}.
\end{aligned}$$

By employing standard methods, we find

$$\begin{aligned}
|\mathrm{Re(II)}_3| &\le \eta\kappa\Delta t\| \div \boldsymbol{E}_h^n\|_{L^2(\Omega)} \|\zeta_h^n\|_{L^2(\Omega)}, \\
|\mathrm{Re(II)}_4| &\le C\Delta t\| \div (\boldsymbol{A}_I^n - \boldsymbol{A}^n)\|_{L^2(\Omega)} \|\zeta_h^n\|_{L^2(\Omega)} \\
&\le Ch\Delta t \|\boldsymbol{A}^n\|_{H^2(\Omega)} \|\zeta_h^n\|_{L^2(\Omega)}, \\
|\mathrm{Re(II)}_5| &\le C\Delta t^{3/2}\Big[\int_\Omega \|\frac{\partial}{\partial t}(\div \boldsymbol{A}\psi)\|_{L^2(\Omega)}^2 dt\Big]^{1/2} \|\zeta_h^n\|_{L^2(\Omega)} \\
&\le C\Delta t^{3/2}\Big[\int_\Omega (\|\frac{\partial \boldsymbol{A}}{\partial t}\|_{H^1(\Omega)}^2 + \| \div \boldsymbol{A}\|_{L^4(\Omega)}^4 \\
&\quad + \|\frac{\partial \psi}{\partial t}\|_{L^4(\Omega)}^4)\Big]^{1/2} \|\zeta_h^n\|_{L^2(\Omega)}.
\end{aligned}$$

Thus we can bound (II) by

$$|\mathrm{Re(II)}| \le \Delta t\|\zeta_h^n\|_{L^2(O)}^2 + \eta\kappa\Delta t\| \div \boldsymbol{E}_h^n\|_{L^2(\Omega)} \|\zeta_h^n\|_{L^2(\Omega)} + C(h^2 + \Delta t)\Theta_1^n, \tag{4.7}$$

where

$$
\begin{aligned}
\Theta_1^n &= \Delta t\big[\| \div \boldsymbol{A}_h^n\|_{L^2(\Omega)} + \|\boldsymbol{A}^n\|^2_{H^2(\Omega)} \\
&+ \int_{I^n}\big[\|\frac{\partial \boldsymbol{A}}{\partial t}\|^2_{H^1(\Omega)} + \| \div \boldsymbol{A}\|^4_{L^4(\Omega)} + \|\frac{\partial \psi}{\partial t}\|^4_{L^4(\Omega)}\big]dt.
\end{aligned}
$$

We shall estimate the term (III) by the method developed in [CHo 94]. To this purpose, denote $g(\psi) = (1-|\psi|^2)\psi$ and notice that the function $g : C \to C$ has following properties

$$
\begin{aligned}
&\mathrm{Re}[g(\psi) - g(\phi)\overline{(\psi-\phi)} \le |\psi-\phi|^2, \quad \forall \phi,\psi\in C,\\
&|g(\psi)-g(\phi)| \le (3K^2+1)\phi - \psi|, \quad \forall\phi,\psi\in C,\ |\phi|,|\psi|\le K.
\end{aligned}
$$

Since

$$
\begin{aligned}
\text{(III)} &= \int_\Omega [g(\psi_h^n) - g(\psi_1^n)]\omega dx + \int_\Omega [g(\psi_1^n) - g(\psi^n)]\omega dx \\
&+ \big[\int_\Omega g(\psi^n)\omega dx - \frac{1}{\Delta t}\int_{I^n}\int_\Omega g(\psi)\omega dx dt\big] \\
&= \sum_{j=1}^3 \text{(III)}_j
\end{aligned}
$$

and

$$
\begin{aligned}
|\mathrm{Re(III)}_1| &\le \Delta t\|\psi_h^n - \psi_I^n\|^2_{L^2(O)} = \Delta t\|\zeta_h^n\|^2_{L^2(\Omega)},\\
|\text{(III)}_2| &\le C\Delta t\|\psi^n_{(\mathrm{I})} - \psi^n\|_{L^2(\Omega)}\|\zeta_h^n\|_{L^2(\Omega)}\\
&\le Ch^2\Delta t\|\psi^n\|_{H^2(\Omega)}\|\zeta_h^n\|_{L^2(\Omega)},\\
|\text{(III)}_3| &\le C\Delta t^{3/2}\big[\int_{I_n}\|\frac{\partial}{\partial t}g(\psi)\|^2_{L^2(\Omega)}dt\big]^{1/2}\|\zeta_h^n\|_{L^2(\Omega)}\\
&\le C\Delta t^{3/2}\big[\int_{I_n}\|\frac{\partial\psi}{\partial t}\|^2_{L^2(\Omega)}dt\big]^{1/2}\|\zeta_h^n\|_{L^2(\Omega)},
\end{aligned}
$$

we have

$$
|\mathrm{Re(III)}|A \le \frac{3}{2}\Delta t\|\zeta_h^n\|^2_{L^2(\Omega)} + C(h^2+\Delta t^2)\Theta_2^n, \tag{4.8}
$$

where

$$
\Theta_2^n = \Delta t\|\psi^n\|^2_{H^2(\Omega)} + \int_{I^n}\|\frac{\partial\psi}{\partial t}\|^2_{L^2(\Omega)}dt.
$$

Denote by $\xi_h^n = -\frac{i}{\kappa}\nabla\zeta_h^n - \boldsymbol{A}_h^n\zeta_h^n$ and note that $\omega = \Delta t\overline{\zeta}_h^n$, we can write (IV) as follows.

$$
\begin{aligned}
\text{(IV)} &= \Delta t\int_\Omega\big[-\frac{i}{\kappa}\nabla(\psi_I^n - \psi^n) - \boldsymbol{A}_h^n(\psi_I^n-\psi^n)\big]\xi_h^n dx\\
&+\Delta t\int_\Omega(\boldsymbol{A}^n - \boldsymbol{A}_I^n)\psi^n\xi_h^n dx + \Delta t\int_\Omega \boldsymbol{E}_h^n\psi^n\overline{\xi}_h^n dx
\end{aligned}
$$

$$
\begin{aligned}
&+\Delta t\Big[\int_\Omega(-\frac{i}{\kappa}\nabla\psi^n - \boldsymbol{A}^n\psi^n)\overline{\xi}_h^n dx - \frac{1}{\Delta t}\int_{I^n}\int_\Omega(-\frac{i}{\kappa}\nabla\psi - \boldsymbol{A}\psi)\overline{\xi}_h^n dxdt\Big]\\
&+\int_{I^n}\int_\Omega(-\frac{i}{\kappa}\nabla\psi - \boldsymbol{A}\psi)(\boldsymbol{A}-\boldsymbol{A}^n)\overline{\zeta}_h^n dxdt\\
&+\int_{I^n}\int_\Omega(-\frac{i}{\kappa}\nabla\psi - \boldsymbol{A}\psi)(\boldsymbol{A}^n-\boldsymbol{A}_I^n)\overline{\zeta}_h^n dxdt\\
&+\int_{I^n}\int_\Omega(-\frac{i}{\kappa}\nabla\psi - \boldsymbol{A}\psi)(\boldsymbol{A}_{(\mathrm{I})}^n-\boldsymbol{A}_h^n)\zeta_h^n dxdt\\
=&\ \sum_{j=1}^{7}(\mathrm{IV})_j.
\end{aligned}
$$

In view of (2.6) and by using inverse estimates and Theorem 3.1, we know that

$$
\begin{aligned}
|\mathrm{Re}(\mathrm{IV})_1 &\le C\Delta t\| -\frac{i}{\kappa}\nabla(\psi_I^n-\psi^n) - \boldsymbol{A}_h^n(\psi_I^n-\psi^n)\|_{L^2(\Omega)}\|\xi_h^n\|_{L^2(\Omega)}\\
&\le C\Delta t(h\|\psi^n\|_{H^2(\Omega)} + h^2\|\boldsymbol{A}_h^n\|_{L^\infty}\|\psi^n\|_{H^2(\Omega)}\|\xi_h^n\|_{L^2(\Omega)}\\
&\le Ch\Delta t(\|\psi^n\|_{L^2(\Omega)} + \|\boldsymbol{A}_h^n\|_{L^2(\Omega)}\|\psi^n\|_{H^2(\Omega)}\|\xi_h^n\|_{L^2(\Omega)}\\
&\le Ch\Delta t\|\psi^n\|_{H^2(\Omega)}\|\xi_h^n\|_{L^2(\Omega)}.
\end{aligned}
$$

It is easy to see that

$$
\begin{aligned}
&|\mathrm{Re}(\mathrm{IV})_2| + |\mathrm{Re}(\mathrm{IV})_3|\\
\le\ & Ch^2\Delta t\|\boldsymbol{A}^n\|_{H^2(\Omega)}\|\xi_h^n\|_{L^2(\Omega)} + \Delta t\|\boldsymbol{E}_h^n\|_{L^2(\Omega)}\|\xi_h^n\|_{L^2(\Omega)},\\
&|\mathrm{Re}(\mathrm{IV})_6| + |\mathrm{Re}(\mathrm{IV})_7|\\
\le\ & Ch\Delta t\|\boldsymbol{A}^n\|_{H^2(\Omega)}\|\zeta_h^n\|_{L^2(\Omega)} + C\Delta t\|\boldsymbol{E}_h^n\|_{L^4(\Omega)}\|\zeta_h^n\|_{L^2(\Omega)},
\end{aligned}
$$

where in the estimation of $|\mathrm{Re}(\mathrm{IV})_7|$ we have used that fact that

$$
\| -\frac{i}{\kappa}\nabla\psi - \boldsymbol{A}\psi\|_{C(0,T;L^4(\Omega))} \le C.
$$

The other two terms can be bounded by standard methods

$$
\begin{aligned}
\mathrm{Re}(\mathrm{IV})_4 &\le \Delta t^{3/2}\Big[\int_{I_n}\|\frac{\partial}{\partial t}(-\frac{i}{\kappa}\nabla\psi - \boldsymbol{A}\psi)\|_{L^2(\Omega)}^2 dt\Big]^{1/2}\|\xi_h^n\|_{L^2(\Omega)}\\
&\le \Delta t^{3/2}\Big[\int_{I_n}(\|\frac{\partial\psi}{\partial t}\|_{H^1(\Omega)}^2 + \|\frac{\partial \boldsymbol{A}}{\partial t}\|_{L^2(\Omega)}^2)dt\Big]^{1/2}\|\xi_h^n\|_{L^2(\Omega)},\\
\mathrm{Re}(\mathrm{IV})_5 &\le \int_{I_n}\|\boldsymbol{A}-\boldsymbol{A}^n\|_{L^4(\Omega)}\|\zeta_h^n\|_{L^2(\Omega)}dt\\
&\le C\Delta t^{3/2}\Big[\int_{I_n}\|\frac{\partial \boldsymbol{A}}{\partial t}\|_{H^1(\Omega)}^2 dt\Big]^{1/2}\|\zeta_h^n\|_{L^2(\Omega)}.
\end{aligned}
$$

Therefore, we have

$$
\begin{aligned}
|\mathrm{Re}(\mathrm{IV}) \le\ & \frac{\Delta t}{2}\|\xi_h^n\|_{L^2(\Omega)} + \Delta t\|\zeta_h^n\|_{L^2(\Omega)}^2 + \Delta t\|\boldsymbol{E}_h^n\|_{L^2(\Omega)}^2\\
&+C\Delta t\|\boldsymbol{E}_h^n\|_{L^4(\Omega)}^2 + C(h^2+\Delta t^2)\Theta_3^n, \qquad (4.9)
\end{aligned}
$$

where

$$\Theta_3^n = \Delta t(\|\psi^n\|_{H^2(\Omega)}^2 + \|\boldsymbol{A}^n\|_{H^2(\Omega)}^2 + \int_{I_n} (\|\frac{\partial \psi}{\partial t}\|_{H^1(\Omega)}^2 + \|\frac{\partial \boldsymbol{A}}{\partial t}\|_{H^1(\Omega)}^2) dt.$$

Note that by Lemma 1.1 we have

$$\begin{aligned} \|\boldsymbol{E}_h^n\|_{L^4(\Omega)}^2 &\leq C\|E_h^n\|_{H^1(\Omega)}\|E_h^n\|_{L^2(\Omega)} \\ &\leq C|E_h^n|_{H_n^1(\Omega)}\|E_h^n\|_{L^2(\Omega)} + C\|E_h^n\|_{L^2(\Omega)}^2, \end{aligned} \tag{4.10}$$

we can substitute (4.6)–(4.9) into (4.3) and get

$$\begin{aligned} & \frac{\eta}{2}\|\zeta_h^n\|_{L^2(\Omega)}^2 - \frac{\eta}{2}\|\zeta_h^{n-1}\|_{L^2(\Omega)}^2 + \frac{\Delta t}{2}\|\xi_h^n\|_{L^2(\Omega)}^2 \\ \leq\; & \frac{\Delta t}{16}[\| \div \boldsymbol{E}_h^n\|_{L^2(\Omega)}^2 + \|\mathbf{curl}\boldsymbol{E}_h^n\|_{L^2(\Omega)}^2] + C\Delta t(\|\boldsymbol{E}_h^n\|_{L^2(\Omega)}^2 + \|\zeta_h^n\|_{L^2(\Omega)}^2) \\ +\; & C(h^2 + \Delta t^2)(\Theta_1^n + \Theta_2^n + \Theta_3^n). \end{aligned} \tag{4.11}$$

We now turn to the estimates of the terms (V), ..., (IX) in (4.5). At first, analogue to (I), we have

$$|(\mathrm{V})| \leq Ch\Delta t\Big[\int_{I_n} \|\frac{\partial \boldsymbol{A}}{\partial t}\|_{H^1(\Omega)}^2 dt\Big]^{1/2}\|\boldsymbol{E}_h^n\|_{L^2(\Omega)}.$$

By standard argument we can obtain

$$|(\mathrm{VI})| \leq Ch\Delta t\|\boldsymbol{A}^n\|_{H^2(\Omega)}|\boldsymbol{E}_h^n|_{\boldsymbol{H}_n^1(\Omega)} + C\Delta t^{3/2}\Big[\int_{I_n} \|\frac{\partial \boldsymbol{A}}{\partial t}\|_{H^1(\Omega)}^2 dt\Big]^{1/2}|\boldsymbol{E}_h^n|_{\boldsymbol{H}_n^1(\Omega)}$$

and

$$|(\mathrm{IX})| \leq C\Delta t^{3/2}\Big[\int_{I_n} \|\frac{\partial H}{\partial t}\|_{L^2(\Omega)}^2 dt\Big]^{1/2}|\boldsymbol{E}_h^n|_{\boldsymbol{H}_n^1(\Omega)}.$$

Therefore,

$$|(\mathrm{V})| + |(\mathrm{VI})| + |(\mathrm{IX})| \leq \frac{\Delta t}{4}|\boldsymbol{E}_h^n|_{\boldsymbol{H}_n^1(\Omega)}^2 + \Delta t\|\boldsymbol{E}_h^n\|_{L^2(\Omega)}^2 + C(h^2 + \Delta t^2)\Theta_4^n, \tag{4.12}$$

where

$$\Theta_4^n = \Delta t\|\boldsymbol{A}^n\|_{H^2(\Omega)}^2 + \int_{I_n} [\|\frac{\partial \boldsymbol{A}}{\partial t}\|_{H^1(\Omega)}^2 + \|\frac{\partial H}{\partial t}\|_{L^2(\Omega)}^2] dt.$$

Since

$$\begin{aligned} & \|\boldsymbol{A}_I^n - \boldsymbol{A}_I^{n-1}\|_{L^4(\Omega)} \\ \leq\; & \|\boldsymbol{A}^n - \boldsymbol{A}^{n-1}\|_{L^4(\Omega)} + \|\boldsymbol{A}^n - \boldsymbol{A}_I^n\|_{L^4(\Omega)} + \|\boldsymbol{A}^{n-1} - \boldsymbol{A}_I^{n-1}\|_{L^4(\Omega)} \end{aligned}$$

$$
\begin{aligned}
&\leq \int_{I_n} \|\frac{\partial \boldsymbol{A}}{\partial t}\|_{L^4(\Omega)} dt + Ch(\|\boldsymbol{A}^n\|_{H^2(\Omega)} + \|\boldsymbol{A}^{n-1}\|_{H^2(\Omega)} \\
&\leq Ch(\|\boldsymbol{A}^n\|_{H^2(\Omega)} + \|\boldsymbol{A}^{n-1}\|_{H^2(\Omega)}) + C\Delta t^{3/4}\Big[\int_{I_n} \|\frac{\partial \boldsymbol{A}}{\partial t}\|^4_{L^4(\Omega)} dt\Big]^{1/4},
\end{aligned}
$$

we can apply the same method as that in (3.5) to get

$$
\begin{aligned}
(\mathrm{VIII}) \;&=\; -\Delta t \int_\Omega |\psi_h^{n-1}|^2 (\boldsymbol{E}_h^n - \boldsymbol{E}_h^{n-1})\boldsymbol{E}_h^n dx \\
&\quad -\Delta t \int_\Omega |\psi_h^{n-1}|^2 (\boldsymbol{A}_I^n - \boldsymbol{A}_I^{n-1})\boldsymbol{E}_h^n dx \\
&= -\Delta t \int_\Omega |\psi_h^{n-1}|^2 (\boldsymbol{E}_h^n - \boldsymbol{E}_h^{n-1})^2 dx \qquad (4.13) \\
&\quad -\Delta t \int_\Omega |\psi_h^{n-1}|^2 (\boldsymbol{E}_h^n - \boldsymbol{E}_h^{n-1})\boldsymbol{E}_h^{n-1} dx \\
&\quad -\Delta t \int_\Omega |\psi_h^{n-1}|^2 (\boldsymbol{A}_I^n - \boldsymbol{A}_I^{n-1})(\boldsymbol{E}_h^n - \boldsymbol{E}_h^{n-1}) dx \\
&\quad -\Delta t \int_\Omega |\psi_h^{n-1}|^2 (\boldsymbol{A}_I^n - \boldsymbol{A}_I^{n-1})\boldsymbol{E}_h^{n-1} dx \\
&\leq -\frac{\Delta t}{2} \int_\Omega |\psi_h^{n-1}|^2 (\boldsymbol{E}_h^n - \boldsymbol{E}_h^{n-1})^2 dx + \Delta t \|\psi_h^{n-1}\|^2_{L^4(\Omega)} \|\boldsymbol{E}_h^{n-1}\|^2_{L^4(\Omega)} \\
&\quad +\Delta t \|\boldsymbol{A}_I^n - \boldsymbol{A}_I^{n-1}\|^2_{L^4(\Omega)} \|\psi_h^{n-1}\|^4_{L^4(\Omega)} \\
&\quad +\Delta t \|\psi_h^{n-1}\|^2_{L^4(\Omega)} \|\boldsymbol{A}_I^n - \boldsymbol{A}_I^{n-1}\|_{L^4(\Omega)} \|\boldsymbol{E}_h^{n-1}\|_{L^4(\Omega)} \\
&\leq -\frac{\Delta t}{2} \int_\Omega |\psi_h^{n-1}|^2 (\boldsymbol{E}_h^n - \boldsymbol{E}_h^{n-1})^2 dx + C\Delta t \|\psi_h^{n-1}\|^2_{L^4(\Omega)} \|\boldsymbol{E}_h^{n-1}\|^2_{L^4(\Omega)} \\
&\quad +C\Delta t \|\psi_h^{n-1}\|^2_{L^4(\Omega)} \|\boldsymbol{A}_I^n - \boldsymbol{A}_I^{n-1}\|^2_{L^4(\Omega)} \\
&\leq -\frac{\Delta t}{2} \int_\Omega |\psi_h^{n-1}|^2 (\boldsymbol{E}_h^n - \boldsymbol{E}_h^{n-1})^2 dx + \frac{\Delta t}{8} |\boldsymbol{E}_h^{n-1}\|^2_{\boldsymbol{H}_n^1(\Omega)} \\
&\quad +C\Delta t (\|\psi_h^{n-1}\|^2_{L^4(\Omega)} + \|\psi_h^{n-1}\|^4_{L^4(\Omega)} \|\boldsymbol{E}_h^{n-1}\|^2_{L^2(\Omega)}) \\
&\quad +C(h^2 + \Delta t^2)\Theta_5^n, \qquad (4.14)
\end{aligned}
$$

where

$$
\begin{aligned}
\Theta_5^n &= \Delta t \|\psi_h^{n-1}\|^2_{L^4(\Omega)} + \Delta t^{1/2} \Big[\int_{I_n} \|\frac{\partial \boldsymbol{A}}{\partial t}\|^4_{L^4(\Omega)} dt\Big]^{1/2} \|\psi_h^{n-1}\|^2_{L^4(\Omega)}] \\
&\leq \Delta t (\|\psi_h^{n-1}\|^2_{L^4(\Omega)} + \|\psi^{n-1}\|^4_{L^4(\Omega)}) + \int_{I_n} \|\frac{\partial \boldsymbol{A}}{\partial t}\|^4_{L^4(\Omega)} dt.
\end{aligned}
$$

In order to estimate (VII) we first write it as follows.

$$
\begin{aligned}
&(\mathrm{VII}) \\
= \;&\mathrm{Re} \int_\Omega (-\frac{i}{\kappa} \nabla \zeta_h^{n-1} - \boldsymbol{A}_h^{n-1} \zeta_h^{n-1}) \overline{\psi}_h^{n-1} B dx
\end{aligned}
$$

$$
\begin{aligned}
&+\mathrm{Re}\int_\Omega [-\frac{i}{\kappa}\nabla(\psi_I^{n-1}-\psi^{n-1})-\boldsymbol{A}_h^{n-1}(\psi_I^{n-1}-\psi^{n-1})]\overline{\psi}_h^{n-1}\boldsymbol{B}dx\\
&+\mathrm{Re}\int_\Omega (\boldsymbol{A}_h^{n-1}-\boldsymbol{A}_I^{n-1})\psi^{n-1}\overline{\psi}_h^{n-1}\boldsymbol{B}dx\\
&+\mathrm{Re}\int_\Omega (\boldsymbol{A}_I^{n-1}-\boldsymbol{A}^{n-1})\psi^{n-1}\overline{\psi}_h^{n-1}\boldsymbol{B}dx\\
&+\mathrm{Re}\int_\Omega (-\frac{i}{\kappa}\nabla\psi^{n-1}-\boldsymbol{A}^{n-1}(\overline{\psi}_h^{n-1}-\overline{\psi}_I^{n-1}))\boldsymbol{B}dx\\
&+\mathrm{Re}\int_\Omega (-\frac{i}{\kappa}\nabla\psi^{n-1}-\boldsymbol{A}^{n-1}\psi^{n-1})(\overline{\psi}_{(\mathrm{I})}^{n-1}-\overline{\psi}^{n-1})\boldsymbol{B}dx\\
&+\mathrm{Re}[\int_\Omega (-\frac{i}{\kappa}\nabla\psi^{n-1}-\boldsymbol{A}^{n-1}\psi^{n-1})\overline{\psi}^{n-1}\boldsymbol{B}dx\\
&-\frac{1}{\Delta t}\int_{I_n}\int_\Omega(-\frac{i}{\kappa}\nabla\psi-\boldsymbol{A}\psi)\overline{\psi}\boldsymbol{B}dxdt]\\
= &\sum_{j=1}^{7}(\mathrm{VII})_j.
\end{aligned}
$$

Note that $\xi_h^{n-1}=-\frac{i}{\kappa}\nabla\zeta_h^{n-1}-\boldsymbol{A}_h^{n-1}\zeta_h^{n-1}$ and $\boldsymbol{B}=\Delta t\boldsymbol{E}_h^{n-1}$. We can bound $(\mathrm{VII})_1$ by

$$
\begin{aligned}
(\mathrm{VII})_1 \le\ & \Delta t\,|\int_\Omega \xi^{n-1}\psi_h^{n-1}(\boldsymbol{E}_h^n-\boldsymbol{E}_h^{n-1})dx\,|+\Delta t\,|\int_\Omega \xi_h^{n-1}\psi_h^{n-1}\boldsymbol{E}_h^{n-1}dx\,|\\
\le\ & \frac{\Delta t}{2}\int_\Omega |\psi_h^{n-1}|^2(\boldsymbol{E}_h^n-\boldsymbol{E}_h^{n-1})^2dx+\Delta t\|\xi_h^{n-1}\|^2_{L^2(\Omega)}\\
&+C\Delta t\|\psi_h^{n-1}\|^2_{L^4(\Omega)}\|\boldsymbol{E}_h^{n-1}\|^2_{L^4(\Omega)}\\
\le\ & \frac{\Delta t}{2}\int_\Omega |\psi_h^{n-1}|^2(\boldsymbol{E}_h^n-\boldsymbol{E}_h^{n-1})^2dx+\Delta t\|\xi_h^{n-1}\|^2_{L^2(\Omega)}\\
&+\frac{\Delta t}{8}|\boldsymbol{E}_h^{n-1}|^2_{\boldsymbol{H}_n^1(\Omega)}+C\Delta t\|\psi_h^{n-1}\|^4_{L^4(\Omega)}\|\boldsymbol{E}_h^{n-1}\|^2_{L^2(\Omega)},
\end{aligned}
$$

where we have used (4.10) for $\boldsymbol{E}_h^{n-1}$. Similar to $(\mathrm{IV})_1$ we have

$$
\begin{aligned}
|(\mathrm{VII})_2| \le\ & Ch\Delta t\|\psi_h^{n-1}\|_{H^2(\Omega)}\|\psi_h^{n-1}\|_{L^4(\Omega)}\|\boldsymbol{E}_h^n\|_{L^4(\Omega)}\\
\le\ & Ch\Delta t\|\psi_h^{n-1}\|_{L^4(\Omega)}\|\boldsymbol{E}_h^n\|_{H^1(\Omega)}.
\end{aligned}
$$

The other terms in (IV) can be bounded by standard methods as follows.

$$
\begin{aligned}
|(\mathrm{VII})_3| \le\ & C\Delta t\|\boldsymbol{E}_h^{n-1}\|_{L^2(\Omega)}\|\psi_h^{n-1}\|_{L^4(\Omega)}\|\boldsymbol{E}_h^n\|_{H^1(\Omega)},\\
|(\mathrm{VII})_4| \le\ & C\Delta t\|\boldsymbol{A}^{n-1}-\boldsymbol{A}_I^{n-1}\|_{L^2(\Omega)}\|\psi_h^{n-1}\|_{L^4(\Omega)}\|\boldsymbol{E}_h^n\|_{L^4(\Omega)}\\
\le\ & Ch^2\Delta t\|\psi_h^{n-1}\|_{L^4(\Omega)}\|\boldsymbol{E}_h^n\|_{H^1(\Omega)},\\
|(\mathrm{VII})_5| \le\ & C\Delta t\|\zeta_h^{n-1}\|_{L^2(\Omega)}\|\boldsymbol{E}_h^n\|_{H^1(\Omega)},\\
|(\mathrm{VII})_6| \le\ & C\Delta t\|\psi_I^{n-1}-\psi^{n-1}\|_{L^4(\Omega)}\|\boldsymbol{E}_h^n\|_{L^2(\Omega)}\\
\le\ & Ch\Delta t\|\psi^{n-1}\|_{H^2(\Omega)}\|\boldsymbol{E}_h^n\|_{L^2(\Omega)},\\
|(\mathrm{VII})_7| \le\ & C\Delta t^{3/2}[\int_{I^n}\|\frac{\partial}{\partial t}\,|\,(-\frac{i}{\kappa}\nabla\psi-\boldsymbol{A}\psi)\overline{\psi}]\|^2_{L^2(\Omega)}dt]^{1/2}\|\boldsymbol{E}_h^n\|_{L^2(\Omega)}
\end{aligned}
$$

$$
\begin{aligned}
&\leq C\Delta t^{3/2}\Big[\int_{I^n}(\|\frac{\partial\psi}{\partial t}\|^2_{H^1(\Omega)}+\|\frac{\partial \boldsymbol{A}}{\partial t}\|^2_{L^2(\Omega)})dt\Big]^{1/2}\|\boldsymbol{E}_h^n\|_{L^2(\Omega)}\\
&+C\Delta t^{3/2}\Big[\int_{I^n}(\|\frac{\partial\psi}{\partial t}\|^4_{L^4(\Omega)}+\|\nabla\psi\|^4_{L^4(\Omega)})dt\Big]^{1/2}\|\boldsymbol{E}_h^n\|_{L^2(\Omega)}.
\end{aligned}
$$

Hence,

$$
\begin{aligned}
|(\mathrm{VII})| \;\leq\;& \frac{\Delta t}{2}\|\psi_h^{n-1}|^2(\|\boldsymbol{E}_h^n-\boldsymbol{E}_h^{n-1})^2dx+\frac{\Delta t}{4}|\boldsymbol{E}_h^n|^2_{\boldsymbol{H}^1(\Omega)}+\Delta t\|\boldsymbol{E}_h^n\|^2_{L^2(\Omega)}\\
+\;& \Delta t\|\xi^{n-1}\|^2_{L^2(\Omega)}+\frac{\Delta t}{8}|\boldsymbol{E}_h^{n-1}|^2_{\boldsymbol{H}^1-n(\Omega)}+C\Delta t\|\zeta_h^{n-1}\|^2_{L^2(\Omega)}\\
+\;& C\Delta t(\|\psi_h^{n-1}\|^4_{L^4(\Omega)}+\|\psi_h^{n-1}\|^2_{L^4(\Omega)}\|\boldsymbol{E}_h^{n-1}\|^2_{L^2(\Omega)}\\
+\;& C(h^2+\Delta t^2)\Theta_6^n, \qquad\qquad (4.15)
\end{aligned}
$$

where

$$
\begin{aligned}
\Theta_6^n \;=\;& \Delta t(\|\psi_h^{n-1}\|^4_{L^4(\Omega)}+\|\psi^{n-1}\|^2_{L^2(\Omega)}\\
&+\int_{I^n}(\|\frac{\partial\psi}{\partial t}\|^2_{H^1(\Omega)}+\|\frac{\partial \boldsymbol{A}}{\partial t}\|^2_{L^2(\Omega)}+\|\frac{\partial\psi}{\partial t}\|^4_{L^4(\Omega)}+\|\Delta\psi\|^4_{L^4(\Omega)})dt.
\end{aligned}
$$

Substituting (4.12)–(4.14) into (4.5) we obtain

$$
\begin{aligned}
&\frac{1}{2}\|\boldsymbol{E}_h^n\|^2_{L^2(\Omega)}-\frac{1}{2}\|\boldsymbol{E}_h^{n-1}\|^2_{L^2(\Omega)}+\frac{\Delta t}{2}\big[\|\div\boldsymbol{E}_h^n\|^2_{L^2(\Omega)}+\|\mathbf{curl}\boldsymbol{E}_h^n\|^2_{L^2(\Omega)}\big]\\
\leq\;& 2\Delta t\|\boldsymbol{E}_h^n\|^2_{L^2(\Omega)}+C\Delta t(\|\psi_h^{n-1}\|^4_{L^4(\Omega)}+\|\psi_h^{n-1}\|^2_{L^4(\Omega)})\|\boldsymbol{E}_h^{n-1}\|^2_{L^2(\Omega)}\\
+\;& \frac{\Delta t}{4}\big[\|\div\boldsymbol{E}_h^n\|^2_{L^2(\Omega)}+\|\mathbf{curl}\boldsymbol{E}_h^n\|^2_{L^2(\Omega)}\big]+\Delta t\|\xi_h^{n-1}\|^2_{L^2(\Omega)}+C\Delta t\|\zeta_h^{n-1}\|^2_{L^2(\Omega)}\\
+\;& C(h^2+\Delta t^2)(\Theta_4^n+\Theta_5^n+\Theta_6^n). \qquad\qquad (4.16)
\end{aligned}
$$

Multiplying (4.15) by 1/4 and adding to (4.11), we end up with

$$
\begin{aligned}
&\big[\frac{\eta}{2}\|\zeta_h^n\|_{L^2(\Omega)}+\frac{1}{8}\|\boldsymbol{E}_h^n\|^2_{L^4(\Omega)}\big]-\big[\frac{\eta}{2}\|\zeta_h^{n-1}\|_{L^2(\Omega)}+\frac{1}{8}\|\boldsymbol{E}_h^{n-1}\|^2_{L^4(\Omega)}\\
&+\frac{1}{2}\Delta t\|\xi_h^n\|^2_{L^2(\Omega)}+\frac{1}{8}\Delta t\big[\|\div\boldsymbol{E}_h^n\|^2_{L^2(\Omega)}+\|\mathbf{curl}\boldsymbol{E}_h^n\|^2_{L^2(\Omega)}\\
\leq\;& C^*\Delta t(\|\boldsymbol{E}_h^n\|^2_{L^2(\Omega)}+\|\zeta_h^n\|^2_{L^2(\Omega)})+C\Delta t\|\zeta_h^{n-1}\|^2_{L^2(\Omega)}+\frac{1}{4}\Delta t\|\xi_h^{n-1}\|^2_{L^2(\Omega)}\\
&+C\Delta t(\|\psi_h^{n-1}\|^4_{L^4(\Omega)}+\|\psi_h^{n-1}\|^2_{L^4(\Omega)}\|\boldsymbol{E}_h^{n-1}\|^2_{L^2(\Omega)}+C(h^2+\Delta t^2)\sum_{j=1}^{6}\Theta_j^n,
\end{aligned}
$$

where $C^*$ is a constant independent of $h$, $\Delta t$ and $M$.

Now choosing $\Delta t$ sufficiently small such that $C^*\Delta t<\min(\eta/4,1/8)$, applying Lemma 3.1 to above inequality and noticing that by Theorem 1.1 and Theorem 3.1,

$$
\sum_{n=1}^{M}\Delta t(\|\psi_h^{n-1}\|^4_{L^4(\Omega)}+\|\psi_h^{n-1}\|^2_{L^4(\Omega)})\leq C \;\text{ and }\; \sum_{j=1}^{6}\Theta_j^n\leq C,
$$

we get

$$\max_{1\le n\le M}\left[\|\zeta_h^n\|^2_{L^2(\Omega)}\right]+\sum_{n=1}^{M}\Delta t\left[\|\xi_h^n\|^2_{L^2(\Omega)}+|\boldsymbol{E}_h^n|^2_{\boldsymbol{H}^1_n(\Omega)}\right]\le C(h^2+\Delta t^2).$$

By applying the triangle inequality we have

$$\max_{1\le n\le M}\left[\|\psi^n-\psi_h^n\|^2_{L^2(\Omega)}+\|\boldsymbol{A}^n-\boldsymbol{A}_h^n\|^2_{L^2(\Omega)}\right]+\sum_{n=1}^{M}\Delta t\|\boldsymbol{A}^n-\boldsymbol{A}_h^n\|^2_{H^1(\Omega)}\le C(h^2+\Delta t^2).$$

Moreover, since

$$\sum_{n=1}^{M}\Delta t\|\nabla\zeta_h^n\|^2_{L^2(\Omega)}\le\kappa\sum_{n=1}^{M}\Delta t\|\xi_h^n\|^2_{L^2(\Omega)}+\kappa\sum_{n=1}^{M}\Delta t\|\boldsymbol{A}_h^n\zeta_h^n\|^2_{L^2(\Omega)}$$

and

$$\begin{aligned}\|\boldsymbol{A}_h^n\zeta_h^n\|^2_{L^2(\Omega)} &\le \|\boldsymbol{A}_h^n-\boldsymbol{A}^n\|^2_{L^\infty(\Omega)}\|\zeta_h^n\|^2_{L^2(\Omega)}+\|\boldsymbol{A}^n\|^2_{L^\infty(\Omega)}\|\zeta_h^n\|^2_{L^2(\Omega)}\\ &\le Ch^{-2}\|\boldsymbol{A}^n-\boldsymbol{A}_h^n\|^2_{L^2(\Omega)}\|\zeta_h^n\|^2_{L^2(\Omega)}+C\|\zeta_h^n\|^2_{L^2(\Omega)}\\ &\le Ch^{-2}(h^2+\Delta t^2)^2+C(h^2+\Delta t^2)\\ &\le C(h^2+\Delta t^2)\end{aligned}$$

when we assume that $\Delta t\le Ch$ for some constant $C>0$, we have

$$\sum_{n=1}^{M}\Delta t\|\nabla\zeta_h^n\|^2_{L^2(\Omega)}\le C(h^2+\Delta t^2).$$

Now (5.2) follows by using triangle inequality. This completes the proof.

We make now some remarks about the theorem.

**Remark 10.4.1** *The assumption that $\Delta t$ should be sufficiently small in the theorem results from that we have used a semi-implicit scheme in the discrete problem (DP). If instead we use the totally implicit scheme to approximate the continuous problem (P), namely, for $n=1,2,\dots,M$, we seek $\{\psi_h^n,\boldsymbol{A}_h^n\}\in\mathcal{S}_h\times\boldsymbol{V}_h$ such that*

$$\psi_h^0=(\psi_0)_I,\quad \boldsymbol{A}_h^n=(\boldsymbol{A}_0)_I, \tag{4.17}$$

*and*

$$\begin{aligned}&\eta\int_\Omega\partial\psi_h^n\omega dx-i\eta\kappa\int_\Omega\div\boldsymbol{A}_h^n\psi_h^n\omega dx\\ &+\int_\Omega\Big(-\frac{i}{\kappa}\nabla\psi_h^n-\boldsymbol{A}_h^n\psi_h^n\Big)\Big(\frac{i}{\kappa}\nabla\omega-\boldsymbol{A}_h^n\omega\Big)dx\\ &+\int_\Omega(|\psi_h^n|^2-1)\psi_h^n\omega dx=0\quad\forall\omega\in\mathcal{S}_h,\end{aligned} \tag{4.18}$$

$$\begin{aligned}
&\int_\Omega \partial \boldsymbol{A}_h^n \boldsymbol{B} dx + \int_\Omega (\div \boldsymbol{A}_h^n \div \boldsymbol{B} + \mathbf{curl} \boldsymbol{A}_h^n \mathbf{curl} \boldsymbol{B}) dx \\
&+ \int_\Omega \frac{i}{\kappa} (\overline{\psi}_h^n \nabla \psi_h^n - \psi_h^n \nabla \overline{\psi}_h^n) \boldsymbol{B} dx + \int_\Omega |\psi_h^n|^2 \boldsymbol{A}_h^n \cdot \boldsymbol{B} dx \\
= &\int_\Omega H^n \mathbf{curl} \boldsymbol{B} dx dt, \quad \forall \boldsymbol{B} \in \boldsymbol{V}_h \qquad (4.19)
\end{aligned}$$

*then we can show the same error estimates (4.1)–(4.2) without the assumption that $\Delta t$ should be small. The advantage in using the semi-implicit scheme (2.8)–(2.10) lies in the fact that in each time step we only need to solve a non-linear system of equations for $\psi$, while in the scheme (4.16)–(4.18) we must solve a non-linear system of equations for $\psi$ and $\boldsymbol{A}$ which certainly will cost more time in practical computations.*

**Remark 10.4.2** *For convenience we have restricted ourselves in this chapter to the use of linear finite element approximations in space. It is easy to see from the proof that under proper regularity assumptions concerning $(\psi, \boldsymbol{A})$, one can demonstrate the error estimates of order $O(h^r + \Delta t)$ if the finite element space $\mathcal{S}$, $\boldsymbol{V}_h$ have following approximation properties*

$$\|\phi - \phi_I\|_{H^j(\Omega)} \le C h^{r+1-j} \|\phi\|_{H^{r+1}(\Omega)}, \quad \forall \phi \in \mathcal{H}^{r+1}(\Omega),\ j = 0, 1$$

*and*

$$\|\boldsymbol{Q} - \boldsymbol{Q}_I\|_{H^j(\Omega)} \le C h^{r+1-j} \|\boldsymbol{Q}\|_{H^{r+1}(\Omega)}, \quad \forall \boldsymbol{Q} \in \boldsymbol{H}^{r+1}(\Omega),\ j = 0, 1.$$

*We note that the error estimates on the $H^1$-norm in (4.1)–(4.2) are of optimal order. But we are not able to prove optimal error estimates in $L^2$ norm by the method in Theorem 4.1 because of the non-linear coupling effects coming from the terms such as $\div \boldsymbol{A}\psi$, $\psi \nabla \psi$ and $\boldsymbol{A} \nabla \psi$. However, our numerical example in the next section indicates that the method in this paper may still preserve optimal convergence order in $L^2$ norm. The question of how to prove the optimal error estimates in the $L^2$ norm for the scheme (2.8)–(2.10) is still open and waiting for further research.*

**Remark 10.4.3** *One other alternative to treat the boundary conditions $\boldsymbol{A} \cdot \boldsymbol{n} = 0$ on $\partial\Omega$ is to use a penalty method. To this end, one should add an additional term $(1/\epsilon) \int_{\partial\Omega} (\boldsymbol{A}_h^n \cdot \boldsymbol{n})(\boldsymbol{B} \cdot \boldsymbol{n}) dx$ in the left side of (2.10). One can also deduce an error estimate by employing the method in this paper. We also note that this method can be used to deal with the case when $\partial\Omega$ is smooth.*

## 10.5 A numerical example

In order to test the numerical method proposed in this paper, we add a term $g$ (respectively $f$) on the right side of the equation (1.1) (respectively (1.2)). More precisely, we are going to solve the following equations

$$\eta\frac{\partial\psi}{\partial t} - i\eta\kappa \div \boldsymbol{A}\psi + \left(-\frac{i}{\kappa}\nabla - \boldsymbol{A}\right)^2\psi + (|\psi|^2 - 1)\psi = g, \quad \text{in } Q_T, \tag{5.1}$$

$$\frac{\partial \boldsymbol{A}}{\partial t} - \Delta\boldsymbol{A} + \frac{i}{2\kappa}(\overline{\psi}\nabla - \psi\nabla\psi) + |\psi|^2\boldsymbol{A} = \mathbf{curl}H + \boldsymbol{f}, \quad \text{in } Q_T \tag{5.2}$$

with the initial and boundary conditions (1.3), (1.4) and gauge condition 1.5.

Let $\Omega = (0,1)\times(0,1)$ and $T = 1$. We take $\eta = \kappa = 1$. The exact solution $(\psi, \boldsymbol{A})$ of (5.1)–(5.2) is defined by

$$\begin{aligned}\psi &= \exp(t/2)(\cos(\pi x) + i\cos(\pi y)),\\ \boldsymbol{A} &= (1+t^2)(\exp(y)\sin(\pi x), \exp(x)\sin(\pi y))^T.\end{aligned}$$

The external magnetic field $H$ is given by

$$H = \mathbf{curl}\boldsymbol{A} = (1+t^2)(\sin(\pi y) - \sin(\pi x))$$

and the functions $\boldsymbol{f}$, $g$ in (6.1)–(6.2) are set down through the equations (6.1)–(6.2).

We use the standard bilinear finite elements with uniform mesh to construct the finite element spaces $\mathcal{S}_h$ and $\boldsymbol{V}_h$. Let $\{P_j\}_{j=1}^N$ be the nodes of the triangulation $\Delta_h$, we define the discrete $L^2$-norm by

$$\|u\|_h = \Big[\sum_{j=1}^{N} |u(P_j)|^2 \cdot h^2\Big]^{1/2} \quad \forall u \in C(\overline{\Omega}).$$

It is well known that the norms $\|\cdot\|_h$ and $\|\cdot\|_{L^2(\Omega)}$ are equivalent on spaces $\mathcal{S}_h$ and $\boldsymbol{V}_h$.

Let $E_1^h = \max_{1\le n\le M}\|\psi^n - \psi_h^n\|_h$, then from the relation $E_1^h = Ch^{p_1}$ it follows that $p_1 = \log[E_1^{h_1}/E_1^{h_2}]/\log[h_1/h_2]$. In the same way we have $E_2^h = \max_{1\le n\le M}\|\boldsymbol{A}^n - \boldsymbol{A}_h^n\|_h$ and the corresponding convergence order $p_2$.

The following tables demonstrate the results computed by taking $\Delta t = h$ and $\Delta t = h^2$, respectively. In the computations, we used Newton's method defined in Section 3 to solve the non-linear equations (3.4).

Table 6.1 shows that the numerical results here are in good agreement with the theoretical analysis in section 5. Table 6.2 indicates that the method proposed in this paper may also preserve optimal convergence order in $L^2$ norm.

In our computations the software system used was "Finite Element Program Automatic Generator" which was created by G.-P. Liang. The computations were performed on a SGI workstation.

| $h$ | $E_1^h$ | $p_1$ | $E_2^h$ | $p_2$ |
|---|---|---|---|---|
| 1/5 | $2.20E-1$ | | $2.88E-1$ | |
| 1/10 | $7.23E-2$ | 1.61 | $1.21E-1$ | 1.25 |
| 1/20 | $2.79E-2$ | 1.37 | $5.72E-2$ | 1.08 |
| 1/30 | $1.74E-2$ | 1.16 | $3.78E-2$ | 1.02 |
| 1/40 | $1.27E-2$ | 1.09 | $2.82E-2$ | 1.02 |

Table 6.1. Numerical results computed by taking $\Delta t = h$

| $h$ | $E_1^h$ | $p_1$ | $E_2^h$ | $p_2$ |
|---|---|---|---|---|
| 1/5 | $1.91E-1$ | | $1.12E-1$ | |
| 1/10 | $5.07E-2$ | 1.91 | $2.47E-2$ | 2.18 |
| 1/20 | $1.28E-2$ | 1.99 | $5.79E-3$ | 2.09 |
| 1/30 | $5.67E-3$ | 2.01 | $2.52E-3$ | 2.05 |
| 1/40 | $3.23E-3$ | 1.96 | $1.41E-3$ | 2.02 |

Table 6.2. Numerical results computed by taking $\Delta t = h^2$

## 10.6 Discretization using variable step length

The weak formulation of the problem remains the same as in Section 2.1. The discretization we introduce here is using non-uniform mesh sizes.

We now introduce the discretized problem, which uses continuous piecewise linear finite elements in space with semi-implicit finite differences in time. We denote by $\tau_n$ the $n$-th time step and set

$$t^n := \sum_{i=1}^{n} \tau_i, \quad u^n(\cdot) := u(\cdot, t^n)$$

for any function $u$ continuous in $(t^{n-1}, t^n]$. Let $N$ be the total number of time steps, that is $t^N \geq T$.

We denote by $\mathcal{M}^n$ a uniformly regular partition of $\Omega$ into simplexes (see [Cia 78]). The mesh $\mathcal{M}^n$ is obtained by refinement/coarsening of $\mathcal{M}^{n-1}$, and thus $\mathcal{M}^n$ and $\mathcal{M}^{n-1}$ are *compatible*. Given a triangle $S \in \mathcal{M}^n$, $h_S$ stands for its diameter and $\rho_S$ for its sphericity and they satisfy $h_S \leq 2\rho_S/\sin(\alpha_S/2)$, where $\alpha_S$ is the minimum angle of $S$; $h_n$ denotes the mesh density function $h_n|_S = h_S$ for all $S \in \mathcal{M}^n$. Uniform regularity of the family of triangulations is equivalent to $\alpha_S \geq \alpha^* > 0$, with $\alpha^*$ independent of $n$. We also denote by $\mathcal{B}^n$ the collection of interior boundaries or sides $e$ of $\mathcal{M}^n$ in $\Omega$ and $\bar{\mathcal{B}}^n$ the collection of all sides of $\mathcal{M}^n$; $h_e$ stands for the size of $e \in \bar{\mathcal{B}}^n$.

Let $V^n$ indicate the usual space of $C^0$ piecewise linear finite elements over $\mathcal{M}^n$ and $\mathbf{V}_0^n = \mathbf{V}^n \cap \mathbf{H}_n^1(\Omega)$. Let $I^n : C(\bar{\Omega}) \to V^n$ be the usual Lagrange interpolation operator, then for any $S \in \mathcal{M}^n$ or $e \in \mathcal{B}^n$, the following local approximation properties hold [Cia 78]

$$\| u - I^n u \|_{L^2(S)} \le C^* h_S^2 \| u \|_{H^2(S)}, \quad \| u - I^n u \|_{L^2(e)} \le C^* h_e^{3/2} \| u \|_{H^2(\tilde{S})}, \tag{6.1}$$

where $\tilde{S}$ is any element in $\mathcal{M}^n$ with $e \in \mathcal{B}^n$ as part of its boundary. The constant $C^*$ depends only on the minimum angle of the mesh $\mathcal{M}^n$.

Let $\varphi^0 = I^0\psi_0$ and $\mathbf{D}^0 = I^0 \boldsymbol{A}_0$. Denote by $\mathcal{P}_n : \mathcal{L}^2(\Omega) \to \mathcal{V}^n$ and $\mathbf{P}_n : \mathbf{L}^2(\Omega) \to \mathbf{V}_0^n$ the $L^2$-projection operators. The discrete problem approximating (2.1)–(2.3) is defined as follows.

**Discrete Problem.** *Given* $(\varphi^{n-1}, \mathbf{D}^{n-1}) \in \mathcal{V}^{n-1} \times \mathbf{V}_0^{n-1}$, *then* $\mathcal{M}^{n-1}$ *and* $\tau_{n-1}$ *are modified as described below to get* $\mathcal{M}^n$ *and* $\tau_n$ *and thereafter* $(\varphi^n, \mathbf{D}^n) \in \mathcal{V}^n \times \mathbf{V}_0^n$ *computed according to the following prescription*

$$\eta\Big\langle \frac{\varphi^n - \mathcal{P}_n\varphi^{n-1}}{\tau_n}, \tilde{\omega}\Big\rangle + \frac{1}{\kappa^2}\langle \nabla\varphi^n, \nabla\tilde{\omega}\rangle + \langle f(\varphi^n, \mathbf{D}^n), \tilde{\omega}\rangle = 0 \quad \forall \tilde{\omega} \in \mathcal{V}^n, \tag{6.2}$$

$$\begin{aligned} &\Big\langle \frac{\mathbf{D}^n - \mathbf{P}_n\mathbf{D}^{n-1}}{\tau_n}, \tilde{\mathbf{B}}\Big\rangle + \langle \div\mathbf{D}^n, \div\tilde{\mathbf{B}}\rangle + \langle \mathbf{curl}\mathbf{D}^n, \mathbf{curl}\tilde{\mathbf{B}}\rangle \\ &+\langle \mathbf{P}_n[\mathbf{g}(\varphi^{n-1}, \mathbf{D}^{n-1})], \tilde{\mathbf{B}}\rangle = \langle\!\langle I^n H^n, \tilde{\mathbf{B}}\cdot\boldsymbol{\tau}\rangle\!\rangle \quad \forall \tilde{\mathbf{B}} \in \mathbf{V}_0^n. \end{aligned} \tag{6.3}$$

Hereafter, $\langle\cdot,\cdot\rangle$ and $\langle\!\langle\cdot,\cdot\rangle\!\rangle$ stand for the $L^2$-scalar products in $\mathcal{L}^2(\Omega)$ and $\mathbf{L}^2(\partial\Omega)$, respectively.

We remark that at each time step $n$, (6.3) is a linear system of equations with positive definite coefficient matrix, which can be solved by standard methods. As soon as we know $\mathbf{D}^n$ from (6.3), we substitute it into (6.2) and solve the non-linear system of equations to obtain $\varphi^n$. Any standard iterative method for solving non-linear system of equations such as simple iterative method or Newton iterative method may be used to solve (6.2) (see for example, [CHo 95]).

We define the interior residuals as follows:

$$R_\psi^n := \eta\frac{\varphi^n - \mathcal{P}_n\varphi^{n-1}}{\tau_n} + f(\varphi^n, \mathbf{D}^n), \quad R_{\mathbf{A}}^n := \frac{\mathbf{D}^n - \mathbf{P}_n\mathbf{D}^{n-1}}{\tau_n} + \mathbf{P}_n[\mathbf{g}(\varphi^{n-1}, \mathbf{D}^{n-1})].$$

Let the jump of $\nabla\varphi^n$ across $e \in \mathcal{B}^n$ be

$$[\nabla\varphi^n]_e := (\nabla\varphi^n_{|S_1} - \nabla\varphi^n_{|S_2})\cdot\mathbf{n}_e. \tag{6.4}$$

Note that with the convention that the unit normal vector $\mathbf{n}_e$ to $e$ points from $S_2$ to $S_1$, the jump $[\nabla\varphi^n]_e$ is well defined. Similarly, we define the jumps

$$[\div\mathbf{D}^n]_e := \div\mathbf{D}^n_{|S_1} - \div\mathbf{D}^n_{|S_2}, \tag{6.5}$$

$$[\mathbf{curl}\mathbf{D}^n]_e := \mathbf{curl}\mathbf{D}^n_{|S_1} - \mathbf{curl}\mathbf{D}^n_{|S_2}. \tag{6.6}$$

Let $\varphi$ and $\hat{\varphi}$ denote the piecewise linear and piecewise constant extensions of $\{\varphi^n\}$, that is $\varphi(\cdot,0) = \hat{\varphi}(\cdot,0) = \varphi^0(\cdot)$ and, for all $t^{n-1} < t \le t^n$,

$$\hat{\varphi}(\cdot,t) := \varphi^n(\cdot) \in \mathcal{V}^n, \quad \varphi(\cdot,t) := \frac{t^n - t}{\tau_n}\varphi^{n-1}(\cdot) + \frac{t - t^{n-1}}{\tau_n}\varphi^n(\cdot).$$

Similarly, we can define $\widehat{\mathbf{D}} \in \mathbf{V}_0^n$ and $\mathbf{D}$. Finally, for any $\gamma > 0, k \geq 0$ and $\mathcal{D} \subset \bar{\Omega}$ we introduce the mesh dependent norms

$$|||\, h_n^\gamma \phi \,|||_{H^k(\mathcal{D})} := \Big( \sum_{e\subset\mathcal{D}, e\in\bar{\mathcal{B}}^n} h_e^{2\gamma} \| \phi \|_{H^k(e)}^2 \Big)^{1/2},$$

$$\| h_n^\gamma \phi \|_{H^k(\mathcal{D})} := \Big( \sum_{S\subset\mathcal{D}, S\in\mathcal{M}^n} h_S^{2\gamma} \| \phi \|_{H^k(S)}^2 \Big)^{1/2}.$$

## 10.7 A dual problem

In this section we introduce and study a dual problem which is the dual of the linearization of (2.1)–(2.3) at $(\psi, \boldsymbol{A})$. We first formulate the linearization of (2.1)–(2.3) at $(\psi, \boldsymbol{A})$ as follows: given $(\psi, \boldsymbol{A})$ be the solution of (2.1)–(2.3), find $(\psi^*, \boldsymbol{A}^*) \in \mathcal{W}(0,T) \times \mathbf{W}_n(0,T)$ such that

$$\psi^*(\cdot, 0) = \psi_0^*(\cdot), \quad \boldsymbol{A}^*(\cdot, 0) = \boldsymbol{A}_0^*(\cdot) \tag{7.1}$$

and

$$\begin{aligned} & \eta \int_0^T \int_\Omega \partial_t \psi^* \omega dx dt - i\eta\kappa \int_0^T \int_\Omega (\div \boldsymbol{A}^* \psi + \div \boldsymbol{A} \psi^*) \omega dx dt \\ + & \int_0^T \int_\Omega \Big( \frac{i}{\kappa} \nabla \psi^* + \boldsymbol{A}^* \psi + \boldsymbol{A} \psi^* \Big) \Big( -\frac{i}{\kappa} \nabla \omega + \boldsymbol{A} \omega \Big) dx dt \\ + & \int_0^T \int_\Omega \Big( \frac{i}{\kappa} \nabla \psi + \boldsymbol{A} \psi \Big) \boldsymbol{A}^* \omega dx dt \\ + & \int_0^T \int_\Omega \Big[ (2|\psi|^2 - 1)\psi^* + \psi \bar{\psi}^* \psi \Big] \omega dx dt = 0, \quad \forall \omega \in L^2(0,T; \mathcal{H}^1(\Omega)), \end{aligned} \tag{7.2}$$

$$\begin{aligned} & \int_0^T \int_\Omega \partial_t \boldsymbol{A} \mathbf{B} dx dt + \int_0^T \int_\Omega (\div \boldsymbol{A}^* \div \mathbf{B} + \mathbf{curl} \boldsymbol{A}^* \mathbf{curl} \mathbf{B}) dx dt \\ + & \int_0^T \int_\Omega \mathrm{Re} \Big[ \Big( \frac{i}{\kappa} \nabla \psi^* + \boldsymbol{A}^* \psi + \boldsymbol{A} \psi^* \Big) \bar{\psi} + \Big( \frac{i}{\kappa} \nabla \psi + \boldsymbol{A} \psi \Big) \bar{\psi}^* \Big] \mathbf{B} dt dt \\ = & \ 0, \qquad \forall \mathbf{B} \in L^2(0,T; \mathbf{H}_n^1(\Omega)). \end{aligned} \tag{7.3}$$

Now we are in the position to define the following dual parabolic problem which is dual to (7.1)–(7.3).

**Dual Problem.** Given $(\theta^*, \mathbf{W}^*) \in \mathcal{L}^2(\omega) \times \mathbf{L}^2(\Omega)$ and an arbitrary $t^* \in (0, T]$, find $(\theta, \mathbf{W}) \in \mathcal{W}(0,T) \times \mathbf{W}_n(0,T)$ such that

$$\theta(\cdot, t^*) = \theta^*(\cdot), \quad \mathbf{W}(\cdot, t^*) = \mathbf{W}^*(\cdot) \tag{7.4}$$

and

$$
\begin{aligned}
&-\eta\int_0^{t^*}\int_\Omega \partial_t\theta\omega dxdt - i\eta\kappa\int_0^{t^*}\int_\Omega \div \boldsymbol{A}\theta\omega dxdt \\
&+ \int_0^{t^*}\int_\Omega \Big(-\frac{i}{\kappa}\nabla\theta+\boldsymbol{A}\theta\Big)\Big(\frac{i}{\kappa}\nabla\omega+\boldsymbol{A}\omega\Big)dxdt \\
&+ \int_0^{t^*}\int_\Omega \Big[\Big(-\frac{i}{\kappa}\nabla\bar\psi+\boldsymbol{A}\bar\psi\Big)\mathbf{W}\omega+\bar\psi\mathbf{W}\Big(\frac{i}{\kappa}\nabla\omega+\boldsymbol{A}\omega\Big)\Big]dxdt \\
&+ \int_0^{t^*}\int_\Omega \Big[(2|\psi|^2-1)\theta+\bar\psi\bar\theta\bar\psi\Big]\omega dxdt = 0, \quad \forall\omega\in L^2(0,t^*;\mathcal{H}^1(\Omega)), \qquad (7.5) \\
&-\int_0^{t^*}\int_\Omega \partial_t\mathbf{W}\mathbf{B}dxdt + \int_0^{t^*}\int_\Omega (\div\mathbf{W}\div\mathbf{B}+\mathbf{curl}\mathbf{W}\mathbf{curl}\mathbf{B})dxdt \\
&+ \int_0^{t^*}\int_\Omega \mathrm{Re}\Big[i\eta\kappa\nabla(\psi\theta)\Big]\mathbf{B}dxdt + \int_0^{t^*}\int_\Omega |\psi|^2\mathbf{W}\mathbf{B}dxdt \\
&+ \int_0^{t^*}\int_\Omega \mathrm{Re}\Big[\Big(\frac{i}{\kappa}\nabla\psi+\boldsymbol{A}\psi\Big)\theta \\
&+ \Big(-\frac{i}{\kappa}\nabla\theta+\boldsymbol{A}\theta\Big)\psi\Big]\mathbf{B}dxdt = 0, \quad \forall\mathbf{B}\in L^2(0,t^*;\mathbf{H}_n^1(\Omega)). \qquad (7.6)
\end{aligned}
$$

The linearization problem (7.1)–(7.3) and its dual problem (7.4)–(7.6) for the G-L system have been first introduced in [CHo 96] in the context of exploring the possibility of controlling the motion of vortices in the superconductors through the external magnetic field. The purpose in this section is two-fold: we first derive strong stability estimates for (7.4)–(7.6) which will be used in next section in the a posteriori error analysis; and second we derive an explicit error representation formula for the error $\|\psi-\varphi\|_{L^2(\Omega)}$ and $\|\boldsymbol{A}-\mathbf{D}\|_{L^2(\Omega)}$ based on the backward parabolic problem (7.4)–(7.6). Throughout we denote by $C$ the generic constant which may depend on $\eta,\kappa,\Omega,T$, and the norms of $\psi_0,\boldsymbol{A}_0,H$ indicated in the hypotheses (H1)–(H2).

### 10.7.1 Stability estimates

We extend the method in [LRa 81] for linear heat equation to derive strong stability estimates for $(\theta,\mathbf{W})$ under weak regularity assumption $(\theta^*,\mathbf{W}^*)\in\mathcal{L}^2(\Omega)\times\mathbf{L}^2(\Omega)$. First we note that by Lemma 2.1 we have

$$
\|\psi\|_{W_2^{2,1}(Q_T)}+\|\boldsymbol{A}\|_{W_2^{2,1}(Q_T)}\le C, \quad \|\nabla\psi\|_{L^4(Q_T)}+\|\nabla\boldsymbol{A}\|_{L^4(Q_T)}\le C. \qquad (7.7)
$$

The latter estimate in (7.7) follows from the former one and the following Nirenberg-Gagliardo inequality

$$
\|u\|_{L^4(\Omega)}\le C\|u\|_{L^2(\Omega)}^{1/2}\|u\|_{H^1(\Omega)}^{1/2} \quad \forall u\in H^1(\Omega). \qquad (7.8)
$$

**Lemma 10.7.1** *The following stability estimates are valid for all* $0 \le t^* \le T$

$$\max_{0\le t\le t^*}(\| \theta \|^2_{L^2(\Omega)} + \| \mathbf{W} \|^2_{L^2(\Omega)}) + \int_0^{t^*} (\| \nabla\theta \|^2_{L^2(\Omega)} + \| \div \mathbf{W} \|^2_{L^2(\Omega)}$$
$$+\| \mathbf{curl W} \|^2_{L^2(\Omega)})dt \le C(\| \theta^* \|^2_{L^2(\Omega)} + \| \mathbf{W}^* \|^2_{L^2(\Omega)}).$$

*Proof.* Denote by $\chi_{(t,t^*]}$ the characteristic function of the interval $(t, t^*]$. We let $\omega = \bar{\theta}\chi_{(t,t^*]}$ in (7.5) and take the real part of the obtained equation, to get

$$\frac{\eta}{2}\| \theta \|^2_{L^2(\Omega)} - \frac{\eta}{2}\| \theta^* \|^2_{L^2(\Omega)} + \int_t^{t^*} \Big\| -\frac{i}{\kappa}\nabla\theta + \boldsymbol{A}\theta \Big\|^2_{L^2(\Omega)} dt$$
$$= -\int_t^{t^*}\int_\Omega \mathrm{Re}\Big[\Big(-\frac{i}{\kappa}\nabla\bar{\psi} + \boldsymbol{A}\bar{\psi}\Big)\mathbf{W}\bar{\theta} + \bar{\psi}\mathbf{W}\Big(\frac{i}{\kappa}\nabla\bar{\theta} + \boldsymbol{A}\bar{\theta}\Big)\Big]dxdt$$
$$-\int_t^{t^*}\int_\Omega \mathrm{Re}\Big[(2|\psi|^2 - 1)|\theta|^2 + \bar{\psi}^2\bar{\theta}^2\Big]dxdt. \tag{7.9}$$

By using (1.6) and (7.8) we can bound the first term on the right-hand side as follows

$$-\int_t^{t^*}\int_\Omega \mathrm{Re}\Big[\Big(-\frac{i}{\kappa}\nabla\bar{\psi} + \boldsymbol{A}\bar{\psi}\Big)\mathbf{W}\bar{\theta}\Big]dxdt$$
$$\le C\int_t^{t^*} \Big\| \frac{i}{\kappa}\nabla\psi + \boldsymbol{A}\psi \Big\|_{L^4(\Omega)} \| \mathbf{W} \|_{L^2(\Omega)} \| \theta \|_{L^2(\Omega)} dt$$
$$+C\int_t^{t^*} \Big\| \frac{i}{\kappa}\nabla\psi + \boldsymbol{A}\psi \Big\|_{L^4(\Omega)} \| \mathbf{W} \|^{1/2}_{L^2(\Omega)} (\| \div \mathbf{W} \|^{1/2}_{L^2(\Omega)}$$
$$+\| \mathbf{curl W} \|^{1/2}_{L^2(\Omega)}) \| \theta \|_{L^2(\Omega)} dt$$
$$\le \delta\int_t^{t^*} (\| \div \mathbf{W} \|^2_{L^2(\Omega)} + \| \mathbf{curl W} \|^2_{L^2(\Omega)})dt + \frac{C}{\delta}\int_t^{t^*} \| \mathbf{W} \|^2_{L^2(\Omega)} dt$$
$$+C\int_t^{t^*} \Big\| \frac{i}{\kappa}\nabla\psi + \boldsymbol{A}\psi \Big\|^2_{L^4(\Omega)} \| \theta \|^2_{L^2(\Omega)} dt, \quad \forall\delta > 0.$$

By using the fact that $|\psi| \le 1$ a.e. in $Q_T$ we get easily

$$-\int_t^{t^*}\int_\Omega \mathrm{Re}\Big[\bar{\psi}\mathbf{W}\Big(\frac{i}{\kappa}\nabla\bar{\theta} + \boldsymbol{A}\bar{\theta}\Big)\Big]dxdt$$
$$\le \frac{1}{2}\int_t^{t^*} \Big\| -\frac{i}{\kappa}\nabla\theta + \boldsymbol{A}\theta \Big\|^2_{L^2(\Omega)} dt + C\int_0^{t^*} \| \mathbf{W} \|^2_{L^2(\Omega)},$$

and

$$-\int_t^{t^*}\int_\Omega \mathrm{Re}\Big[(2|\psi|^2 - 1)|\theta|^2 + \bar{\psi}^2\bar{\theta}^2\Big]dxdt \le C\int_t^{t^*} \| \theta \|^2_{L^2(\Omega)} dt.$$

Inserting these estimates into (7.9) we get

$$\begin{aligned}
&\frac{\eta}{2}\| \theta \|^2_{L^2(\Omega)} - \frac{\eta}{2}\| \theta^* \|^2_{L^2(\Omega)} + \frac{1}{2}\int_t^{t^*} \left\| -\frac{i}{\kappa}\nabla\theta + \boldsymbol{A}\theta \right\|^2_{L^2(\Omega)} dt \\
\leq\ & \delta \int_t^{t^*} (\| \div \mathbf{W} \|^2_{L^2(\Omega)} + \| \mathbf{curlW} \|^2_{L^2(\Omega)})dt + \frac{C}{\delta}\int_t^{t^*} \| \mathbf{W} \|^2_{L^2(\Omega)} dt \\
+\ & C\int_t^{t^*} \left(1 + \left\| \frac{i}{\kappa}\nabla\psi + \boldsymbol{A}\psi \right\|^2_{L^4(\Omega)}\right) \| \theta \|^2_{L^2(\Omega)} dt. \qquad (7.10)
\end{aligned}$$

Similarly, by letting $\mathbf{B} = \mathbf{W}\chi_{(t,t^*]}$ in (7.6) we can obtain that

$$\begin{aligned}
&\frac{1}{2}\| \mathbf{W} \|^2_{L^2(\Omega)} - \frac{1}{2}\| \mathbf{W}^* \|^2_{L^2(\Omega)} + \frac{1}{2}\int_t^{t^*} (\| \div \mathbf{W} \|^2_{L^2(\Omega)} + \| \mathbf{curlW} \|^2_{L^2(\Omega)})dt \\
\leq\ & \delta \int_t^{t^*} \left\| -\frac{i}{\kappa}\nabla\theta + \boldsymbol{A}\theta \right\|^2_{L^2(\Omega)} dt + \frac{C}{\delta}\int_t^{t^*} \| \mathbf{W} \|^2_{L^2(\Omega)} dt \\
+\ & C\int_t^{t^*} \left(1 + \left\| \frac{i}{\kappa}\nabla\psi + \boldsymbol{A}\psi \right\|^2_{L^4(\Omega)}\right) \| \theta \|^2_{L^2(\Omega)} dt. \qquad (7.11)
\end{aligned}$$

Now observe that by (7.7)

$$\int_0^T \left\| \frac{i}{\kappa}\nabla\psi + \boldsymbol{A}\psi \right\|^2_{L^4(\Omega)} dt \leq C,$$

we conclude after adding (7.10) and (7.11), taking $\delta$ appropriately small and using Gronwall inequality that

$$\begin{aligned}
&\max_{0\leq t\leq t^*} \| \theta \|^2_{L^2(\Omega)} + \int_0^{t^*} \left\| -\frac{i}{\kappa}\nabla\theta + \boldsymbol{A}\theta \right\|^2_{L^2(\Omega)} dt \\
+\ & \max_{0\leq t\leq t^*} \| \mathbf{W} \|^2_{L^2(\Omega)} + \int_0^{t^*} (\| \div \mathbf{W} \|^2_{L^2(\Omega)} + \| \mathbf{curlW} \|^2_{L^2(\Omega)})dt \\
\leq\ & C(\| \theta^* \|^2_{L^2(\Omega)} + \| \mathbf{W}^* \|^2_{L^2(\Omega)}).
\end{aligned}$$

Now the lemma follows by observing that

$$\begin{aligned}
\int_0^{t^*} \| \nabla\theta \|^2_{L^2(\Omega)} dt &\leq C\int_0^{t^*} \left\| -\frac{i}{\kappa}\nabla\theta + \boldsymbol{A}\theta \right\|^2_{L^2(\Omega)} dt \\
&\quad + C\int_0^{t^*} \| \boldsymbol{A} \|^2_{L^\infty(\Omega)} \| \theta \|^2_{L^2(\Omega)} dt \\
&\leq C\int_0^{t^*} \left\| -\frac{i}{\kappa}\nabla\theta + \boldsymbol{A}\theta \right\|^2_{L^2(\Omega)} dt + C\max_{0\leq t\leq t^*} \| \theta \|^2_{L^2(\Omega)} dt.
\end{aligned}$$

This completes the proof.

**Corollary 10.7.1** *The following stability estimates are valid for all* $0 < t^* \leq T$

$$\int_0^{t^*} (\| \theta \|_{L^4(\Omega)}^4 + \| \mathbf{W} \|_{L^4(\Omega)}^4) dt \leq C(\| \theta^* \|_{L^2(\Omega)}^2 + \| \mathbf{W}^* \|_{L^2(\Omega)}^2)^2.$$

*Proof.* This is a direct consequence of Lemma 4.1 and (7.8).

To proceed, we set

$$\begin{aligned} f_1(\psi, \boldsymbol{A}; \theta, \mathbf{W}) &= -i\eta\kappa \div \boldsymbol{A}\theta - \frac{i}{\kappa} \div (\boldsymbol{A}\theta) \\ &\quad + \boldsymbol{A}\Big( -\frac{i}{\kappa}\nabla\theta + \boldsymbol{A}\theta \Big) + \Big( -\frac{i}{\kappa}\nabla\bar{\psi} + \boldsymbol{A}\bar{\psi} \Big)\mathbf{W} \\ &\quad + \Big( -\frac{i}{\kappa}\nabla + \boldsymbol{A} \Big)(\bar{\psi}\mathbf{W}) + (2|\psi|^2 - 1)\theta + \bar{\psi}^2\bar{\theta} \end{aligned}$$

and

$$\begin{aligned} \mathbf{g}_1(\psi, \boldsymbol{A}; \theta, \mathbf{W}) &= \operatorname{Re}\Big[ i\eta\kappa\nabla(\psi\theta) + \Big(\frac{i}{\kappa}\nabla\psi + \boldsymbol{A}\psi\Big)\theta \\ &\quad + \Big( -\frac{i}{\kappa}\nabla\theta + \boldsymbol{A}\theta \Big)\psi \Big] + |\psi|^2\mathbf{W}. \end{aligned}$$

Then it is easy to see that (7.4)–(7.6) is the weak formulation of the following linear parabolic equations

$$-\eta\partial_t\theta - \frac{1}{\kappa^2}\Delta\theta + f_1(\psi, \boldsymbol{A}; \theta, \mathbf{W}) = 0 \quad \text{in } Q^*, \tag{7.12}$$

$$-\partial_t\mathbf{W} - \Delta\mathbf{W} + \mathbf{g}_1(\psi, \boldsymbol{A}; \theta, \mathbf{W}) = 0 \quad \text{in } Q^*, \tag{7.13}$$

$$\nabla\theta \cdot \mathbf{n} = 0, \quad \mathbf{W} \cdot \mathbf{n} = 0, \quad \mathbf{curl W} = 0 \quad \text{on } \Gamma^*, \tag{7.14}$$

$$\theta(\cdot, t^*) = \theta^*(\cdot), \quad \mathbf{W}(\cdot, t^*) = \mathbf{W}^*(\cdot) \quad \text{on } \Omega, \tag{7.15}$$

where $Q^* = \Omega \times (0, t^*)$ and $\Gamma^* = \partial\Omega \times (0, t^*)$.

**Lemma 10.7.2** *The following stability estimates are valid for all* $0 < t^* \leq T$

$$\begin{aligned} &\max_{0 \leq t \leq t^*} (t^* - t)\Big( \| \nabla\theta \|_{L^2(\Omega)}^2 + \| \div \mathbf{W} \|_{L^2(\Omega)}^2 + \| \mathbf{curl W} \|_{L^2(\Omega)}^2 \Big) \\ &+ \int_0^{t^*} (t^* - t)\Big( \| \partial_t\theta \|_{L^2(\Omega)}^2 + \| \partial_t\mathbf{W} \|_{L^2(\Omega)}^2 \Big) dt \\ &\leq C(\| \theta^* \|_{L^2(\Omega)}^2 + \| \mathbf{W}^* \|_{L^2(\Omega)}^2). \end{aligned}$$

*Proof.* First we know from Lemma 4.1 that there exists a sequence $t_j^* \nearrow t^*$ such that

$$(t^* - t_j^*)\Big( \| \nabla\theta(t_j^*) \|_{L^2(\Omega)}^2 + \| \div \mathbf{W}(t_j^*) \|_{L^2(\Omega)}^2 + \| \mathbf{curl W}(t_j^*) \|_{L^2(\Omega)}^2 \Big) \to 0. \tag{7.16}$$

We multiply (7.12) by $-(t^* - t)\partial_t\bar{\theta}$, integrate over $\Omega \times (s, t_j^*]$ and then take the real part of the equation to obtain that

$$\eta \int_s^{t_j^*} (t^* - t) \| \partial_t\theta \|_{L^2(\Omega)}^2 dt = \frac{1}{2\kappa^2} \int_s^{t_j^*} (t^* - t) \frac{d}{dt} \| \nabla\theta \|_{L^2(\Omega)}^2$$
$$+ \int_s^{t_j^*} \int_\Omega (t^* - t) \mathrm{Re}\Big[ f_1(\psi, \boldsymbol{A}; \theta, \mathbf{W}) \partial_t\bar{\theta} \Big] dt. \tag{7.17}$$

Integrating by parts and using Lemma 4.1 we have

$$\begin{aligned}
& \int_s^{t_j^*} (t^* - t) \frac{d}{dt} \| \nabla\theta \|_{L^2(\Omega)}^2 dt \\
= \;& \int_s^{t_j^*} \| \nabla\theta \|_{L^2(\Omega)}^2 dt + (t^* - t_j^*) \| \nabla\theta(t_j^*) \|_{L^2(\Omega)}^2 - (t^* - s) \| \nabla\theta(s) \|_{L^2(\Omega)}^2 \\
\leq \;& (t^* - t_j^*) \| \nabla\theta(t_j^*) \|_{L^2(\Omega)}^2 - (t^* - s) \| \nabla\theta(s) \|_{L^2(\Omega)}^2 \\
+ \;& C(\| \theta^* \|_{L^2(\Omega)}^2 + \| \mathbf{W}^* \|_{L^2(\Omega)}^2).
\end{aligned} \tag{7.18}$$

Next by Lemma 4.1 we get easily that

$$\begin{aligned}
& \int_s^{t_j^*} \int_\Omega (t^* - t) \mathrm{Re}\big[ f_1(\psi, \boldsymbol{A}; \theta, \mathbf{W}) \partial_t\bar{\theta} \big] dt \\
\leq \;& \frac{\eta}{2} \int_s^{t_j^*} (t^* - t) \| \partial_t\theta \|_{L^2(\Omega)}^2 dt + C \int_0^{t^*} (t^* - t) \| f_1(\psi, \boldsymbol{A}; \theta, \mathbf{W}) \|_{L^2(\Omega)}^2 dt \\
\leq \;& \frac{\eta}{2} \int_s^{t_j^*} (t^* - t) \| \partial_t\theta \|_{L^2(\Omega)}^2 dt + C(\| \theta^* \|_{L^2(\Omega)}^2 + \| \mathbf{W}^* \|_{L^2(\Omega)}^2).
\end{aligned} \tag{7.19}$$

For example, we can estimate the first term in $f_1(\psi, \boldsymbol{A}; \theta, \mathbf{W})$ by using (7.7) and Corollary 4.1 as follows

$$\begin{aligned}
\int_0^{t^*} \| i\eta\kappa \div \boldsymbol{A}\theta \|_{L^2(\Omega)}^2 dt \;&\leq\; C\Big( \int_0^T \| \div \boldsymbol{A} \|_{L^4(\Omega)}^4 dt \Big)^{1/2} \Big( \int_0^{t^*} \| \theta \|_{L^4(\Omega)}^4 dt \Big)^{1/2} \\
&\leq\; C(\| \theta^* \|_{L^2(\Omega)}^2 + \| \mathbf{W}^* \|_{L^2(\Omega)}^2).
\end{aligned}$$

Substituting (7.18)–(7.19) into (7.17), letting $t_j^* \nearrow t^*$ and using (7.16), we get

$$\frac{\eta}{2} \int_s^{t^*} \| \partial_t\theta \|_{L^2(\Omega)}^2 dt + (t^* - s) \| \nabla\theta(s) \|_{L^2(\Omega)}^2 \leq C(\| \theta^* \|_{L^2(\Omega)}^2 + \| \mathbf{W}^* \|_{L^2(\Omega)}^2).$$

Similarly, we obtain from (7.13) that

$$\begin{aligned}
& \frac{1}{2} \int_s^{t^*} (t^* - t) \| \partial_t\mathbf{W} \|_{L^2(\Omega)}^2 + (t^* - s)(\| \div \mathbf{W} \|_{L^2(\Omega)}^2 + \| \mathbf{curl} \mathbf{W} \|_{L^2(\Omega)}^2) \\
\leq \;& C(\| \theta^* \|_{L^2(\Omega)}^2 + \| \mathbf{W}^* \|_{L^2(\Omega)}^2).
\end{aligned}$$

This completes the proof of Lemma 4.2.

**Lemma 10.7.3** *The following stability estimates are valid for all* $0 < t^* \leq T$

$$\int_0^{t^*} (t^* - t)\Big(\| \theta \|^2_{H^2(\Omega)} + \| \mathbf{W} \|^2_{H^2(\Omega)}\Big) dt \leq C(\| \theta^* \|^2_{L^2(\Omega)} + \| \mathbf{W}^* \|^2_{L^2(\Omega)}).$$

*Proof.* This is a direct consequence of Lemma 4.2. For instance, since

$$\int_0^{t^*} \| \mathbf{g}_1(\psi, \boldsymbol{A}; \theta, \mathbf{W}) \|^2_{L^2(\Omega)} dt \leq C(\| \theta^* \|^2_{L^2(\Omega)} + \| \mathbf{W}^* \|^2_{L^2(\Omega)}),$$

we know from (7.13)–(7.14) and Lemma 4.2 that

$$\int_0^{t^*} (t^* - t)\| \Delta \mathbf{W} \|^2_{L^2(\Omega)} \leq C(\| \theta^* \|^2_{L^2(\Omega)} + \| \mathbf{W}^* \|^2_{L^2(\Omega)}).$$

This, by (H3), yields

$$\int_0^{t^*} (t^* - t)\| \mathbf{W} \|^2_{H^2(\Omega)} dt \leq C(\| \theta^* \|^2_{L^2(\Omega)} + \| \mathbf{W}^* \|^2_{L^2(\Omega)}).$$

This completes the proof.

### 10.7.2 Error representation formula

We now derive an explicit representation formula for the error $\| \psi - \varphi \|_{L^2(\Omega)}$ and $\| \boldsymbol{A} - \mathbf{D} \|_{L^2(\Omega)}$ based on the linear dual problem (7.4)–(7.6). We first multiply (7.12) by $(\psi - \hat{\varphi})$, (7.13) by $(\boldsymbol{A} - \widehat{\mathbf{D}})$, and integrate in space and in time from 0 to $t^*$. We examine the various contributions in turn. Since $\hat{\varphi}, \widehat{\mathbf{D}}$ are piecewise constant in time, we have

$$\begin{aligned} -\eta \int_0^{t^*} \langle \partial_t \theta, \psi - \hat{\varphi} \rangle dt &= \eta \int_0^{t^*} \Big( \langle \theta, \partial_t(\psi - \varphi) \rangle + \langle \partial_t \theta, \hat{\varphi} - \varphi \rangle \Big) dt \\ &+ \eta \langle \theta^0, \psi_0 - \varphi^0 \rangle - \eta \langle \theta^*, \psi(t^*) - \varphi(t^*) \rangle, \end{aligned}$$

and

$$\begin{aligned} -\int_0^{t^*} \langle \partial_t \mathbf{W}, \boldsymbol{A} - \widehat{\mathbf{D}} \rangle dt &= \int_0^{t^*} \Big( \langle \mathbf{W}, \partial_t(\boldsymbol{A} - \mathbf{D}) \rangle + \langle \partial_t \mathbf{W}, \widehat{\mathbf{D}} - \mathbf{D} \rangle \Big) dt \\ &+ \langle \mathbf{W}^0, \boldsymbol{A}_0 - \mathbf{D}^0 \rangle - \langle \mathbf{W}^*, \boldsymbol{A}(t^*) - \mathbf{D}(t^*) \rangle. \end{aligned}$$

Integrating by parts we get

$$-\int_0^{t^*} \frac{1}{\kappa^2} \langle \Delta \theta, \psi - \hat{\varphi} \rangle) dt = \int_0^{t^*} \frac{1}{\kappa^2} \langle \nabla \theta, \nabla (\psi - \hat{\varphi} \rangle dt,$$

$$\begin{aligned} &-\int_0^{t^*} \langle \Delta \mathbf{W}, \boldsymbol{A} - \widehat{\mathbf{D}} \rangle dt \\ &= \int_0^{t^*} \Big( \langle \div \mathbf{W}, \div(\boldsymbol{A} - \widehat{\mathbf{D}}) \rangle + \langle \mathbf{curl} \mathbf{W}, \mathbf{curl}(\boldsymbol{A} - \widehat{\mathbf{D}}) \rangle \Big) dt, \end{aligned}$$

Collecting these equalities and making use of (2.4)–(2.5), we easily end up with

$$\begin{aligned}
&\eta\| \psi(t^*) - \varphi(t^*) \|_{L^2(\Omega)} + \| \boldsymbol{A}(t^*) - \mathbf{D}(t^*) \|_{L^2(\Omega)} \\
\leq\ & 2 \sup_{(\theta^*,\mathbf{W}^*)\in\mathcal{L}^2(\Omega)\times\mathbf{L}^2(\Omega)} \frac{\mathrm{Re}\Big[\eta\langle\theta^*, \psi(t^*) - \varphi(t^*)\rangle + \langle\mathbf{W}^*, \boldsymbol{A}(t^*) - \mathbf{D}(t^*)\rangle\Big]}{\| \theta^* \|_{L^2(\Omega)} + \| \mathbf{W}^* \|_{L^2(\Omega)}} \\
=\ & 2 \sup_{(\theta^*,\mathbf{W}^*)\in\mathcal{L}^2(\Omega)\times\mathbf{L}^2(\Omega)} \frac{\mathrm{Re}[\mathcal{R}(\theta, \mathbf{W})]}{\| \theta^* \|_{L^2(\Omega)} + \| \mathbf{W}^* \|_{L^2(\Omega)}}, \qquad (7.20)
\end{aligned}$$

where $\mathcal{R}$, the parabolic residual, is the following distribution

$$\begin{aligned}
\mathcal{R}(\theta, \mathbf{W}) &= \eta\langle\theta^0, \psi_0 - \varphi^0\rangle + \langle\mathbf{W}^0, \boldsymbol{A}_0 - \mathbf{D}^0\rangle \\
&+ \int_0^{t^*} \Big(\eta\langle\partial_t\theta, \hat{\varphi} - \varphi\rangle + \langle\partial_t\mathbf{W}, \widehat{\mathbf{D}} - \mathbf{D}\rangle\Big)dt \\
&- \int_0^{t^*} \Big(\eta\langle\partial_t\varphi, \theta\rangle + \frac{1}{\kappa^2}\langle\nabla\hat{\varphi}, \nabla\theta\rangle + \langle f(\hat{\varphi}, \widehat{\mathbf{D}}), \theta\rangle\Big)dt \\
&- \int_0^{t^*} \Big(\langle\partial_t\mathbf{D}, \mathbf{W}\rangle + \langle\div\widehat{\mathbf{D}}, \div\mathbf{W}\rangle \\
&+ \langle\mathbf{curl}\widehat{\mathbf{D}}, \mathbf{curl}\mathbf{W}\rangle + \langle\mathbf{g}(\hat{\varphi}, \widehat{\mathbf{D}}), \mathbf{W}\rangle\Big)dt \\
&+ \int_0^{t^*} \langle\!\langle H, \mathbf{W}\cdot\boldsymbol{\tau}\rangle\!\rangle dt \\
&+ \int_0^{t^*} \Big(\langle f(\hat{\varphi}, \widehat{\mathbf{D}}) - f(\psi, \boldsymbol{A}), \theta\rangle \\
&+ \langle f_1(\psi, \boldsymbol{A}; \theta, \mathbf{W}), \psi - \hat{\varphi}\rangle\Big)dt \\
&+ \int_0^{t^*} \Big(\langle\mathbf{g}(\hat{\varphi}, \widehat{\mathbf{D}}) - \mathbf{g}(\psi, \boldsymbol{A}), \mathbf{W}\rangle \\
&+ \langle\mathbf{g}_1(\psi, \boldsymbol{A}; \theta, \mathbf{W}), \boldsymbol{A} - \widehat{\mathbf{D}}\rangle\Big)dt. \qquad (7.21)
\end{aligned}$$

The first five terms in (7.21) which depend solely on the discrete quantities and data will yield the desired a posteriori error estimates for the discrete solution $(\varphi, \mathbf{D})$ of (6.2)–(6.3) upon using the stability bounds in Section 3.1. The last two terms in $\mathcal{R}$ which depend both on the discrete and continuous solutions will be shown that they are indeed of higher order and thus can be controlled by the error in the left side of (7.20) via (H5) to be introduced in next section.

## 10.8 A posteriori error analysis

We first denote by the error functions

$$e_\psi := \psi - \varphi, \quad \hat{e}_\psi = \psi - \hat{\varphi}; \quad e_{\mathbf{A}} := \boldsymbol{A} - \mathbf{D}, \quad \hat{e}_{\mathbf{A}} := \boldsymbol{A} - \widehat{\mathbf{D}}.$$

Let $h_{\max} = \max_{1\le n\le N} \bar{h}_n$, where $\bar{h}_n = \max_{S\in\mathcal{M}^n} h_S$, and $\tau_{\max} = \max_{1\le n\le N} \tau_n$. We need the following assumptions on the discrete solutions.

(H4) There exists a constant $C$ independent of the mesh $\mathcal{M}^n$ and the time step $\tau_n$ such that

$$\max_{1\le n\le N} (\| \varphi^n \|_{L^2(\Omega)} + \| \mathbf{D}^n \|_{L^2(\Omega)}) \le C.$$

(H5) The quantity $\Lambda = \max_{0\le t^*\le T} \Lambda(t^*)$, where

$$\Lambda(t^*) = \frac{\left(\int_0^{t^*} (\| \hat{e}_\psi \|^2_{H^1(\Omega)} + \| \hat{e}_{\mathbf{A}} \|^2_{H^1(\Omega)})dt\right)^3}{\max_{0\le t\le t^*} (\| \hat{e}_\psi \|^2_{L^2(\Omega)} + \| \hat{e}_{\mathbf{A}} \|^2_{L^2(\Omega)})},$$

tends to zero as $(h_{\max}, \tau_{\max}) \to 0$.

These two assumptions deserve further remarks. The a priori bound (H4), which clearly achieved experimentally, can indeed be proved under suitable restriction on mesh changes. The assumption (H5) is in fact a non-degeneracy assumption. We remark that by the a priori error analysis in [CHo 95], the following optimal energy a priori error estimates hold under suitable restriction on the mesh changes:

$$\begin{aligned} \max_{0\le t\le t^m} (\| \hat{e}_\psi \|^2_{L^2(\Omega)} + \| \hat{e}_{\mathbf{A}} \|^2_{L^2(\Omega)}) \quad &+ \quad \int_0^{t^m} (\| \hat{e}_\psi \|^2_{H^1(\Omega)} + \| \hat{e}_{\mathbf{A}} \|^2_{H^1(\Omega)})dt \\ &\le \quad C \max_{1\le n\le m} (\bar{h}_n^2 + \tau_n^2). \end{aligned} \tag{8.1}$$

Now (H5) is a direct consequence of the following non-degeneracy assumption: There exists a constant $C_*$ independent of $1 \le m \le N$ such that, for any $t^* \in (t^{m-1}, t^m]$,

$$\max_{0\le t\le t^*} (\| \hat{e}_\psi \|_{L^2(\Omega)} + \| \hat{e}_{\mathbf{A}} \|_{L^2(\Omega)}) \ge C_* \max_{1\le n\le m} (\bar{h}_n^2 + \tau_n). \tag{8.2}$$

We note that $\max_{1\le n\le m}(\bar{h}_n^2 + \tau_n)$ is the expected optimal $L^\infty - L^2$ error estimate for $\hat{e}_\psi$ and $\hat{e}_{\mathbf{A}}$. Moreover, following from (8.1)–(8.2) we have now

$$\Lambda(t^*) \le C \max_{1\le n\le m} (\bar{h}_n^2 + \tau_n^2) \quad \forall t^* \in (t^{m-1}, t^m],$$

which clearly tends to zero as $(h_{\max}, \tau_{\max}) \to 0$.

**Theorem 10.8.1** *Let* (H1)–(H5) *be satisfied. Then there exist two positive constants* $h^*, \tau^*$ *and a constant* $C$ *depending only on* $h^*, \tau^*, \eta, \kappa, \Omega, T$, *the norms of* $\psi_0, \boldsymbol{A}_0, H$ *indicated in the hypotheses* (H1)–(H2), *and the minimum angle of the mesh* $\mathcal{M}^n$ *such that for* $h_{\max} \le h^*$ *and* $\tau_{\max} \le \tau^*$ *the following a posteriori error estimates hold for any* $0 \le t^m \le T$

$$\| \psi^m - \varphi^m \|_{L^2(\Omega)} + \| \boldsymbol{A}^m - \mathbf{D}^m \|_{L^2(\Omega)} \le C\mathcal{E}_0 + C\Big(1 + \log \frac{t^m}{\tau_m}\Big)^{1/2} \sum_{i=1}^{6} \mathcal{E}_i,$$

*where* $\mathcal{E}_i = \max_{1 \le n \le m} \mathcal{E}_i^n$ *and the error indicators* $\mathcal{E}_0$ *and* $\mathcal{E}_i^n$ *are given by*

$$\begin{aligned}
\mathcal{E}_0 \;&:= \| \psi_0 - I^0\psi_0 \|_{L^2(\Omega)} + \| \boldsymbol{A}_0 - I^0 \boldsymbol{A}_0 \|_{L^2(\Omega)}\\
&\textit{initial error},\\
\mathcal{E}_1^n \;&:= \| \varphi^n - \mathcal{P}_n\varphi^{n-1} \|_{L^2(\Omega)} + \| \mathbf{D}^n - \mathbf{P}_n\mathbf{D}^{n-1} \|_{L^2(\Omega)}\\
&\quad + \| \mathbf{g}(\varphi^n, \mathbf{D}^n) - \mathbf{P}_n[\mathbf{g}(\varphi^{n-1}, \mathbf{D}^{n-1})] \|_{L^2(\Omega)}\\
&\textit{time residual},\\
\mathcal{E}_2^n \;&:= \Big\| \frac{h_n^2}{\tau_n}\Big(\varphi^{n-1} - \mathcal{P}_n\varphi^{n-1}\Big) \Big\|_{L^2(\Omega)} + \Big\| \frac{h_n^2}{\tau_n}\Big(\mathbf{D}^n - \mathbf{P}_n\mathbf{D}^{n-1}\Big) \Big\|_{L^2(\Omega)}\\
&\quad + \| \varphi^{n-1} - \mathcal{P}_n\varphi^{n-1} \|_{L^2(\Omega)} + \| \mathbf{D}^{n-1} - \mathbf{P}_n\mathbf{D}^{n-1} \|_{L^2(\Omega)}\\
&\textit{coarsening},\\
\mathcal{E}_3^n \;&:= \| h_n^2 R_\psi^n \|_{L^2(\Omega)} + \| h_n^2 R_{\mathbf{A}}^n \|_{L^2(\Omega)}\\
&\textit{interior residual},\\
\mathcal{E}_4^n \;&:= |\!|\!| h_n^{3/2}[\nabla\varphi^n]_e |\!|\!|_{L^2(\Omega)} + |\!|\!| h_n^{3/2}[\div \mathbf{D}^n]_e |\!|\!|_{L^2(\Omega)}\\
&\quad + |\!|\!| h_n^{3/2}[\mathbf{curl}\mathbf{D}^n]_e |\!|\!|_{L^2(\Omega)}\\
&\textit{jump residual},\\
\mathcal{E}_5^n \;&:= |\!|\!| h_n^{3/2}\nabla\varphi^n \cdot \mathbf{n} |\!|\!|_{L^2(\partial\Omega)} + |\!|\!| h_n^{3/2}(\mathbf{curl}\mathbf{D}^n - I^n H^n) |\!|\!|_{L^2(\partial\Omega)}\\
&\textit{boundary residual},\\
\mathcal{E}_6^n \;&:= \max_{t^{n-1} \le t \le t^n} [\![ H - I^n H^n ]\!] L^2(\partial\Omega)\\
&\textit{boundary error}.
\end{aligned}$$

### 10.8.1 Residuals

The error analysis depends on the crucial estimate (7.20). We first derive the so-called Galerkin orthogonality property by rewriting the discrete problem (7.5)–(7.6), for $t^{n-1} < t \le t^n$, $(\omega, \mathbf{B}) \in \mathcal{H}^1(\Omega) \times \mathbf{H}_n^1(\Omega)$ and $(\widetilde{\omega}, \widetilde{\mathbf{B}}) \in \mathcal{V}^n \times \mathbf{V}_0^n$, as follows:

$$\begin{aligned}
&\eta\langle \partial_t\varphi, \omega\rangle + \frac{1}{\kappa^2}\langle \nabla\hat{\varphi}, \nabla\omega\rangle + \langle f(\varphi^n, \mathbf{D}^n), \omega\rangle\\
&= \eta\Big\langle \frac{\mathcal{P}_n\varphi^{n-1} - \varphi^{n-1}}{\tau_n}, \omega\Big\rangle + \langle R_\psi^n, \omega - \widetilde{\omega}\rangle + \frac{1}{\kappa^2}\langle \nabla\varphi^n, \nabla(\omega - \widetilde{\omega}) \rightarrow, \qquad (8.3)\\
&\langle \partial_t\mathbf{D}, \mathbf{B}\rangle + \langle \div\widehat{\mathbf{D}}, \div\mathbf{B}\rangle + \langle \mathbf{curl}\widehat{\mathbf{D}}, \mathbf{curl}\mathbf{B}\rangle + \langle \mathbf{P}_n[\mathbf{g}(\varphi^{n-1}, \mathbf{D}^{n-1})], \mathbf{B}\rangle\\
&= \Big\langle \frac{\mathbf{P}_n\mathbf{D}^{n-1} - \mathbf{D}^{n-1}}{\tau_n}, \mathbf{B}\Big\rangle + \langle R_{\mathbf{A}}^n, \mathbf{B} - \widetilde{\mathbf{B}}\rangle + \langle \div\mathbf{D}^n, \div(\mathbf{B} - \widetilde{\mathbf{B}})\rangle\\
&+ \langle \mathbf{curl}\mathbf{D}^n, \mathbf{curl}(\mathbf{B} - \widetilde{\mathbf{B}})\rangle + \langle\!\langle I^n H^n, \widetilde{\mathbf{B}} \cdot \boldsymbol{\tau}\rangle\!\rangle. \qquad (8.4)
\end{aligned}$$

By taking $(\omega, \mathbf{B}) = (\theta, \mathbf{W})$ in (8.3)–(8.4) and selecting $(\widetilde{\omega}, \widetilde{\mathbf{B}})$ to be

$$\widetilde{\omega}(\cdot,t) = I^n\theta(\cdot,t), \quad \widetilde{\mathbf{B}}(\cdot,t) = I^n\mathbf{W}(\cdot,t) \quad \forall t^{n-1} < t \leq t^n,$$

we then obtain an explicit expression for the residual $\mathcal{R}(\theta, \mathbf{W}) = \sum_{i=0}^{8} \mathcal{R}_i(\theta, \mathbf{W})$ of (7.21), where, for any $t^* \in (t^{m-1}, t^m]$,

$$\mathcal{R}_0(\theta, \mathbf{W}) = \eta\langle \theta^0, \psi_0 - I^0\psi_0\rangle + \langle \mathbf{W}^0, \boldsymbol{A}_0 - I^0\boldsymbol{A}_0\rangle,$$

$$\mathcal{R}_1(\theta, \mathbf{W}) = \int_0^{t^*} \Big(\eta\langle \partial_t\theta, \hat{\varphi} - \varphi\rangle + \langle \partial_t\mathbf{W}, \widehat{\mathbf{D}} - \mathbf{D}\rangle\Big)dt,$$

$$\mathcal{R}_2(\theta, \mathbf{W}) = \sum_{n=1}^{m} \int_{t^{n-1}}^{t^n\wedge t^*} \Big(\eta\Big\langle \frac{\varphi^{n-1} - \mathcal{P}_n\varphi^{n-1}}{\tau_n}, \theta\Big\rangle$$
$$+\Big\langle \frac{\mathbf{D}^{n-1} - \mathbf{P}_n\mathbf{D}^{n-1}}{\tau_n}, \dot{\mathbf{W}}\Big\rangle\Big)dt,$$

$$\mathcal{R}_3(\theta, \mathbf{W}) = \sum_{n=1}^{m} \int_{t^{n-1}}^{t^n\wedge t^*} \Big(\langle R_\psi^n, I^n\theta - \theta\rangle + \langle R_{\mathbf{A}}^n, I^n\mathbf{W} - \mathbf{W}\rangle\Big)dt,$$

$$\mathcal{R}_4(\theta, \mathbf{W}) = \sum_{n=1}^{m} \int_{t^{n-1}}^{t^n\wedge t^*} \Big(\frac{1}{\kappa^2}\langle \nabla\varphi^n, \nabla(I^n\theta - \theta)\rangle\Big)dt$$
$$+ \sum_{n=1}^{m} \int_{t^{n-1}}^{t^n\wedge t^*} \Big(\langle \div\mathbf{D}^n, \div(I^n\mathbf{W} - \mathbf{W})\rangle$$
$$+\langle \mathbf{curl}\mathbf{D}^n, \mathbf{curl}(I^n\mathbf{W} - \mathbf{W})\rangle\Big)dt,$$

$$\mathcal{R}_5(\theta, \mathbf{W}) = \sum_{n=1}^{m} \int_{t^{n-1}}^{t^n\wedge t^*} \langle \mathbf{P}_n[\mathbf{g}(\varphi^{n-1}, \mathbf{D}^{n-1})] - \mathbf{g}(\varphi^n, \mathbf{D}^n), \mathbf{W}\rangle dt,$$

$$\mathcal{R}_6(\theta, \mathbf{W}) = \sum_{n=1}^{m} \int_{t^{n-1}}^{t^n\wedge t^*} \Big(\langle\!\langle H - I^nH^n, \mathbf{W}\cdot\boldsymbol{\tau}\rangle\!\rangle$$
$$-\langle\!\langle I^nH^n, (I^n\mathbf{W} - \mathbf{W})\cdot\boldsymbol{\tau}\rangle\!\rangle\Big)dt,$$

$$\mathcal{R}_7(\theta, \mathbf{W}) = \int_0^{t^*} \Big(\langle f(\hat{\varphi}, \widehat{\mathbf{D}}) - f(\psi, \boldsymbol{A}), \theta\rangle + \langle f_1(\psi, \boldsymbol{A}; \theta, \mathbf{W}), \psi - \hat{\varphi}\rangle\Big)dt,$$

$$\mathcal{R}_8(\theta, \mathbf{W}) = \int_0^{t^*} \Big(\langle \mathbf{g}(\hat{\varphi}, \widehat{\mathbf{D}}) - \mathbf{g}(\psi, \boldsymbol{A}), \mathbf{W}\rangle + \langle \mathbf{g}_1(\psi, \boldsymbol{A}; \theta, \mathbf{W}), \boldsymbol{A} - \widehat{\mathbf{D}}\rangle\Big)dt.$$

Here $t^n \wedge t^* = \min(t^n, t^*)$. The rest of the arguments consists of estimating the terms $\mathcal{R}_i$ separately. For convenience, we define in the following

$$\Phi^* = \| \theta^* \|_{L^2(\Omega)} + \| \mathbf{W}^* \|_{L^2(\Omega)}.$$

First by using Lemma 4.1 we get

$$|\mathcal{R}_0(\theta, \mathbf{W})| + |\mathcal{R}_5(\theta, \mathbf{W})| \leq C\Phi^*(\mathcal{E}_0 + \mathcal{E}_1).$$

Next we observe by Lemmas 4.2–4.3 that

$$\begin{aligned} &\int_0^{t^{m-1}} \Big(\| \partial_t\theta \|_{L^2(\Omega)} + \| \partial_t \mathbf{W} \|_{L^2(\Omega)} + \| \theta \|_{H^2(\Omega)} + \| \mathbf{W} \|_{H^2(\Omega)}\Big) dt \\ \leq\ & C\Big(\log \frac{t^*}{t^* - t^{m-1}}\Big)^{1/2} \Phi^*. \end{aligned} \tag{8.5}$$

Denote by $l(t)$ the piecewise linear function $l(t) := \tau_n^{-1}(t^n - t)$, then

$$\hat{\varphi} - \varphi = l(t)(\varphi^n - \varphi^{n-1}), \quad \widehat{\mathbf{D}} - \mathbf{D} = l(t)(\mathbf{D}^n - \mathbf{D}^{n-1}) \quad \forall t^{n-1} < t \leq t^n.$$

Thus, we have

$$\begin{aligned} \mathcal{R}_1(\theta, \mathbf{W}) &= \sum_{n=1}^{m-1} \int_{t^{n-1}}^{t^n} l(t) \Big(\eta\langle \partial_t\theta, \varphi^n - \varphi^{n-1}\rangle + \langle \partial_t \mathbf{W}, \mathbf{D}^n - \mathbf{D}^{n-1}\rangle\Big) dt \\ &+ \int_{t^{m-1}}^{t^*} \frac{t^m - t}{\tau_m} \Big(\eta\langle \partial_t\theta, \varphi^m - \varphi^{m-1}\rangle + \langle \partial_t \mathbf{W}, \mathbf{D}^m - \mathbf{D}^{m-1}\rangle\Big) dt \\ &:= \mathcal{R}_{1a}(\theta, \mathbf{W}) + \mathcal{R}_{1b}(\theta, \mathbf{W}). \end{aligned}$$

By (8.5), we have

$$\begin{aligned} |\mathcal{R}_{1a}(\theta, \mathbf{W})| &\leq C\Big(\log \frac{t^*}{t^* - t^{m-1}}\Big)^{1/2} \Phi^* \\ &\quad \max_{1\leq n\leq m-1} (\| \varphi^n - \varphi^{n-1} \|_{L^2(\Omega)} + \| \mathbf{D}^n - \mathbf{D}^{n-1} \|_{L^2(\Omega)}). \end{aligned}$$

Integrating by parts and using Lemma 4.1 we get

$$\begin{aligned} \mathcal{R}_{1b}(\theta, \mathbf{W}) &= \frac{1}{\tau_m} \int_{t^{m-1}}^{t^*} \Big(\eta\langle \varphi^m - \varphi^{m-1}, \theta\rangle + \langle \mathbf{D}^m - \mathbf{D}^{m-1}, \mathbf{W}\rangle\Big) dt \\ &+ \frac{t^m - t^*}{\tau_m} \Big(\eta\langle \theta^*, \varphi^m - \varphi^{m-1}\rangle + \langle \mathbf{W}^*, \mathbf{D}^m - \mathbf{D}^{m-1}\rangle\Big) \\ &- \Big(\eta\langle \theta^{m-1}, \varphi^m - \varphi^{m-1}\rangle + \langle \mathbf{W}^{m-1}, \mathbf{D}^m - \mathbf{D}^{m-1}\rangle\Big) \\ &\leq C\Phi^*(\| \varphi^m - \varphi^{m-1} \|_{L^2(\Omega)} + \| \mathbf{D}^m - \mathbf{D}^{m-1} \|_{L^2(\Omega)}). \end{aligned}$$

Thus

$$|\mathcal{R}_1(\theta, \mathbf{W})| \leq C\Big(1 + \log \frac{t^*}{t^* - t^{m-1}}\Big)^{1/2} \Phi^*(\mathcal{E}_1 + \mathcal{E}_2).$$

Next denote by

$$\theta_{\text{int}}^n = \int_{t^{n-1}}^{t^n \wedge t^*} \theta dt, \quad \mathbf{W}_{\text{int}}^n = \int_{t^{n-1}}^{t^n \wedge t^*} \mathbf{W} dt.$$

**Lemma 10.8.1** *The following estimate*

$$\sum_{n=1}^{m}(\| \theta^n_{\text{int}} \|_{H^2(\Omega)} + \| \mathbf{W}^n_{\text{int}} \|_{H^2(\Omega)}) \leq$$
$$C\Big(1+\log\frac{t^*}{t^*-t^{m-1}}\Big)^{1/2}(\| \theta^* \|_{L^2(\Omega)} + \| \mathbf{W}^* \|_{L^2(\Omega)})$$

*is valid for any* $t^* \in (t^{m-1}, t^m]$.

*Proof.* We only prove the estimate for $\theta^n_{\text{int}}$ since the estimate for $\mathbf{W}^n_{\text{int}}$ is similar. First from (8.5), Lemma 4.1 and elliptic regularity theory (see [Gri 85]), we have that

$$\begin{aligned}
&\sum_{n=1}^{m} \| \theta^n_{\text{int}} \|_{H^2(\Omega)} \\
\leq\ & \int_0^{t^{m-1}} \| \theta \|_{H^2(\Omega)} dt + \Big\| \int_{t^{m-1}}^{t^*} \theta dt \Big\|_{H^2(\Omega)} \\
\leq\ & \int_0^{t^{m-1}} \| \theta \|_{H^2(\Omega)} dt + C\Big\| \int_{t^{m-1}}^{t^*} \theta dt \Big\|_{L^2(\Omega)} + C\Big\| \int_{t^{m-1}}^{t^*} \Delta\theta dt \Big\|_{L^2(\Omega)} \\
\leq\ & C\Big(1+\log\frac{t^*}{t^*-t^{m-1}}\Big)^{1/2}\Phi^* + C\Big\| \int_{t^{m-1}}^{t^*} \Delta\theta dt \Big\|_{L^2(\Omega)}. \qquad (8.6)
\end{aligned}$$

But from (7.12) and Lemma 4.1 we have

$$\begin{aligned}
&\Big\| \int_{t^{m-1}}^{t^*} \Delta\theta dt \Big\|_{L^2(\Omega)} \\
\leq\ & C\Big\| \int_{t^{m-1}}^{t^*} \partial_t\theta dt \Big\|_{L^2(\Omega)} + C\Big\| \int_{t^{m-1}}^{t^*} f_1(\psi, \boldsymbol{A}; \theta, \mathbf{W}) dt \Big\|_{L^2(\Omega)} \\
\leq\ & C\Phi^* + C\int_{t^{m-1}}^{t^*} \| f_1(\psi, \boldsymbol{A}; \theta, \mathbf{W}) \|_{L^2(\Omega)} dt \\
\leq\ & C\Phi^*. \qquad (8.7)
\end{aligned}$$

For example, the first term in $f_1(\psi, \boldsymbol{A}; \theta, \mathbf{W})$ can be estimated by using (7.7) and Corollary 4.1 as follows

$$\int_{t^{m-1}}^{t^*} \| i\eta\kappa \div \boldsymbol{A}\theta \|_{L^2(\Omega)} dt \leq C\| \div \boldsymbol{A} \|_{L^4(Q_T)} \| \theta \|_{L^4(Q^*)} \leq C\Phi^*.$$

Inserting (8.7) into (8.6) we finally obtain that

$$\sum_{n=1}^{m} \| \theta^n_{\text{int}} \|_{H^2(\Omega)} \leq C\Big(1+\log\frac{t^*}{t^*-t^{m-1}}\Big)^{1/2}\Phi^*.$$

This completes the proof.

Now we continue our efforts in estimating $\mathcal{R}_i(\theta, \mathbf{W})$. By Lemma 4.1, the definition of the projection operator $\mathcal{P}_n : \mathcal{L}^2(\Omega) \to \mathcal{V}^n$ and the interpolation estimate (6.1), we have

$$\begin{aligned}
&\sum_{n=1}^{m} \int_{t^{n-1}}^{t^n \wedge t^*} \Big\langle \frac{\varphi^{n-1} - \mathcal{P}_n \varphi^{n-1}}{\tau_n}, \theta \Big\rangle dt \\
= &\sum_{n=1}^{m} \tau_n^{-1} \langle \varphi^{n-1} - \mathcal{P}_n \varphi^{n-1}, \theta_{\text{int}}^n - I^n \theta_{\text{int}}^n \rangle \\
\leq &\; C \max_{1 \leq n \leq m} \Big\| \frac{h_n^2}{\tau_n} (\varphi^{n-1} - \mathcal{P}_n \varphi^{n-1}) \Big\|_{L^2(\Omega)} \sum_{n=1}^{m} \| \theta_{\text{int}}^n \|_{H^2(\Omega)} \\
\leq &\; C \Big( 1 + \log \frac{t^*}{t^* - t^{m-1}} \Big)^{1/2} \Phi^* \mathcal{E}_2.
\end{aligned}$$

The same argument then leads to

$$|\mathcal{R}_2(\theta, \mathbf{W})| \leq C \Big( 1 + \log \frac{t^*}{t^* - t^{m-1}} \Big)^{1/2} \Phi^* \mathcal{E}_2.$$

Again, by Lemma 4.1 and (6.1), we obtain

$$\begin{aligned}
|\mathcal{R}_3(\theta, \mathbf{W})| &\leq \sum_{n=1}^{m} \Big( \| h_n^2 R_\psi^n \|_{L^2(\Omega)} \| \theta_{\text{int}}^n \|_{H^2(\Omega)} + \| h_n^2 R_{\mathbf{A}}^n \|_{L^2(\Omega)} \| \mathbf{W}_{\text{int}}^n \|_{H^2(\Omega)} \Big) \\
&\leq C \Big( 1 + \log \frac{t^*}{t^* - t^{m-1}} \Big)^{1/2} \Phi^* \mathcal{E}_3.
\end{aligned}$$

We decompose the integral $\langle \nabla \varphi^n, \nabla (I^n \theta - \theta) \rangle$ over all elements $S \in \mathcal{M}^n$ and next integrate by parts to obtain the equivalent expression

$$\langle \nabla \varphi^n, \nabla (I^n \theta - \theta) \rangle = \sum_{e \in \mathcal{B}^n} \langle\!\langle [\nabla \varphi^n]_e, \theta - I^n \theta \rangle\!\rangle_e + \langle\!\langle \nabla \varphi^n \cdot \mathbf{n}, I^n \theta - \theta \rangle\!\rangle,$$

where $\langle\!\langle \cdot, \cdot \rangle\!\rangle_e$ denotes the $L^2$-scalar product on $e \in \mathcal{B}^n$, and $[\nabla \varphi^n]_e$ is defined in (6.4). In view of the interpolation estimate (6.1) we obtain

$$\sum_{n=1}^{m} \int_{t^{n-1}}^{t^n \wedge t^*} \langle \nabla \varphi^n, \nabla (I^n \theta - \theta) \rangle dt \leq \sum_{n=1}^{m} ||| h_n^{3/2} [\nabla \varphi^n]_e |||_{L^2(\Omega)} \| \theta_{\text{int}}^n \|_{H^2(\Omega)}.$$

Since $\mathbf{W}, I^n \mathbf{W} \in \mathbf{H}_n^1(\Omega)$, we have, after decomposing the integral and integrating by parts, that

$$\langle \div \mathbf{D}^n, \div (I^n \mathbf{W} - \mathbf{W}) \rangle = \sum_{e \in \mathcal{B}^n} \langle\!\langle [\div \mathbf{D}^n]_e, (\mathbf{W} - I^n \mathbf{W}) \cdot \mathbf{n}_e \rangle\!\rangle_e, \tag{8.8}$$

$$\begin{aligned}
\langle \mathbf{curl} \mathbf{D}^n, \mathbf{curl} (I^n \mathbf{W} - \mathbf{W}) \rangle &= \sum_{e \in \mathcal{B}^n} \langle\!\langle [\mathbf{curl} \mathbf{D}^n]_e, (\mathbf{W} - I^n \mathbf{W}) \cdot \boldsymbol{\tau}_e \rangle\!\rangle_e \\
&+ \langle\!\langle \mathbf{curl} \mathbf{D}^n - I^n H^n, (I^n \mathbf{W} - \mathbf{W}) \cdot \boldsymbol{\tau} \rangle\!\rangle + \langle\!\langle I^n H^n, (I^n \mathbf{W} - \mathbf{W}) \cdot \boldsymbol{\tau} \rangle\!\rangle,
\end{aligned} \tag{8.9}$$

where $\boldsymbol{\tau}_e = (-n_2^e, n_1^e)^T$ is the unit tangent of $e \in \mathcal{B}^n$. Since the last term in (8.9) cancels out a similar term in $\mathcal{R}_6$, we add $\mathcal{R}_4$ and $\mathcal{R}_6$, employ (6.1) and Lemma 4.1, and get

$$|\mathcal{R}_4(\theta, \mathbf{W}) + \mathcal{R}_6(\theta, \mathbf{W})| \le C\Big(1 + \log \frac{t^*}{t^* - t^{m-1}}\Big)^{1/2} \Phi^*(\mathcal{E}_4 + \mathcal{E}_5 + \mathcal{E}_6).$$

Now we turn to the estimate of $\mathrm{Re}[\mathcal{R}_7(\theta, \mathbf{W}) + \mathcal{R}_8(\theta, \mathbf{W})]$, which is different in nature. Recall that $\hat{e}_\psi = \psi - \hat{\varphi}$ and $\hat{e}_{\mathbf{A}} = \boldsymbol{A} - \widehat{\mathbf{D}}$, we obtain by simple calculations that

$$\begin{aligned}
\mathrm{Re}[\mathcal{R}_7(\theta, \mathbf{W}) + \mathcal{R}_8(\theta, \mathbf{W})] = & \int_0^{t^*} \int_\Omega \mathrm{Re}\Big[\Big(\frac{i}{\kappa} - i\eta\kappa\Big) \div \hat{e}_{\mathbf{A}} \hat{e}_\psi \theta\Big] dx dt \\
& + \int_0^{t^*} \int_\Omega \mathrm{Re}\Big[\frac{2i}{\kappa} \hat{e}_{\mathbf{A}} \nabla \hat{e}_\psi \theta\Big] dx dt \\
& + \int_0^{t^*} \int_\Omega \mathrm{Re}\Big[\Big(\hat{e}_{\mathbf{A}} \hat{e}_\psi (\boldsymbol{A} + \widehat{\mathbf{D}}) + |\hat{e}_{\mathbf{A}}|^2 \psi + \hat{e}_\psi^2 \bar{\psi} \\
& + \hat{\varphi} |\hat{e}_\psi|^2 + \psi |\hat{e}_\psi|^2\Big)\theta\Big] \\
& + \int_0^{t^*} \int_\Omega \mathrm{Re}\Big[\frac{i}{\kappa} \nabla \hat{e}_\psi \bar{\hat{e}}_\psi - 2\hat{e}_{\mathbf{A}} \bar{\hat{e}}_\psi \psi + \widehat{\mathbf{D}} |\hat{e}_\psi|^2\Big] \mathbf{W} dx dt \\
:= & \ (\mathrm{I}) + \cdots + (\mathrm{IV}). \qquad (8.10)
\end{aligned}$$

By Hölder inequality, (7.8) and Corollary 4.1, we get

$$\begin{aligned}
|(\mathrm{I})| \le & \int_0^{t^*} \| \div \hat{e}_{\mathbf{A}} \|_{L^2(\Omega)} \| \hat{e}_\psi \|_{L^4(\Omega)} \| \theta \|_{L^4(\Omega)} dt \\
\le & \ C\Big(\int_0^{t^*} \| \theta \|_{L^4(\Omega)}^4 dt\Big)^{1/4} \Big(\int_0^{t^*} \| \div \hat{e}_{\mathbf{A}} \|_{L^2(\Omega)}^{4/3} \| \hat{e}_\psi \|_{H^1(\Omega)}^{2/3} \| \hat{e}_\psi \|_{L^2(\Omega)}^{2/3} dt\Big)^{3/4} \\
\le & \ C\Phi^* \max_{0 \le t \le t^*} \| \hat{e}_\psi \|_{L^2(\Omega)}^{1/2} \Big(\int_0^{t^*} \| \div \hat{e}_{\mathbf{A}} \|_{L^2(\Omega)}^{4/3} \| \hat{e}_\psi \|_{H^1(\Omega)}^{2/3} dt\Big)^{3/4} \\
\le & \ C\Phi^* \max_{0 \le t \le t^*} \| \hat{e}_\psi \|_{L^2(\Omega)}^{1/2} \Big(\int_0^{t^*} \| \hat{e}_{\mathbf{A}} \|_{H^1(\Omega)}^2\Big)^{1/2} \Big(\int_0^{t^*} \| \hat{e}_\psi \|_{H^1(\Omega)}^2 dt\Big)^{1/4} \\
\le & \ C\Phi^* \Lambda(t^*)^{1/4} \max_{0 \le t \le t^*} (\| \hat{e}_\psi \|_{L^2(\Omega)} + \| \hat{e}_{\mathbf{A}} \|_{L^2(\Omega)}), \qquad (8.11)
\end{aligned}$$

where

$$\Lambda(t^*) = \Big(\int_0^{t^*} (\| \hat{e}_\psi \|_{H^1(\Omega)}^2 + \| \hat{e}_{\mathbf{A}} \|_{H^1(\Omega)}^2) dt\Big)^3 \Big/ \max_{0 \le t \le t^*} (\| \hat{e}_\psi \|_{L^2(\Omega)}^2 + \| \hat{e}_{\mathbf{A}} \|_{L^2(\Omega)}^2).$$

Similarly, we have

$$\begin{aligned}
|(\mathrm{II})| \le & \ C\Phi^* \max_{0 \le t \le t^*} \| \hat{e}_{\mathbf{A}} \|_{L^2(\Omega)}^{1/2} \Big(\int_0^{t^*} \| \hat{e}_\psi \|_{H^1(\Omega)}^2\Big)^{1/2} \Big(\int_0^{t^*} \| \hat{e}_{\mathbf{A}} \|_{H^1(\Omega)}^2 dt\Big)^{1/4} \\
\le & \ C\Phi^* \Lambda(t^*)^{1/4} \max_{0 < t < t^*} (\| \hat{e}_\psi \|_{L^2(\Omega)} + \| \hat{e}_{\mathbf{A}} \|_{L^2(\Omega)}). \qquad (8.12)
\end{aligned}$$

The first term in (III) can be estimated by (7.7), Corollary 4.1 and (7.8) as follows:

$$
\begin{aligned}
&\int_0^{t^*}\int_\Omega \mathrm{Re}[\hat{e}_{\mathbf{A}}\hat{e}_\psi(\boldsymbol{A}+\widehat{\mathbf{D}})\theta]dxdt \\
= & \int_0^{t^*}\int_\Omega \mathrm{Re}[2\hat{e}_{\mathbf{A}}\hat{e}_\psi \boldsymbol{A}\theta - |\hat{e}_{\mathbf{A}}|^2\hat{e}_\psi\theta]dxdt \\
\leq & \int_0^{t^*} \| \hat{e}_{\mathbf{A}} \|_{L^4(\Omega)} \| \hat{e}_\psi \|_{L^4(\Omega)} \| \boldsymbol{A} \|_{L^4(\Omega)} \| \theta \|_{L^4(\Omega)} dt \\
& + \int_0^{t^*} \| \hat{e}_{\mathbf{A}} \|^2_{L^4(\Omega)} \| e_\psi \|_{L^4(\Omega)} \| \theta \|_{L^4(\Omega)} dt \\
\leq & \ C\Phi^*\Big(\int_0^{t^*} \| \hat{e}_{\mathbf{A}} \|^2_{L^4(\Omega)} \| \hat{e}_\psi \|^2_{L^4(\Omega)} dt\Big)^{1/2} \\
& + C\Phi^*\Big(\int_0^{t^*} \| \hat{e}_{\mathbf{A}} \|^{8/3}_{L^4(\Omega)} \| \hat{e}_\psi \|^{4/3}_{L^4(\Omega)} dt\Big)^{3/4} \\
\leq & \ C\Phi^* \max_{0\leq t\leq t^*}(\| \hat{e}_{\mathbf{A}} \|^{1/2}_{L^2(\Omega)} \| \hat{e}_\psi \|^{1/2}_{L^2(\Omega)})\Big(\int_0^{t^*} \| \hat{e}_\psi \|^2_{H^1(\Omega)}\Big)^{1/4} \\
& \Big(\int_0^{t^*} \| \hat{e}_{\mathbf{A}} \|^2_{H^1(\Omega)} dt\Big)^{1/4} \\
& + C\Phi^* \max_{0\leq t\leq t^*}(\| \hat{e}_{\mathbf{A}} \|_{L^2(\Omega)} \| \hat{e}_\psi \|^{1/2}_{L^2(\Omega)})\Big(\int_0^{t^*} \| \hat{e}_\psi \|^2_{H^1(\Omega)}\Big)^{1/4} \\
& \Big(\int_0^{t^*} \| \hat{e}_{\mathbf{A}} \|^2_{H^1(\Omega)} dt\Big)^{1/2} \\
\leq & \ C\Phi^*\Lambda(t^*)^{1/6} \max_{0\leq t\leq t^*}(\| \hat{e}_\psi \|_{L^2(\Omega)} + \| \hat{e}_{\mathbf{A}} \|_{L^2(\Omega)})^{4/3} \\
+ & \ C\Phi^*\Lambda(t^*)^{1/4} \max_{0\leq t\leq t^*}(\| \hat{e}_\psi \|_{L^2(\Omega)} + \| \hat{e}_{\mathbf{A}} \|_{L^2(\Omega)})^{3/2}.
\end{aligned}
$$

The other terms in (III), (IV) can be estimated similarly to obtain

$$
|(\mathrm{III})| + |(\mathrm{IV})| \leq C\Phi^*\Lambda(t^*)^\alpha \max_{0\leq t\leq t^*}(\| \hat{e}_\psi \|_{L^2(\Omega)} + \| \hat{e}_{\mathbf{A}} \|_{L^2(\Omega)})^\beta, \tag{8.13}
$$

for some $\alpha > 0$ and $\beta \geq 1$. Substituting (8.11)–(8.13) into (8.10) we finally obtain that

$$
|\mathrm{Re}[\mathcal{R}_7(\theta, \mathbf{W}) + \mathcal{R}_8(\theta, \mathbf{W})]| \leq C\Phi^*\Lambda(t^*)^\alpha \max_{0\leq t\leq t^*}(\| \hat{e}_\psi \|_{L^2(\Omega)} + \| \hat{e}_{\mathbf{A}} \|_{L^2(\Omega)})^\beta
$$

for some $\alpha > 0$ and $\beta \geq 1$.

### 10.8.2 Proof of Theorem 4.1

Collecting the above estimates for $\mathcal{R}_i$, and inserting them into (7.20), we obtain that

$$\begin{aligned}
&\eta \| \psi(t^*) - \varphi(t^*) \|_{L^2(\Omega)} + \| \boldsymbol{A}(t^*) - \mathbf{D}(t^*) \|_{L^2(\Omega)} \\
\leq\ & C\mathcal{E}_0 + C\Big(1 + \log \frac{t^*}{t^* - t^{m-1}}\Big)^{1/2} \sum_{i=1}^{6} \mathcal{E}_i \\
& + C\Lambda(t^*)^\alpha \max_{0 \leq t \leq t^*} (\| \hat{e}_\psi \|_{L^2(\Omega)} + \| \hat{e}_{\mathbf{A}} \|_{L^2(\Omega)})^\beta
\end{aligned} \tag{8.14}$$

for some $\alpha > 0$ and $\beta \geq 1$. But

$$\begin{aligned}
\| \hat{e}_\psi \|_{L^2(\Omega)} &\leq \| \psi - \varphi \|_{L^2(\Omega)} + \| \varphi - \hat{\varphi} \|_{L^2(\Omega)} \\
&\leq \| \psi - \varphi \|_{L^2(\Omega)} + \max_{1 \leq n \leq m} \| \varphi^n - \varphi^{n-1} \|_{L^2(\Omega)} \\
&\leq \| \psi - \varphi \|_{L^2(\Omega)} + (\mathcal{E}_1 + \mathcal{E}_2) \quad \forall t \in (0, t^*],
\end{aligned}$$

and

$$\| \hat{e}_{\mathbf{A}} \|_{L^2(\Omega)} \leq \| \boldsymbol{A} - \mathbf{D} \|_{L^2(\Omega)} + (\mathcal{E}_1 + \mathcal{E}_2) \qquad \forall t \in (0, t^*],$$

we use (H4)–(H5) to conclude that for sufficiently small $h_{\max}$ and $\tau_{\max}$, the right-most term in (8.14) can be absorbed into the left-hand side, and thus

$$\| \psi^m - \varphi^m \|_{L^2(\Omega)} + \| \boldsymbol{A}^m - \mathbf{D}^m \|_{L^2(\Omega)} \leq C\mathcal{E}_0 + C\Big(1 + \log \frac{t^m}{\tau_m}\Big)^{1/2} \sum_{i=1}^{6} \mathcal{E}_i.$$

This completes the proof.

## 10.9 Numerical implementation

In this section, we follow the work of [Bau 99] to present our computational results. Two numerical schemes are use: one scheme with non-adaptive grids and the other scheme with adaptive grids. The adaptive scheme is controlled by the a posteriori error-estimator. The comparison shows that – not surprisingly, – that the adaptive scheme is a good choice for a transient simulation of the GL-model for the following reasons:
1. The number of degrees of freedom required to achieve a good approximation is much smaller than that in the non-adaptive case.
2. The computation is much more stable, which means that

2.1: there are fewer convergence problems with the Newton-Method used to resolve the non-linearity (even for higher values of $\kappa$).

2.2: the approximation at the boundary is better: numerical artefacts in the corners of the boundary can be avoided.

### 10.9.1 Comparison of the schemes

In both schemes, the implicit-Euler method is used for time discretization and continuous, linear elements on triangles for the discretization in space. To proceed in time equidistant steps of length 0.1 are used. The Lorentz gauge guarantees the uniqueness of the solution.

In the adaptive scheme the grid is adapted every four time steps by applying the recursive bisection method [Mit 89]. The values of the unknowns at new grid points are interpolated. This simple "explicit" adoption strategy is suitable since the non-linearity of the problems restricts the length of the time steps anyway.

#### Non-adaptive scheme

Figure 5 shows the effect on $|\psi|^2$ when switching on an external magnetic field. The domain involved in the computation is the rectangle with associate corners at (0, 0) and (1, 1.1). We take the values of the parameters $\kappa = 10$, $\eta = 1.2$. We take the initial values at t=0 as the perfect superconducting state with $\psi(x, 0) = 0.6 + i0.8$ and $\boldsymbol{A} = 0$. The external magnetic field is initially $H = 3$ and is increased linearly to $H = 8$ over the time interval $(0, 5)$. When $t > 5$, $H$ takes the constant value 8. The pictures below show the values of $|\psi|^2$ at different times where dark region indicating superconductivity region (where $|\psi| \sim 1$) and white region indicating non-supercondictivity region (where $|\psi| \sim 0$).

The last picture shows the stationary formation of the eight vortices.

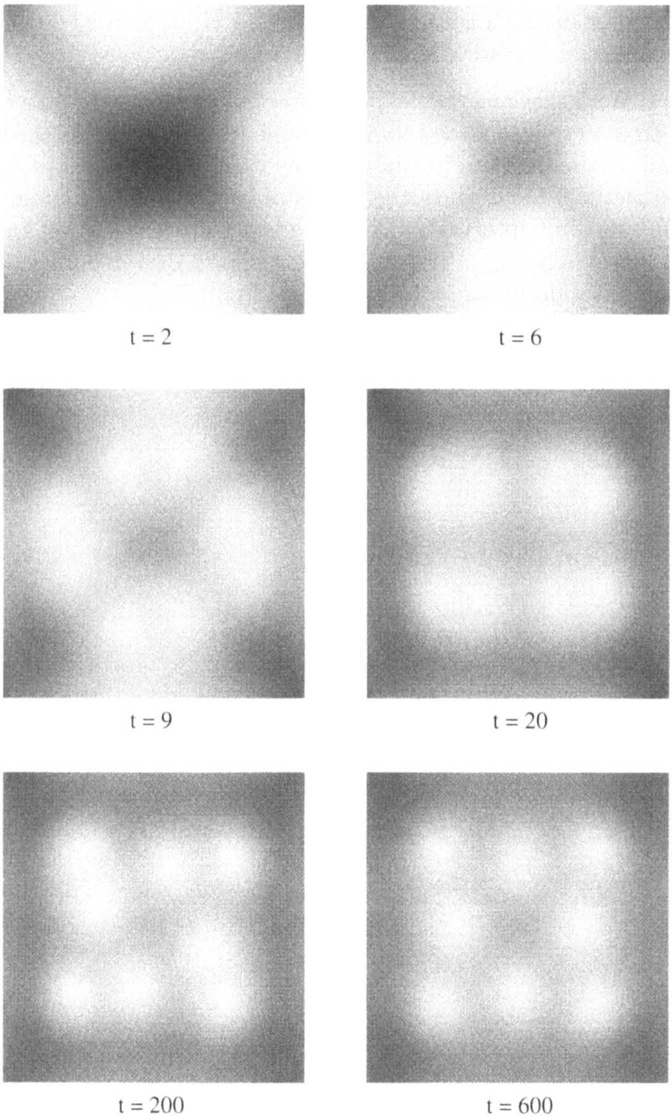


Figures 5: Non-adaptive, t= 2, 6, 9, 20, 200, 600

**Adaptive Scheme**

A rectangular domain with corners at (0, 0) and (1, 1) is used in the programs. The initial values are the same as in the previous part.

**Experiment 1:** With the same model parameter as before but we take the magnetic field to vary from $H = 3$ to $H = 4$ over the same time interval. As the final applied magnetic field is smaller than in the previous case, only 4 vortices are formed in the end. The pictures below show the values of $|\psi|^2$ at different times. Due to the fact that a shorter time interval is taken, the last picture is not yet the stationary solution but reasonably close. The experiment shows that using an adaptive grid makes the computation much more stable (see Figure 6).

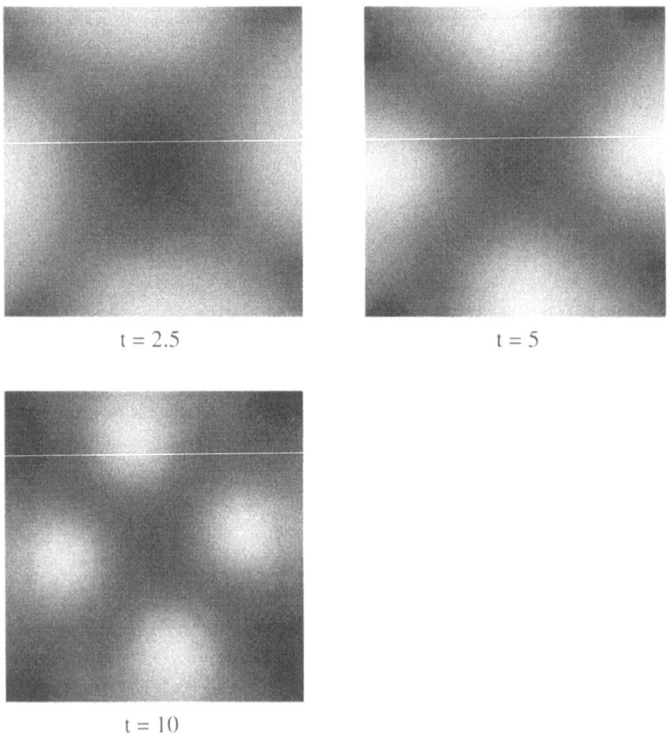


Figure 6: Adaptive 1, t= 2.5, 5, 10

**Experiment 2:** The model parameter $\kappa$ is increased to the value of 15 which will lead to "steeper singularities". This makes the adaptive method even more suitable. The magnetic field is increased from 4.0 at the time $t = 0$ linearly to 6.0 at $t = 5$. The picture below shows that the diameter of the vortices is smaller (see Figure 7).

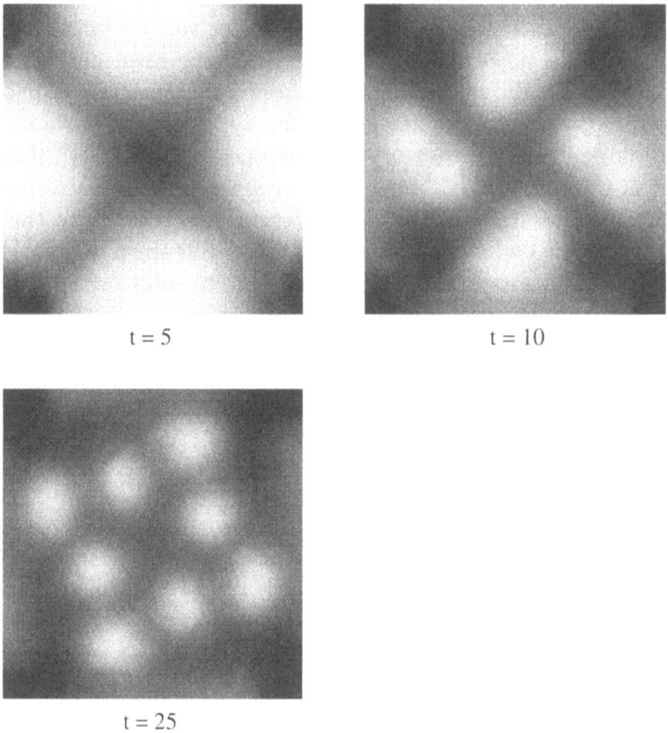


Figure 7: Adaptive 2, t= 5, 10, 25

## 10.10 Bibliographical remarks

A posteriori error estimates are computable quantities that measure the actual errors without knowledge of the limit solution. They are essential in designing algorithms for mesh and time-step modification which equidistribute the computational effort and so optimize the computations. Since the seminal paper [BRh 78] on elliptic problems there has been an ever-increasing interest on the development of reliable and efficient adaptive algorithms for various linear and non-linear PDEs. In particular, a posteriori error estimates have been derived in [EJo 91], [EJo 95] for linear and mildly non-linear parabolic problems, and in [Noc 99], [CNS 99] for degenerate parabolic problems of Stefan type with or without convection. The main tool in deriving a posteriori error estimates in [EJo 91], [EJo 95], [NSV 99], [CNS 99] is the analysis of linear dual problems of the corresponding *error* equations. This method has been extended recently in [Iva 98] to the G-L model under gauge choice (1.5) when using implicit finite difference in time. However, in this situation, the resulting dual problem has discrete solutions in the coefficients which, in turn, do *not* provide necessary smoothness to obtain the strong stability estimates similar to those used in [EJo 91] for linear parabolic problems.

# References


[ABe 98] L. Almeida and F. Bethuel; Topological methods for the Ginzburg-Landau equations, J. Math. Pures Appl., **77** (1998), 1–49

[Abr 57] A. Abrikosov; On the magnetic properties of superconductors of the 2nd type, Soviet Phys. JETP, **5** (1957), 1174–1182.

[ACi 97] A. Assous and P. Ciarlet Jr.; Une caractérisation de l'orthogonal de $\Delta(H^2(\Omega) \bigcap H_0^1(\Omega))$ dans $L^2(\Omega)$, C. R. Acad. Sci. Paris, **325**, Série I (1997) 605–610.

[ASh 95] N. André and I. Shafrir; Minimization of the Ginzburg-Landau functional with weight, C.R. Acad. Sci. Paris, Série I, **321** (1995), 999–1004.

[Ada 75] R. Adams; Sobolev Spaces, Academic Press, New York, 1975.

[Aft 97] A. Aftalion; On the minimizers of the Ginzburg-Landau energy for high kappa: the axially symmetric case, preprint.

[ASo 98] L. Ambrosio and H.M. Soner; A measure theoretic approach to higher codimension mean curvature flows, preprint.

[Ban 91] E. Bänsch; Local mesh refinement in 2 and 3 dimensions, Impact of Computing in Science and Engineering **3**, (1991), pp. 181–191.

[Bat 67] G.K. Batchelor; An Introduction to Fluid Mechanics, Cambridge University Press, 1967.

[Bau 99] H.-J. Bauer; Objektorientierter Entwurf und Implementierung multilevel-adaptiver Finite-Element-Methoden, Dissertation, Technische Universität München, 1999.

[BBC 94] H. Berestycki, A. Bonnet and J. Chapman; A semi-elliptic system arising in the theory of type-II superconductivity, Comm. Appl. Nonlinear Anal., **3** (1994), 1–21.

[BBH 92] F. Bethuel, H. Brezis and F. Hélein; Limite singulière pour la minimization de fonctionnelles du type Ginzburg-Landau, C.R.Acad.Sci.Paris, **314**, Ser I, 1992, pp. 891–895.

[BBH 93] F. Bethuel, H. Brezis and F. Hélein; Asymptotics for the minimization of a Ginzburg-Landau functional, Calculus of Variations and PDE, **1** (1993), 123–148.

[BBH 94] F. Bethuel, H. Brezis & F. Hélein; Ginzburg-Landau Vortices, Birkhäuser (1994).

[BCh 89] M. Berger and Y. Chen; Symmetric vortices for Ginzburg-Landau equations of superconductivity and nonlinear desingularization, J. Func. Anal., **82** (1989) 259–295.

[BCE 99] A.J. Briggs, J.R. Claisse and C.M. Elliott, Finite difference approximation of a non-local Hamilton-Jacobi equation arising in superconductivity, to appear, Preprint of CMAIA, University of Sussex.

[BCP 93] P. Bauman, N. Carlson and D. Phillips; On the zeros of solutions to Ginzburg-Landau type systems, SIAM J. Math. Anal., **24** (1993), 1283–1293.

[BCPS 94] P. Bauman, C.N. Chen, D. Phillips and P. Sternberg; Annihilation of vortices for the Ginzburg-Landau heat flow, European J. Appl. Math. 6 (1995), no. 2, 115–126.

[BCS 57] J. Bardeen, L.N. Cooper and J.R. Schreiffer; Theory of Superconductivity, Phy. Rev., **108** (1957), p1175.

[BDG 81] K.J. Brown, P.C. Dunne & R.A. Gardner; A Semilinear Parabolic System Arising in the Theory of Superconductivity, J. of Diff. Equa., **40**, (1981) 232–252.

[BGT 92] E. Barany, M. Golubitsky and J. Turski; Bifurcation with local gauge symmetries in the Ginzburg-Landau equations, Physica D, **56** (1992) 36–56.

[BHa 95] A. Beaulieu and R. Hadiji; Asymptotics for minimizers of a class of Ginzburg-Landau equations with weight, C.R. Acad. Sci. Paris, Série I, **320** (1995), 181–186.

[BHe 93]. C. Bolley and B. Helffer; An application of semi-classical analysis to the asymptotic study of the supercooling field of a superconducting material, Ann. Inst. Henri Poincaré, Anal. Non Linéaire, **58** (1993), 189–233.

[BHe 96] C. Bolley and B. Helffer; Upper bounds for the solutions of the Ginzburg-Landau equations in a semi-infinite superconducting film and applications to the superheating field in the large $\kappa$ regime, preprint ECN.

[BMi 87] I. Babuska and A. Miller; A feedback finite element method with a posteriori error estimation: Part I. The finite element method and some basic properties of a posteriori error estimator, Computer Meth. Appl. Mech. Engineer. **61** (1987), pp. 1–40.

[BMR 94] H. Brezis, F. Merle and T. Rivière; Quantization effects for $-\Delta u = u(1-|u|^2)$ in $\mathbb{R}^2$, Arch. Rat. Mech. Anal., **126** (1994), 35–58.

[BNe 97] J. Bak and D.J. Newman; Complex Analysis, 2nd edition, Springer, 1997

[Bol 92] C. Bolley; Modélisation de champ de retard à la condensation d'un supraconducteur par un problème de bifurcation, $M^2AN$, **26** (1992), 235–287.

[BPT 98] P. Baumann, D. Phillips and Q. Tang; Stable nucleation for the Ginzburg-Landau system with an applied magnetic field, Arch. Rat. Mech. Anal., 142 (1998), 1–43.

[BRh 78] I. Babuska and C. Rheinboldt; Error estimates for adaptive finite element computations, SIAM J. Numer. Anal. **15** (1978), pp. 736–754.

[BRi 95] F. Bethuel and T. Rivière; A variational problem related to superconductivity, Annales Inst. H. Poincaré, Analyse Nonlinéaire, **12** (1995), 243–303.

[BRu 98] J. Berger and J. Rubinstein; Bifurcation and phase transitions in nonuniform superconducting rings, SIAM J. Appl. Math., **58** (1998) 103.

[CDa 99] Z. Chen and S. Dai; Adaptive Galerkin methods with error control for a dynamical Ginzburg-Landau model in superconductivity, to appear.

[CDG 96] S.J. Chapman, Q. Du and M.D. Gunzburger; A model for variable thickness superconducting thin films, Z. angew. Math. Phys., **47** (1996), 410–431.

[CDGP 95] S. Chapman, Q. Du, M. Gunzburger and J. Peterson; Simplified Ginzburg-Landau models for superconductivity valid for high Kappa and high fields, Advances in Applied Math., **5** (1995) 193–218.

[CET 94] X. Chen, C.M. Elliott and Q. Tang; Shooting method for the vortex solutions of a Ginzburg-Landau equation, Proc. Roy. Soc. Edinburgh, 124A, (1994) 1075–1088.

[CET 98] Z. Chen, C.M. Elliott and Q. Tang; Justification of a two dimensional evolutionary Ginzburg-Landau superconductivity model, $M^2AM$, Vol 32, No.1, (1998).

[CFT 85] P. Constantin, C. Foias anf R. Temam; Attractors representing turbulent flows, Memoirs of AMS **53** 314 (1985).

[Cha1 95] S.J. Chapman; Nucleation of superconductivity in a decreasing field, Euro. J. Appl. Math., **6** (1995) 97–114.

[Cha2 95] S.J. Chapman; A mean-field model of superconducting vortices in three dimensions, SIAM J. Appl. Math., **55** (1995) 1259–1274.

[Che 90] Y.Y. Chen; Vortices for the Ginzburg-Landau equations – the nonsymmetric case in unbounded domain, Contemporary Mathematics, **108** (1990), 19–32.

[Che 97] Z. Chen; Mixed finite element methods for a dynamical Ginzburg-Landau model in superconductivity, Numer. Math., **76** (1997), pp. 323–353.

[CHL 93] Z. Chen, K.H. Hoffmann and J. Liang; On a non-stationary Ginzburg-Landau superconductivity model, Math. Meth. Appl. Sci., **16** (1993), 855–875.

[CHo 94] Z. Chen and K.-H. Hoffmann, An error estimate of a finite element scheme for a phase field model, IMA J. Numer. Anal. 14 (1994), 243–255.

[CHo 95] Z. Chen and K.-H. Hoffmann; Numerical studies of a non-stationary Ginzburg-Landau model for superconductivity, Adv. Math. Sci. Appl., **5** (1995), 363–389.

[CHo 96] Z. Chen and K.-H. Hoffmann; Optimal control of dynamical Ginzburg-Landau vortices in superconductivity, Numer. Func. Anal. and Optimiz. **17** (1996), pp. 241–258.

[CHO 92] S.J. Chapman, S.D. Howison, J.R. Ockendon; Macroscopic models for superconductivity, SIAM Review, **34** No 4 (1990), 529–560.

[Cia 78] P.G. Ciarlet; The Finite Element Method for Elliptic Problems, North-Holland, Amsterdam, 1978.

[CNS 99] Z. Chen, R.H. Nochetto and A. Schmidt; A characteristic Galerkin method with adaptive error control for the continuous casting problem, to appear.

[CRi 95] S.J. Chapman and G. Richardson; Motion of vortices in type II superconductors, SIAM J. Appl. Math., **55** (1995) 1275–1296.

[Cro 80] C. Cronström; A simple and complete Lorentz-covariant gauge condition, Phys. Lett., **90B** (1980), 267–269.

[CSW 97] M. Chipot, I. Shafrir and G. Wolansky; On the solutions of Liouville systems, Journal of Differential Equations, **140** 1 (1997) 59–105.

[Cti 78] A.J. Callegari and L. Ting; Motion of a curved vortex filament with decaying vortical core and axial velocity, SIAM J. Appl. Math., **35** (1978), 148–175.

[DGP 92] Q. Du, M.D. Gunzburger and J.S. Peterson; Analysis and Approximation of the Ginzburg-Landau Model of Superconductivity, SIAM Review, **34**, (1992), 54–81.

[DGP 95] Q. Du, M.D. Gunzburger and J.S. Peterson; Computational simulations of type-II superconductivity including pinning mechanisms, Phys. Rev. B., **3** (1995) 16194–16203.

[DGu 93] Q. Du and M.D. Gunzburger; A model for superconducting thin films having variable thickness, Phys. D., **69** (1993) 215–231.

[Dor 92] A. Dorsey; Vortex motion and the Hall effect in type II superconductors: a time dependent Ginzburg-Landau theory approach, Phys. Rev. B. **46** (1992) 8376–8368.

[DHe 93] M. Dauge and B. Helffer; Eigenvalues variation, I, Neumann problem for Sturm-Liouville operators, J. Diff. Eq., **104** (1993), 243–262.

[DKN 97] P. Drábek, A. Kufner and F. Nicolosi; Quasilinear Elliptic Equations with Degenerations and Singularities, de Gruyter Series in Nonlinear Analysis and Applications, Vol. 5, Berlin-New York, 1997.

[DLi 76] G. Duvaut, J.L. Lions; Inequalities in Mechanics and Physics, Springer, 1976.

[DLi 97A] S. Ding and Z. Liu; Pinning of vortices for a variational problem related to the superconducting thin films having variable thickness, J.PDEs, 1997,vol.10, 174–192

[DLi 1] S. Ding and Z. Liu; Asymptotic behaviour for minimizers of anistropic G-L functional, to appear in Diff. Int. Eqs..

[DLi 97B] S. Ding and Z. Liu; Asymptotics for a class of G-L functionals, Chin. Ann. Math., 1997, 18A, 437–444.

[DLi 98] S. Ding and Z. Liu; A remark on "Asymptotics for a class of G-L functionals", Chin. Ann.Math., 1998, 19A, 621–628.

[DLY 97] S. Ding, Z. Liu and W. Yu; Pinning of vortices for the G-L functional with variable coefficient, Appl. Math. -JCU, **12B** (1997), 77–88.

[Du 94] Q. Du; Global existence and uniqueness of solutions of the time-dependent Ginzburg-Landau model for superconductivity, Applicable Analysis, **52** (1994), 1–17.

[Du2 94] Q. Du: Finite element methods for the time-dependent Ginzburg-Landau model for superconductivity, Comput. Math. Appl. **27** (1994), pp. 119–133.

[Du 98] Q. Du; Studies of a Ginzburg-Landau model for d-wave superconductors, SIAM J. Appl. Math **59** (1999), no 4, 1225–1250.

[E1 94] W. E; Dynamics of vortices in Ginzburg-Landau Theories and applications to superconductivity, Phys. D., **77** (1994) 383–404.

[E2 94] W. E; Dynamics of vortex liquid in Ginzburg-Landau theories, with application to superconductivity, Phys. Rev. B., **50** (1994) 1126–1135.

[EJo 91] K. Eriksson and C. Johnson; Adaptive finite element methods for parabolic problems I: A linear model problem, SIAM J. Numer. Anal. **28** (1991), pp. 43–77.

[EJo 95] K. Eriksson and C. Johnson; Adaptive finite element methods for parabolic problems IV: Nonlinear problems, SIAM J. Numer. Anal. **32** (1995), pp. 1729–1749.

[Ell 97] C.M. Elliott, Approximation of curvature dependent interface motion, State of the Art in Numerical Analysis, ed. I.S.Duff and G.A.Watson, Clarendon Press, Oxford, (1997), pp. 407–440

[EMT 93] C.M. Elliott, H. Matano and Q. Tang; Zeros of a complex Ginzburg-Landau order parameter with applications to superconductivity, Eur. J. Appl. Math., **5** 7 (1994); 437–448.

[ESS 98] C.M. Elliott, R. Schatzle and B. Stoth, Viscosity solutions of a degenerate parabolic elliptic system arising in the mean-field theory of superconductivity, Arch. Rat. Mech. Anal., 145 (1998), 99–127.

[ESt1 99] C.M. Elliott and V. Styles, Flux pinning and boundary nucleation of vorticity in a mean field model of superconducting vortices, Interfaces and Free Boundaries, 2 (2000), 143–180.

[ESt2 99] C.M. Elliott and V. Styles, Numerical analysis of a mean field model of superconducting vortices, to appear in I.M.A.J. Numerical Analysis.

[ETe 75] I. Ekeland, R. Temam; Convex Analysis and Variational Problems, North-Holland, 1976.

[Fed 69] H. Federer; Geometric Measure Theory, Springer, 1969.

[FGa 95] I. Fonseca and W. Gangbo; Degree Theory in Analysis and Application, Clarendon Press, 1995.

[FKa 64] H. Fujita & T. Kato; On the Navier-Stokes Initial Value Problem, I, Arch. Rat. Mech. Anal., **16**, (1964), 269–315.

[GEl 68] L.P. Gor'kov & G.M. Éliashberg; Generalization of the Ginzburg-Landau Equations for Non-Stationary Problems in the Case of Alloys with Paramagnetic Impurities, Soviet Physics JETP, **27**, (1968), 328–334.

[Gen 66] P.D. de Gennes; Superconductivity of Metals and Alloys, W.A. Benjamin, 1966; Reprinted edition, Addison-Wesley, 1996.

[Ghi 84] J.M. Ghidaglia; Régularité de solutions de certains problèmes aux limites linéaires liés aux équations d'Euler, Comm. in Partial Differential Equations, **9** 13, (1984), 1265–1298.

[Gia 83] M. Giaquinta; Multiple Integrals in the Calculus of Variations and Nonlinear Elliptic Systems, Princeton Univ. Press, Princeton, 1983.

[GLa 50] V. Ginzburg & L. Landau; On the Theory of Superconductivity, Zh. Èksper. Teoret. Fiz., **20**, (1950), 1064–1082. (In Russian.). Man of Physics: L. D. Landau, I. D. ter Haar, ed. Pergamon. Oxford, (1965), 138–167. (In English.)

[GPh 99] Giogio and D. Phillips; The breakdown of superconductivity due to strong fields for the Ginzburg-Landau model, SIAM J. Math. Anal. 30 (1999), no. 2, 341–359.

[GRa 86] V. Girault & P.-A. Raviart; Finite Element Methods for Navier-Stokes Equations, Springer-Verlag, Berlin, (1986).

[Gri 85] P. Grisvard; Elliptic Problems in Nonsmooth Domains, Pitman, 1985.

[GTr 77] D. Gilbarg, N.S. Trudinger; Elliptic Partial Differential Equations, 2nd ed.; Springer, 1983.

[GYu 98] B. Guo and G. Yuan; Cauchy problem of the Ginzburg-Landau equation for the superconductivity model, Preprint.

[Hen 81] D. Henry; Geometric Theory of Semilinear Parabolic Equations, Springer-Verlag Berlin Heidelberg New York, (1981).

[Hon 96] M.C. Hong; On a problem of Bethuel, Brezis and Hélein concerning the Ginzburg-Landau functional, C.R. Acad. Sci. Paris. Ser. I Math. **320** (1995) No.6, 679–684.

[HSh 95] Z.-C. Han and I. Shafrir; Lower bounds for the energy of $S^1$-valued maps in perforated domains, J. Analyse Math., **66** (1995), 295–306.

[Iva 98] J. Ivarsson; A posteriori error analysis of a finite element method for the time-dependent Ginzburg-Landau equations, Preprint, Chalmers University of Technology, 1998.

[JMo 96] S. Jimbo and Y. Morita; Ginzburg-Landau equations and stable solutions in a rotational domain, SIAM Math. Anal., **27** 5 (1996); 1360–1385.

[JTa 80] A. Jaffe, C. Taubes; Vortices and Monopoles, Birkhäuser, 1980.

[JZh 98] S. Jimbo and J. Zhai; Ginzburg-Landau equation with magnetic effect: non-simply-connected domains, J.Math.Soc.Japan, **50** 3 (1998) 663–684.

[Kel 58] J.B. Keller; Propagation of a magnetic field into a superconductor, Phys. Rev., **111** (1958) 1497.

[Kis 95] Kissling; Conformally invariant systems of nonlinear PDEs of Liouville type, Geometric and Functional Analysis, **5** 6 (1995) 924–947.

[Lad 69] O.A. Ladyzhenskaya; The Mathematical Theory of Viscous Incompressible Flow, Gordon And Breach, Science Publishers, New York London Paris, (1969).

[Lan 97] S. Lang; Complex Analysis, Addison-Wesley, 1977.

[LDu 97] F.H. Lin and Q. Du; Ginzburg-Landau vortices: dynamics, pinning and hysteresis, SIAM J. Math. Anal., **28** (1997), 1265–1293.

[Lin 95] F.H. Lin; Mixed vortex and anti-vortex solutions of Ginzburg-Landau equations, Arch. Rat. Mech. Anal., **133** (1995) 103–127.

[Lin2 95] F.H. Lin; Solutions of Ginzburg-Landau equations and critical points of the renormalized energy, Ann. Inst. H.Poincaré Anal. Non Lin., 12(5), 1995, 599–622.

[Lin 96] F.H. Lin; Some dynamic properties of Ginzburg-Landau vortices, Comm. Pure. Appl. Math., **49** (1996) 323–364.

[LRa 81] M. Luskin and R. Rannacher; On the smoothing property of the Galerkin method for parabolic equations, SIAM J. Numer. Anal. **19** (1981), pp. 93–113.

[LSU 68] O.A. Ladyzhenskaya, V.A. Solonnikov and N.N. Uralceva; Linear and Quasilinear Equations of Parabolic Type, American Mathematical Society, Providence, 1968.

[LTa 95] J. Liang and Q. Tang; Asymptotic behavior of the solutions of an evolutionary Ginzburg-Landau superconductivity model, J. Math. Anal. Appl. **195**, (1995); 92–107.

[MGP 91] A.B. Monvel-Berthier, V. Georgescu and R. Purice; A boundary value problem related to the Ginzburg-Landau Model, Comm. Math. Phys., **142** (1991), 1–23.

[Mit 89] W.F. Mitchell; A Comparison of Adaptive Refinement Techniques for Elliptic Problems, ACM Trans. Math. Software, **15(4)** (1989), 326–347.

[MKe 69] M.H. Millman and J.B. Keller; Perturbation theory of nonlinear boundary value problems, J.Math.Phys., **10** (1969), 342–361.

[Mod 87] L. Modica; The gradient theory of phase transitions and the minimal interface criterion, Arch. Rat. Mech. Anal., **98** 1 1987, pp. 123–142

[Mor 66] C.B. Morrey; Multiple Integrals in the Calculus of Variations, Springer, 1966

[MSa 71] D.W. Moore and P.G. Saffman, Structure of a line vortex in an imposed strain, Aircraft Wake Turbulence, eds Olsen, Goldburg, Rogers, Plenum, 1971.

[MTY 93] S. Muller, Q. Tang and B.S. Yan; On a new class of elastic deformations not allowing for cavitations, Ann. Inst. H. Poincaré, Analyse Non Linear, **11** (1994), 217–243.

[Mu 99] M. Mu; A linearized Crank-Nicolson-Galerkin method for the Ginzburg-Landau model, SIAM J. Sci. Comput., **18** (1997), No.4, 1028–1039.

[Nec 67] J. Necas; Les Méthodes Directes en Théorie des Equations Elliptique, Masson, 1967.

[Neu 90] J. Neu; Vortices in complex scalar fields, Physica D, **43** (1990) 385–406.

[Noc 93] R.H. Nochetto; Removing the saturation assumption in a posteriori error analysis, Istit. Lombardo Accad. Sci. Lett. Rend. **A 127** (1993), pp. 67–82.

[Noc 95] R.H. Nochetto; Pointwise a posteriori error estimates for elliptic problems on highly graded meshes, Math. Comp. **64** (1995), pp. 1–22.

[NSV 99] R.H. Nochetto, A. Schmidt and C. Verdi; A posteriori error estimation and adaptivity for degenerate parabolic problems, Math. Comp. (to appear).

[Pri 96] L. Prigozhin; Sandpiles, river networks and type-II superconductors, preprint.

[Pri 97] L. Prigozhin; On the Bean critical-state model in superconductivity, preprint.

[PRu 93] L. Peres and J. Rubinstein; Vortex dynamics for $U(1)$ Ginzburg-Landau models, Physica D, **64** (1993) 299–309.

[RSt 95] J. Rubinstein and P. Sternberg; On the slow motion of vortices in the Ginzburg-Landau heat flow, Siam J. Math. Anal., **26** (1995), no.6, 1452–1466.

[Rub 95] J. Rubinstein; On the equilibrium position of Ginzburg-Landau vortices, A. angew. Math. Phys., **46** (1995) 739–751.

[Saf 93] P.G. Saffman; Vortex dynamics, Cambridge University Press, Cambridge, UK, 1993

[Ser1 98] S. Serfaty; Local minimizers for the Ginzburg-Landau energy near critical magnetic field, preprint.

[Ser2 98] S. Serfaty; Stable configurations in superconductivity: uniqueness, multiplicity and vortex nucleation, preprint.

[SFi 73] G. Strang and G. Fix; Analysis of the Finite Element Method, Prentice-Hall Englewood Cliffs (1973).

[SGe 62] D. Saint-James and P.G. de Gennes; Onset of superconductivity in decreasing fields, Phys. Rev. Lett., **9** (1962) 266.

[SGe 63] D. Saint-James and P.G. de Gennes; Onset of superconductivity in decreasing fields, Phys. Letters, **7** (1963) 306–308.

[Sim 83] L. Simon; Asymptotics for a class of non-linear evolution equations with applications to geometric problems, Ann. of Math, **118** (1983), pp. 525–572

[Sim1 83] L. Simon; Lectures on Geometric Measure Theory. Proceedings of the Centre for Mathematical Analysis, Australian National University, Centre for Mathematical Analysis, Canberra, 1983.

[SLU 68] O.A. Ladyzhenskaya, V.A. Solonnikov and N.N. Uralceva, Linear and Quasilinear Equations of Parabolic Type, American Mathematical Society, Providence (1968).

[SST 69] D. Saint-James, G. Sarma and E.J. Thomas; Type-II Superconductivity, Pergamon Press, (1969).

[SSt 99] R. Schatzle and V. Styles, Analysis of a mean-field model of superconducting vortices, EJAM, **10**(1999), pp. 319–352.

[Str 94] M. Struwe; On the asymptotic behaviour of minimizers of the Ginzburg-Landau model in 2-dimensions, Differential Integral Equations, **7** (1994), 1613–1624; Erratum, Differential Integral Equations, **8** (1995), 224.

[Tan 93] Q. Tang; Remark on a relation between the magnetic field and the topological degree of the order parameter in superconductivity, Physica D; **69** (1993), 320–326.

[Tan 95] Q. Tang; On an evolutionary system of Ginzburg-Landau equations with fixed total magnetic flux, Commun. in Partial Differential Equations, **20** (1 and 2) (1995); 1–36.

[Tau 80] C.H. Taubes; Arbitrary N-vortex solutions to the first order Ginzburg-Landau equations, Commun. Math. Phy.; **72** (1980), 277–292.

[Tem 79] R. Temam; Navier-Stokes Equations, North-Holland, (1979).

[Tem 83] R. Temam; Navier-Stokes and nonlinear functional analysis, CBMS-NSF Regional Conference Series in Applied Mathematics (SIAM, Philadelphia, 1983).

[Tem 88] R. Temam; Infinite-Dimensional Dynamical Systems in Mechanics and Physics, Springer-Verlag, 1988.

[Tem 85] R. Temam; Mathematical Problems in Plasticity, Trans-Inter Scientia, 1985.

[Tin 75] M. Tinkham; Introduction to superconductivity, McGraw-Hill, New York, 1975

[TTi 86] D. Tilley and J. Tilley; Superfluidity and Superconductivity, 2nd edition, Adam Hilger Ltd., Bristol, 1986.

[TWa 95] Q. Tang and S. Wang; Time dependent Ginzburg-Landau equations of superconductivity, Physica D, **8**, (1995); 139–166.

[Wlo 82] J. Wloka, Partielle Differentialgleichungen, B.G.Teubner, Stuttgart, (1982).

[WWi 92] D. Waxman and G. Williams; Estimate of the core size of a vortex in an extreme type II superconductor, Annals of Physics, **220** 2 (1992), 274–299.

[Yan 90] Y. Yang; Boundary value problems of the Ginzburg-Landau equations, Proc. Roy. Soc. Edinburgh, **114A** (1990), 355–365.

[Yan 98] Y. Yang; Coexistence of vortices and antivortices in an abelian gauge theory, Phys. Rev. Lett., **80** 1 (1998), 26–29.

[Yos 78] K. Yosida, Functional analysis, Springer-Verlag, 1978.

# Index

MIX
Papier aus verantwortungsvollen Quellen
Paper from responsible sources
FSC® C105338

If you have any concerns about our products,
you can contact us on
**ProductSafety@springernature.com**

In case Publisher is established outside the EU,
the EU authorized representative is:
**Springer Nature Customer Service Center GmbH**
**Europaplatz 3, 69115 Heidelberg, Germany**

Printed by Libri Plureos GmbH
in Hamburg, Germany